PRACTICAL ENVIRONMENTAL FORENSICS

PRACTICAL ENVIRONMENTAL FORENSICS

Process and Case Histories

Patrick J. Sullivan, Ph.D
Franklin J. Agardy, Ph.D
Richard K. Traub, J.D.

John Wiley & Sons, Inc.
New York • Chichester • Weinheim • Brisbane • Singapore • Toronto

This book is printed on acid-free paper. ♾

Published simultaneously in Canada.

Library of Congress Cataloging-in-Publication Data:

Sullivan, Patrick J., Ph. D.
Practical environmental forensics : process and case histories / Patrick J. Sullivan, Franklin J. Agardy, Richard K. Traub.
p. cm.
Includes bibliographical references and index.
ISBN 0-471-35398-1 (cloth : alk. paper)
1. Environmental forensics. I. Agardy, Franklin J. II. Traub, Richard K. III. Title

TD193.4.S85 2000
628.5—dc21 00-036803

Printed in the United States of America.

10 9 8 7 6 5 4 3 2 1

Contents

Preface **xix**

1 The Truth, the Whole Truth, and Nothing but the Truth **1**

A QUESTION OF ETHICS 4

A DIFFERENCE OF OPINION 5

DEFENDING THE FAITH 10

A TORTUOUS PATH 12

2 A Legacy of Waste and Chemical Pollution **19**

CHEMICAL-HANDLING PRACTICE 20

A HISTORY OF WASTE AND POLLUTION 25

3 Introduction to Environmental Law **49**

THE EVOLUTION OF THE REGULATORY ENVIRONMENT 49

REAL ESTATE AND PROPERTY TRANSFERS AND THE ROLE OF THE ENVIRONMENTAL AUDIT 67

4 An Introduction to Insurance Coverage for Environmental Losses **69**

INTRODUCTION 69

OCCURRENCE-BASED POLICIES 77

ALLOCATION OF DEFENSE AND INDEMNITY COSTS 86

POLLUTION EXCLUSIONS 91

CLAIMS-MADE POLICIES 99
SUMMARY 101

5 Site Histories: The Paper Trail How and Why to Follow It 103

SITE HISTORIES: WHAT THEY ARE 105
PHASE I ENVIRONMENTAL SITE ASSESSMENTS: A DIFFERENT SORT OF HISTORY 110
POTENTIALLY RESPONSIBLE PARTY SEARCHES: THE USE AND LIMITS OF HISTORY IN ENVIRONMENTAL FORENSICS 115

6 The Forensic Application of Contaminant Transport Models 123

THE CONTAMINANT MODELING PROCESS 124
PROBLEM STATEMENT: CONVECTION–DISPERSION–ADSORPTION MODEL 128
SOLUTION OF CDA EQUATION FOR CONTINUOUS INJECTION OF CONTAMINANT 134
SOLUTION OF CDA EQUATION FOR A SLUG SOURCE OF CONTAMINANT 136
NUMERICAL INTEGRATION IN STIFF PROBLEMS 139
MODEL IMPLEMENTATION 140
MODEL LIMITATIONS 140
THE BORDEN SITE FIELD EXPERIMENT 141
INVERSE CALCULATIONS WITH THE INSTANTANEOUS POINT SOURCE SOLUTION 145
FORWARD CALCULATION OF THE BORDEN PLUMES WITH THE DISTRIBUTED SLUG SOURCES 147
SUMMARY 147

7 Chemical Fingerprinting 151

CHEMICAL FINGERPRINTING TEST METHODS 153
FINGERPRINTING APPLICATIONS 179

REVIEW OF CHEMICAL FINGERPRINTING EVALUATIONS 201

8 Forensic Applications of Environmental Health Sciences 213

CASE STUDIES 220

SUMMARY 230

9 Risk Assessments 223

DATA COLLECTION 233

DATA GAPS 234

EXPOSURE 235

RECEPTORS 236

EXPOSURE PATHWAYS 236

ROUTES OF EXPOSURE 237

TOXICITY 237

CALCULATION OF RISK 239

RISK COMMUNICATION 239

LITIGATION 240

10 Why Use Visuals 245

A PICTURE IS WORTH . . . 245

DO YOU SEE WHAT I SEE? 246

TIME IS OF THE ESSENCE 247

ONE FROM COLUMN A AND ONE FROM COLUMN B 247

VARIETY IS THE SPICE OF LIFE 248

TWO TIN CANS AND A STRING 259

11 Evidence Issues: Getting Expert Opinions Past the Judicial Gatekeeper and Into Evidence 261

HISTORY OF ADMISSIBILITY OF EXPERT TESTIMONY IN FEDERAL COURT 263

FRYE "GENERAL ACCEPTANCE" TEST 263
FEDERAL RULES OF EVIDENCE 264
DAUBERT RELEVANCE AND RELIABILITY 265
GENERAL ELECTRIC COMPANY V. JOINER 266
KUMHO TIRE COMPANY V. CARMICHAEL 268
EXPERT ENVIRONMENTAL TESTIMONY UNDER *DAUBERT, JOINER,* AND *KUMHO* 269
AN EXPERT SHOULD RELY ON MORE THAN "EXPERIENCE" IN REACHING A CONCLUSION 269
SCIENTIFIC METHODOLOGY MUST BE FOLLOWED 271
COURTS MAY OR MAY NOT RELY ON THE SPECIFIC FACTORS DESCRIBED IN *DAUBERT* 272
THE *DAUBERT* INQUIRY IS FLEXIBLE AND TIED TO THE SPECIFIC FACTS OF EACH PARTICULAR CASE 273
OPINIONS SHOULD NOT BE DEVELOPED SOLELY FOR PURPOSES OF TESTIFYING 275
A REVIEWING COURT MUST APPLY AN ABUSE OF DISCRETION STANDARD 278
EXPERT CONCLUSIONS BASED SOLELY ON SPECULATION AND POSSIBILITY ARE INSUFFICIENT 279
MOTIONS TO EXCLUDE EXPERT TESTIMONY PURSUANT TO *DAUBERT* 282
ADDITIONAL CHALLENGES TO EXPERT TESTIMONY 288

12 Alternative Dispute Resolution Techniques 301

DESCRIPTION OF ADR METHODS 302
HISTORICAL ROLE FOR ADR IN ENVIRONMENTAL DISPUTE RESOLUTION 308
TECHNICAL NEUTRALS IN ADR 314
CASE STUDIES 317
MANAGING THE PROCESS 322
SUMMARY 323

13 **Forensic Case Management** **327**

TIMING IS EVERYTHING 327

ELEMENTS OF FORENSIC CASE MANAGEMENT 330

SUMMARY 348

14 **Environmental Extortion** **349**

A PROPERTY SALE GONE WRONG 349

FILLING IN THE DATA GAPS 362

EXPLAINING THE PIECES OF THE PUZZLE 365

A DIFFERENCE OF OPINION: ROUND 2 398

DEFINING THE EXPERTS 398

THE MEDIATION 400

SOLD AT LAST 400

LESSONS LEARNED 400

15 **A Lesson in Communications** **403**

AN OPPORTUNITY TO IMPROVE PROPERTY VALUES 403

A "DYSFUNCTIONAL" MEDIATION 407

A NEW BEGINNING AND A SHORT FUSE 408

AN ABSENCE OF REGULATORY INPUT 412

DEVELOPING A MEDIATION STRATEGY 413

THE FINAL MEDIATION 414

THE SETTLEMENT 418

LESSONS LEARNED 420

16 **Allocation, Allocation, Allocation** **421**

A POLLUTION-FREE INDUSTRY AND GROUNDWATER CONTAMINATION 421

THE INDUSTRIAL PARK INVESTIGATIONS 422

ALLOCATION AMONG PARCELS 427

ALLOCATION AMONG PARCEL OCCUPANTS 434

DECREASING REMEDIAL COSTS 439

LESSONS LEARNED 439

17 Deflating Environmental Costs **441**

THE PARTIES AND THEIR PROBLEMS 441
THE FORENSIC INVESTIGATION 444
AN EXAGGERATED CLAIM 450
SETTLEMENT PREPARATION 451
THE FIRST SETTLEMENT MEETING 452
A MORE REALISTIC SETTLEMENT 453
LESSONS LEARNED 454

18 Guilt by Association **455**

THE HISTORY OF MANUFACTURED GAS IN GRASS HILLS COUNTY 455
THE COUNTY'S SITE INVESTIGATIONS 458
THE FORENSIC INVESTIGATION 458
AN ARCHEOLOGY INVESTIGATION 465
LESSONS LEARNED 468

19 An Issue of Disclosure **469**

CLAIM BACKGROUND 470
THE LITIGATION 472
SETTLEMENT NEGOTIATIONS 476
LESSON LEARNED 476

20 Settling a Claim **477**

A SMALL FAMILY BUSINESS 477
AN INSURANCE CLAIM 480
AN ATTEMPTED SETTLEMENT 482
LESSONS LEARNED 484

21 A Cost of Doing Business **485**

SOLID WASTE DISPOSAL 485

THE CITY'S HISTORY OF SOLID WASTE MANAGEMENT 487
THE FORENSIC INVESTIGATION 493
A SETTLEMENT MEETING 496
LESSONS LEARNED 496

22 A Sudden and Accidental Event 497

BACKGROUND TO A PROPERTY SALE 497
GROUNDWATER REMEDIATION: PHASE I 501
LAWSUIT 1 503
GROUNDWATER REMEDIATION: PHASE II 504
THE INSURANCE LITIGATION 504
LESSONS LEARNED 513

23 An Act of God 515

SITE HISTORY 515
A CGL INSURANCE CLAIM 518
AN OPPOSING OPINION 519
DEPOSITIONS 533
LESSONS LEARNED 534

Appendix A 537

Appendix B 547

Index 577

Contributing Authors

Chapter 5

Shelley Bookspan, Ph.D.
Julie Corley
PHR Environmental
Consultants, Inc.
5290 Overpass Road, Suite 220
Santa Barbars, CA 93111-2051

Chapter 6

Tadeusz W. Patzek, Ph.D.
University of California
577 Evans Hall, #1760
Berkeley, CA 94720-1760

Chapter 7

James E. Bruya, Ph.D.
Friedman & Bruya, Inc.
3012 16th Avenue West
Seattle, WA 98119-2029

Chapter 8

Paul C. Chrostowski, Ph.D.
Sarah A. Foster
CPF Associates, Inc.
7708 Takoma Avenue
Takoma Park, MD 20912

Chapter 9

Susan L. Mearns, Ph.D.
Mears Consulting Corporation
521 Pier Avenue, Suite 5
Santa Monica, CA 90405

Chapter 10

John Muir Whitney
Mark S. Greenberg
Think Twice, Inc.
Pier 9 on the Embarcadero
San Francisco, CA 64111

Chapter 11

Timothy A. Colvig, Esq.
Wulfsberg, Reese & Sykes
300 Lakeside Drive, 24th Floor
Oakland, CA 94612-3524

Chapter 12

Michael Kavanaugh, Ph.D., P.E.
Malcoml Pirnie, Inc.
1800 Grand Avenue,
Suite 1000
Oakland, California 94612
Honorable Daniel Weinstein
(retired)
JAMS/Endispute
2 Embarcadero Center,
Suite 1100
San Francisco, CA 94111

Chapter 14

Tad Patzek (see chapter 6)
Doug Zunkel
A.D. Zunkel Consultants
8020 N.E. 71st Loop
Vancouver, WA 98662

Preface

This book addresses the comprehensive environmental forensic process. It is not based solely on chemical and investigative methods that are used as forensic tools, although these tools are presented within the text and also illustrated within the context of environmental forensic case histories. The primary goal of this book is not only to help engineering and scientific professionals understand the forensic process as it pertains to environmental problems, but also to assist any potential expert (e.g., experts involved with construction, intellectual property, patents, medicine, product liability or personal injury litigation) in gaining a "common insight" into the forensic process. A secondary goal is to provide both technical professionals and attorneys an in-depth treatise on both environmental law and engineering/scientific issues commonly encountered in environmental cases.

Practical Environmental Forensics has an immediate application for engineering and scientific professionals who currently provide or anticipate providing expert opinions or environmental litigation consulting services, as well as attorneys practicing in the environmental field. The case history chapters present not only "real-world" examples of environmental cases, but also "illustrative problem solving" as a teaching resource for use in graduate-level engineering/environmental courses.

As a result of the authors' combined 50-plus years of experience in basic research, teaching, and consulting—as well as more than 30 years of environmental litigation experience—this book leans heavily upon knowledge gained from the review of historic industrial and commercial business practices as contrasted with the published literature. Thus, in order to protect the confidential nature of this information, the names of all the sites and their locations, companies, and consultants used in this book are fictitious. Any similarity to actual sites, companies, and consultants is a coincidence. Although case descriptions are necessarily generic, site facts pertaining to facility operations and analytical data that are reported in public records (e.g., regulatory agency files) have been presented in sufficient detail to correctly represent the technical information found therein.

Based on the authors' experience, this book has been organized into four areas. Chapters 1 through 4 comprise the first section of the text and provide an introduction to the process of environmental forensics, as well as the historical basis of chemical handling and solid and liquid waste management, representing a "core" for understanding environmental contamination issues. A review of environmental law and insurance issues is presented in order to establish the bases of environmental litigation. Each of the chapters in the first section is intended to provide a general understanding of the key components associated with the environmental forensic process.

Recognizing that forensic opinions must be based on fundamental scientific/engineering principles, specific technical chapters on the most commonly employed principles and methods utilized in environmental forensic investigations are presented in Chapters 5 through 9, which constitute the second section of the book.

The third section of the book combines, in Chapters 10 through 13, an introduction to essential environmental forensic support functions. These functions address the development and presentation of scientific information, understanding the admissibility of evidence, the application of mediation methods for settling cases, and forensic case management.

The fourth section of the book, Chapters 14 through 23, provides, through the vehicle of case studies, examples of the environmental forensic process in which the authors have participated. One might normally assume that the genesis of an environmental forensics case begins with site contamination. In actuality, it begins much earlier with the historical study of chemicals, hazardous substances, waste management and disposal practices, and regulations. Litigation, however, is usually based on the actual and potential costs associated with a cleanup or health-related damages. Environmental litigation in many cases is also the result of insurance claim disputes. As a consequence, environmental forensic experts are retained to provide their services in "typical" forensic cases that can be grouped into the following areas.

Minimizing the Damage In private-party disputes, prior to the point where a regulatory agency establishes a final remedial action plan (i.e., the cost of the cleanup), a client may retain experts to provide a realistic assessment of the future environmental cost. In these cases, a client might be asked to pay what appears to be an unjustifiably high remedial cost. Thus, it is the objective of the forensic scientist to provide factual data and testimony to support a more realistic and, hopefully, reduced cleanup cost. Examples of minimizing site damages are given in Chapters 14 and 15.

Allocation of Damages (Cost) Once a final remedial action plan (RAP) has been adopted for a site and there are no disputes over the RAP, a client may retain experts to assist with the allocation of the cost, as projected in the RAP, between the responsible parties. In these cases, it is the objective of the forensic scientist to provide factual data and testimony to support a minimum percentage allocation for his or her client and, at times, to aid the search for other potentially responsible parties. Allocation examples are described in Chapters 16 through 18.

Insurance-Related Litigation Insurance litigation usually involves either environmental impairment liability (EIL) or comprehensive general liability (CGL) insurance coverage. The differences between these two types of insurance are detailed in Chapter 4. As a forensic expert hired by either an insured or an insurer, various policy-dependent issues must be addressed.

In EIL cases, the forensic experts are expected to provide facts and testimony regarding knowledge of the occurrence of site contamination, the cause or causes of the contamination, and the damage, which is presumably covered by the EIL policy.

Conversely, CGL cases require not only a study of facts prior to contamination, but also a detailed forensic analysis of the fate and transport of the contaminant, over time, as well as the effect of any subsequent contamination.

In general, the typical insurance cases that require expert testimony focus on (1) when the event giving rise to the contamination first took place, and in the case of EIL coverage, the date a claim for money or damage was actually made (these examples are given in Chapters 19 and 20); (2) the cost of the remedial work versus normal business costs (this example is illustrated in Chapter 21); (3) if the contamination was the result of a sudden and accidental event (this example is described in Chapter 22); and (4) if the pollution was expected and intended (this example is given in Chapter 23).

Special thanks go to the contributing chapter authors as well as to Dr. Nelson L. Nemerow, who provided a constructive review of the text. We owe a sincere debt of gratitude to Paula Massoni for her patience and skill in editing the text. A thank you is also extended to John R. Kiefer for his assistance with analytical cost and engineering analyses.

It is anticipated that a second edition of this text will be based on additional case histories. In this regard, if any reader would like the authors to consider a case history contribution, please contact the senior author at: http://www.psfmaenv@pacbell.net.

Finally, the authors are hopeful that the information, guidance, and history provided in this text will assist the reader in understanding the multidisciplinary focus of environmental litigation where science, engineering, and law are united in the forensic process to bring about practical solutions to complex disputes.

Patrick J. Sullivan
Franklin J. Agardy
Richard K. Traub

Chapter 1

The Truth, the Whole Truth, and Nothing but the Truth

Forensics, simply stated, is the art of debate occurring in a legal forum.[1] Environmental forensics is the study, analysis, and evaluation of environmental issues in a legal dispute. Environmental forensics is not just chemical analyses or modeling, it is a much more complex process. In the legal context, the objective of the forensic scientist/engineer is to establish the "truth"[2] for each and every technical and often-competing element at issue in litigation. Although it may sound simple, it is not.

Most environmental issues are multifaceted. As a result, even factual "truth" is complex. For example, chemical wastes placed on a property by several different owners/operators over decades of occupancy may necessitate a cleanup of both polluted soil and underlying groundwater. The ability to determine which disposal practice(s) by each owner/operator resulted in the need for cleanup activities usually involves historical information and often requires detailed evaluations of industry specific knowledge and practice employing a number of scientific and engineering disciplines: organic and inorganic chemistry, toxicity, biology, solid waste and wastewater treatment practices and methods, soil science, geology, hydrology, groundwater contaminate modeling, and air emissions and particulate modeling. "Simple" has taken on added meaning. The forensic investigation now includes not only the horizontal dimension of expanded science and engineering, but also the vertical dimension of time.

[1]Not unlike the Roman Colosseum, forensic experts play the role of the Christians, while the lawyers play the lions (a polite euphemism).

[2]The actual state of a matter or reality.

In addition to determining the factual "truth" of who is actually responsible for a cleanup, the determination also must be made as to who pays and how much is paid.[3] Reflecting on the presumption that more than one party will be identified as a "contributor" of pollution, a percentage allocation also must be determined. Once an allocation has been made, it then must be determined if the identified "contributor" has the ability to pay the proportionate share; in other words, which of the identified parties possesses the financial resources and/or insurance coverage to pay for a cleanup. If there is the potential for insurance coverage of environmental pollution, then another layer of complicated issues will be added to the forensic process. Finally, in some instances, pollution cases also may require a determination of whether employees or other individuals may have been impacted by environmental pollution and suffered either acute or chronic health effects. Enter the toxic tort (the most complex form of environmental forensics).

These issues, when commingled within the litigation process, generate a vast array of technical and legal views of the "truth." Adding to this complexity is the fact that factual "truth" is not always the same as legal "truth." An example of this dichotomy is given in Sidebar 1.1.

Because environmental pollution is the result of historical practices, inevitably critical information often is missing or "unavailable" (e.g., chemical analyses, geologic/soil boring logs, design and construction documents, as-builts, process and unit operations, etc.). However, even when such information is available, it still may not include the complete level of detail so necessary for today's environmental evaluations.[4] This dilemma, combined with the technical aspects of environmental pollution, dictates the need for forensic scientists/engineers to provide information, insights, and opinions in order to explain the complex puzzle of data and their related details and facts.

Like most puzzles, the entertaining part of determining the "truth" is in finding and placing the individual pieces together. Again, like most puzzles, rarely are all of the pieces found. That is why the forensic scientist/engineer may be required either to explain the whole "truth" based on the identified information or to just define and address the missing pieces. It is at this point that in some cases the forensic scientist/engineer may em-

[3]In litigation, money is often confused with responsibility.

[4]Analytical site data collected for a remedial investigation are not collected for forensic purposes. As a result, specific analytical forensic data may have to be collected (if possible).

Sidebar 1.1. Divergence of the Factual "Truth" and "Legal Truth"[5]

By the very nature of the profession, an attorney is an advocate.[6] Herein lies the predicament as well as the source of a forensic expert's income. Because advocacy requires the best possible argument to support a client's interest, a case initially may be based on and presented with more spin than fact, thus, the need for a forensic expert to counter the advocate's position. For example, the following simplified case illustrates the difference between an "advocate position" (i.e., a client's belief, opinion, or desire) and a factual or "technical position" (i.e., a forensic scientist's/engineering opinion).

A company that manufactured pharmaceuticals wanted to prove that the groundwater contamination on its property resulted from a spill of tetrachloroethylene (PCE). If the company could prove that the spill (an accident) caused the contamination, then the groundwater cleanup would be covered by its insurance as a "fortuitous" event.

The company used PCE in various buildings. It stored drums of pure PCE on an outdoor asphalt pad. Interviews of the company employees revealed that a drum of PCE had been spilled. However, when considering all of the technical facts, the site information demonstrated that the vast majority of the PCE groundwater contamination did not originate with the reported PCE drum spill. In spite of these known facts, the attorney representing the pharmaceutical company presented the position that all of the PCE groundwater contamination at the facility was the result of a spilled drum of PCE.

In this case, it was the role of forensics to demonstrate that the facts could not support the attorney's "spin" of the "truth." However, no matter how well the facts may support an environmental case, the issues are extremely complex. As a result, a good "spin doctor" can dazzle an environmentally uneducated/uninitiated judge or jury. This is one reason why most environmental cases seldom go to trial.

[5]A legal argument designed with a suitable "spin."

[6]Experts should not be advocates. However, once experts' opinions are developed, they can be advocates of their opinions.

ploy tools such as aerial photography, process mineralogy, chemical fingerprinting, age dating, focused analytical studies, and water/chemical transport modeling to fill in data gaps. However, in many instances, there are no appropriate tools with which to answer site-specific questions. Or, in some cases, the litigation will be at such a point that no further studies and/or data collection are possible (i.e., discovery is closed). Consequently, some information simply cannot be determined and, therefore, remains unknown.

Once the technical "truth" has been defined, the most difficult process still remains: convincing someone who is not familiar with the pieces of the puzzle, and has never before even seen the shapes of the pieces, that your explanation of the puzzle is indeed the "truth," or at least a complete enough picture so that it is better understood than that of the opposing expert presenting his or her version of the "truth."

Finally, in building on the technical "truth," the ultimate goal of environmental forensics is the persuasive presentation of the whole puzzle. This last statement begs the question: If "truth" is "truth," why do we usually end up with at least two versions of the "truth"?

The very nature of having an incomplete data set describing pollution is the basis of environmental forensics; that is, it permits different scientists and engineers to have conflicting views of the "truth." During the environmental litigation process, prior to settlement or trial, experts may provide their "truth" as opinions in the form of affidavits, expert reports, rebuttal opinions, and deposition testimony. When all of the testimony has been completed and reviewed, there are occasions when divergent expert opinions are clearly valid. However, on other occasions, a review of the testimony, specifically the factual support for that testimony, will show that only segments of the underlying "truth," if any, have been used to support an opinion.

A QUESTION OF ETHICS

If two or more opposing experts have differing opinions, does this present an ethical conflict? No, because there can be legitimate reasons. However, an expert can ignore ethical practice and misrepresent his or her professional qualifications, facts, information, test methods, data, conclusions, or opinions. If this occurs, one available defense is to provide a well-documented and referenced rebuttal to (1) discredit the opposing expert's opinions and (2) provide the necessary facts for examining the expert during the deposition process.

Ethical issues aside, opposing experts generally have differing opinions. Thus, it is important to understand how this can occur. Based on the authors' experience with opposing scientific and engineering opinions, we offer several examples of how some experts using the same facts, for the most part, can arrive at totally different conclusions, views, or opinions.

A DIFFERENCE OF OPINION

The first phase of the forensic process is the development of opinions. This process usually requires the expert to review documents provided by counsel and to review documents collected by the expert. In some cases, it may also require the expert to undertake additional chemical evaluations or simulate environmental conditions for a site (e.g., model the transport of a chemical in soil and/or groundwater). Based on the site-specific information and the expert's background and experience, the expert will form opinions.

Generally, scientists and engineers who conduct research and address environmental problems tend to have a broad academic background and work experience. This diversity can result in expertise based on a unique experience and understanding of the "truth" relative to specific issues. More often than not, if there is a difference of opinions, the differences will exist because the experts did not review and/or rely on the same documents, data, or model. In most cases, such differences are legitimate. However, when site facts do not support the client's "truth," some experts (or expert's counsel) have been known to use the following methods in order to express an opinion compatible with their client's wishes.

What You Don't Know Can't Hurt You

An expert can be intentionally biased by the client, whereby the client only gives the expert selected documents that support the client's version of the "truth." For example, in a case involving one of the largest chemical companies in the United States, attorneys for the chemical company failed to give their own environmental experts copies of the company's internal publications (which were actually available in local libraries). These publications contained factual and historical environmental information that was contrary to the company experts' reports. Unbelievable, but true. One has to wonder why the client's experts did not investigate and find these materials on their own. Intellectual curiosity alone should have been enough stimulus to warrant further research; or perhaps this was an example of the expert deciding to "play along" with the client's "truth."

Backing Into the "Truth"

In addition to documents given to experts, experts select and/or rely on other documents/data that specifically support their opinions but ignore documents/data that would contradict their opinions.

Unless specifically instructed by the client, a credible expert should include all documents and data containing both "favorable" and "unfavorable" information relative to his or her client's litigation objectives. The client's "truth," if it exists, will become evident through the weight of evidence. It should also be noted that there is an extreme variation on this theme when the expert witness gives an opinion with no supporting literature because "I have 40 years of experience." In other words, if the facts are on your side, argue the facts. If not, argue the experience. This approach is valid in some cases but should not be attempted unless the expert really has the legitimate experience and can persuade the parties, or unless it is the only remaining choice.

Alzheimer's[7]

The experts fail to remember the basic principles learned in school or forget what they themselves wrote in journal articles and books. Because much of the information in environmental forensics deals with historical knowledge, some experts choose to leave out relevant facts in the hope that an opposing expert or attorney may not be able to find historical support for an opposing opinion or simply in an attempt to confuse the issue.[8]

An example of this approach occurred in a case wherein the expert held that it was an acceptable practice in the 1950s to dispose of waste in a landfill that was in direct contact with groundwater. This practice would result in the immediate pollution of the groundwater. Since the case at issue dealt specifically with the landfilling of hazardous waste, this opinion was intentionally misleading. Was it the intent of the expert to convince a judge or jury that it was accepted practice to dispose of toxic chemicals directly into groundwater? Indeed, it was.

[7]In litigation, Alzheimer's may be self-induced or client-induced.

[8]One may legitimately wonder how this section in any way fits into the concept of "truth."

There are technical journal articles from 1940 to 1950 describing municipal waste (including inert construction debris) being placed in swampy or wet landfill environments. Municipal waste and inert waste, however, are not the same as hazardous waste! There are no historical documents reporting, as acceptable procedure, the disposal of hazardous and toxic waste into landfills that have a direct connection to groundwater. In fact, historical literature clearly makes a distinction between landfill characteristics that are acceptable for municipal waste disposal and those designed for hazardous waste disposal. Thus, in this case, the opposing expert's deceptive opinion was overly broad and subsequently exposed as being false.

In some cases, experts may not be required to give a complete list of all their publications (e.g., only publications from the most recent five years). However, it is important to complete a search for all relevant papers, reports, articles, and published speeches attributed to an expert. In many cases, experts present written opinions, occasionally in the form of an affidavit, that are directly contradicted by their own prior articles (articles often published 10 or more prior years). Occasionally, a prior affidavit is found which contradicts currently expressed opinions.

Differences in expert opinions also occur when experts review the same documents but interpret the information using different approaches (a natural result of education, experience, and litigation objectives). Several of these examples are given next.

The History of the World: Part I

An expert selects a date at which some environmentally relevant event (i.e., laws, regulations, enforcement, chemicals of concern, management practices) occurred, but ignores all historical data prior to that date.

A typical illustration references the establishment of the U.S. Environmental Protection Agency, in 1970, as the "new testament" of the environmental era while totally ignoring the "old testament" as exemplified by the U.S. Public Health Service, whose history dates back to the 19th century, and individual state pollution regulations.

A Rose by Any Other Name

An expert is ignorant of a historical environmental condition because it has a different name today. For example, an expert once provided an opinion that prior to the passage of the Resource Conservation and Recovery Act (i.e., an act regulating hazardous waste) in 1976, there were no university classes that taught the management of "hazardous waste" (i.e., treatment, storage, and disposal). This opinion, as stated, essentially is correct. However, the opinion fails to mention that university classes taught the management of "industrial waste,"[9] as far back as the 1940s, which dealt with the same materials.

In another case, an expert expressed an opinion that prior to the mid 1980s engineers and scientists were not aware of DNAPLs (dense non-aqueous phase liquids, i.e., organic liquids that are heavier than water). Simply stated, a DNAPL will sink rather than float on groundwater. Prior to the use of the term DNAPL, these liquids were simply called "sinkers." Again, the "science" did not change—only the name changed.

Ignorance Is Bliss

An expert uses only the data that defines his or her client's "truth." For example, an industrial site (with a long history of various industrial uses) was used from the 1930s through the 1960s for the manufacture of paint that utilized sulfuric acid as one of the chemicals in its process. In the 1980s, it was discovered that the groundwater was contaminated with sulfuric acid. Remedial investigation experts representing the current landowner firmly placed the blame (and, thus, a demand to clean up the groundwater) for the sulfuric acid contamination on the paint manufacturer (the only prior owner with a deep pocket). However, the remedial investigation experts failed to identify the fact that a chemical company, at the exact same location but earlier in time, manufactured sulfuric acid. Without commenting on which operation might have contributed to the pollution, and by ignoring the former chemical company, a bias clearly was demonstrated and documented. Subsequently, that expert opinion was discredited.

Shuck and Jive

An expert gives an opinion that ignores the obvious. In a case that involved the use of large lagoons to hold wastewater from fertilizer manufacturing, seepage from the unlined lagoons contaminated the groundwater. An ex-

[9] As distinguished from municipal waste or inert waste mentioned previously.

pert in this case gave the opinion that the contamination actually occurred from wastewater overflowing the lagoon berms due to a storm (i.e., an accident for which insurance may pay for the cleanup) and not through seepage (i.e., the location of the lagoon on sandy soil with groundwater at only 10 feet below ground surface). Ironically, it was later shown that the "storm" incident date, which allegedly supported contamination by overtopping, was found to be in error and did not support the expert's conclusion. The expert, in his deposition, blamed this error on inadequate staff support.

Nothing but Blue Sky

An expert gives an opinion that defines the client's "truth" without scientific or engineering basis or support. A company that manufactured gas from coal began operations in the 1850s. These activities created groundwater contamination that began during the same time period. In a 1998 litigation, this company retained an expert to model the historic groundwater contamination to show that during a specified time period there was "more" contamination—a formidable task since no facts were available and no groundwater monitoring began until the mid-1990s.

Undaunted, the expert provided a model that was based completely on unsupported assumptions. The model was uncalibrated, unvalidated, and, according to the expert deposition testimony, "did not predict anything." However, the model was still offered as proof of the historic groundwater contamination. The often-expressed and aptly phrased "garbage in, garbage out" defines this type of modeling.

In fact, the one area of environmental forensics in which expert opinions are the most diverse is contaminate modeling of air, water, and soil. Because models of the environment by their very nature are complex, their use requires expert analysis. Thus, the ability of a model to provide accurate predictions requires:

- An expert who can select the appropriate model (if one exists) for the given problem to be modeled
- The availability of appropriate input data
- The analysis of the input data to select appropriate model parameters and boundaries (especially important if actual site data are unavailable)
- The ability to calibrate and verify model results against site-specific information

As a result, modeling of the environment provides opposing experts ample opportunity to disagree. This process is described in greater detail in Chapter 6.

Illustrating these examples of how experts express different opinions is critical, because it is during the evolution of the forensic process that the truth is eventually defined. Although some opposing experts may use one of many different approaches for avoiding the "truth," a forensic expert must be aware of these techniques in order to help counsel define and expose misleading, unsupported, and/or deceptive opinions during the deposition process.

DEFENDING THE FAITH

The second most important phase in the forensic process is the deposition. In addition to providing "truthful" deposition testimony in support of expert opinions, it is also important that an expert provide guidance to counsel when taking the deposition of opposing experts.

When taking the deposition testimony of an expert, it may be the objective of opposing counsel to simply determine the expert's opinion in a straightforward approach. In other cases, the attorney also may try to establish how the expert supports each opinion by determining the factual basis for the opinion. When an expert offers an opinion that is not fully supported, misleading, or deceptive, the deposition process can be used to expose the "truth" and/or deficiencies. It is critical that the expert provide counsel with the necessary questions and documents with which to examine the opposing expert.

When experts offering unsupported, misleading, or deceptive opinions are deposed, some of the following tactics have been employed in an attempt to avoid having their "truth" challenged. In order to assist counsel in the examination of experts, all forensic experts should be aware of some of these techniques.

The "Clinton Principle"

Because environmental issues are complex, an expert often knows and appreciates the subtlety associated with the definitions of technical words. Thus, an expert can "truthfully" answer any question based on *his or her* definitions (if not properly followed up and "impeached" by counsel). For example, counsel is looking for a "yes" answer to the following question:

MR. JACKSON: Q. Dr. Lee, is it true that the leaching of absorbed arsenic from the soil is the source of the groundwater contamination?

THE WITNESS: A. No.

If counsel had asked if either "adsorbed" or "sorbed" arsenic was a source of the groundwater contamination, the answer would have been "yes."

Data Overload

An expert confirms a fact but cannot find the reference because there are so many documents and information that, "I do not remember where I saw that fact, but I know it is in there somewhere." Nevertheless, the opinion will ultimately be brought into question if the reference is not found.

The Best Defense Is a Good Offense

Answer the question by incorporating other scientific terms or principles that are tangentially related, but for which the attorney is most likely unprepared to ask follow-up questions. All fields of science and engineering afford this opportunity in a deposition, especially when dealing with a poorly prepared or neophyte attorney.

A good attorney will make the expert work through all new terms, principles, and definitions in order to understand the relevant relationships or issues. Such a diversion can, however, take a considerable amount of time and may deflect further questions from the area counsel was probing. When an expert uses this tactic, counsel should contact his or her own expert in order to define the proper follow up questions. In complex cases, it is highly recommended that a forensic expert attend an opposing expert's deposition (if permitted).

If Not Asked, Do Not Tell

If counsel's question is not specific, then there is no requirement to give a specific answer. The operative rule is not to provide any information that is *not* specifically solicited (usually at the direction of counsel). If the attorney is not clever enough to ask the right question, do not volunteer the answer he or she is struggling to resolve. The successful application of this rule, of course, presumes that the expert is smarter—or at least better prepared—than the attorney. In the expert's mind, this is usually the case, but almost never the case if you ask the attorney. The downside of using this method is that the deposition process will be much longer.

The "Theory of Relativity"

When trying to discount historical scientific information, an expert stated that in the 1940s there may have been a vast amount of information on a given subject, but because scientists know much more today, a scientist in the 1940s really had only a comparatively small amount of information and therefore did not know very much.[10]

It Is Hard to Find Good Help

The expert cannot answer the question because the relevant information was collected by staff personnel. One standard excuse is the failure to recall just who collected the information, or who lost it (begging the question as to why it was lost). An expert utilizing this response will have to consult with his or her staff to find the relevant information prior to trial.

A Head Case

The ability to support an opinion is an important test of an expert's credibility. As a consequence, experts under the pressure of deposition examination may offer an unlikely foundation for a shaky opinion. For example, one expert, when asked the basis of an opinion concerning specific requirements of a federal regulation, replied that the information was in "his head." Needless to say, this so called regulation in "his head" did not exist. This type of response is both unnecessary and damaging to an expert's credibility.

A TORTUOUS PATH

Because there are different technical opinions in environmental litigation, the path to explaining, defending, and resolving case issues can take a tortuous path. The legal path undertaken by a forensic scientist/engineer to define the "truth" (as illustrated in Sidebar 1.2) is an iterative process whereby scientific and engineering information is collected, analyzed, and

[10]Tell that to the scientists who developed the atomic bomb.

Sidebar 1.2. Forensic Path to the "Truth"

The time at which an attorney determines that an expert forensic investigation and/or opinion is necessary varies anywhere from the beginning of a case to immediately before experts are to be declared and discovery (i.e., collection of data and information) ends. Unfortunately, in many cases, it tends to be the latter. Regardless of the time frame, however, the process of developing a technical opinion will follow the same general path, to the degree that time and money allow.

The first step onto this path begins with either a meeting or a phone call with counsel/client to understand the case history, review the technical litigation objectives, and determine if one's expert qualifications/experience allow a creditable technical opinion.

Once chosen to provide an expert opinion, the usual practice is for the client either to provide an initial set of documents for review or, more often, simply to ask the expert what he or she requires. A review of the initial documents almost always results in a discussion with the client to request more specific documents, if available, in order to further assess the technical issues.

Developing an Opinion

Based on the initial technical data that are available, a meeting or discussion with the client is held in order to determine if the facts are sufficient for the expert to support the client's "truth." It should be understood at this point that the facts may or may not support the client's legal objectives. If the technical facts and opinions do not support the client's legal objectives, the expert should so inform the client and suggest that the client attempt to settle the case or find another expert who may provide a more suitable opinion.[11]

In a perfect world, opinions cannot be "purchased" (i.e., have the expert say anything the client wants). However, in the world of litigation, "purchasing" an opinion should not be confused with paying an expert to

(*continues*)

[11]It should be noted that an adverse early opinion by an expert need not be disclosed if the expert has not been declared.

(Sidebar 1.2, continued)

develop an opinion. Obviously, all experts are paid for their work product and the opinions they develop. Just because an expert is paid for an opinion does not mean that the opinion has been or can be "purchased." Thus, the neophyte expert should not be surprised if some opposing client/counsel incorrectly compares one's practice of "selling" opinions with the practice of prostitution.

Beyond this point, the next phase of the work deals strictly with further review and collection of relevant information. This information is collected either from parties in the litigation or from public sources. Documents from parties may include both plaintiff/defendant site documents, as well as documents from remedial/analytical firms that have investigated the site(s) at issue.

In addition to these documents, information also may be obtained by site visits and/or by conducting site-specific sampling and analysis to answer technical issues not addressed by the remedial studies (e.g., process mineralogy, chemical fingerprinting, contaminant transport modeling, etc.). In fact, remedial investigations rarely address the technical issues that an expert would like to have. Their missions are most often different. As a consequence, additional site-specific investigations may be necessary, if still feasible.

Information also should be obtained from all of the relevant opposing expert's deposition testimony (from all cases in which the expert may have participated), expert reports, referenced documents, and published papers.[12]

A vast amount of pertinent information can be collected from public sources, which can often prove to be quite relevant. For example:

- Site/file information from city, county, state, and federal agencies relating to the operational activities at the site
- Relevant literature (i.e., state of the art, geology, hydrology, soils, chemicals, etc.) from libraries, universities, institutes (private and federal), trade associations, and government archives
- Map data from the U.S. Geological Survey, Sanborn maps (i.e., historic fire insurance maps), and aerial and satellite photos

[12] . . . and be aware that what you have published may in some cases come back to haunt you.

- Climatic data from the U.S. Weather Service and Universities (if necessary)
- Newspapers

Armed with the available documents, an expert may (1) form general opinions that are verbally communicated to the client, (2) write detailed opinions, if asked, (3) write an expert report that provides opinions and the basis of each opinion, and (4) prepare an affidavit. Based on the collected and reviewed informational database, and buttressed by the expert's personal experience and education, the expert, at the request of the client, may also perform the following tasks:

- Collect and analyze the opposing expert(s) articles so as to develop deposition questions.
- Directly develop deposition questions to be put to opposing expert(s).
- Attend depositions of the opposition's experts to Insure that appropriate questions, and follow-up questions, are asked.
- Attend meetings with regulatory agencies to discuss the appropriateness of remedial alternatives.

Pretrial Activities

Once opinions have been developed, the expert then can assume that a deposition will be forthcoming. For both the experienced and the inexperienced expert, the process of providing deposition testimony is acknowledged to be the most difficult aspect of forensics.[13]

For example, the following exchange occurred at the beginning of an expert's deposition:

MR. O'CONNEL: If there's ever a point in time where you don't understand a question, please let me know. If there's ever a time you want to break—particularly today, I understand that you had some recent dental problems.

(*continues*)

[13] Although trial testimony also is stressful, opposing counsel is generally polite and less antagonistic in the presence of a judge and/or jury.

(Sidebar 1.2, continued)

THE WITNESS: I've always said I'd rather have a root canal than give a deposition. Somehow I've managed to put them both together.

This response is shared by many experts who associate a deposition more closely with an inquisition. Unfortunately, this is the price an expert pays in order to participate in a legal debate.

Obviously, if an expert's opinions are well supported, the deposition process will be fundamentally easier than when offering poorly supported opinions. In either case, a deposition can be made less stressful by completing the following steps prior to a deposition:

- Review your résumé and work experience.
- Determine if your deposition will be videotaped (if it is, make sure that you are dressed appropriately and prepared for video testimony).
- Review all of your prior relevant testimony that may be related to the opinions that will be the subject of the deposition.
- Review all publications that you have written that may be related to the opinions that will be the subject of the deposition.
- Review the opposing expert's opinions, supporting documents, and case-specific deposition testimony relative to the opinions that will be the subject of the deposition (if applicable).
- Prepare a binder(s) of relevant documents that support your opinions that may be readily referred to during the deposition (i.e., the deposition is not a test of your memory).
- Based on this review, determine the specific areas where your opinions differ and understand the basis of this difference.
- Understand the litigation objectives of both the plaintiffs and the defendants.

It is not possible to anticipate every question that will be asked by opposing counsel. However, the number of unanticipated questions and, thus, the level of stress can be minimized with proper preparation.

In many cases, the parties settle[14] and no deposition is required. Settle-

[14]As litigation progresses, each side in a dispute learns more about the technical issues, and the costs of litigation rise. Thus, at some point, common sense often dictates a monetary solution or settlement.

ment meetings can be held any time during the litigation process. These meetings may require expert presentations to explain and clarify technical issues and address remediation cost elements. In the authors' experience, well over 95% of environmental cases settle prior to trial. However, this usually is after extensive investigation, research, and formal discovery.

Even though the majority of cases settle, experts must prepare for trial testimony early. Failure to do so often leads to disaster when suddenly confronted with unexpected demands or changes in schedules (which almost always occur). An expert should prepare an organized layout of his or her trial testimony so that key documents and information can be selected and incorporated into the testimony. The expert also needs to develop demonstrative exhibits to clearly explain site issues and scientific principles that will be understandable to a layperson. If possible, this process should be implemented at least 45 to 60 days before the trial date, if known, given the time required to produce quality graphics and to make the modifications almost always requested by the client.

Telling the Whole Truth

Once trial begins, a testifying expert should expect one of the following scenarios to unfold:

- Provide one's testimony as planned.
- Based on the opposing expert's testimony, provide only that portion of the planned testimony requested by trial counsel.
- Based on the opposing expert's testimony and subsequent successful cross examination, give no testimony.
- Provide no testimony because the case settled during the trial.

reanalyzed in order to uncover facts that support opinions truthfully describing the situation being studied.

The final and perhaps the most important phase of the forensic process is the presentation of these same facts and opinions either in a settlement meeting or in court. The presentation of opinions in their most simplified form is the single most fundamental component of a convincing forensic analysis. Because of the complex nature of environmental problems, the ability to both educate and entertain, while presenting the "truth," in many cases can prove more important than the factual argument.

Environmental forensics is not an academic discipline per se, nor is it a process that is easily taught. This arises out of the complicated nature of both litigation and environmental science and engineering. In defining the "truth," virtually all successful environmental forensics investigations depend on an experienced environmental attorney as well as a multidisciplinary team of professionals (e.g., civil, chemical, environmental, and petroleum engineers; chemists; process mineralogists; geologists; soil scientists; hydrologists; biologists; economists; toxicologists; physicists; etc.).

Environmental forensics is an interactive process taking place between scientific professionals and attorneys. It can only be effectively understood by participating in the process. The dynamics of the forensic investigation and the strategies that evolve on the basis of legal objectives are unique to each and every case.

Because this is a complex process, and in order to provide a sense of the process used to define the "truth," this text has been structured to not only provide an introduction to the scientific/engineering and legal aspects of environmental forensics, but also to illustrate the process using actual case histories.

Chapter 2

A Legacy of Waste and Chemical Pollution

Most individuals recognize that a multitude of factors determine their quality of life. For example, to live in an environment that provides clean air, clean water, and a home/workplace that does not expose the owners or occupants to a chemical hazard is considered (at least by citizens of most industrialized nations) a given right.

When the United States was founded, environmental factors were dependent, in large part, on one's neighbors (e.g., agriculture, foundries, shipyards, tanneries, lumber mills, etc.). Today, neighbors have an even greater impact on the quality of life due to increased numbers, size, and diversity (e.g., refineries, petrochemical plants, gas stations, agribusiness, utilities, freeways, etc.). More important, however, today's quality of life in a selected environment (i.e., the property you may own/occupy or your place of business) is sometimes dictated by the actions of prior owners, thus the historical use of land and water. In other words, past industrial practices and associated chemical use may have degraded the quality of a land's present and future use.

During the transition from a colonial agricultural society through an industrial age and now into an information-based technological community, progress meant utilizing natural resources (i.e., petroleum, coal, metals, radioactive elements), developing energy technologies (i.e., gas from coal, electricity, natural gas, petroleum, nuclear, solar), developing and expanding mechanized transportation, and manufacturing an ever-expanding list of synthetic products and chemicals. Each developmental phase resulted in increasing consumption, expansion of production, advancement of and expanded use of chemicals, and subsequently an increasing necessity for disposing of waste products, often referred to as "residuals."[1]

Thus, over decades of use, a given property may have had one or more owners that manufactured or used chemicals or disposed of chemical wastes. The land and water impacted by these practices often retains the layered and commingled environmental legacy of each owner or neighbor. Because all property owners share the land use, a property owner may be an involuntary recipient of potential environmental hazards and associated "dislocations" (both health and economic related). For those unknowingly bequeathed with the environmental consequences of chemical pollution, it should be no surprise that litigation often is the selected remedy.

As a result, the focus of environmental forensics often is the unraveling of man's historical storage and handling practices associated with raw materials and products and disposal practices that may have contaminated soil and groundwater, as well as surface waters and air. The methods used to untangle past practices include common sense, review of historic records, application of scientific/engineering principles, and, in some cases, chemical/geochemical forensics.[2] Examples of these approaches are developed in detail within individual case histories in subsequent technical chapters. However, prior to reading these chapters, it is strongly recommended that the following introduction to the historical industrial practices associated with the handling of hazardous chemicals and the disposal of chemical waste products be reviewed so as to gain the appropriate foundation and insight leading to a greater appreciation of the environmental forensic process.

CHEMICAL-HANDLING PRACTICE

The use and storage of chemicals (both solids and liquids) by households, commercial businesses, utilities, government agencies, and manufacturing facilities has been and continues to be a common practice. The methods by

[1]"Residuals" often take on many lives. On various occasions, one company's residuals become another company's raw materials. Additionally, many residuals can be recycled, such as the use of shredded rubber as a component of asphalt. Companies have "stored" their wastes or residuals in the hope that at a later date there may be some future value.

[2]Chemical/geochemical forensics is the use of chemical fingerprinting of complex organic chemical assemblages, inorganic chemical properties and ratios, chemical age dating methods, determination of chemical isotopes, statistical analysis of chemical data, and process mineralogy to define and separate sources of pollution.

which chemicals are used and stored obviously depend on whether the chemicals are solids or liquids.[3] When these materials are allowed to be discharged into the environment (i.e., air, land, and water), environmental damage may occur.

Geologic Materials

A vast amount of geologic materials are mined, processed, and utilized by various industries throughout the United States. As a consequence, stored raw minerals and their associated waste products disposed of at the earth's surface are subject to the actions of both wind and water. When these materials are allowed to be eroded by the wind (i.e., air emissions) and contact water, both soil and water resources will be contaminated. The degree of environmental damage that may occur, however, is dependent on the toxicity of the chemical elements contained in specific mineral product or waste. Some of the more common "geologic contaminants" are discussed next.

In the 1800s as well as into the 1900s, geologic materials were utilized for the production of both metals/metal-based products and nonmetal products. For example, aluminum, iron, barium, copper, lead, and zinc ores were and continue to be extracted, crushed, and sorted into selected metal feedstocks. These feedstocks, if stored on soil and allowed to be eroded by rainfall and/or wind, can contaminate both soil and surface water with a wide range of toxic metals (i.e., major metals such as barium, copper, lead, and zinc with trace metals such as arsenic, cadmium, chromium, cobalt, nickel, silver, and strontium).

Waste rock and mill tailings from these processes have accumulated at mining sites around the United States. These materials contaminate soil and have been routinely discharged directly into surface bodies of water (i.e., creeks, streams, rivers, and lakes). The disposal of these materials at the earth's surface, without controlling wind and water erosion, results in extensive soil and water pollution.

A special case associated with metal ores and their wastes occurs when sulfides are present. When sulfides are exposed to water and air, sulfide oxidation causes the release of metals into the surface water and groundwater. If iron sulfides are present, extreme acid [1] conditions also can occur (which can increase metal transport in the environment).

[3]Although gases can be responsible for both soil and water pollution, their impact (compared to both liquids and solids) is less significant.

Since the 1940s, the processing and storage of radioactive feedstocks created an added burden on soil and surface waters. Radioactive waste disposal is a particularly large-scale problem associated with the processing of uranium and the generation of uranium mill tailings (excluding the disposal of spent radioactive materials). In addition to radioactive ores and their associated waste rock and tailings, the manufacture of metal alloys also can generate radioactive waste (e.g., magnesium-thorium). These materials, which were usually stored/disposed of at the earth's surface, are an extensive source of pollution to soil, air, and surface water resources.

The extraction and storage of nonmetal ores can also cause widespread soil and surface water pollution by toxic compounds (if stored on soil and allowed to be eroded by rainfall and/or wind). For example, nonmetal ores such as phosphate may contain radioactive elements; coal may contain trace concentrations of lead, zinc, copper, cadmium, nickel, arsenic, and selenium; sulfur ores may contain arsenic; and limestone may contain lead. Coal and coal refuse also contain iron and trace metal sulfides.

As discussed previously, this will also generate acid mine drainage. In addition, solids handling, which may include grinding and bagging, can result in fugitive releases of dust for all of the feedstocks previously listed. However, since the 1930s, one of the most toxic sources of fugitive dust results from the processing and handling of solid herbicides and pesticides that are attached to clays (i.e., clay as a carrier). When these solids are allowed to contaminate soil and surface water by direct contact or by wind and water erosion, extensive areas can be affected.

However, the greatest pollution threat is from the storage and handling of finely ground powders and liquids. The fact that these products are concentrated is an important factor. It is also important to realize, however, that a good portion of the toxic pollution that occurs today is from both organic liquids and solids and that these organic compounds have evolved through time.[4]

The Handling of Chemicals

Within most manufacturing facilities, liquids are moved from process unit (tanks, retorts, etc.) to process unit through pipes. Dry or powdered materials are transported by conveyor belts and air driven systems. It is from

[4]This does not minimize the hazard and damage caused by liquids containing inorganic compounds (e.g., toxic metals).

these types of systems that we find spills, leaks, and the airborne distribution of particulate solids. Control of these releases often depends on the assumed health hazard of the materials (worker safety) [2], general concern over cleanliness, regulatory requirements, and ultimately cost (i.e., loss of product can be expensive).

Consider for a moment the situation of leaks and spills of liquids. A release may occur directly onto and/or into soil and rock, thereby contaminating not only the soil but potentially groundwater as well (i.e., depending on site-specific conditions). When a liquid release occurs within a building, the spill may be totally contained with no contact with the outside environment.

However, when a liquid is released within a building, it may seep through cracks in the floor, discharge to a floor drain,[5] or be channeled to a sump. If discharged to a sump, the sump may be lined (made of concrete) but sometimes simply may be a concrete box with a soil bottom. In some cases, the sumps have discharge lines to sanitary or storm drains, or to ditches, ponds, or lagoons (causing pollution if the lines leak or the ditches, ponds, and lagoons are unlined).

A liquid also can be an air contaminate when the release occurs as an aerosol. These types of emissions commonly occur from cooling towers, scrubbers, and spray evaporation systems. The released pollutants are then distributed onto adjacent properties in the direction of the prevailing winds.[6]

Particulate matter released within a building either can eventually escape from the building (e.g., from a bag house) or be washed from surfaces and floors into a sump. When particulate matter is released into the air (e.g., leaks and spills from conveyor belts and air driven systems, loading and unloading operations, emissions from buildings, sources of combustion, and wind erosion of storage piles), the particles will be distributed into the environment based on particle and environmental conditions (as discussed previously).

Airborne pollutants, either liquid or solid, eventually fall to the ground or into a surface body of water. Once these pollutants are at the earth's

[5]Whose pipes may leak to the soil (etc.).

[6]For example, depending on the properties of the aerosol, the height of the release, and the local wind magnitude and direction, a distribution pattern can be modeled and then confirmed by the actual concentrations of the pollutant on the ground. Air pollution modeling, commonly referred to as "plume modeling," had its origins long before "contaminant hydrogeologists" invented groundwater plume modeling.

surface, they can be solubilized or entrained into surface runoff (i.e., eventually discharge to surface waters). As a result, some soluble fraction of the air pollution may also eventually contaminate groundwater.

Although the transport of both liquids and solids provides sources of pollution to the environment, the simple practice of the storage of liquids has been the greatest potential threat to the environment.

The Storage of Liquids

The most important liquid first stored by humans was water. Thus, engineers have dealt with the storage of water for thousands of years. For over a hundred years, water reservoirs were made watertight by employing linings such as clay, concrete, asphalt, combinations of concrete/asphalt, and subsequently synthetic fibers and chemically treated soil. Other types of storage vessels also were commonly employed to store smaller amounts of liquids such as chemicals. These vessels, again by the turn of the century, included wood, metal (iron or steel), and concrete. The degree of care relative to storage was "engineered" according to the nature of the chemical (i.e., corrosive, reactive, highly volatile, flammable, etc.). Containers were lined with glass or brick, kept cool, kept under pressure, and occasionally double vaulted—each designed to maintain chemical integrity.

Storage units were also constructed both above and below the ground surface. By their very nature, vessels that store liquids and the systems used to transport liquids into and out of storage units are always a potential source of pollution (i.e., poor or no maintenance, noncompatible liquid/tank material, soil/water corrosion, spills during transfers, etc.). Thus, there has always been the need to store liquids without loss to the environment.

Industries have consistently demonstrated an excellent capability to store chemicals and liquid products (each of which had value), but seemed to lack the desire (or bear the related cost) to employ these same types of containment structures when storing liquid wastes. Again, it is almost axiomatic that the greater the value of the chemicals, the greater is the care given to their safe storage. It is one thing to "lose waste liquids;" it is quite another thing to "lose chemicals of value." For example, products that had value would not be stored in earthen pits. However, waste liquids and solids were commonly discharged to earthen pits, ponds, and lagoons.

In spite of this practice, earthen pits, ponds, and lagoons could have been lined in order to prevent the loss of waste liquids. Since the early 1900s, there have been engineering options available to limit seepage into soil and groundwater using liners. A brief historical overview of liner technology is found in Table 2.1.

The handling of inorganic and organic compounds, either liquid or solid, is a threat not only to the workers who are in direct contact with these materials but also to the environment. As a result, it is necessary to understand the toxic characteristics of industrial chemicals.

Classification of Chemicals

Although there are those who are fond of "finger pointing" at industry regarding the hazardous nature of its products, chemical manufacturers were the first to initiate descriptions of their chemicals [3, 4], describing safe handling methods. As more emphasis was placed on safety, the descriptive forms generated by industry became more comprehensive and ultimately took the form of material safety data sheets (MSDSs), which were subsequently made a requirement by the federal government. Thus, since the early 1970s, an MSDS has accompanied all chemical products that contain hazardous compounds.

The U.S. Department of Transportation also developed requirements for the safe transportation of hazardous substances, including their fire hazard, toxicity, and reactivity. Ultimately, the Comprehensive Environmental Response, Compensation, and Liability Act (CERCLA) required that if a listed hazardous chemical (based on the compound and the amount released) was discharged into the environment, that release must be reported to the U.S. Environmental Protection Agency (i.e., a reportable quantity).

Each of these steps can be considered as evolutionary, in that as more was learned about hazardous materials and as regulations became more "focused," the detail paid by the federal government to worker safety and potential harm to the environment increased progressively. Along with this knowledge, came the gradual regulation of the disposal of wastes derived from the production and use of hazardous chemicals. However, it cannot be stressed often enough that long before laws were passed, industries producing chemical waste were well aware of the hazards.

A HISTORY OF WASTE AND POLLUTION

A Definition of Waste

Waste simply can be defined as any gas, liquid, or solid residual that no longer has a perceived value. These wastes may be generated from many

Table 2.1 Evolution of Liner Technology[a]

Year	Event
1907	University of California at Berkeley Agricultural and Experiment Station published a bulletin regarding the lining of ditches and reservoirs to prevent seepage.
1914	U.S. Department of Agriculture discussed the use of concrete linings for irrigation canals.
1942	U.S. Bureau of Reclamation discussed the use of compacted earth liners in canals.
1948	U.S. Bureau of Reclamation reported on low-cost liners, including concrete, asphalt, clay (bentonite), and plastic.
1952	A buried asphalt-membrane lining was used to line Columbia Basin canals.
1954	Soil dispersants were employed to seal an International Paper Company sulfite lagoon.
1954	Gulf States Asphalt Company advertised its prefabricated asphalt linings in the *Journal of the American Water Works Association*.
1957	Portland Cement Association published a pamphlet summarizing the uses of cement as liner material.
1958	Dow Chemical Company advertised its chemically resistant plastic liners in *Sewage and Industrial Wastes*.
1960	Bentonite, asphalt coatings, and clay blankets were used to seal waste stabilization lagoons.
1960	Staff Industries, Inc., advertised its vinyl plastic liners in *Civil Engineering* on the same subject. The brochure stated that this type of liner has "been under test at a number of locations for the past ten years, and has prevented seepage completely when properly installed."
1962	Agricultural Plastics Company advertised its vinyl plastic liners in the *Journal of the American Water Works Association*.
1963	Agricultural Plastics Company advertised its vinyl plastic liners in the *Journal of the Water Pollution Control Federation*.
1963	W. R. Meadows advertised its asphalt-plastic liners in the *Industrial Water and Wastes Journal*.
1963	U.S. Bureau of Reclamation reported that some 3000 miles of canal linings had been installed, including 355 miles of asphalt, plastic, and bentonite linings.
1966	Asphalt Institute described the different types of asphalt linings that could be used to seal waste ponds.
1966	Gulf Seal Corporation advertised its asphalt liners in the *Journal of the Water Pollution Control Federation*.

1967	Hydraulic Linings, Inc., advertised asphalt, rubber, and plastic linings in the *Journal of the American Water Works Association.*
1967	Enjay Chemical Company advertised polyvinyl chloride (PVC) liners in the *Journal of the Water Pollution Control Federation.*

[a] Every technology has its "history," and liner technology is no exception. The need to maintain "integrity," beginning with reservoirs, aqueducts, and canals, goes back as far as recorded history. However, the examples in this table begin with the 20th century.

sources, including homes, commercial and industrial businesses, agriculture, mining, transportation, and energy. The wastes from all of these sources are often categorized as sewage or liquid/solid wastes. These liquid/solid wastes can be classified as having originated from municipal sources[7] (i.e., municipal waste from homes, schools, commercial businesses), industrial sources (i.e., waste from manufacturing and processing facilities), agricultural/feedlot operations, and stormwater runoff (often classified as non-point pollution).

It is important to remember that each of these waste categories is characteristically different both chemically and biologically. Each waste can be further characterized by its own unique hazards (with industrial waste usually being considered the most chemically hazardous). Therefore, methods for municipal/domestic waste disposal should not be considered as being equally acceptable for the disposal of industrial wastes (i.e., hazardous wastes in today's terminology).

The practices utilized in the management of industrial/hazardous waste have not always relied on state-of-the-art science and engineering (primarily due to cost considerations). As a result, the historic pollution of air, soil, surface water, and groundwater from industrial facilities has been widespread throughout the United States.

The Historic Management of Waste

Since recorded history, wastes have been disposed of on land, to the water, or to the air. The ancient Greeks, Romans, and Egyptians, and, more recently, major commercial centers all had "sewers" of some type to convey liquid wastes "out of the cities." Classic solid and liquid waste disposal meth-

[7]Also known as domestic wastes.

ods included the direct discharge of waste to nearby bodies of water, burning of solids, and disposal of refuse[8] and garbage in open dumps (i.e., no soil cover), into bodies of water, or by burial. The discharge of waste into the environment has resulted in many of the illnesses that have plagued the world. The best example in recent times was the cholera epidemic in London in 1854, commonly referred to as the Broad Street Pump cholera epidemic. A drinking water well was polluted by wastes from a nearby privy. The resulting pattern of sickness led investigators to determine that cholera was being spread through groundwater. This was the first direct proof that disease was indeed spread by water and that a contaminated water resource represented a serious health issue. At about this time, a French mathematician named Darcy developed the theory and equation (Darcy's law) that could be used to describe the flow of water through porous media (e.g., groundwater). This was the beginning of humanity's "focused" understanding of the science of groundwater pollution from waste disposal. However, it was not until the U.S. Geological Survey published a report on the movement of groundwater in 1902 [5] and in its subsequent publication in 1911 [6] that the "anecdotal" references to groundwater pollution became "factual." Even without the help of science, our ancestors somehow, someway, generally seemed to understand that one did not locate a privy upstream from one's drinking water well or stream.

Because human and animal wastes were discharged directly to surface bodies of water, scientists and engineers developed technologies designed to treat liquid domestic wastes prior to their discharge. As population centers grew, the obvious "stench and sights" in major cities such as London demanded "a better way;" the commonly expressed environmental phrase that the "solution to pollution is dilution" already was being taxed in many metropolitan environments. Thus, the need to maintain clean water resources led to the development of the "sanitary engineering" discipline as an outgrowth of civil engineering.

Chicago was the first U.S. city to plan an urban sewer system, in 1855. The idea (and engineering) of filtering wastes from water and using naturally occurring bacteria to biodegrade (i.e., purify and reduce the volume of waste) is of a more recent time. One of the earliest biological units for the treatment of liquid domestic sewage was developed in Germany by, and named after, Karl Imhoff—the Imhoff Tank. Even before the turn of the 19th century,

[8]Refuse is solid waste that usually does not contain garbage (i.e., organic materials that will decompose).

sand filtration was being employed to remove impurities from water and, by 1908, chlorination was being practiced in Jersey City, New Jersey.

Although "new technology" was catching on in large population centers, one should remember that at the turn of the last century, even the most industrialized of nations still had the majority of their populations living on farms and in villages and small towns. In these locations, typical waste disposal practice included cesspools (with earth bottoms) and septic tanks discharging to leach fields (commonly referred to as tile fields). The objective of the septic tank/leach field system was to separate waste solids from liquids and allow the liquids to infiltrate through unsaturated soil. If there was a sufficient thickness of unsaturated soil above groundwater, it was assumed that the waste liquid would be "cleansed" (bio-oxidized) so that pathogens were destroyed before their seepage into the groundwater [7].

It does not take much thought to recognize that the problem of waste and waste disposal is a consequence of population density. More people generate more industrial activity, greater demand for food, greater utilization of natural resources, and more waste. Thus, with an increasing population, the visible discharge of waste to lakes and rivers becomes more obvious and therefore more objectionable.

By the mid-1920s, the discharge of human and animal wastes to soil, surface water, and groundwater was an obvious health problem, even in the rural areas of the United States. By this time, potential health effects from toxic compounds associated with the disposal of industrial wastes also were becoming more prolific.

Beginning in the 1800s, huge volumes of solid waste generated from the processing of mineral ores (ore dressing) and coal throughout the United States began to contaminate soil, surface water, and groundwater with toxic metals and acid. By the 1850s, the development of manufactured gas (from coal and then oil) added organics such as coal tars and lampblack as contaminates. With the advent of electricity from coal burning utilities, mountains of ash containing trace metals and leachable salts were being disposed of on land.

In addition to solid wastes from ores/coal utilization, both urban centers and rural communities were disposing of municipal solid waste into open dumps and surface bodies of water. The first "sanitary landfill"[9] for municipal garbage and refuse did not occur until 1934.

[9]A sanitary landfill was created by covering garbage with at least 1 foot of soil to eliminate insect and rodent problems.

Near the turn of the 19th century, a wide range of liquid and solid wastes were also being generated by major industries that did not manufacture or utilize synthetic materials. Examples include:

- Canning
- Cement manufacturing/products
- Coal chemicals
- Dairies and feedlots
- Distilling
- Fertilizer production
- Foundries
- Glass manufacturing
- Iron and steel
- Inorganic chemicals
- Metal plating
- Nonferrous metals
- Petroleum refining
- Pulp and paper
- Paint formulation
- Meat packing and rendering
- Rubber processing
- Tanning
- Textiles
- Timber products

An increasing population and technological advances through the 1930s and 1940s resulted not only in the production and expansion of existing industries, but created major new industries such as:

- Battery manufacturing
- Electroplating and metal finishing
- Electrical equipment and electronics
- Fibers (natural and synthetic)
- Herbicides and pesticides
- Pharmaceuticals
- Plastics
- Synthetic chemicals

During this period of technological advancements and expansion, industries were burying or burning their solid wastes and discharging liquid wastes into streams and rivers. In some cases, they also discharged liquids directly to groundwater using leaching pits, ponds, dry wells, shallow groundwater wells, and deep injection wells [8]. Given this ever increasing problem [9], it was inevitable that regulations would be proposed and implemented to control the discharge of all forms of waste [10].

A Need for Regulation

In tracing the history of U.S. laws addressing pollution, we can begin with the Rivers and Harbors Act of 1899, commonly referred to as the Refuse

Act, which was designed to prevent the discharge of solids to navigable bodies of water. The main thrust was to remove impediments to navigation so that commerce and capitalism could flourish unabatedly. Interestingly, no waste discharge was allowed without a permit, except for sewage and surface water runoff from the land. However, the discharge of toxic materials was banned.

Concurrent with this regulation, investigations were under way regarding the effects of pollution on water quality. This included the identification of both inorganic and organic pollutants, which by their presence in water either represented some degree of harm or were considered as "indicators" of pollution. In many cases, it was common sense for people to recognize that if water had on unpleasant odor (sulfur) or was discolored (iron and silt), it was of poor quality (without the need to conduct chemical analyses). Bad taste and odor also could be caused by dissolved organic compounds such as phenols (water-soluble compounds formed by the chemical conversion of coal and wood). Phenols impart a bad taste as well as an odor even at very low parts per billion (ppb) ranges but are not considered toxic until they reach parts per million (ppm) levels. However, most chemical compounds that occur in water have no taste, odor, or visual impact. Thus, their detection must be determined by chemical analyses.

The U.S. Public Health Service (PHS) published its first drinking water standards in 1914, addressing only bacterial contamination. A brief list of the historical federal water quality criteria is given in Appendix A. From the introduction of the first drinking water quality standards, it has been implied that waste discharges should not degrade water resources to the point where the published criteria are exceeded. However, just because there may not be a published criterion or standard does not mean that it was acceptable to discharge "unregulated" compounds onto the land or water.

For example, a common definition of "pollution" is any impairment of water by sewage or industrial waste to a degree that does not create an actual hazard to the public health but that does adversely and unreasonably affect such water for domestic, industrial, agricultural, navigational, recreational, or other beneficial use. This type of definition, which was commonly recognized and used by many states in the 1940s, demonstrates that damage to water resources is not based strictly on criteria or toxicity. Therefore, the discharge of any chemical was a potential threat to the beneficial use of water resources.

The next major piece of federal legislation regulating waste disposal took place in 1948 with passage of the Federal Water Pollution Control Act. As a result of this act, many states implemented programs to regulate wastewater discharges to water resources. This act was further expanded in 1956 to include broad water quality standards and placed greater emphasis on enforcement. The Solid Waste Disposal Act followed in 1965 and was instrumental in the closing of open-dump operations and implementing more stringent sanitary landfill practices [11] (how long does it take the government to recognize a problem and respond accordingly?).

In 1970, the U.S. Environmental Protection Agency (EPA)[10] was formed along with passage of the Water Quality Improvement Act, which firmly established standards commonly referred to as effluent limitations. The limits were employed when issuing National Pollutant Discharge Elimination System (NPDES) permits. Thus, discharge of waste to surface waters without a permit became illegal. This act also addressed the issue of liability (i.e., the polluter pays). By 1972, with passage of the Federal Water Pollution Control Act (commonly referred to as the Clean Water Act), the NPDES permit system was expanded to include effluent limits tied to available treatment technology. At this point in time, the EPA recognized that standards for surface water discharge could only be met in stages and initial compliance was dependant on the treatment technologies available to each industry sector. Because of this need, studies were conducted by consultants familiar with various industry sectors and their recommendations relative to available treatment technology were matrixed against proposed effluent discharge standards (in the early years, the EPA demonstrated great enthusiasm, innovation, leadership, and an uncommon lack of constraining bureaucracy).

In 1976, a new federal era of waste management began with passage of the Resource Conservation and Recovery Act (RCRA), which defined "hazardous waste" and established the often-expressed concept of "cradle to grave" responsibility for hazardous waste. This regulation addressed those facilities that were active businesses and that wished to continue generating, storing, transporting, treating, or disposing of hazardous waste (i.e., get a permit under RCRA or cease hazardous waste operations). Finally, in 1980, regulations pertaining to the prevention, containment, and remediation of pollution from abandoned hazardous waste areas or chemical spills came into place with the passage of the Comprehensive Envi-

[10]An often overlooked federal agency, and the precursor to the EPA, was the Federal Water Pollution Control Administration.

ronmental Response, Compensation and Liability Act (CERCLA), commonly referred to as Superfund.

So as not to be misled, the federal government was not the only government addressing problems of pollution. The states also expressed their concerns and passed their own laws—often referred to by those of little faith (or interest in complying) as "nuisance laws." Each state has its own unique history of environmental regulation.

For example, in 1929, the state of Michigan created the Stream Control Commission to control the pollution of surface waters. In 1949, by Public Act 117, Michigan amended its 1929 law extending its authority to control pollution of underground waters. The state of Washington passed its first water pollution law in 1909 and established its Pollution Control Commission in 1945, granting it control over the pollution of all waters of the state—including underground waters. In 1951, the State of Pennsylvania Sanitary Water Board adopted a resolution that by law underground waters were part of the waters of the Commonwealth and the pollution of groundwater was subject to the provisions of the antipollution law. In California, the Dickey Act resulted from legislative activities that began in early 1947. The resolution adopted by the California Assembly in June of 1947 specifically addressed pollution resulting from sewage and industrial waste disposal and its impact on both surface and underground waters. Suffice it to say, virtually every state established laws to protect its environment, often with language more specific and detailed than the federal laws in effect at the time. Several examples of these state regulations are given in Appendix B.

Science and Industry

Not to be ignored, industry, both in the United States and in Europe, recognized the pollutional nature of its wastes and, indeed, developed methods of waste control to prevent contamination. However, this information was rarely shared with the public at large (see Sidebar 2.1).

For example, even before the cholera epidemic associated with the Broad Street Pump, industry recognized that its wastes were reaching the groundwater. A civil engineer by the name of Thomas S. Peckston published a "Practical Treatise on Gas-Lighting" in which he wrote in 1841, "It is of the utmost importance that the tar cistern should be quite tight, not only on account of preventing waste of the tar and ammoniacal liquor it may receive, but also to prevent an escape of either of these fluids, which

Sidebar 2.1 Industry-Public Sector "Dysfunctional Family"

Industry always has been and continues to be the most knowledgeable about its own waste and its pollution potential. However, this patriarchy of institutional knowledge for the most part failed to share its information with its public sector relatives. In other words, little of the evolutionary knowledge of the industrial family was provided to the public sector family (citizens, government, and regulators) that was trying to define and address pollution problems.

The public sector, as represented by average citizens and their elected officials, had its primary concern with pollution of drinking water and, within that sphere, bacterial pollution was of primary concern. Added to that, what people saw and smelled ranked high on the list of concerns. Thus, the pollution of streams, rivers, and lakes—more commonly referred to as surface waters—was addressed "head on" along with "smoke" (i.e., air pollution). Although garbage dumps and early landfills were recognized as problem areas, health issues such as dust, odor, and fly and rodent infestation were the impetus for control measures. Having said all that, why were soil and groundwater not addressed more aggressively? Simply stated, what could not be seen did not appear on the radar screen of concern. Since much of the domestic pollution still came from septic tanks and leach fields, and since industrial wastes were being disposed of behind fences on private property (i.e., lagoons and landfills)—each having its path to the groundwater through the soil—and each being relatively "invisible," the extent and magnitude of the problems were not universally recognized by governmental agencies until many of our underground aquifers had been polluted by wastes, which then began to manifest themselves in drinking-water wells. Somewhat analogous to a termite infestation, what goes on "under the stucco" often goes undetected until the structure begins to crumble.

would percolate through the ground, particularly if of a gravelly nature, and, if by such means it found its way into any adjacent spring or well, would soon render the water unfit for use."[11]

By the early 1920s, industries manufacturing and using chemicals began to study and address waste disposal, as well as the effect of chemicals discharged to surface water and groundwater, thereby developing an extensive history and body of knowledge regarding pollution [12–14]; a knowledge that often ran in advance on what was known and understood by professionals outside the industrial community, including the government (i.e., regulators). Certainly, this knowledge was never presented in any detail to the public at large. Again, common sense supports the thesis that those who manufacture and use chemicals know and understand the most about these chemicals, including their potential effect on the environment.

Industries (including their employees), however, did belong to professional organizations and societies. These groups represented a forum for the discussion and distribution of industrial knowledge, which included waste management and pollution control. As a result, a brief history of professional organizations and societies delving into the issues of water and pollution is in order.

The plain fact is that much that was known about the subject of water quality, waste management, and pollution control was considered common knowledge within the professional community. In no small measure, the sharing of information within the professional community came through such organizations as the American Society of Civil Engineers (founded in 1852), the American Chemical Society (1872), the Manufacturing Chemists Association (1872), the American Water Works Association (1881), the Water Pollution Control Federation (1928), and the Purdue (University) Industrial Waste Conferences, which began in 1944.

Information distributed by these organizations through their meetings and publications, including surface water and groundwater pollution issues, began as far back as the 1920s. This information included the 1923 report of pollution of groundwater supplies (wells) by industrial wastes. By the early 1930s, a significant body of knowledge was being developed and reported regarding the character of domestic and industrial wastes and wastewaters, including such parameters as biochemical oxygen demand (BOD) [15], chemical oxygen demand (COD), phenolic content [16],

[11]It should come as no surprise that there was a lengthy history of damages to groundwater supplies by industries long before the federal government recognized the problem.

and the concentration of inorganics such as arsenic and cyanide. An explanation of the application of both BOD and COD is given in Sidebar 2.2.

Other issues being reported and addressed by these associations during the 1920s and 1930s included the land disposal of waste and the subsequent pollution of groundwater (e.g., from garbage dumps [17]); the properties of soil relative to the passage of wastes through soil and the "capacity" of soil as either a waste treatment medium or a "waste storage" medium; and the relationship and proximity of waste disposal areas to groundwater.

Advanced methods of treating domestic wastes (with a greater emphasis on biological treatment, such as activated sludge) as well as methods to treat specific industrial wastes (such as phenolic wastes) by biological treatment [18], solvent extraction, distillation, condensation, and adsorption (activated carbon) were commonly discussed and reported on—usually by industry. By the mid- to late 1930s, manufacturers of chemicals also were reporting on the physical properties of their chemicals, including their toxicities.

World War II triggered a near explosion of manufactured chemicals, including many of the solvents that have contaminated sites throughout the United States. Along with the increased manufacture and use of chemicals came greater concerns over their use and their disposal. Needless to say, the manufacturers and users of these chemicals continued to be the most knowledgeable about their chemical and physical properties. For example, in both 1938 and 1952, Dow Chemical Company published the chemical and physical properties of solvents and chemicals it manufactured. In addition to understanding product chemistry, industry also appreciated the chemical and physical properties of its wastes, and had knowledge of, and often developed, relevant state-of-the-art methods of waste treatment.

By the late 1940s and 1950s, advanced biological treatment, as well as such processes as ion exchange and incineration [19, 20], were being used to treat difficult-to-dispose-of industrial wastes, and these technological applications were appropriately reported by companies through their trade and professional associations. The emphasis continued to be on minimizing the impact (of wastes, including their toxicity [21]) on receiving waters (particularly those used for drinking water), as well as reducing or eliminating the practice of open burning. To accomplish these objectives, and in keeping with a cost minimization (profits and bottom line considerations) strategy, industry turned to more extensive land disposal of their wastes, there-

Sidebar 2.2 BOD/COD Relationship

Microorganisms can transform some organic chemicals into other compounds (i.e., biodegradation) that may be less toxic than their parent compounds. One measure of an organic chemical's ability to be biodegraded is the biological oxygen demand (BOD). Because oxygen is used by microorganisms during (aerobic) biodegradation, there will be a resulting depletion of oxygen in water or wastewater, over time. Thus, a high BOD value usually indicates the presence of a large amount of biodegradable organics dissolved in water. However, a very low or no BOD could indicate very little organic matter or that the chemicals contained therein are toxic to microorganisms and biodegradation thereby is limited. In this latter case, chemicals that do not biodegrade will persist in the environment and may pose a long-term hazard to the environment. It is for this reason that the chemical oxygen demand (COD) is often run in conjunction with the BOD test. An additional indicator of the presence of toxic chemicals is to compare the ratio of COD to BOD. If the ratio is large, then a fair assumption is that toxic chemicals may be present and further analyses are called for.

by further contaminating both the ground (soil) and, ultimately, the underlying groundwater.[12]

However, while industry turned to the land for disposal, the body of knowledge concerning the presence, fate, transport and pollutional effect of chemicals in the environment grew rapidly during this same period. For example, tracing groundwater pollution by using dyes, which dates back to 1896, became more widely employed during the 1940s and 1950s. Indeed, trichloroethylene (TCE) pollution of groundwater was reported by 1949 [22], and the toxicity of phenolic wastes to fish was reasonably defined by 1942.[13] Groundwater pollution from industrial waste lagoons was frequently reported in the 1930s and 1940s, and the distances traveled by chemicals

[12]Parenthetically, the regulatory emphasis on surface waters also may have contributed to the reason that so many industries turned to land disposal.

[13]Phenol toxicity to fish was known and understood by the turn of the last century, but the cause-and-effect relationships as measured by concentration and time of exposure were not defined until the early 1940s.

through the groundwater were published by 1932 [23]. The persistence of industrial chemicals in both soil and groundwater, particularly pesticides and agricultural chemicals, was being investigated and reported on during the 1940s. Many of these findings and reports, as well as their implications, were well recognized and discussed by professional associations.

As early as 1947, publications were recommending legislation to protect groundwater. As a result, from 1948 to 1950, a number of chemical companies turned to incineration as a more foolproof technology by which to treat hard-to-dispose-of chemical wastes, including solvents, thereby avoiding land disposal and associated groundwater pollution. In a similar fashion, by the early 1950s, new ground rules were being proposed by professional groups for the design of sanitary landfills,[14] which included addressing concerns regarding the pollution of groundwater [24]. In parallel, concerns over synthetic organic chemicals and their effect on the environment received considerable attention in the professional literature [25–27].

By 1956, several professional publications reported that impoundments and lagoons were responsible for groundwater pollution in some 20 states. In 1957, even the World Health Organization expressed concerns over groundwater pollution [28]. Early on, it was recognized and reported that groundwater cleanup was both difficult and expensive when compared to cleaning up surface waters.[15] It has often been stated that once the groundwater regime has been contaminated, no amount of remediation will ever bring back the resource to its original pristine state.

By the early 1960s, the groundwater pollution issue was so apparent that the PHS conducted meetings, as well as training sessions, specifically dealing with this problem, including discussions concerning the behavior of pollutants in groundwater [29]. Research and publications on groundwater pollution caused by industrial chemicals and waste disposal practices continued to expand through the 1970s to the 1990s. A brief history of groundwater pollution from waste disposal activities is given in Table 2.2.

How Did We Get Here?

Having established that the scientific and engineering professions have had a progressive and increasing understanding of pollution issues, and

[14]These same proposed rules also apply (at a minimum) to industrial landfills and lagoons.

[15]One of the earliest (industrial) documented examples of extraction of groundwater coupled with treatment to remove pollutants contained therein (commonly referred to as pump and treat) was reported in 1954.

have had the knowledge, experience, and technology for dealing with these problems for the better part of 100 years, a fair question to ask at this juncture is if industry and the allied professions were so well aware of the nature of pollutants, waste disposal methods, and their negative effect on the environment, why was so little done to prevent the problems we are presented with today?

The answer is both simple and complex. The simple answer is cost. It simply was cheaper and easier to hide wastes than to treat them. Bear in mind that the more readily treatable wastes, both liquids and solids, were treated. In most cases, these wastes also were the least costly to treat. On the other hand, the most difficult to deal with residuals were the pollutants that were often "hidden" in and under the ground—why treat a waste when one can dispose of it cheaply? The more complex answer requires a more detailed explanation of the regulatory environment. Most laws and regulations dealing with pollutants were, in effect, self-regulating and -reporting. Thus, if an industry applied for a permit to discharge wastes to a nearby body of water, the industry performed analyses on its waste streams, usually analyzing only those constituents specified by the local permitting agency. If the levels met the standards, permits were issued. On an ongoing basis, unless something obvious occurred—such as taste or odor in water supplies or excessive chlorine demand at sewage treatment plants or water treatment facilities—regulatory agencies relied on reports from waste dischargers (i.e., the industries) to define the nature of their wastes. Thus, pollution of unregulated compounds was, de facto, allowed to be discharged (i.e., even if industry knew there were potential hazards associated with their discharge).

Regulatory agencies had to catch up to the industry/professional knowledge base regarding waste characteristics and toxicity before they could implement new regulatory controls. In some cases, this process took decades. This is understandable since the generation of new regulations requires both public support and governmental funding.

Industries also recognized that if a portion of their wastes was discharged onto their property (emphasis here on their property), and subsequently to the groundwater beneath it, so long as local wells (which were usually some distance away from industrial facilities) were not obviously polluted,[16] no one would be analyzing the groundwater environment to determine if industry-specific contaminates were present and, therefore,

[16]For example, impacted by taste or odor.

Table 2.2 History of Groundwater and Waste Disposal Practice

1923	American Water Works Association reported on the pollution of water supplies by industrial wastes.
1931	First use of a deep injection well to dispose of industrial brine.
1932	American Water Works Association reported on impounded garbage pollution of groundwater eights months after disposal.
1936	American Water Works Association report indicated that the closer the water table is to a source of pollution, the more likely that groundwater will be contaminated.
1940	American Water Works Association report discussed chemical and biological groundwater pollution thousands of feet from disposal pits.
1943	American Water Works Association reported on tracing groundwater pollution using tracer dye.
1945	American Water Works Association reported on groundwater pollution by refinery and industrial waste with attempted cleanup.
1947	American Water Works Association report listed nine cases of industries polluting groundwater (one by dichlorophenol).
1948	American Water Works Association report warned that once a groundwater reservoir is polluted the pollution may persist almost indefinitely.
1949	Manufacturing Chemists' Association Stream Pollution Abatement Committee voted that the name of the committee be changed to Water Pollution Abatement Committee, members feeling that the change was desirable in view of the inclusion of groundwater in committee activities.
1950	*Sewage and Industrial Waste* article discussed industrial waste lagoons that can pollute groundwater if constructed in porous soils.
1952	American Water Works Association article described groundwater pollution from industrial wastes that do not biologically degrade; a long-term problem has been phenols and a charcoal-iron production plant.
1953	American Water Works Association article reported on groundwater pollution from industrial sources such as phenols, petroleum, and cleaning fluids, stating that the "proper time to control underground pollution is before it occurs.
1954	American Water Works Association article reported that chemical movement through soil into groundwater can be expected with percolating liquids.
1955	*Sewage and Industrial Waste* article reported on groundwater pollution from industrial sources, stating that soil filtration does not appreciably reduce the concentration of many chemical compounds.
1956	*Sewage and Industrial Waste* article recommended that in areas of groundwater use lagoons should have impermeable bottoms.

1957	American Water Works Association article reported that groundwater pollution, impoundments, and lagoons were responsible for groundwater pollution in 20 states.
1958	Purdue Industrial Waste Conference reported that recommended surface impoundments for oil refinery wastes be located in areas of impervious geological formation and very little groundwater.
1959	American Water Works Association article indicated that the cleanup of contaminated groundwater is far more difficult and time consuming than pollution removal in surface water.
1960	American Water Works Association articles on groundwater pollution stressed that ponding of liquids is a hazard to groundwater, as well as septic systems and cesspools.
1961	U.S. Public Health Service reported on the hydrogeologic aspects of groundwater pollution and types of groundwater pollution.
1962	American Water Works Association article described groundwater pollution from liquids poured into pits.
1964	American Water Works Association article on siting waste disposal facilities considered depth to the water table and soil permeability for the prevention of groundwater pollution.
1967	American Water Works Association article reported on the results of groundwater pollution from sanitary landfills in direct contact with groundwater.

wastes would go—and often did go—unnoticed. Let us never forget that the public only "gets what it pays for." If funding for regulatory agencies is limited, then policing also is limited. One can only expect so much from public agencies, which were usually underbudgeted, understaffed, and underequipped.

Pollution also occurred because regulatory agencies often "managed" industrial waste discharges by the threat of withholding or revoking permits. For example, a regulatory agency would issue a waste disposal permit based on the characteristics of the waste and compliance with stated discharge standards. Once the permit was issued, the facility would be allowed to operate unless the facility was found to be out of compliance with the permit conditions. When facilities were in noncompliance, they normally were given a set time period to address and remedy the identified problem. However, in some cases, the time period would be continually extended, often for years, with the anticipation that the industry would ultimately comply because without a remedy the permit would be revoked.[17] Thus, when confronted with serious pollution problems, industry commonly asserted that "we had a permit; we were never told that we couldn't discharge our waste, therefore, we were in compliance."

Finally, many industries that had access to state-of-the-art technology failed to employ that technology (see Table 2.3 for examples of state-of-the-art technologies). Thus, when industries were confronted with pollution problems, they argued that "we did what everyone else did so how could we be responsible?"[18] As a result, the question then becomes, what is the difference between "state of the art" and "state of practice"? State of the art is what you know you can do if you desire to do so (i.e., spend the money). State of practice is what you have been doing and continue to do (because it is usually cheaper) and there is no apparent regulatory objection. For example, in the 1940s and 1950s, state of practice might have been dumping liquid wastes into an unlined lagoon. On the other hand, state of the art called for an "engineered" lined lagoon, thereby eliminating, or at least greatly curtailing, leakage. Similarly, state of practice meant dumping solvents in a pit, while state of the art was to incinerate these solvents.

A Well-Kept Secret

It is obvious that the scientific and engineering professions have had a progressive understanding of pollution issues and have had the knowledge, experience, and technology for dealing with these problems for the better part of 100 years, perhaps longer. Given this scientific and engineering understanding, why are we having to contend with (historic) widespread pollution and at such a huge cost? There is no simple answer. Each pollution problem has its own set of unique circumstances that must be unraveled and analyzed. Perhaps pollution, especially industrial pollution, was a well-kept professional secret just beyond the public's awareness. Conversely, domestic wastes, collected in sewers and handled at municipal treatment plants, were much more visible and subject to much greater public scrutiny. Many industries have continued to play the "shell game,"

[17]It should be noted that regulatory agencies, in some instances, were hesitant to revoke a permit because of the potential economic impact. In addition, sometimes it was difficult, if not impossible, to obtain public permission for new or expanded waste disposal facilities such as industrial landfills.

[18]This "common practice" argument parallels the argument that "I may have been speeding but so was everyone else."

Table 2.3 State-of-the-Art Technologies[a]

Year	Event
1935	DuPont established a laboratory to investigate the toxic effects of chemicals.
1937	Dow Chemical Company completed the first large-scale biological treatment plant in the United States to destroy phenols in its wastewaters.
1945	Hoffman-La Roche operated a modified garbage incinerator to dispose of organic liquid wastes.
1946	Lederle Laboratories employed incineration to dispose of wastes, including solvents.
1948	Dow Chemical Company began deep-well injection of industrial liquid waste.
1952	DuPont used a catalytic oxidation unit for destruction of organic wastes.
1957	Dow Chemical Company built a high-temperature incinerator to burn highly toxic waste materials, including plastics.
1957	Upjohn Company employed incineration to destroy waste solvents.
1957	Koppers Company employed incineration to destroy organic wastes.
1957	Monsanto incinerated liquid wastes from vinyl chloride production.

[a] Examples of when several state-of-the-art technologies were employed by industry in the United States.

hiding their wastes to the degree that regulatory circumstances allowed, thereby deferring "waste-related costs" and protecting the "bottom line."

The concept "out of sight, out of mind" was the prevalent historic waste management approach at the headquarters of many industries and, coincidentally, in many municipalities. So, where we are today, in no small measure, resulted from the inability to get a powerful enough message out so that the public, becoming aware of the problem, would bring pressure on the politicians to write new legislation and to fund the enforcement of existing laws and regulations.

A perfect example of this process is illustrated by Rachel Carson's book *Silent Spring*, which was published in 1961. Although the hazards of dichlorodiphenyl trichloroethane (DDT) were reported by Carson, it still took some nine years for the government to ban the use of DDT. In effect, Carson's book was the "microscope" that brought the problem into focus. Unfortunately, relative to industrial waste practices and pollution, there was no similar (public domain) microscope. The issue of bringing public focus cannot be overstated. Even when rivers were running yellow, pink, green, and brown; even when gas bubbles routinely broke the surface; even when fish floated to the surface and washed up on shores; even when a river "caught fire," corrective actions only came about when the public outcry became too loud to ignore. Needless to say, when these same chemicals, which often were so obvious "on the surface," ended up in the groundwater—their impact went unnoticed—a time bomb of immense proportion and cost was created.

Pollution of natural resources, regardless of its form (i.e., pollution of air, soil, surface water, and/or groundwater), resulted from decisions to defer waste-related costs, to either protect the bottom line or increase production.

An Issue of Deferred Maintenance

The United States continues to be viewed by many as "utopia," the ultimate example of dynamic growth, the land of opportunity. Without a doubt, technological innovation, whether measured by agribusiness, manufactured goods, abundant energy, transportation, and communication, has been and continues to be the economic platform of this nation. Had industry directed more of its resources to addressing issues of pollution, perhaps we would have decelerated our growth and, indeed, might have lost some of our initiative.

Fundamental to any development is the investment of available capital coupled with the ability to pass costs on to the consumer. Profits or "surplus capital" directed toward pollution control measures can (and indeed in certain industries did) stifle investment in new technologies and plant modernization and expansion, as well as job creation. Similarly, the increased cost of finished products has given advantages to overseas competitors with the result that both facilities and jobs have moved "offshore."

While this argument is no excuse for intentional (or for that matter, unintentional) environmental damage, it can be argued that industrial growth, in special cases, is a beneficial social tradeoff to pollution. Consider for a moment just how much pollution was created during World War II, when the government essentially ignored problems of pollution in order to win the war. Was this a good tradeoff? We won the war, but we are still fighting the residuals—such as health problems from asbestos, as well as pollution that resulted from the widespread and often indiscriminate use of solvents, pesticides, and weapons manufacture chemicals.

Today, the federal government, as well as many state and local governments, are committed to spending billions of dollars on addressing our decaying infrastructure—an infrastructure that is vital to our nation. How did this happen? Simply stated, available funds were directed to other programs, resulting in a massive "delayed maintenance" issue with associated increased costs. Thus, many industries and companies that deferred pollution abatement are paying a greater price today.

It is reasonable to state that, in today's environment, addressing pollution issues, regardless of whether we focus on municipal pollution, agricultural pollution, or industrial pollution, requires a careful investigation of each "singular" situation. Indeed, each situation should be evaluated on its own merits. While the reasons for deferred maintenance can be argued, the necessity of dealing with the problem is obvious. The only open issue, as always, is who pays.

REFERENCES

1. W. Hodge, "Pollution of Streams by Coal Mine Drainage," *Industrial and Engineering Chemistry*, 29(9): 1048–1055 (September 1937).

2. M. Bowditch et al., "Code for Safe Concentrations of Certain Common Toxic Substances Used in Industry," *Journal of Industrial Hygiene and Toxicology*, 22(6): 251 (June 1940).

3. Dow Chemical Company, *Dow Industrial Chemicals and Dyes* (1938).

4. Dow Chemical Company, *Dow Organic Solvents* (1938).

5. C. Slichter, "The Movement of Underground Waters," *U.S. Geological Survey Water-Supply and Irrigation Paper* No. 67 (1902).

6. G. Matson, "Pollution of Underground Waters in Limestone," in M. Fuller et al., Eds., Underground Water Papers, *U.S. Geological Survey Water-Supply Paper* No. 258 (1911).

7. C. Stiles and H. Croshurst, "The Principles Underlying the Movement of Bacillus Coli in Ground-Water, With Resulting Pollution of Wells," *Public Health Reports*, 38(4): 1350 (June 15, 1923).

8. C. Stiles et al., "Experimental Bacterial and Chemical Pollution of Wells via Ground Water, and the Factors Involved," *Hygienic Laboratory Bulletin*, No. 147 (June 1927).

9. W. Brown, "Industrial Pollution of Ground Waters," *Water Works Engineering*, 88(4): 171 (February 20, 1935).

10. B. Doll, "Formulating Legislation to Protect Ground Water from Pollution," *Journal of American Water Works Association*, 39(10): 1003 (October 1947).

11. C. Calvert, "Contamination of Ground Water by Impounded Garbage Waste," *Journal of American Water Works Association*, 24(2): 266 (February 1932).

12. Anonymous, "Stream Pollution Becomes a Problem for Chemical Engineers," *Chemical and Metallurgical Engineering*, 45(3): 138 (March 1938).

13. I. Harlow, "Waste Problems of a Chemical Company," *Industrial and Engineering Chemistry*, 31(11): 1346 (November 1939).

14. R. Goudey, "Symposium—Disposal of Liquid Industrial Wastes, 1. The Industrial Waste Problem," *Sewage Works Journal*, 16(6): 1177 (November 1944).

15. F. W. Mohlman, "The Biochemical Oxidation of Phenolic Wastes," *American Journal of Public Health*, 19(2): 145 (February 1929).

16. J. Baylis, "An Improved Method for Phenol Determinations," *Journal of the American Water Works Association*, 19(5): 597 (May 1928).

17. C. Calvert, "Contamination of Ground Water by Impounded Garbage Plant Wastes," *Water Works and Sewerage*, 78(12): 371 (December 1931).

18. W. Rudolfs, "A Survey of Recent Developments in the Treatment of Industrial Wastes," *Sewage Works Journal*, 9(6): 998 (November 1937).

19. Anonymous, "Incineration Solves a Waste Disposal Problem," *Chemical Engineering*, 55(3): 110 (March 1948).

20. C. Arbogast, "Incineration of Wastes from Large Pharmaceutical Establishments," *Proceedings of the Fourth Purdue Industrial Waste Conference*, Extension Series No. 68: 255 (September 21–22, 1948).

21. W. Rudolfs et al., "Review of Literature on Toxic Materials Affecting Sewage Treatment Processes, Streams, B.O.D. Determinations," *Sewage and Industrial Wastes*, 22(9): 1157 (September 1950).

22. F. Lyne and T. McLachlan, "Contamination of Water by Trichloroethylene," *The Analyst*, 74(882): 513 (September 1949).

23. F. Veatch, "Tracing Underground Flows—Special Reference to Pollution," *Water Works and Sewerage*, 81(11): 379 (November 1934).

24. American Water Works Association, "Findings and Recommendations on Underground Waste Disposal: Task Group Report," *Journal of American Water Works Association*, 45(12): 1295 (December 1953); presented by Task Group E4–C (Underground Waste Disposal and Control) at May 14, 1953 Annual Conference.

25. M. O'Neal and T. Wier, "Mass Spectrometry of Heavy Hydrocarbons," *Analytical Chemistry*, 23(6): 830 (June 1951).

26. H. Braus, et al., "Systematic Analysis of Organic Industrial Wastes," *Analytical Chemistry*, 24(12): 1872 (December 1952).

27. C. Lamb and G. Jenkins, "B.O.D. of Synthetic Organic Chemicals," *Proceedings of Seventh Purdue Industrial Waste Conference*, Extension Series No. 79: 326 (May 7–9, 1952).

28. World Health Organization, "Pollution of Ground Water: World Health Organization Report," *Journal of American Water Works Association*, 49(4): 392 (April 1957).

29. U.S. Public Health Service, *Ground Water Contamination: Proceedings of the 1961 Symposium*, Robert A. Taft Sanitary Engineering Center, Cincinnati, OH (April 5–7, 1961).

Chapter 3

Introduction to Environmental Law

THE EVOLUTION OF THE REGULATORY ENVIRONMENT

To understand the evolution of the regulatory environment, one must first have a general understanding of what is meant by the term *environmental law*. It is a concept that means different things to different people. It is a term that evades an agreed upon meaning. To environmentalists, it may be perceived as a champion of the fundamental right to clean water, clean land, and clean air. To politicians, environmental law may be a vehicle of public popularity or public scorn. To governmental agencies, it may be viewed as a powerful regulatory tool. To industry, environmental law may be considered a myriad of vague regulations, a necessary evil of doing business, or an albatross of unwanted expense. To academicians, environmental law is viewed as an intellectual endeavor, a body of law incorporating constitutional law, tort law, property law, administrative law, and more. To realists, it is probably seen as a scattered attempt at addressing developing problems, missing the necessary foresight to create a sustainable solution. To lawyers, it is another in a long line of "in vogue" "cottage industries." To Captain Hazelwood,[1] it may be seen, in triplicate, as a well-oiled machine careening out of control, only to be run aground and halted by an unseen, unanticipated condition.

Whatever view one has of the definition of environmental law, most will agree that it is necessary. As noted by David Sive, the "grandfather of en-

[1]Captain (former), *Exxon Valdez*.

vironmental law," "there is only one Earth"[1]. At least that we know of in this endless universe. That reason alone may be deemed sufficient for the development of the massive body of law that is environmental law, to avert wandering down the lonely path of the dinosaur.

Rivers and Harbors Act of 1899

For as long as history has been recorded, men have sailed the seven seas, pitched their tents next to bodies of water, and haphazardly dumped whatever refuse existed, overboard. With the industrialization of the world, this refuse and waste became much more prevalent and, for that matter, a bit more aromatic. Our great leaders of the day decided to put an end to this reckless practice. In 1899, in what may be considered the initial statutory step toward environmental protection, Congress passed the Rivers and Harbors Act of 1899 (RHA). The RHA prohibits the discharge of refuse matter "of any kind or description whatever" into the navigable waters of the United States without a permit from the U.S. Army Corps of Engineers.[2] However, as is often the case with a strong lobby or a conduct that screams for regulation, the RHA was difficult to police and, therefore, rarely used for enforcement. It was ultimately replaced, for the most part, by the Clean Water Act of 1972.[3]

Faustian Pact

While the Rivers and Harbors Act and other random pollution control statutes were an early attempt at addressing environmental problems, in truth, Americans continued to utilize the environment for economic well-being, doing whatever was possible to avoid addressing the environmental consequences of those actions. Pursuant to this "credit card" theory or Faustian pact,[4] the United States, in actuality, cleverly avoided addressing the massive environmental degradation caused by years of technological advancement, with environmental impunity, until the past three decades.

The evolution of the environmental regulatory scheme coincided with the

[2]Rivers and Harbors Act of 1899, 33 U.S.C. §407.

[3]Federal Water Pollution Control Act (Clean Water Act [CWA]) 101–607, 33 U.S.C. 1251–1387 (1994). The CWA was a substantial change to previous water pollution control, which was left primarily to states and was deemed inadequate for a number of reasons, including the fact that water quality standards were ineffective by themselves, the lack of a permitting process, and inefficient administration.

[4]Warren Freedman, *Hazardous Waste Liability* 75 (1987). Goethe's Faust sold his soul to the devil for short-term fame and success.

first Earth Day in 1970, when society began to reevaluate its perception of environment and economy, with 53% of Americans choosing the reduction of air and water pollution as the second greatest problem facing government.[5] It was at this time, in response to growing public concern over the quality of the nation's air and water, that Congress created the "command-and-control" environmental regulatory scheme that predominantly exists today. With these newly enacted "command-and-control" statutes, "environment" became a household name, and, to some, a four-letter word.

Command and Control

Command and control generally describes regulation that encompasses the theory of government controlling the behavior of the regulated community [2]. In the environmental context, this means that Congress and the EPA set forth in detail rules, such as technology and operating standards, that businesses must follow regarding various contamination sources, such as containers, tanks, trucks, impoundments, wells, pipes, and landfills, and the contaminants, such as organic wastes or inorganic wastes, that may be distributed by industry through the air, soil, water, or sewer systems. The regulators use these standards to develop a permitting system, under each statute, for various facilities. This permitting system is ultimately used to enforce compliance. These sweeping changes, mandated by Congress, have had and will continue to have an impact on business practices no less than that created by the New Deal of the 1930s [3]. Many have debated whether this hard line regulatory approach is needed; however, very few have resisted maximizing their gain from the regulations or falling into the tragedy of the commons.[6]

Many of these command-and-control statutes on their face require a "Fantasy Island" elimination of the harmful pollution the act addresses. In reality, the promulgated standards reduce the levels of pollution to an acceptable level of protection. For example, Section 101(a)(1) of the Clean Water Act states that "it is the national goal that the discharge of pollutants into the navigable waters be eliminated by 1985"; yet, today any regulated entity possessing a National Pollutant Discharge Elimination System

[5]Jackson B. Battle, *Environmental Decisionmaking and NEPA* 3 (1986). Crime was considered the first greatest problem facing government.

[6]Garrett Hardin, The Tragedy of the Commons, Science 162:1243 (1968). "Ruin is the destination toward which all men rush, each pursuing his own best interest in a society that believes in the freedom of the commons. Freedom in a commons brings ruin to all." Id.

(NPDES) or State Pollutant Discharge Elimination System (SPDES) permit may lawfully discharge pollutants into navigable waters under certain controlled circumstances.[7]

Environmental Legislation: Examples

Clean Water Act

Many command-and-control schemes evolved in the 1970s. One of the earliest was the Clean Water Act (CWA),[8] a substantial supplement to the Federal Water Pollution Control Act and the Clean Air Act (CAA).[9] The Clean Water Act was deemed necessary to reduce the introduction of pollutants, such as organic wastes, toxic chemicals, and hazardous substances, into the water bodies of the United States. The CWA is composed of a system of standards and permits designed to utilize technology to reduce water pollution, requiring a permit for the discharge of pollutants from a point source into the navigable waters of the United States. To this end, the EPA set forth technological standards in the form of effluent limitations that must be met to avoid federal, state, or civil enforcement.

Clean Air Act

The Clean Air Act is another example of one of the powerhouse command-and-control environmental statutes that were enacted in the 1970s. Increased levels of carbon monoxide (CO), nitrogen oxide (NO_x), sulfur oxide (SO*x*), particulate matter (PM-10), and hydrocarbons (HCs) in the nation's air led to the CAA's enactment. These pollutants were found to cause or contribute to cancer, emphysema, bronchitis, asthma, eye and lung irritation, damage to crops, and smog. To address the introduction of these pollutants into the air, National Ambient Air Quality Standards (NAAQSs), both primary and secondary, were established, by the EPA, under Section 109 of the Clean Air Act. The primary standards were required to protect public health, while the secondary standards were implemented to protect public welfare. In turn, states were required to submit state implementation plans (SIPs) that set forth each state's plan for attaining the NAAQSs. As with the CWA, technological standards were set and integrated into a permit sys-

[7]Conversely, the government will not decertify the Sheboyagan Harbour from being navigable, a waterway that has not been actually navigable for more than 30 years, thereby delaying a significantly less expensive cleanup process.

[8]Federal Water Pollution Control Act (Clean Water Act [CWA]) 101-607, 33 U.S.C. 1251-1387 (1994).

[9]Clean Air Act 101-618, 42 U.S.C. 7401-7671q.

tem. The standards varied, depending on the classification of the source—stationary source, mobile source, existing source, new source, and so on. Also similar to the CWA, a civil enforcement provision was adopted in the CAA, to complement state and federal enforcement. In other words, it was time to clean up your act, if you were a polluter, legal or otherwise.

We have not attempted to provide in this introductory setting anything more than a basic introduction[10] to the history of environmental legislation. It is, however, already evident from this basic introduction how government, recognizing the growing concern for environmental protection and the lack of successful tools to implement existing statutes, took command and control and imposed a vast array of intrusive rules and regulations on business. Government has now told business what to do, how to do it, when to do it, and will force it to do it, if necessary.

Soon after effective use of command-and-control measures in connection with the enactment of the CWA and CAA, the government moved into the hazardous waste business with the enactment of hazardous waste legislation. RCRA and CERCLA are two of the most intrusive and broad-based command-and-control statutes ever passed. No other legislation has probably had a greater impact on business in the short but impressive history of the United States.

Introduction to RCRA and CERCLA

RCRA[11]

The Resource Conservation and Recovery Act (RCRA), first known as the Solid Waste Disposal Act (SWDA), was established to confront the problems that hazardous waste was causing to human health and the environment, by encouraging the management of wastes via conservation and reduction or elimination. A cradle-to-grave statute, RCRA applies to the generation of waste, the transportation of that waste anywhere regardless of the distance, and the storage and disposal of hazardous waste. Once a waste is classified as a hazardous waste,[12] it falls under the purview of Subtitle C, where the responsibility of handling the waste is divided among three groups: generators, transporters, and the owners/operators of TSDFs

[10]Read the footnotes to quench that thirst for more.

[11]Resource Conservation and Recovery Act (RCRA) 1002-11012, 42 U.S.C. 6901-6992k (1991).

[12]RCRA hazardous wastes can be "characteristic wastes" (as distinguished from listed wastes), depending on their ignitability, corrosivity, and toxicity. Hazardous wastes can also be listed hazardous wastes, which can be found in 40 C.F.R. 261.

(treatment, storage, and disposal facilities). A brief listing of what is required from each follows:

Generator[13]

- Obtaining EPA ID# for tracking
- Preparation of transportation for safe transport
- Accumulation and storage requirements to prevent leaking manifest
- Recordkeeping and reporting requirements

Transporter[14]

- Obtaining EPA ID#
- Complying with manifest handling hazardous waste discharges

TSDF owner/operator[15]

- Administrative/nontechnical requirements and technical requirements
- Waste analyses
- Security
- Inspections
- Training
- Management of ignitable, reactive, or incompatible wastes-location standards
- Contingency plan, emergency procedures, corrective action manifest
- Recordkeeping and reporting
- Groundwater monitoring and protection
- Closure/postclosure (see following discussion) and financial requirements (see following discussion)

Closure & Postclosure Closure is the period during which owners or operators of TSDFs complete their operations and apply the final covers or cap landfills. Postclosure is the 30-year period following closure, during which the owners or operators maintain and monitor the closed disposal system for leaching, cap punctures, and other contamination problems. A plan must be crafted for closure to include how the facility will be closed, the estimated year of closure, a schedule for the closure, and an estimate

[13]Requirements for generators are located at 40 C.F.R. pt. 262.
[14]Transporters' requirements are located at 40 C.F.R. pts. 171-179 and 40 C.F.R. pt. 263.
[15]TSDF regulations are located at 40 C.F.R. pts. 264-265.

of the maximum amount of waste the site will hold. A similar plan must be created for postclosure. The postclosure plan usually provides for groundwater monitoring and reporting, maintenance and monitoring of waste containment systems, and security.

Financial Assurance of Closure Pursuant to RCRA Sections 3004(a)(6) and 3004(t), owners or operators of TSDFs must demonstrate evidence of financial responsibility for the closure of a facility, for postclosure activities, and for compensation of third parties damaged by accidents occurring in relation to the facility's operation.

One of the primary objectives of the financial responsibility provision of CERCLA is to prevent the TSDFs from becoming Superfund sites by ensuring adequate funding for proper closure. The six mechanisms of demonstrating financial responsibility are:

- Trust fund
- Surety bond
- Letter of credit
- Closure/postclosure insurance
- Corporate guarantee
- Financial test

Any one or a combination of these mechanisms may prove adequate.

A Glance at a State RCRA Statute and Its Requirements Most states have their own hazardous waste statutes. As an example, we will review that found in New York under Article 27, Title 9, of the Environmental Conservation Law (ECL). New York's RCRA equivalent is also a cradle-to-grave statute, with its purpose being "to regulate the management of hazardous waste (from its generation, storage, transportation, treatment and disposal) in this state . . . "[4]. Hazardous waste is defined by the statute as:

> a waste or combination of wastes, which because of its quantity, concentration, or physical, chemical or infectious characteristics may: a. [c]ause, or significantly contribute to an increase in mortality or an increase in serious irreversible, or incapacitating reversible illness; or b. [p]ose a substantial present or potential hazard to human health or the environment when improperly treated, stored, transported, disposed, or otherwise managed.[16]

[16]N.Y. Envtl. Cons. L. 27-0901(3) - (3)(b). Identification and listing of hazardous wastes can be found at 6 N.Y.C.R.R. 366.

Similar to RCRA, in order to achieve the purpose of the statute, New York regulates three specific groups: generators; transporters; and treatment, storage, and disposal facilities. For specific statutory standards, the New York Code of Rules and Regulations should be consulted. General requirements are reviewed next.

Generators General requirements for generators of hazardous wastes, pursuant to ECL Section 27-0907, include:

- recordkeeping to identify quantities generated, the constituents of the wastes which are significant in quantity or potential harm, and the disposition of such wastes;
- labeling of containers to identify the wastes;
- use of appropriate containers;
- use of a manifest; and
- submission of an annual report to the commissioner.

Transporters General standards for transporters, located at ECL Section 27-0909, include:

- recordkeeping to identify the wastes, their source, and their destination;
- only may transport and store properly labeled wastes;
- compliance with manifest, including transportation of such wastes only to the TSDF listed on the manifest; and
- bond requirement, as a condition to issuance of a permit, in case of a release.

TSDFs The statute, in ECL Section 27-0911, sets forth that the standards for owners/operators of TSDFs are to be consistent with comparable standards promulgated pursuant to RCRA.

Financial Requirements Similar to RCRA, Section 27-0917 of the ECL, requires financial assurance:

> . . . to be included as conditions in hazardous waste facility permits for the remediation of failures during operation and after facility closure, for facility closure, and for pre-closure facility monitoring and maintenance.[17]

Methods for meeting the financial requirements are also similar: through trust funds, surety or performance bonds, letters of credit, liability insur-

[17]N.Y. Envtl. Cons. L. 27-0917(1).

ance or annuities, or corporate guarantees or guarantees by other legal or financial affiliates of the owner or operator.

Closure and Postclosure Section 27-0918 of the ECL imposes requirements for closure and postclosure plans upon owners and operators of hazardous waste facilities. Pursuant to this provision, owners and operators must submit closure plans and plans for postclosure monitoring and maintenance to the Department of Environmental Conservation. In addition to these plans, owners and operators must also submit closure and postclosure cost estimates, supplementing them any time that a change in the regulatory requirements would affect the cost estimates.

CERCLA[18]

The Comprehensive Environmental Response Compensation and Liability Act (CERCLA), enacted at the close of the Carter Administration, may be considered the most daunting, powerful piece of environmental legislation ever enacted. By way of example, under CERCLA, a dry-cleaning business owner who had been lawfully storing hazardous substances in the 1940s may, in the 1990s, find himself being taken to the cleaners with his own hefty cleaning bill. Along with its reputation as one of the most powerful pieces of legislation, CERCLA is also one of the most highly criticized pieces of environmental legislation. Critics characterize CERCLA as slow, expensive, and unfair. Some of these characteristics will become apparent in the following discussion.

CERCLA was intended to remedy the developing hazardous waste nightmares such as Love Canal[19] and Valley of the Drums.[20] These are not B movies or Spectrovision, but rather, the likely impetus for Congress for the enactment of this sweeping law. In fact, CERCLA was enacted primarily to finance the cleanup of hazardous substance contaminated sites and, secondarily, to recover those costs from the parties responsible for the conta-

[18]Comprehensive Environmental Response, Compensation and Liability Act (CERCLA) 101-405, 42 U.S.C. 9601-9675 (1993).

[19]Love Canal, located in Niagara Falls, New York, was a toxic waste dump site and an impetus for the enactment of CERCLA. In the 1970s, during heavy periods of rainfall, residents surprisingly found barrels rising up into their yards as the groundwater level rose. In the 1990s, some families were finally allowed to move back into the area; however, the site must be maintained and monitored indefinitely to prevent cap punctures and other potential problems.

[20]Valley of the Drums, along with Love Canal, was a hazardous waste dump full of thousands of rusting, leaking drums, this one located in Bullitt County, Kentucky, that helped lead to the enactment of CERCLA.

mination. Toward this end, Congress created the Superfund, a multibillion-dollar fund, for the EPA to utilize in remediating contaminated sites.

As RCRA focuses on the effective management of hazardous wastes with all of its elusive definitions, CERCLA maintains broad jurisdiction over hazardous substances,[21] which are statutorily defined to include:

- CWA toxic pollutants and hazardous substances
- CAA hazardous air pollutants (HAPs)[22]
- RCRA hazardous wastes (listed and characteristic)[23]
- Toxic substances control act (TSCA) hazardous chemical substances or mixtures[24]

CERCLA equips the EPA with four methods of dealing with contaminated sites:

- Clean the site, pursuant to Section 104, and obtain reimbursement, under Section 107, from the potentially responsible parties (PRPs).
- Use Section 106 to issue a unilateral order to the PRPs requiring the PRPs to conduct an investigation and cleanup.
- Use Section 106 to seek a court order compelling the PRPs to undertake the response activities.
- Engage in a settlement with one or more of the PRPs that will require the settling PRPs to perform the investigation and cleanup.

Remediation Before the EPA may undertake an extensive remediation, the site to be remediated must be listed on the National Priorities List (NPL),[25] which is, oddly enough, a listing of contaminated sites prioritized by their relative risk, as determined by the Hazardous Ranking System (HRS).

[21]CERCLA 101(14), 42 U.S.C. 9601(14). In addition to the other statutory substances that fall under the purview of CERCLA, 40 C.F.R. 302 lists CERCLA-specific hazardous substances.

[22]The regulations listing CAA HAPs are located at 40 C.F.R. 61.01.

[23]RCRA listed hazardous wastes can be found at 40 C.F.R. 261. Characteristic hazardous wastes, under RCRA, depend on their ignitability, corrosivity, and toxicity.

[24]TSCA toxic substances are indexed, as the Toxic Substances Chemical—CAS Number Index, at the end of 40 C.F.R. 700-789.

[25]The number-one site on the National Priorities List, the most hazardous of all sites, is the Lipari Landfill—a peach farm.

A site will be listed on the NPL if it is equal to or greater than a set score in the HRS.[26] The HRS has specific criteria for determining whether a site should be listed on the NPL. The criteria are based on:

> . . . relative risk or danger to public health or welfare or the environment . . . taking into account to the extent possible the population at risk, the hazard potential of the hazardous substances at such facilities, the potential for contamination of drinking water supplies, the potential for direct human contact, the potential for destruction of sensitive ecosystems, the damage to natural resources which may affect the human food chain and which is associated with any release or threatened release, the contamination or potential contamination of the ambient air which is associated with the release or threatened release, State preparedness to assume State costs and responsibilities, and other appropriate factors.[27]

Any EPA cleanup of a contaminated site must be completed in a manner not inconsistent with the National Contingency Plan (NCP), pursuant to CERCLA Section 105. The NCP provides for an array of studies to be performed and criteria that must be applied to the studies.

First, the NCP requires scoping, which is, essentially, an early review of data to determine future steps to be taken. Scoping includes such acts as developing health and safety plans, evaluating existing data, identifying potential cleanup methods, notifying natural resource trustees, and identifying potential Applicable or Relevant and Appropriate Requirements (ARARs).[28]

Once scoping has been completed, a remedial investigation (RI) must be performed. A remedial investigation helps in forming a remedy selection by assessing the site and evaluating alternative forms of remediation—often leading to the question: How clean is clean? The remedial investigation involves a baseline risk assessment that is intended to determine the current and potential threats to human health and the environment and to determine acceptable exposure levels.

[26]The score to beat is 28.5. More information on the HRS can be found in Appendix A to 40 C.F.R. 300.

[27]CERCLA 105(a)(8)(A), 42 U.S.C. 9605(a)(8)(A).

[28]ARARs refer to the application of all federal standards and more stringent state standards to the cleanup of a site. For example, a remedial cleanup that involves transferring the hazardous substances off site may only send those substances to a RCRA permitted facility. An applicable standard is a standard that specifically deals with hazardous substances, pollutants, remedial actions, or other circumstances at a site, whereas relevant and appropriate standards are those standards that are not "applicable" but address "problems sufficiently similar to those encountered at a CERCLA site." See William H. Rodgers, *Environmental Law* 741-743 (2nd ed. 1994 and Supp. 1996); see also 40 C.F.R. 300.5 for the NCP definition of ARARs.

The feasibility study (FS) generally occurs concurrent to the remedial investigation. Its purpose is to develop and analyze alternatives for an appropriate response. This involves a consideration of a wide variety of remedies, followed by deleting those options that may be too expensive or ineffective. An expensive remedy may be eliminated when a cheaper, equally effective remedy exists.

Upon completion of the RI/FS, the remaining cleanup options are evaluated in detail, applying the following criteria:

- overall protection of the public and environment;
- compliance with ARARs;
- long-term effectiveness;
- reduction of toxicity and mobility;
- short-term effectiveness;
- implementability;
- cost;
- acceptability to the affected state or Indian tribe; and
- acceptability to the public.[29]

Generally, the first two criteria must always be met, the middle five are balancing criteria, and the last two are simply additional criteria to be considered. Once a remedy is chosen, it must be presented to the public for comment and acceptance.

The CERCLA Liability Scheme CERCLA identifies four groups of PRPs that may encounter liability under the act[30]:

- current owners and operators;
- owners and operators at the time of disposal;
- generators who arranged for the disposal or arranged with the transporter for the disposal; and
- transporters.

This on its face seems fair enough. However, the liability scheme under CERCLA is extremely broad. A current owner or operator who never touched a hazardous material or an owner or operator at the time of disposal can and has included, within the long arm of CERCLA, a plant supervisor who had authority over the disposal of wastes; one who had the authority to prevent the disposal, even though that person never engaged

[29] 40 C.F.R. 300.430(e)(9).

[30] See CERCLA 107(a)(1)-(4), 42 U.S.C. 9607(a)(1)-(4).

in the disposal activities; a lending institution with a security interest in the facility; a landlord that had no involvement in a tenant's waste disposal; and countless other individuals and entities.[31] This broad liability scheme is the primary reason that CERCLA is often criticized for its unfairness. Many found liable under CERCLA were, at the time of disposal, acting in accordance with the law and/or acting pursuant to their occupational requirements. Operators are, among others, those that have sufficient involvement in the management and operations of the site; those with control over the pollution-causing disposal system; those positioned to prevent or abate the pollution, or those responsible for the environmental controls at the facility. They face CERCLA liability as a result of their control over the sites' management and operation.[32] Owners are generally liable under CERLCA because they are deemed to have control over how the property will be utilized. An owner can be, among others, a lessee, a parent corporation that maintains substantial control over the subsidiary, or a fee owner.[33] Generators or arrangers for disposal are those that create the waste and arrange for its disposal, or those that did not create the waste, but determined where the waste would be disposed. These parties face CERCLA liability because they made the decision to send the contaminants to the site that must now be cleaned.[34] A transporter is generally a

[31]*United States v. Maryland Bank*, 632 F. Supp. 573 (D. Md. 1986) (construing the CERCLA 107(a)(1) phrase "owner and operator" disjunctively); *Ecodyne Corp. v. Shah*, 718 F. Supp. 1454 (N.D. Cal. 1989) (holding prior owner, who introduced chromium to the property, liable, and not subsequent interim owners, by inferring an active conduct requirement pursuant to the definition of disposal); but see *Nurad, Inc. v. Hooper & Sons Co.*, 966 F.2d 837 (4th Cir.), *cert. denied*, 113 S. Ct. 377 (1992) (finding that leaking, found in the definition of disposal, typically occurs without any active conduct); see *Edward Hines Lumber v. Vulcan Materials*, 861 F.2d 155 (7th Cir. 1988) (holding, in a contribution action against designer of plant and supplier of chemicals, that CERCLA does not set owner/operator liability on "slipshod architects, clumsy engineers, poor construction contractors, and negligent suppliers of on-the-job training"); but see *FMC Corp. v. U.S. Dept. of Commerce*, 29 F.3d 833 (3rd Cir. 1994) (holding, with four judges dissenting, the United States to be an operator of the plaintiff's facility due to its extensive involvement in the facility's operations—the United States had (1) required the facility to assist in the war effort and manufacture rayon, (2) built a new plant and supervised employees, (3) exercised a significant degree of control over the operations, (4) supplied equipment, and (5) controlled marketing and pricing); see *CPC Intl., Inc. v. Aerojet-General Corp.*, 731 F. Supp. 783 (W.D. Mich. 1989) (noting government liability as an operator when government undertakes a cleanup operation and its subsequent mistakes cause and increase in the ultimate cost of cleanup).

[32]Id.

[33]*U.S. v. Monsanto*, 858 F.2d 160 (4th Cir. 1988), *cert. denied*, 490 U.S. 1106 (1989) (noting that lessors are generally liable for any contamination problems caused by their lessees); *U.S. v. A&N Cleaners & Launderers, Inc.*, 788 F. Supp. 1317 (S.D. N.Y. 1992) (finding lessee, who subleased property to party causing contamination, as an owner under Section 107(a).

[34]*U.S. v. Wade*, 577 F. Supp. 1326 (E.D. Pa. 1983) (requiring a causal nexus between cleanup costs incurred and the generator's waste—such a link need only be that hazardous substances like those

hauling company that picks up waste from a generator and transports it to a disposal facility. Transporters of hazardous substances also face liability under CERCLA because they were the party that actually took the wastes to the contaminated site. However, due to the concern that transporters will not transport hazardous substances if CERCLA's strict liability applies to them, Congress fashioned CERCLA so that transporters will be found liable only if they were involved in the decision regarding the selection of the site of disposal.[35] CERCLA also has extended its reach to "second tier" parties as well. This reach has found directors and officers, parent corporations, and successor corporations liable for loss.[36]

The liability to which these parties may be subject includes any removal or remedial costs incurred by the government "not inconsistent" with the

---

found in the defendant's hazardous wastes be present at the site); see also *U.S. v. Alcan Aluminum Corp.*, 964 F.2d 252 (3rd Cir. 1992); see *U.S. v. Ward*, 618 F. Supp. 884 (E.D. N.C. 1985) (finding that a generator does not have to select the site for liability to hold—doing so would allow generators to evade liability by closing their eyes to the method of disposal); *New York v. General Electric Co.*, 592 F. Supp. 291 (N.D. N.Y. 1984) (finding liability where defendant sold contaminated waste oil to a drag strip for its use—defendant argued that the sale could not be characterized as a disposal); *Florida Power & Light Co. v. Allis Chalmers Corp.*, 893 F.2d 1313 (11th Cir. 1990) (holding that to impose liability upon a manufacturer of a product, the evidence must show that the manufacturer arranged for the disposal—the mere sale of a product will not suffice).

[35]*B.F. Goodrich v. Betkoski*, 99 F.3d 505 (2nd Cir. 1996) (requiring actual selection of the disposal site or an active participation in the selection in order to find transporter liability); *Tippins, Inc. v. USX Corp.*, 37 F.3d 87 (3rd Cir. 1994) (determining that a transporter is liable under CERCLA when it ultimately selects the disposal site or when the transporters' active participation in the selection of the site is substantial); *U.S. v. Western Processing Co., Inc.*, 756 F. Supp. 1416 (W.D. Wash. 1991) (granting summary judgment to transporters as to claims based on waste disposal, for which they did not select the site); *U.S. v. New Castle County*, 727 F. Supp. 854, 875 (D. Del. 1989) (noting that "[i]t is generally accepted that, in order to find liability as a transporter under section 107(a)(4) of CERCLA, there must be a finding that the site was selected by the transporter.").

[36]Officers, Directors, Employees:
U.S. v. Northeastern Pharmaceutical & Chemical Co., Inc., 810 F.2d 726 (8th Cir. 1986), *cert. denied*, 484 U.S. 848 (1987) (finding vice president of corporation liable for approving the arrangement of waste disposal, and finding president liable due to his ultimate authority over all operations, even though plaintiff could not show any knowledge of the activities on the president's part); *Kelley v. Thomas Solvent Co.*, 727 F. Supp. 1532 (W.D. Mich. 1989) (requiring a rigorous fact-specific analysis of various circumstances weighing toward the corporate individual's degree of authority to prevent or abate the hazardous waste disposal that led to the CERCLA cleanup).

Parent Corporations:
U.S. v. Kayser-Roth Corp., 910 F.2d 24 (1st Cir. 1990), *cert. denied*, 498 U.S. 1084 (1991) (holding parent corporation liable as an operator because it exercised pervasive control over the subsidiary corporation by controlling environmental matters, approving large capital expenditures, placing its own personnel in the subsidiary's director positions, controlling operational structure, and other measures); *Joslyn Manufacturing Co. v. T.L. James & Co., Inc.*, 893 F.2d 80 (5th Cir. 1990), *cert. denied*, 498 U.S. 1108 (1991) (declining to extend liability to the parent of an offending wholly owned subsidiary).

NCP, "necessary" costs incurred by any other person "consistent with" the NCP, and any damages to natural resources that result from the release of hazardous substances.[37] These response costs, which can and have become the liability of a retired automobile service station owner and a manufacturing plant supervisor, are almost always figured in the millions of dollars and, more often, billions.

Lender Liability Lenders also fall under the broad umbrella of CERCLA's liability scheme. The statute provides an exemption from liability for any person who, "without participating in the management of a vessel or facility, holds indicia of ownership primarily to protect his security interest in the vessel or facility."[38] Although this exemption exists, lenders must tread carefully, due to the *Fleet Factors* decision.[39]

In *Fleet Factors*, a lender agreed to provide a loan to a cloth company on the condition that the lender, Fleet Factors, receive a security interest in the facility.[40] Upon default, Fleet Factors foreclosed on the equipment and contracted to auction the collateral, but did not foreclose on the property.[41] Subsequently, the EPA found toxic chemicals and asbestos-containing materials on the property and, thereafter, sued the cloth company and Fleet Factors to recover response costs.[42] The court determined that a secured party may incur CERCLA liability if it participates in the "financial management of a facility to a degree indicating a capacity to influence the corporation's treatment of hazardous wastes."[43]

In response to lending agencies' increasing concern over potential CERCLA liability, especially in light of the *Fleet Factors* decision, the EPA promulgated a lender liability rule. This rule was subsequently invalidated by the D.C. Circuit in *Kelley v. United States EPA*, which held that the EPA

Successor Corporations:
In re Acushnet River & New Bedford Harbor Proceedings Re Alleged PCB Pollution, 712 F. Supp. 1010 (D. Mass. 1989) (holding successor corporation liable when successor corporation is really a continuation of the seller, as evidenced by (1) the continued manufacturing of the same product, (2) the continued sale of products under the same name, (3) the same officers holding the same offices, (4) those in middle management positions being the same, (5) the employees being substantially the same, (6) using the same physical facilities, and (7) using the same bank and insurance company).

[37]CERCLA 107, 42 U.S.C. 107.

[38]CERCLA 101(20)(A), 42 U.S.C. 9601(20)(A).

[39]*United States v. Fleet Factors, Corp.*, 901 F.2d 1550 (11th Cir. 1990).

[40]Id., at 1552.

[41]Id.

[42]Id., at 1553.

[43]Id., at 1557.

had exceeded its authority.[44] Today, most courts will find in favor of the lender, absent a showing of significant participation in the management that would affect the disposal practices of the allegedly offending company, but care must be taken by the lender to protect itself.

CERCLA Defenses[45] Basically, there are no defenses to CERCLA. If they want you, they got you. However, if you demonstrate that the contamination was caused by an act of God or by an act of war, you have a chance. Further, the Superfund Amendments and Reauthorization Act (SARA) of 1986, amending CERCLA, has allowed in rare situations for what is known as the "innocent landowner defense."[46]

An "innocent landowner" may escape CERCLA liability if the owner can establish, by a preponderance of the evidence, that the release or threat of release involved and the resulting damages were caused by "an act or omission of a third party other than an employee or agent of the defendant, or other than one whose act or omission occurs in connection with a contractual relationship . . . ," and that he or she exercised due care with respect to the hazardous substances, and precautions were taken against foreseeable acts or omissions of third parties.[47]

When this defense involves the purchase of property, the contract of sale or transfer of the property will be considered a contractual relationship negating the defense, unless the defendant, acquiring the property after contamination, did not know and had no reason to know of the contamination.[48] A failure to investigate is *not* a defense. In order to show that the defendant "had no reason to know," he or she "must have undertaken, at the time of acquisition, all appropriate inquiry into the previous ownership and uses of the property consistent with good commercial or customary practice. . . ."[49]

Unfortunately, the careful consumer will be hard pressed to find a court that has granted the "innocent landowner" defense or one that has established a bright line interpretation of "all appropriate inquiry." However, a 1996 Second Circuit Court of Appeals decision may just prove to be the light at the end of the tunnel. In *New York v. Lashins Arcade*, the court of ap-

[44]*Kelley v. United States EPA*, 15 F.3d 1100 (D.C. Cir. 1994).
[45]CERCLA 107(b), 42 U.S.C. 9607(b).
[46]CERCLA 107(b)(3), 42 U.S.C. 9607(b)(3).
[47]CERCLA 107(b)(3), 42 U.S.C. 9607(b)(3).
[48]CERCLA 101(35)(A)(I), 42 U.S.C. 9601(35)(A)(I).
[49]CERCLA 101(35)(B), 42 U.S.C. 9601(35)(B).

peals held that a purchaser of land had established the third-party defense to CERCLA liability.[50] There, New York brought suit against the owner of a shopping plaza to recover costs incurred while investigating and cleaning up a release of perchloroethylene (PCE), resulting from a previous dry-cleaning operation, into the groundwater around the shopping center.[51] One of the defendants, Lashins Arcade, the current shopping center owner, asserted the CERCLA Section 107(b)(3) defense and was granted summary judgment.[52]

The facts presented showed that, prior to purchasing the property, Lashins contacted the town of Bedford and was assured that there were no past or present problems with the property. It even went so far as to interview other tenants of the property. After purchasing the shopping center and being apprised of the contamination, Lashins maintained the granular activated-carbon (GAC) filters that were already in place, took water samples, instructed tenants not to discharge hazardous substances into the waste and septic systems, incorporated this tenant requirement into the leases, and conducted periodic inspections of the tenants' premises to ensure compliance. However, the state of New York was not impressed and contended that Lashins never inquired specifically about groundwater contamination.[53]

After reviewing the case, the district court found that the defendants did not have a direct or indirect contractual relationship with the dry cleaners that released the PCE, nor did they have a contractual relationship with the owners of the shopping center at the time the dry-cleaning companies operated and the pollution occurred.[54] The court also noted that Lashins had done "everything that could reasonably have been done to avoid or correct the pollution."[55] In its review of the case, the court of appeals examined the elements of the "innocent landowner" defense. First, the court noted that the only contractual relationship Lashins had with the previous owner, the straightforward sale of the shopping center, had no relation to the release of hazardous substances.[56] Second, the last release of PCE at

[50]*New York v. Lashins Arcade Co.,* 91 F.3d 353 (2nd Cir. 1996). The success of Lashins Arcade clearly demonstrates that you can amount to something if "you don't stop sitting around and playing video games all day."

[51]*Lashins Arcade*, 91 F.3d at 356.

[52]Id., at 355.

[53]Id., at 357.

[54]Id., at 359.

[55]Id. (quoting *Lashins Arcade Co.,* 856 F.Supp. 157-158).

[56]Id., at 360.

the site occurred more than 15 years prior to the sale to Lashins, and there was nothing that Lashins could have done to prevent the actions that led to the release.[57] Finally, the court found that the actions taken by Lashins, including water sampling, maintaining the GAC filter, and ensuring that tenants did not dispose of hazardous substances, were all the steps Lashins needed to take to demonstrate due care in this fact-specific situation. Thus, the court of appeals affirmed the district court's granting of summary judgment to Lashins Arcade. The significance of this decision will be apparent when we discuss the role of an environmental audit later in this chapter.

Common Law Exposure to Environmental Loss Statutory redress aside, a plaintiff can often find a remedy for environmental claims in the realm of tort law. The availability of common law claims increases the array of liability a defendant may face over such statutory claims, as CERCLA. Some of the more common tort claims brought in the environmental arena follows (however, novel claims, under such theories as battery, emotional distress, and others, may also be brought)[58]:

- *Trespass*. Actions alleging a physical invasion of property through groundwater contamination, air contamination, and surface water contamination. Generally, one can bring a suit under trespass for an invasion of invisible substances.
- *Negligence*. To use negligence in an environmental context, a toxic tort plaintiff must show that the defendant breached a standard of care that has been established in the environmental scheme.
- *Nuisance*. Many toxic tort plaintiffs utilize a private nuisance claim, the interference with one's use and enjoyment of property, or a public nuisance claim, the interference with a right common to the public, to abate environmental harms. For example, a plaintiff may bring a cause of action under a private nuisance theory because contaminated groundwater is interfering with his or her enjoyment of property.

Strict Liability (Evolving from *Rylands v. Fletcher*)[59] Strict liability has become another weapon in the environmental litigation arsenal. This the-

[57]Id.

[58]For an interesting account of a toxic tort case, you may wish to read *A Civil Action* by Jonathan Harr.

[59]*Rylands v. Fletcher*, L.R. 3 H.L. 330 (1868).

ory renders absolute liability on landowners or occupiers of land on which hazardous activities have occurred—including the storage of hazardous chemicals.

REAL ESTATE AND PROPERTY TRANSFERS AND THE ROLE OF THE ENVIRONMENTAL AUDIT

The advent of CERCLA has led to a "chill" in real estate and property transfers due to the potential multimillion-dollar liability for investigation, cleanup, and natural resources damages that a purchaser of property may encounter. Purchasers, sellers, and lenders must all be prudent when entering into property actions, in order to be aware of their respective liabilities. A purchaser should be concerned with:

- Potential common law claims
- CERCLA liability for being a current owner/operator
- Impact of any contamination on the property's value
- Due diligence and investigating into the past uses of the property, including interviews of previous owners and neighbors and research of various records (CERCLIS, NPL, permitting lists, and various CERCLA and RCRA lists)
- Conducting an environmental audit

Sellers will want to receive the highest possible price, thereby wishing to limit an environmental investigation. However, they need to keep in mind their potential liabilities.

Lenders will want a thorough investigation to ensure that the purchaser's potential liabilities for any cleanup will not be greater than the value of the property. Also, lenders should make sure that they do not exert any control over hazardous waste disposal.

The environmental audit is, essentially, an environmental investigation of a piece of property that should be conducted prior to performing a property transaction. An audit has become a valuable tool for identifying potential liabilities, but it does not yet provide for a valid defense in a CERCLA action. However, in light of the *Lashins Arcade* decision detailed previously, an environmental audit can be seen as not only a valuable tool for identifying potential liabilities, but it can also be used to assist in establishing a defense to a CERCLA cost recovery action.

REFERENCES

1. D. Sive, "The Litigation Process in the Development of Environmental Law," *13 Pace Environmental Law Review* 32 (1995).
2. A. Babich, "A New Era in Environmental Law," *Colorado Lawyer* 1 (February 1991).
3. J.B. Battle, *Environmental Decisionmaking and NEPA* 3 (1986).
4. McKinney, *New York Environmental Construction*, L. 27–0900 (1996).

Chapter 4

An Introduction to Insurance Coverage for Environmental Losses

To thoroughly analyze the insurance coverage issues that arise in the context of commercial general liability (CGL) and environmental impairment liability (EIL) policies would require an entire book and, in fact, is the subject of several. Thus, we will use this opportunity to provide only a brief introduction to the basic topics and issues and some very basic conclusions, all of which are subject to state law and the unique factual scenario of each case. It will also give you a break from the excitement of the scientific information stuffed into the other chapters of this book.

INTRODUCTION

Coverage disputes that concern environmental matters differ from most other types of coverage disputes in several ways. One major difference is that the incident that may constitute an "occurrence" under occurrence-based policies often is not a discrete event such as a slip-and-fall accident, a punch in the nose, or a theft, which generally have an identifiable beginning and end as well as a recognizable cause and effect. In contrast, occurrences affecting the environment, such as a fuel leak or a chemical spill, often, although not always, begin slowly, continue for months or even years, and experience a gradual decline in effect that may not be detectable by the unaided senses. Related to the temporal indefiniteness of environmental hazards is the fact that they typically are not confined to a limited space. Rather, pollutants such as carbon monoxide, benzene, chlorinated solvents, and silica and lead dust have a way of penetrating most structures and dispersing throughout the surrounding air, soil, or water.

Because of the characteristics of environmental wrongs mentioned previously, attorneys concerned with the insurance coverage issues that arise in environmental matters usually require assistance with the technical questions that accompany the attempt to resolve the coverage issues. This will come as a surprise, but such assistance is provided by experts in environmental forensics. For unless an attorney has a technical education or background, it will likely be impossible for him or her to effectively analyze the large quantities of data available on the site and event at hand and formulate a conclusion with respect to such questions as the date the release began, the cause and effect of the contamination, and whether and to what extent the toxic materials originated from the operations of the insured. Even if an attorney had the ability to perform such tasks, he or she almost definitely does not have the time. Conversely, the environmental forensic expert will not know without the assistance of legal counsel how to determine which portions of the data accumulated is relevant to legal coverage issues.

The issues that arise in environmental coverage matters flow from the fundamental principles of insurance and contract law and from the specific structure and substantive provisions of insurance policies. A key set of issues concerns whether the incident on which the claim is based was fortuitous as opposed to intentional, known or in progress at the time the insurance was procured. As will be discussed later, the fortuity requirement is usually expressed in policy language in the definition of an "occurrence" in occurrence-based policies as an "accident" resulting in injury or damage that is "neither expected nor intended" from the standpoint of the insured.

Another important issue concerns which types of actions taken against the insured constitute a "suit" that, according to now-standard CGL policy language, obligates the insurer to provide the insured with a defense. While courts virtually unanimously hold that an insured has a defense obligation with respect to a lawsuit filed against the insured seeking damages for covered losses, in the environmental context responses to pollution-causing events by governmental agencies can take many nontraditional forms, such as PRP letters, information requests, and so forth. Courts varied widely in their decisions and reasoning on whether a particular type of action by a regulatory body creates a duty to defend.

Two issues that are inherent in the definition of "occurrence" introduced previously are when the "occurrence" took place and when the injury or damage occurred. This will be relevant to which and how many policies will be "triggered" by any loss. Again, such questions are frequently relatively easy to answer when the event at issue is, for example, an automobile accident or a tree limb breaking off and damaging someone's garage. In an environmental context, however, when a hazardous substance migration began, how many "occurrences" the pollution constitutes, and when a disease the symptoms of which manifest slowly constitutes "bodily injury" are often exceedingly complex issues, which an attorney can competently tackle only with the assistance of experts.

Forensics experts also may provide valuable assistance in determining whether a particular policy provision operates to bar coverage for a loss. Insurers have for approximately the past 25 years included provisions in policies excluding from coverage liability due to loss caused by environmental pollution. The language of pollution exclusions has become increasingly broader over the years, but any such exclusion must employ terms such as "release," "irritant," and "pollutant," which must then be interpreted by attorneys and courts. Another exclusion—the owned property exclusion—excludes from property damage coverage certain categories of property, including property that is owned, rented, or occupied by the insured. Attorneys are often required to consult forensics specialists to construct a sophisticated argument regarding whether a particular chemical should be viewed as an "irritant" or whether some of the damage caused by hazardous wastes was to the insured's own property. In the context of policies providing coverage only for off-site contamination, forensic investigation is often required to determine how much of the pollution occurred offsite and how much onsite.

The purpose of this chapter is to show how forensics will assist in determining the rights and obligations of coverage. This task will be carried out by providing an overview of some of the more frequently seen coverage issues encountered in claims for loss resulting from environmental contamination. Our focus will be on the role forensic experts can—and many times, must—play in the process of framing the issues of a dispute, analyzing and evaluating the data, and developing strategy.

BASIC SIMILARITIES AND DIFFERENCES BETWEEN CGL AND EIL POLICIES

The most common type of third-party coverage for businesses is that provided by CGL policies. The standard CGL coverage provision states that the insurer "will pay those sums that the insured becomes obligated to pay as damages because of bodily injury or property damage . . . to which this insurance applies."[1] Most, although not all, CGL policies provide coverage for bodily injury or property damage caused by an "occurrence" that takes place during the policy period. Because courts have construed the term "damages" found in the previously referenced coverage provision as meaning only tort damages, breach of contract claims are generally not covered.[2] Another justification for the rule that there is no coverage under CGL policies for breach of contract claims is that only a fortuitous event falls within the parameters of the definition of "occurrence"; tort claims satisfy the fortuity requirement, while breach of contract claims do not.[3]

Unlike a CGL policy, an EIL policy provides coverage only for personal injury or property damage caused by an "environmental impairment" that arises out of or in the course of the insured's "operations, installations, or premises." The term "environmental impairment" encompasses most conceivable forms of pollution and contamination, but also includes odors, noises, vibrations, and other phenomena that affect the environment. Another difference between EIL and CGL policies is that the former provide coverage on a claims-made basis, which means that the insurer agrees to pay all amounts that the insured is obligated to pay as damages as a result of all claims first made against the insured during the policy period. "Claim" is defined as a "demand received by the insured for money or services." Coverage issues unique to claims-made policies will be explored more fully later in the chapter.

Note, however, that the coverage language in EIL policies contains the reference to "damages" found in CGL policies. Further, the fundamental limitation of insurance to fortuitous events, which concept will be discussed more fully later, applies to both CGL and EIL policies.

DUTY TO DEFEND

In both CGL and EIL policies, the insurer's duty to defend is distinct from the obligation to indemnify for losses falling within the policy's coverage language. It is a basic premise of insurance law and undisputed that the

[1]Businessowners Liability Coverage Form BP 00 06 01 96.

[2]See, e.g., *Wilmington Liquid Bulk Terminals, Inc. v. Somerset Marine, Inc.,* 53 Cal. App. 4th 186, 61 Cal. Rptr. 2d 727 (Cal. Ct. App. 1997).

[3]See *Stein-Brief Group, Inc. v. Home Indem. Co.,* 66 Cal. App. 4th 364, 76 Cal. Rptr. 2d 3 (Cal. Ct. App. 1998).

insurer's duty to defend is broader than the duty to indemnify, and that the duty to defend is triggered if the allegations of the complaint are such that the insured's conduct is potentially within coverage.[4] In determining whether the claim is potentially covered, courts look to the underlying complaint and the policy itself, liberally construing policy language in favor of the insured.[5] Because environmental coverage disputes often go on for years, the rule that insurers must provide a defense with respect to claims that may be indemnifiable can mean that the insurer is funding the insured's defense for quite some time before sufficient data are gathered and analyzed to allow a determination as to whether the incident(s) on which the claim is based falls is excluded from coverage. Throughout this long process of investigation, forensics experts help the attorney translate mountains of technical data into a characterization and assessment of the contamination to which the policy terms can be meaningfully applied.

Difficulties may arise in the attempt to apply the foregoing principles and rules regarding the duty to defend when the claim against the insured does not take the form of a lawsuit. Thus, we now turn to a discussion of the duty to defend issues in the context of regulatory agencies' responses to environmental pollution other than litigation.

PRP Letters

Unlike a complaint, a letter from a governmental agency informing an insured that it has been determined to be a potentially responsible party (PRP) for environmental contamination discovered at a site typically does not employ language that would clearly constitute a claim for money or services from the insured. Further, it is not obvious that a PRP letter qualifies as a "suit" as that term is contemplated in CGL policies providing that the insurer "shall have the right *and duty to defend any suit* against the insured seeking damages . . . and *may* make such investigation . . . of any *claim or suit* . . ." (emphasis supplied). Because most CGL policies do not define the terms "suit" and "claim," the question thus arises whether such a letter triggers the insurer's duty to defend.

While most courts now find PRP letters to qualify as a "claim," there is a split of authority nationwide on this issue. Some courts have adopted a

[4] E.g., *LaSalle National Trust, N.A. v. Schaffner*, 818 F. Supp. 1161 (N.D. Ill. 1993); *Colon v. Aetna Life & Casualty Ins. Co.*, 66 N.Y.2d 6, 484 N.E.2d 1040, 494 N.Y.S.2d 688 (1985).

[5] E.g., *Seaboard Surety Co. v. Gillette Co.*, 64 N.Y.2d 304, 476 N.E.2d 272, 486 N.Y.S.2d 873 (1984).

"plain meaning" or "literal meaning" approach to the question whether a PRP letter should be construed as constituting a "suit." Illinois courts have held that the term "suit" unambiguously refers to a proceeding in a court of law, and therefore letters from environmental agencies, including PRP letters, are not considered "suits" and therefore do not trigger the duty to defend.[6] The interpretation of the "claim" portion of the policy language is different. This view with regard to "suit" has been advanced by courts in a number of jurisdictions, including Wisconsin,[7] Missouri,[8] and Ohio.[9] Some courts concluding that an insurer does not have a duty to provide a defense when a PRP letter has been issued have also justified their position by referring to the rule that the allegations of the complaint determine whether a duty to defend has been triggered. It naturally follows that where there is no complaint, no such duty is owed.

A further basis for such reasoning is that because CGL policies employ the terms "suit" *and* "claim," they must be construed as having different meanings in order for all of the policy's language to be given effect. If the term "suit" were broadened to include any type of "claim," any distinction between the two would vanish.[10] Related to this justification is the notion that insurers are drawing an unambiguous line to define and delimit their contractual obligations in specifying that only a "suit" and not a "claim" triggers the duty to defend.[11] Because insurers have effected a clear delineation using two terms rather than one, some courts argue, the insured should not be permitted to obtain more benefits from the policy than the insurer bargained for. This is basic contract law: The insurer's obligations should be no more and no less than that for which the insured paid.

In contrast to the position just discussed, courts in some jurisdictions have adopted a "functional" approach to the question whether a PRP letter should be interpreted as triggering the duty to defend. According to this view, the receipt of a PRP letter or any prelitigation environmental agency correspondence constitutes a "suit" for the purpose of construing CGL poli-

[6]See, e.g., *Lapham-Hickey Steel Corp. v. Protection Mut. Ins. Co.,* 166 Ill. 2d 520, 655 N.E.2d 842 (1995).

[7]E.g., *City of Edgarton v. General Casualty Co. of Wisconsin,* 184 Wis. 2d 750, 517 N.W.2d 463 (1994), *cert. denied,* 514 U.S. 1017 (1995).

[8]*Aetna Casualty & Surety Co. v. General Dynamics Corp.,* 783 F. Supp. 1199 (E.D. Mo. 1991), aff'd in relevant part, rev'd in part and remanded, 968 F.2d 707 (8th Cir. 1992).

[9]*Professional Rental, Inc. v. Shelby Ins. Co.,* 75 Ohio App. 3d 365, 599 N.E.2d 423 (1991).

[10]See, e.g., *Lapham-Hickey Steel,* 655 N.E.2d at 847-848; *Ray Industries, Inc. v. Liberty Mutual Ins. Co.,* 974 F.2d 754 (6th Cir. 1992) (applying Michigan law).

[11]See *Foster-Gardner, Inc. v. National Union Fire Ins. Co. of Pittsburgh,* 18 Cal. 3d 857, 77 Cal. Rptr. 2d 107 (1998).

cies.[12] The reasoning behind this stance is that the legal proceeding initiated by the receipt of a PRP letter is the functional equivalent of a lawsuit for the purposes of the "suit" requirement. Courts applying a functional approach view the term "suit" as ambiguous, necessitating an inquiry as to whether the insured would expect a defense of the administrative proceeding.[13] Because a PRP letter will generally force an insured to hire technical experts to protect its interests, courts adhering to this view find that a PRP letter triggers a duty to defend.[14] Nobody ever said that the law was consistent.

Letters from Regulatory Agencies Demanding or Requesting Remediation or Investigation

Actions by the government in response to contamination can take many forms, including action that sometimes does not reach the level of a PRP letter. Frequently, a regulatory agency is not certain of the source of the toxic materials causing the hazardous condition. In such a situation, it may demand or request that PRPs participate in investigatory activities to determine the extent, source, and effects of the pollution. This could take the form of a request of information or RFI. In contrast, when an agency has concluded that an entity is responsible for the contamination, it may issue a notice of a statutory violation without inquiring whether there are other responsible parties. Despite the wide range of possible actions by the government, courts' analysis of whether CGL coverage exists for the particular response at issue is generally guided by their views on the previously discussed issue of whether an action other than a lawsuit constitutes a "suit."[15] It is important, therefore, to understand that in the end the de-

[12]See, e.g., *Mich. Miller Mutual Ins. v. Bronson Plat.*, 445 Mich. 558, 519 N.W.2d 864 (1994); *Coakley v. Maine Bonding and Cas. Co.*, 136 N.H. 402, 618 A.2d 777 (1992); *Spangler Const. v. Industries Crankshaft & Eng. Co.*, 326 N.C. 133, 388 S.E.2d 557 (1990).

[13]*Aetna Cas. Sur. Co. v. Pintlar Corp.*, 948 F.2d 1200 (9th Cir. 1991) (applying Idaho law).

[14]See, e.g., id.; *Broadwell Realty Servs., Inc. v. Fidelity & Cas. Co. of N.Y.*, 218 N.J. Super. 516, 528 A.2d 76 (App. Div. 1987).

[15]See, e.g., *Joslyn Mfg. Co. v. Liberty Mut. Ins. Co.*, 836 F. Supp. 1273 (W.D. La. 1993), aff'd, 30 F.3d 630 (5th Cir. 1994) (concluding that the term "suit" refers to a formal proceeding in a court of law and holding that an insured had no duty to defend request by regulatory agency that it clean up toxic materials at former plant); *SCSC Corp. v. Allied Mutual Ins. Co.*, 536 N.W.2d 305 (Minn. 1995) (holding that insurer had duty to provide defense when insured received information request from state agency, reasoning that the term "suit" included such actions); *C.D. Spangler Constr. Co. v. Industrial Crankshaft & Eng. Co.*, 326 N.C. 133, 388 S.E.2d 557 (1990) (construing "suit" as an "attempt to gain an end by the legal process" and finding duty to defend regarding state agency order to remove hazardous wastes from its property).

termination of whether a duty to defend exists will be based on the *interpretation of the policy language*—that is, the use of the term "suit" in CGL policies—in the context of the facts of the particular case.

Issues Arising Out of CERCLA Claims

In providing federal regulatory agencies with various means by which to carry out their attempts to clean up hazardous waste sites, the Comprehensive Environmental Response, Compensation and Liability Act (CERCLA)[16] raises many issues impacting coverage under all insurance forms. For example, the question has arisen whether CGL policies provide coverage for governmental suits to obtain reimbursement for costs expended for site cleanup under Section 107(a)(4)(A) of CERCLA. Some courts have held because the remedy such actions seek is essentially the equitable remedy of restitution, they do not constitute claims for legal "damages" as required for CGL coverage.[17] One court fleshed out the reasoning behind this rule by remarking that CGL policies provide coverage not for "property damage" per se but for "sums which the insured shall become legally obligated to *pay as damages* because of . . . property damage." The legal mind is indeed a terrible thing to waste.

On the other hand, many courts have taken the position that it does not matter whether the environmental regulators attempt to force the insured to clean up the contamination or carry out the cleanup work themselves and then seek reimbursement; either way, the insured is being required to incur the cost of restoring the natural resource to its precontamination state.[18] According to this theory, the fact that the costs are being sought after the cleanup has been paid for does not alter their status as "damages" under CGL policies.

[16]The Comprehensive Environmental Response, Compensation and Liability Act, 42 U.S.C. 9601 et seq. (hereinafter CERCLA).

[17]See, e.g., *City of Edgarton v. General Cas. Co. of Wisconsin,* 184 Wis. 2d 750, 517 N.W.2d 463 (1994), *cert. denied,* 514 U.S. 1017 (1995); *Continental Ins. Co. v. Northeastern Pharmaceutical & Chem. Co.,* 842 F.2d 977 (8th Cir. 1988) (en banc) (applying Missouri law), *cert. denied,* 488 U.S. 821 (1988); *Maryland Cas. Co. v. Armco, Inc.,* 822 F.2d 1348 (4th Cir. 1987), *cert. denied,* 484 U.S. 1008 (1988).

[18]See, e.g., *Aerojet-General Corp. v. San Mateo Superior Court (Chesire & Cos.),* 209 Cal. App. 3d 973, 257 Cal. Rptr. 621 (1st Dist.), mod. & reh'g denied, 211 Cal. App. 3d 216, 258 Cal. Rptr. 684 (1st Dist. 1989), rev. denied, No. S010409 (Cal. August 10, 1989); *American Motorists Ins. Co. v. Levelor Lorentzgen, Inc. Co.,* No. 88-1994 (D. N.J. October 14, 1988); *United States Aviex Co. v. Travelers Ins. Co.,* 125 Mich. App. 579, 336 N.W.2d 838 (1983).

The Duty to Defend and Costs Incurred Investigating CERCLA Claims

Once an insured establishes a duty to defend, the next question becomes: What constitutes defense and defense cost? Pursuant to Section 106 of CERCLA, an insured may be required to investigate the source, extent, and nature of contamination, as well as alternative remediation methods. Investigative efforts can be extremely costly, involving the retention of experts, paying for feasibility studies, and crafting remedial action plans. Further, the insured will no doubt wish to determine whether the government has valid grounds for holding it liable for the contamination. Courts generally hold that postclaim payments by an insurer to the insured for such investigative efforts should be deemed defense, rather than indemnity, expenses.[19] Why is the issue of characterization of these costs important? Because many policies provide for defense costs in *excess* of limits and as such, in effect, increase available insurance dollars.

The law is less settled regarding the issue of whether insurers have a duty to reimburse the insured for costs incurred in investigation done prior to the claim against the insured. One court has held that if the suit against the insured was nearly certain, and the investigative costs thus virtually inevitable, then the insurer's defense obligation reaches back to such preclaim costs.[20] An alternative view, however, is that no matter how likely it was that a suit would be filed, the fact that the insurer's duty had not yet ripened at the time the investigation was commenced bars recovery for the costs incurred in connection therewith.[21]

OCCURRENCE-BASED POLICIES

Under occurrence-based policies, the insurer's obligation to pay the insured is limited to liability due to an "occurrence," which is defined in CGL policies as an "accident" resulting in injury or damage that is "neither expected nor intended" from the standpoint of the insured. The task of determining how the abstract concepts that serve as the building blocks of

[19]See, e.g., *Ins. Co. of N. Am. v. American Home Assur. Co.,* 391 F. Supp. 1097 (D. Colo. 1975); *Hertzka & Knowles v. Salter,* 6 Cal. App. 3d 325, 86 Cal. Rptr. 23 (Cal. Ct. App. 1970).

[20]See, e.g., *Liberty Mut. Ins. Co. v. Continental Cas. Co.,* 771 F.2d 579 (1st Cir. 1985); *Travelers Ins. Co. v. Chicago Bridge & Iron Co.,* 442 S.W.2d 888 (Tex. Civ. App. 1969), writ of error ref. (Tex. November 12, 1969).

[21]See *Gibbs v. St. Paul Fire & Marine Ins. Co.,* 22 Utah 2d 263, 451 P.2d 776 (1969).

this seemingly simple definition are to be applied to the facts of a particular coverage dispute can be exceptionally complex and technical with respect to environmental matters. In providing an introduction to issues inherent in the definition of an "occurrence," this section will examine how forensics specialists can help the attorney construct a bridge between the abstract and the concrete.

The Concept of Fortuity, the Known-Loss Rule, and the Loss-in-Progress Rule

In order for an incident to qualify as an "occurrence," it must be an accident, that is, a fortuity, which means that the injurious event and the resulting damage must both be unexpected. This principle arises from the essence of insurance, which is to provide coverage for risks as opposed to planned, intended, or expected losses.[22] The Supreme Court of the United States and nearly all lower courts have adopted the position that the fortuity doctrine applies to all insurance policies, regardless of whether their language expresses or is even consistent with it.[23]

In the context of loss due to environmental contamination, it is not always a simple matter to determine whether the incident causing the damage or injury occurred accidentally or naturally or was the product of an intentional act or an expected event. The causes of environmental pollution are frequently multiple and interrelated, involving natural processes operating over time and in several locations at once. Even a sudden event such as an explosion may be the result of the interplay of many factors, and the role of each in causing it may be difficult and costly to identify.

The messiness, as it were, of incidents causing damage to the environment necessitates the consultation of experts to develop an understanding of the nature and causes of the loss. A petrochemical engineer can, for example, advise an attorney with respect to whether a benzene leak was caused by the insured's storage methods or the faulty design of the underground storage tanks it used. An expert on soil absorption of toxic chemicals can advise as to whether compounds being released into a stream will enter groundwater and, if so, what harm would be caused thereby. Information obtained from consultants can be central to an inquiry regarding whether a discharge or escape was "unexpected, unin-

[22]See, e.g., *University of Cincinnati v. Arkwright Mut. Ins. Co.,* 51 F.3d 1277 (3rd Cir. 1995).
[23]See, e.g., *Ins. Co. of N. Am. v. U.S. Gypsum Co.,* 870 F.2d 148 (4th Cir. 1989).

tended, and short-lasting," as one construed the term "accident" in occurrence-based policies.[24]

A corollary to the fortuity requirement is the doctrine that no coverage is unavailable under CGL policies for losses that are known to the insured at the time the policy is issued. The justification for the so-called "known-loss rule" is that once a risk becomes a certainty, the concept of fortuity fundamental to insurance law prohibits the acquisition of valid insurance for that act and losses flowing therefrom.[25] The critical questions in a known-loss analysis are whether the events from which the loss resulted were within the control of the insured, and whether the insured knew, or had reason to know, at the time of the policy's inception that there was a substantial probability that the loss would result in liability.[26] A consultant may be able to provide the information and technical analysis on the basis of which an insurer could argue that the insured had such knowledge and therefore was not entitled to coverage. With respect to whether an insured had reason to know, a consultant is essential, for such questions often boil down to a "battle of the experts."

Related to the known-loss rule is the loss-in-progress rule, which renders coverage unavailable under policies issued after a loss has begun to manifest itself.[27] Coverage is precluded by this rule if the insurer was or should have been aware of an ongoing loss at the time insurance is obtained. Some courts, however, have stated that the loss-in-progress rule, like the known loss rule, may be conceived of as barring coverage for losses about which the insured's liability was not uncertain.[28]

As is the case regarding disputes in which the known-loss rule may be significant, an expert can help an attorney construct an educated argument regarding whether and when the insured knew or should have known that a loss was ongoing. Further, either party to a dispute involving this issue may benefit from a forensic expert's investigation into when

[24]*Ins. Co. of N. Am. v. ASARCO, Inc.*, No. L-6164-93 (N.J. Super. Ct. Law Div. Middlesex County, January 30, 1997).

[25]See, e.g., *Outboard Marine Corp. v. Liberty Mut. Ins. Co.*, 154 Ill. 2d 90, 607 N.E.2d 1204 (1992); *Morton Int'l, Inc. v. General Accid. Ins. Co. of America*, 266 N.J. Super. 300, 629 A.2d 895 (App. Div. 1991), aff'd, 134 N.J. 1, 629 A.2d 831 (1993), *cert. denied*, 512 U.S. 1245 (1994).

[26]*Outboard Marine Corp.*, 154 Ill. 2d at 90, 607 N.E.2d at 1212.

[27]See, e.g., *Continental Ins. Co. v. Beecham, Inc.*, 836 F. Supp. 1027 (D. N.J. 1993); *Prudential-LMI Commercial Ins. Co. v. Superior Court* (Lundberg), 51 Cal. 3d 674, 798 P.2d 1230, 274 Cal. Rptr. 387 (1990).

[28]See, e.g., *Zurich Ins. Co. v. Transamerica Ins. Co.*, 34 Cal. App. 4th 933, 34 Cal. Rptr. 2d 913 (1994); *Pines of La Jolla Homeowners Association v. Industrial Indemnity*, Cal. App. 4th 714, 7 Cal. Rptr. 2d 53 (1992).

the loss began. Nature rarely allows bright lines to be drawn as to causes, beginnings and endings, boundaries between parcels of property, and other phenomena that insurers, insureds, and attorneys often would like to be able to delineate and quantify.[29]

Number of Occurrences

Limits of liability in occurrence-based policies are generally established on a per-occurrence basis. An insurer issuing a policy with limits of $1 million per occurrence will, therefore, have more than a passing interest keeping the number of covered "occurrences" that are found to have taken place as low as possible. Aside from the economic issues at stake, an essential part of the process of determining the respective parties' coverage rights and obligations with respect to a particular loss is ascertaining just how many "occurrences" will have to be dealt with.

Cause Test

In a majority of jurisdictions, the question of the number of "occurrences" an event constitutes is determined by the cause of the resulting injury or damage. The causetest analysis involves determining whether there was one proximate, uninterrupted, and continuing cause that resulted in all of the injuries and damage, even though several discrete instances of damage resulted.[30] The cause test thus yields a finding of one occurrence in situations where one event results in damage to multiple persons or items of property.[31]

In jurisdictions applying the cause test, attorneys often rely on technical consultants to determine the likely cause(s) of the contamination at issue. If the data support the conclusion that there was more than one cause, then a finding of multiple occurrences is justified. In complex cases involving, for example, environmental damage resulting from several chem-

[29]Depending on which position is being adopted in a dispute, one may prefer that causes, sources, effects on natural resources, etc. remain indeterminate and imprecise. Of course, this is usually the preference of the party that is, in fact, responsible for the environmental damage at issue!

[30]See, e.g., *General Accident Insurance Company of America v. Allen,* 708 A.2d 828 (Pa. Super. Ct. 1998); *Endicott Johnson Corp. v. Liberty Mut. Ins. Co.,* 928 F. Supp. 176 (N.D. N.Y. 1996); *Household Mfg., Inc. v. Liberty Mut. Ins. Co.,* No. 85 C 8519 (N.D. Ill. February 11, 1987), mod. on reconsid., No. 85 C 8519 (N.D. Ill. Nov. 16, 1987).

[31]See, e.g., *Home Indem. Co. v. City of Mobile,* 749 F.2d 659 (11th Cir. 1984); *Owens-Illinois, Inc. v. United Ins. Co.,* 264 N.J. Super. 460, 625 A.2d 1 (App. Div. 1993), rev'd in part on other grounds and remanded, 138 N.J. 437, 650 A.2d 974 (1994).

icals released over a long period of time in multiple locations, the issue of how many causes, and thus how many occurrences, may require the efforts of a team of experts from various fields.

Effect Test

The effect test is employed less frequently than the cause test. Under this test, a court will find, in situations involving multiple instances of injury or damage resulting from one cause, a separate accident or occurrence for each instance.[32] In an environmental context, courts will often employ this test in situations involving exposure of multiple persons to a toxin, for example, a factory in which asbestos-containing materials were used.[33] In such cases, courts virtually always hold that each person's injury resulting from the exposure constitutes an "occurrence." Courts have also found multiple occurrences in cases involving disposal of waste at multiple sites.[34] It follows, therefore, that it may be important to know how many sites have been affected by a client's pollution-creating operations, and forensics consultants can be retained to provide advice in this regard.

Trigger of Coverage

when environmental contamination exists over a period of years, or where the injury is such that the toxic conditions were likely present for a time before the pollution manifested itself in the form of injury, the issue of when and if the insurer's defense and indemnity obligations are activated becomes important for claims under occurrence-based policies. The event that activates these obligations is known as the "trigger" of coverage.[35] Which event is seen as triggering coverage determines which policy years, and thus which insurers, will have to respond to the loss. An insurer whose policy or policies have been triggered is said to be "on the risk" for that loss.

Determining which event is the triggering event can be very difficult

[32]See, e.g., *Maurice Pincoffs Co. v. St. Paul Fire & Marine Ins. Co.,* 447 F.2d 204 (5th Cir. 1971); *Slater v. United States Fidel. & Guar. Co.,* 379 Mass. 801, 400 N.E.2d 1256 (1980).

[33]See, e.g., *United States Gypsum Co. v. Admiral Ins. Co.,* 268 Ill. App. 3d 454, 532 N.E.2d 1226 (1994); *Asbestos Ins. Coverage Cases,* No. 1072 (Cal. Super. 1988).

[34]See, e.g., *Southern Pacific Rail Corp. v. Certain Underwriters at Lloyd's,* No. BC 154722 (Cal. Super. 1998); *Illinois Power Co. v. The Home Ins. Co.,* No. 95-L-284 (Ill. Cir Ct. 1997).

[35]See, e.g., *Armstrong World Industries, Inc. v. Aetna Cas. & Sur. Co.,* 45 Cal. App. 4th 1, 39, 52 Cal. Rptr. 690 (1996).

when the contamination involves, for example, toxic chemicals that were present at a site for an extended period of time and resulted in diseases that a person or persons may have contracted sometime before the visible symptoms developed. The same difficulties can exist in the property damage context, where, for example, the damage to a building or the effects on soil are very gradual and are not detected until months or even years after the destructive processes commenced.

For these reasons, courts have developed several methods for determining which policies have been triggered by a hazardous condition: the exposure theory, the manifestation theory, the injury-in-fact theory, and the continuous-trigger theory.

Exposure Theory

Under the exposure theory, coverage is triggered on the date the damaged property or the body of the injured person first comes into contact with the substance or condition that caused the loss.[36] In cases involving allegations of injury due to exposure to harmful substances such as asbestos and lead dust, application of the exposure theory, which is sometimes referred to as the "injurious exposure rule," has been justified on the grounds that it promotes a finding of coverage because it is often less difficult to determine when the injured party was exposed to the substance than when the injury was manifested.[37]

The exposure theory has also been applied in cases involving property damage resulting from hazardous conditions or defective products.[38] For example, if a defective product causes damage to the piece of machinery in which it is being used, a court may find that a CGL policy's property damage coverage was triggered when the machinery was first exposed to the part—that is, when the part was installed therein.[39] Similarly, in cases involving environmental contamination such as seepages or releases of toxic chemicals, courts may hold that an insurer's coverage obligation is triggered at the time the damaged property was first exposed to

[36]See, e.g., *Babcock & Wilcox Co. v. Arkwright-Boston Mfg. Mut. Ins. Co.,* 53 F.3d 762 (6th Cir. 1995) (applying Ohio law); *Boardman Petroleum, Inc. v. Federated Mut. Ins. Co.,* 926 F. Supp. 1566 (S.D. Ga. 1995); *Colonial Gas Co. v. Aetna Cas & Sur. Co.,* 823 F. Supp. 975 (D. Mass. 1993); *Imperial Cas. and Indem. Co. v. Radiator Specialty Co.,* 862 F. Supp. 1437 (E.D. N.C. 1994).

[37]See *Hancock Laboratories, Inc. v. Admiral Ins. Co.,* 777 F.2d 520 (9th Cir. 1985).

[38]See, e.g., *Eljer Mfg., Inc. v. Liberty Mut. Ins. Co.,* 972 F.2d 805 (7th Cir. 1992); *Vermont American Corp. v. American Employers' Ins. Co.,* No. 330-6-95 (Vt. Super. 1997).

[39]Id.

[40]See, e.g., *Town of Peterborough v. Hartford Fire Ins. Co.,* 824 F. Supp. 1102 (D. N.H. 1993).

the hazardous substance.[40] Such reasoning lends itself especially well to situations in which toxic gases or liquids invade a neighbor's property, causing a loss.

Because exposure of a body or item of property to hazardous substances is often discovered long after the fact, attorneys often must consult experts to assist with the inquiry regarding when exposure most likely first took place. Once that is determined, or a plausible argument with respect to this issue is constructed, only the policies whose periods cover those dates may potentially be triggered. Under the exposure theory, multiple policy years can be triggered if it is determined that a continuous or repeated exposure occurred.

Manifestation Theory

Under the manifestation theory, a policy's coverage is triggered at the time the injury or property damage was discovered or manifested.[41] The manifestation theory is applied more frequently in property damage matters than in those involving bodily injury. As the Fourth Circuit Court of Appeals has observed, it is often nearly impossible to determine when property damage caused by hazardous wastes began.[42] For this reason, some courts have adopted the rule that the "occurrence" triggering coverage should be deemed to take place when the damage due to contamination first manifests itself, which is when the effects of the contamination are first discovered.[43]

Although determining the date of the discovery of damage, to either property or body, may seem obvious, this is not always the case. For example, it may be that the only way to determine when the "manifestation" occurs is through the interpretation of technical data relating to the presence of toxic agents in soil or water. One person's damage may be another's wear and tear, so to speak. Similarly, some types of damage, such as soil erosion, occur naturally over periods of time but may be accelerated by conditions present on a neighbor's property. These types of distinctions, which are inherent in the application of the manifestation theory to a particular loss, may be possible only with the assistance of forensic specialists.

[41]See, e.g., *Mraz v. Canadian Universal Ins. Co.,* 804 F.2d 1325 (4th Cir. 1986); *West Am. Ins. Co. v. Tufco Flooring E., Inc.,* 104 N.C. App. 312, 409 S.E.2d 692 (1991); *United States Fidelity & Guar. Co. v. American Ins. Co.,* 169 Ind. App. 1, 345 N.E.2d 267 (1976).

[42]Mraz, 804 F.3d at 1328.

[43]Id.

Injury-in-Fact Theory

Under the injury-in-fact theory, only policies in effect when the injury or damage is shown to have actually taken place will be triggered.[44] Some courts have opted for the injury-in-fact, or actual-injury, approach based on their interpretation of occurrence-based policies; in their opinion, the CGL definition of "occurrence" logically encompasses "only those injuries, sicknesses, or diseases that are proved to have existed" during the policy period.[45] The injury-in-fact theory has also been applied to claims of property damage.[46] For example, in a case involving property damage to buildings caused by the presence therein of asbestos-containing products, the Second Circuit Court of Appeals, relying on New York cases holding that the incorporation of a defective product into another product inflicts property damage, concluded that the "damage-in-fact" occurs at the time of the products' installation in the buildings.[47]

In contrast to the exposure and manifestation theories, the injury-in-fact theory involves inquiring as to when the injury or damage itself occurred, rather than the exposure resulting injury or the manifestation of the injury, respectively. In the context of injuries or damage caused by environmental contamination, which often develop gradually and within the recesses of the body or the soil, one could argue that "injury" or "damage" should realistically be seen as grammatical constructs; the time of exposure or the manifestation are often more easily identifiable and quantifiable, whereas what we refer to with the concepts of injury and damage is actually only the result of a unity of those two ideas. Nevertheless, in a jurisdiction applying this rule, attorneys will need to develop arguments for their client regarding when the "injury" or "damage" actually took place. As with the other tests for determining trigger, this effort will require the assistance of forensics consultants who can construct out of the data and other types of evidence an account of the incident that offers an opinion as to when the harm occurred.

[44]See, e.g., *Trizec Properties, Inc. v. Biltmore Constr. Co.,* 767 F.2d 810 (11th Cir. 1985); *Armstrong World Industries, Inc. v. Aetna Cas. & Sur. Co.,* 45 Cal. App. 4th 1, 39, 52 Cal. Rptr. 690 (1996); *St. Paul Fire & Marine Ins. Co. v. McCormick & Baxter Creosoting Co.,* 324 Ore. 184, 923 P.2d 1200 (1996).

[45]*American Home Products Corp. v. Liberty Mut. Ins. Co.,* 565 F.Supp. 1485, 1497 (S.D. N.Y. 1983), aff'd as modified, 748 F.2d 760 (2nd Cir. 1984).

[46]See, e.g., *Maryland Cas. Co. v. W.R. Grace & Co.,* 23 F.3d 617 (2nd Cir. 1994) (as amended), *cert. denied,* 115 S. Ct. 655 (1994).

[47]Id., at 627.

Continuous-Trigger Theory

Recently, many courts have turned away from the preceding three tests for determining the date of trigger and have adopted the continuous-trigger theory.[48] Under the continuous-trigger theory, all policies that were in effect from the time of the initial exposure to the harmful substance or condition to the manifestation of the effects thereof are triggered.[49] In claims for injury or damage resulting from environmental contamination, the trigger of the insurer's coverage obligations is generally considered to be the date the claimant was first exposed to environmental contamination. Further, the cutoff point under the continuous-trigger theory is commonly deemed to be the date the damages or injuries are manifested or the date of the making of the claim.[50]

Courts applying a continuous trigger to environmental claims sometimes justify their conclusion by pointing out that injury occurs in each phase of environmental contamination. Further, many environmental actions involve claimants who suffered continuous and repeated exposure to a continuing series of loss-producing events.[51] It is thus appropriate, they reason, to require all insurers whose policies are on the risk throughout the entire period of the contaminants' impact to respond to the losses occasioned thereby.[52] Environmental pollution is not restricted to easily determined periods the way, for example, auto accidents are. The uncertainties inherent in attempting to pinpoint the dates of exposure, manifestation, and injury warrant a more expansive theory, courts applying this rule claim.[53]

Because under a continuous-trigger theory courts ask when the initial exposure and the subsequent manifestation of the harm occurred, attorneys must attempt to determine the dates of both of these events in order to properly assess whether a particular policy will be triggered. As discussed pre-

[48]See, e.g., *Montrose Chemical Corp. of Cal. v. Admiral Ins. Co.,* 10 Cal. 4th 645, 42 Cal. Rptr. 2d 324 (1995); *Benoy Motor Sales, Inc. v. Universal Underwriters Ins. Co.,* 287 Ill. App. 3d 942, 679 N.E.2d 414 (1997); *Owens-Illinois, Inc. v. United Ins. Co.,* 264 N.J. Super. 460, 625 A.2d 1 (App. Div. 1993), rev'd in part on other grounds and remanded, 138 N.J. 437, 650 A.2d 974 (1994).

[49]See, e.g., *Carter Wallace v. Admiral Ins. Co.,* No. A-3558-95T3 (N.J. App. Div. 1997).

[50]See *Owens-Illinois, Inc. v. United Ins. Co.,* 264 N.J. Super. 460, 625 A.2d 1 (App. Div. 1993), rev'd in part on other grounds and remanded, 138 N.J. 437, 650 A.2d 974 (1994).

[51]See, e.g., *Zurich Ins. Co. v. Transamerica Ins. Co.,* 34 Cal. App. 4th 933, 34 Cal. Rptr. 913 (Cal. Ct. App. 1994).

[52]Id.

[53]See, e.g., *J.H. France Refractories Co. v. Allstate Ins. Co.,* 534 Pa. 29, 626 A.2d 502 (1993); *Bristol-Myers Squibb Co. v. AIU Ins. Co.,* No. A-145,672 (Tex. Dist. Ct. 1996).

viously, forensics experts can be of great value regarding both of these inquiries. Further, because the application of a continuous theory involves developing an understanding of the progression of the contamination from beginning to the last manifestation of its effects, the need for consultants' advice is not necessarily limited to the issues of exposure and manifestation. Technical experts can assist the coverage attorney in gaining the broad understanding of the site and the loss occurring there, which a continuous-trigger-theory analysis may demand.

ALLOCATION OF DEFENSE AND INDEMNITY COSTS

The need to develop a method for the allocation of losses among various policy years arises when a continuous trigger is applied in response to injury or damage occurring over a period during which more than one policy was in effect. Contamination that occurs over a period of many years may implicate the policies of multiple insurers. In contrast, when a manifestation or exposure trigger is applied, one or more policy years is triggered. Courts have thus been required to develop for continuous- or multiple-trigger claims allocation methods that take into account the contractual language of the policies involved, the interests of the claimants, and general issues of equity. This allocation requirement is complicated by the fact that different insurers provide coverage in different years, with different limits of liability, perhaps a different number of occurrences in a given year, and, at times, different forms of coverage.

Joint and Several Liability

The concept of joint and several liability as applied to insurance coverage disputes means that an insurer is required to provide coverage up to the limits of each of its policies that has been triggered.[54] This method of allocation has also been referred to as a "pick-and-choose" and/or "all-sums" approach because the insured may make a claim against any or all insurers whose policies have been triggered and each would be liable for the insured's injury or damage up to policy limits, regardless of the insurer's degree of responsibility for the loss. One justification for this doctrine is that standard CGL policies require the insurer to pay for "all sums" the insured

[54] See, e.g., *Keene Corp. v. Insurance Co. of N. Am.*, 667 F.2d 1034 (D.C. Cir. 1981), *cert. denied*, 455 U.S. 1007 (1982); *J.H. France Refractories Co. v. Allstate Ins. Co.*, 534 Pa. 29, 626 A.2d 502 (1993).

becomes legally obligated to pay because of personal injury or property damage during the policy period.[55]

Pro Rata Allocation Based on Time on the Risk

In contrast to the joint and several liability method, pro rata allocation apportions liability for covered losses among triggered policies. The question then becomes what factors will be taken into account in determining what each insurer's proportionate share amounts to. Some courts apportion based on the amount of time each insurer was "on the risk," which involves asking what proportion of the duration of the loss the combined duration of a particular insurer's policies totals.[56] Although insurers are likely to view this approach as fairer than imposing joint and several liability, allocating solely based on time on the risk allocation does not take into account the fact that policy limits can vary greatly.

Limits Multiplied by Time on the Risk

A variation on the time-on-the-risk method is that which multiplies the combined duration of an insurer's triggered policies by the combined limits of liability of the triggered policies.[57] In adopting this method, New Jersey's Supreme Court justified its decision by stating that allocating risk in this manner is more consistent with the economic realities of risk retention or risk transfer, especially with respect to periods when there was a year-by-year increase in policy limits.[58]

"Other Insurance" Clauses

A fundamental purpose of "other insurance" clauses is to anticipate situations in which an insured will have obtained other policies for the same time period as the subject policy, or other policies issued to the insured are triggered along with the subject policy with respect to a particular loss. There are three main types of "other insurance" clauses: pro rata, excess,

[55]*Acands, Inc. v. Aetna Cas. & Sur. Co.,* 764 F.2d 968 (3rd Cir 1985).

[56]See, e.g., *Insurance Co. of North America v. Forty-Eight Insulations, Inc.,* 633 F.2d 1212 (6th Cir. 1991), clarified and aff'd on reh'g, 657 F.2d 814 (6th Cir.), *cert. denied,* 454 U.S. 1109 (1981); *Clemtex, Inc. v. Southeastern Fidel. Ins. Co.,* 752 F.2d 976 (5th Cir. 1987).

[57]See, e.g., *Owens-Illinois, Inc. v. United Ins. Co.,* 138 N.J. 437, 650 A.2d 974 (1994).

[58]*Owens-Illinois, Inc.,* 138 N.J. at 475, 650 A.2d at 993.

and escape. Pro rata clauses typically provide for contribution by the subject policy according to the ratio of its applicable limits of liability to the total applicable limits of all insurers. Excess clauses typically state that the insurance provided by the subject policy is excess over any other insurance covering the loss insured under the subject policy, whether such other insurance is collectible or not. Escape clauses typically provide that if the insured has other insurance, whether on an excess, primary, or contingent basis, covering the losses covered under the subject policy, no insurance will be provided under the subject policy.

Courts will at times look to the types of "other insurance" clauses found in triggered policies to assist in determining how to best allocate liability.[59] Courts virtually never, however, base an allocation finding solely on these clauses. Rather, other factors, including equitable considerations, are employed, with the language of "other insurance" clauses merely contributing to the overall analysis. Further, some courts have concluded that "other insurance" clauses provide nothing of value to allocation inquiries.[60]

Apportionment Between Covered and Noncovered Claims

Many times the complaint filed in a lawsuit contains numerous allegations, not all of which are necessarily covered under the defendant-insured's insurance policy. The majority rule is that in such a situation there will be apportionment of defense and indemnity costs between the covered and noncovered claims.[61] However, some courts have refused to order allocation where the claims consist of the same factual core—for example, covered negligence claims and noncovered claims with respect to the same acts, but alleging intentional behavior.[62] Other courts have refused to allocate costs based on a supposed lack of a reasonable manner of effecting proration.[63] When allocation is permitted, the insurer is generally required to provide a defense for covered and noncovered claims alike and then seek

[59]See, e.g., *Uniroyal, Inc. v. Home Ins. Co.,* 707 F. Supp. 1368, (E.D. N.Y. 1988).

[60]See *Owens-Illinois, Inc.,* 138 N.J. at 470-471, 650 A.2d at 991.

[61]See, e.g., *EEOC v. Southern Publishing Co., Inc.,* 894 F.2d 785 (5th Cir. 1990); *Morrone v. Harleysville Mut. Ins. Co.,* 283 N.J. Super. 411, 662 A.2d 562 (App. Div. 1995).

[62]See, e.g., *Public Utility Dist. No. 1 v. International Ins. Co.,* 124 Wash. 2d 789, 881 P.2d 1020 (1994).

[63]See, e.g., *National Steel Constr. Co. v. National Union Fire Ins. Co.,* 14 Wash. App. 573, 543 P.2d 642 (1975); *Burlington Drug Co. v. Royal Globe Ins. Co.,* 616 F. Supp. 481 (D. Vt. 1985).

reimbursement for the costs of defending the noncovered claims after the defense has been concluded.[64]

Allocation Issues Arising when There Are Periods when the Insured Is Self-Insured or Uninsured

As was discussed previously, when a continuous trigger is applied, the triggered policies may span a period of years. In situations where the insured was self-insured or obtained no insurance for a portion of that period, the question arises whether the triggered policies should provide coverage for the loss corresponding to the self-insured or uninsured periods. Courts have split in their decisions on this issue. Some courts have held that the insured is not required to share in the defense and/or indemnity costs for such periods.[65] While this conclusion flows logically from a joint and several liability approach to allocation,[66] some courts refusing to allocate costs to the insured for self-insured or uninsured periods have stated that this rule is justifiable so long as harm is caused at least in part during the policy period.[67] Other courts have, however, rejected this position altogether, holding that with respect to periods of self-insurance or no insurance, an insured is responsible for its share of the costs as if it were an insurer.[68]

Applicability of CGL Personal Injury Coverage to Pollution Claims

The preceding discussion of topics such as trigger and the duty to defend have largely been confined to CGL policies' coverage for liability resulting from bodily injury or property damage. The standard CGL policy also provides coverage, however, for losses due to personal injury, whether under the personal and advertising injury coverage part or a personal injury endorsement. "Personal injury" is generally defined as including, in addition

[64]See, e.g., *Gulf Chemical & Metallurgical Corp. v. Associated Metals & Minerals Corp.,* 1 F.3d 365 (5th Cir. 1993).

[65]See, e.g., *Keene Corp. v. Insurance Co. of North America,* 667 F.2d 1034 (D.C. Cir. 1981); *B&L Trucking & Constr. Co. v. Northern Ins. Co.,* 134 Wash. 2d 413, 951 P.2d 250 (1998); *Aerojet-General Corp. v. Transport Ind. Co.,* 17 Cal. 4th 38, 70 Cal. Rptr. 2d 118 (1997).

[66]See *Keene Corp. v. Insurance Co. of North America,* 667 F.2d 1034 (D.C. Cir. 1981).

[67]See, e.g., *Aerojet-General Corp. v. Transport Ind. Co.,* 17 Cal. 4th 38, 70 Cal. Rptr. 2d 118 (1997).

[68]See, e.g., *Domtar v. Niagara Fire Ins. Co.,* 563 N.W.2d 724 (Minn. 1997); *Outboard Marine Corp. v. Liberty Mut. Ins. Co.,* 283 Ill. App. 3d 630, 670 N.E.2d 740 (1996); *Owens-Illinois, Inc. v. United Ins. Co.,* 138 N.J. 437, 650 A.2d 974 (1994).

to such torts as defamation and unlawful imprisonment, harm in the form of "wrongful entry or eviction, or other invasion of the right of private occupancy." The issue has arisen whether CGL personal injury coverage extends to losses resulting from environmental contamination or pollution. This question is usually addressed with respect to allegations that the seepage or release of a toxic substance onto one's property constitutes the torts of trespass and/or nuisance, which are typically not included in the CGL policy definition of "personal injury."

In considering this question, many courts have focused on the fact that the phrase "or other invasion of the right of private occupancy" immediately follows "wrongful entry or eviction." Applying the principle of *ejusdem generis*, which states that general terms should be interpreted so as to ascribe to them a meaning consistent with that of specific terms preceding them, courts have interpreted the phrase "other invasion of the right of private occupancy" as being limited to interferences with a possessory right, a characteristic shared by "wrongful entry" and "wrongful eviction."[69] In pollution cases, then, the question becomes whether a party whose property has been damaged by the effects of toxic substances has suffered an interference with a possessory right. Many courts have answered this question in the negative, holding that no personal injury coverage existed for such losses.[70] In addition to the previously referenced analysis applying *ejusdem generis*, some courts have justified their holding by stating that pollution exclusions would be nullified if coverage for pollution damage could be recast as personal injury.[71]

Other courts have held that personal coverage is available for losses resulting from contamination, despite the absence of an interference with possessory rights.[72] For example, the Seventh Circuit Court Appeals has held that personal injury coverage existed for a suit alleging damage to ground wells due to chemical releases, concluding that the term "wrongful entry" encompasses allegations of negligent trespass.[73] It has also been held that because a private nuisance involves an invasion of the right to the use and enjoyment of land, personal injury coverage was available for

[69]See, e.g., *Martin v. Brunzelle,* 699 F. Supp. 167 (N.D. Ill. 1988); *National Union Fire Ins. Co. v. Rhone-Poulenc, Inc.,* No. 87C-SE-11 (Del. Super. Ct. New Castle County May 19, 1993).

[70]See, e.g., *Martin v. Brunzelle,* 699 F. Supp. 167 (N.D. Ill. 1988); *W.H. Breshears, Inc. v. Federated Mut. Ins. Co.,* 832 F. Supp. 288 (E.D. Cal. 1993); *Legarra v. Federated Mut. Ins. Co.,* 35 Cal. App. 4th 1472, 42 Cal. Rptr. 2d 101 (Cal. Ct. App. 1995).

[71]*Legarra v. Federated Mut. Ins. Co.,* 35 Cal. App. 4th 1472, 42 Cal. Rptr. 2d 101 (Cal. Ct. App. 1995).

[72]See, e.g., *Titan Holdings Syndicate, Inc. v. City of Keene,* 898 F.2d 265 (1st Cir. 1990).

[73]*Scottish Guar. Ins. Co., Ltd. v. Dwyer,* 19 F.3d 307 (7th Cir. 1994).

pollution-related allegations involving disposal of waste at a landfill.[74] Courts finding in favor of personal injury coverage for claims alleging environmental contamination appear to be increasingly in the minority, however. One can readily see why.

POLLUTION EXCLUSIONS

In reaction to the tremendous costs involved in the investigation, cleanup, and remediation of environmental contamination, insurance companies began in the mid-1970s to include in their policies provisions excluding from coverage injury or damage resulting from pollution, or at least so they thought. The first exclusion adopted on a wide scale was the so-called sudden and accidental pollution exclusion, which, contrary to its name, purported to preclude coverage for all pollution-related liability except that resulting from a sudden and accidental event. In the late 1970s and early 1980s, the need for a broader exclusion became apparent, and insurers adopted the absolute pollution exclusion. This so-called "absolute" provision was proven not to be as seamless as was hoped by the insurers, yielding in the early 1990s the latest incarnation, the total pollution exclusion, which, unlike its name, has also resulted in some decisions unfavorable to insurers finding that "total" does not always mean "total."

While, in theory, any pollution exclusion can be drafted more effectively, whether a court concludes that an exclusion operates to bar coverage in a particular case will nearly always be determined by a fact-intensive application of the exclusion's language to the situation at issue. For this reason, the advice and expertise of technical consultants are often extremely valuable to the attorney who must convince a judge and/or a jury that one or more of the terms of the pollution exclusion found in the policy under examination do or do not apply to the contamination-causing incident at issue. To better understand the issue, we will provide an introduction to the three pollution exclusions found successively in CGL policies during the last two and a half decades. We will also discuss briefly some of the case law on whether a particular substance qualifies as a pollutant under the pollution exclusions.

Sudden and Accidental Pollution Exclusion

The sudden and accidental pollution exclusion typically contains language identical or very similar to the following:

[74] *Royal Ins. Co. v. Xtra Corp., Inc.*, No. 95-CV-43 (Wis. Cir. 1997).

> This insurance does not apply:
> to "bodily injury" or "property damage" arising out of the discharge, dispersal, release or escape of smoke, vapors, soot, fumes, acids, alkalis, toxic chemicals, liquids or gases, waste materials or other irritants, contaminants or pollutants into or upon land, the atmosphere or any water course or body of water; but this exclusion does not apply if such discharge, dispersal, release or escape is sudden and accidental. . . .

Because the sudden and accidental pollution exclusion precludes coverage for pollution that occurs gradually, the key issue in interpreting this provision has been which polluting events qualify as "sudden and accidental." Some courts have interpreted the term "accidental" as meaning unexpected and unintended.[75] With respect to the meaning of "sudden," many courts have interpreted it to involve a temporal aspect or element.[76] These courts tend to view "sudden" as meaning "abrupt" and, accordingly, to hold that pollution damage taking place over a lengthy period of time is barred by the exclusion.[77]

Courts' focus on whether the insured intended or expected that the contaminating event would occur, as well as the importance frequently ascribed to the length of time over which the pollution took place, necessitates in many cases that forensic experts be consulted. It is not always a simple matter to determine how long, for example, PCE was being released into groundwater. With respect to the requirement that pollution be unexpected and unintended for it to be covered, insureds are not always forthcoming about what they expected or intended. It is sometimes possible to determine from data collected at the site or corporate records obtained through discovery whether management knew about or ordered a release, dispersal, or other type of contamination-producing event.

Some enterprising insureds have argued that a polluting event seemingly gradual by any standard is nothing more than a series of discreet sudden and accidental events.

Absolute Pollution Exclusion

The typical absolute pollution exclusion contains the following language, or a variation thereof:

[75]See, e.g., *A-H Plating, Inc. v. American Nat'l Fire Ins. Co.,* 57 Cal. App. 4th 427, 67 Cal. Rptr. 2d 113 (Cal. Ct. App. 1997); *LaSalle National Trust, N.A. v. Schaffner,* 818 F. Supp. 1161 (N.D. Ill. 1993); *Claussen v. Aetna Cas. & Sur. Co.,* 259 Ga. 333, 380 S.E.2d 686 (1989).

[76]See, e.g., *Shell Oil Co. v. Winterthur Swiss Ins. Co.,* 12 Cal. App. 4th 715, 15 Cal. Rptr. 2d 815 (1993); *Dimmitt Chevrolet, Inc. v. Southeastern Fidelity Ins. Co.,* 636 So.2d 700 (Fla. 1993).

[77]See, e.g., *Service Control Corp. v. Liberty Mut. Ins. Co.,* 46 Cal. App. 4th 1047, 54 Cal. Rptr. 2d 74 (Cal. App. 1996).

This insurance does not apply:

(1) to bodily injury or property damage arising out of the actual, alleged or threatened discharge, dispersal, seepage, migration, release or escape of pollutants:

(a) at or from premises owned, rented or occupied by the named insured;

(b) at or from any site or location used by or for the named insured or others for the handling, storage, disposal, processing or treatment of waste;

(c) which are at any time transported, handled, stored, treated, disposed of, or processed as waste by or for the named insured or any person or organization for whom the named insured may be legally responsible; or

(d) at or from any site or location on which the named insured or any contractors or subcontractors working directly or indirectly on behalf of the named insured are performing operations:

(i) if the pollutants are brought on or to the site or location in connection with such operations; or

(ii) if the operations are to test for, monitor, clean up, remove, contain, treat, detoxify or neutralize the pollutants.

(2) to any loss, cost or expense arising out of any governmental direction or request that the named insured test for, monitor, clean up, remove, contain, treat, detoxify or neutralize pollutants.

Pollutants means any solid, liquid, gaseous or thermal irritant or contaminant, including smoke, vapor, soot, fumes, acids, alkalis, chemicals and waste. Waste includes materials to be recycled, reconditioned or reclaimed.[78]

One would certainly think that coverage for pollution was not intended. In fact, courts interpreting the absolute pollution exclusion have not always required that the insured be responsible for the pollution alleged by the claimant in order for the exclusion to apply.[79] This reasoning is based on the exclusion's lack of specification as to which party's activities or operations must cause the loss; the exclusion merely requires that the injury or damage arise out of the pollution. The Ninth Circuit Court of Appeals affirmed a district court decision holding that the absolute pollution ex-

[78]Some versions of the absolute pollution exclusion and the total pollution exclusion also include the following provision: "Subparagraphs (a) and (d) of paragraph (1) of this exclusion do not apply to bodily injury or property damage caused by heat, smoke or fumes from a hostile fire. As used in this exclusion, a hostile fire means one which becomes uncontrollable or breaks out from where it was intended to be."

[79]See, e.g., *Homestead Ins. Co. v. Ryness Co.,* 851 F. Supp. 1441 (N.D. Cal. 1992), aff'd, 15 F.3d 1085 (9th Cir. 1994); *Town of Harrison v. National Union Fire Ins. Co.,* 89 N.Y.2d 308, 675 N.E.2d 829, 653 N.Y.S.2d 75 (1996).

clusion applied to a lawsuit alleging that the insured, a real estate broker, negligently failed to disclose that property adjacent to that which was sold was contaminated.[80]

The abstractness of the exclusion's "arising out of" language has yielded many disputes regarding whether a particular injury or damage arose out of the release, seepage, and so forth, of pollutants so as to render the exclusion applicable. Without input from a person knowledgeable about the technical issues surrounding a particular toxic event, coverage attorneys' attempts to construct arguments regarding whether the loss arose out of the contamination would generally amount to nothing more than a speculative exercise.

Conversely, although the are numerous decisions holding that the exclusion precluded coverage for various types of pollution-related damage, the provision has not been 100% effective. For one thing, some courts have held that certain terms contained in it are ambiguous.[81] For example, a Pennsylvania court concluded that the term "atmosphere" as employed in the exclusion is ambiguous and held that claims arising out of a restaurant's release of carbon monoxide fumes were not excluded thereby.[82] The court also adopted the position that the exclusion only applies to environmental contamination, as opposed to ordinary risks associated with a building.[83]

The New Jersey Appellate Division has also imposed a limitation on the application of the absolute pollution exclusion. *Kimber Petroleum Corp. v. Travelers Indemnity Co.*[84] involved a claim for coverage for various third-party environmental cleanup actions and state directives arising out of the insured's sale and distribution of gasoline to service stations. Evaluating a clause identical to the foregoing, the court found that the absolute pollution exclusion did not preclude coverage for completed operations because "[w]ith respect to operations, the pollution exclusion speaks in the present tense and refers to an exclusion of insurer liability where the named insured or its contractors or subcontractors 'are performing operations.'" Because the completed operations hazard encom-

[80]See *Homestead Ins. Co. v. Ryness Co.,* 851 F. Supp. 1441 (N.D. Cal. 1992), aff'd, 15 F.3d 1085 (9th Cir. 1994).

[81]See, e.g., *Alabama Plating Co. v. United States Fid. and Guar. Co.,* 690 So.2d 331 (Ala. 1996); *American States Ins. Co. v. Kriger,* 622 N.E.2d 945 (Ind. 1996); *Gamble Farm Inn, Inc. v. Selective Ins. Co.,* 440 Pa. Super. 501, 656 A.2d 142 (Pa. App. 1995).

[82]See *Gamble Farm Inn, Inc. v. Selective Ins. Co.,* 440 Pa. Super. 501, 656 A.2d 142 (Pa. App. 1995).

[83]Id.

[84]*Kimber Petroleum Corp. v. Travelers Indemnity Co.,* 689 A.2d 747 (N.J. Super. Ct. App. Div. 1997).

passes the past tense, part (1) of the absolute pollution exclusion was held inapplicable to claims falling under that hazard. *Kimber* did not, however, address the exclusion's application to the mere sale of a product (with no related operation).

Another important issue regarding the applicability of the absolute pollution exclusion has been whether the particular substance causing the damage qualifies as a "pollutant" under the definition contained in the exclusion. Because the total pollution exclusion contains the same definition as is contained in the absolute pollution exclusion, this issue will be addressed after a brief discussion of the total pollution exclusion.

Total Pollution Exclusion

Because the total pollution exclusion has only been in use since approximately 1993, few courts have had the opportunity to interpret it. A small body of case law has begun to develop, however.

The total pollution exclusion typically contains the following language:

> This insurance does not apply to:
> (1) "Bodily injury" or "property damage" which would not have occurred in whole or part but for the actual, alleged or threatened discharge, dispersal, seepage, migration, release or escape of pollutants at any time.
> (2) Any loss, cost or expense arising out of any:
> (a) Request, demand or order that any insured or others test for, monitor, clean up, remove, contain, treat, detoxify or neutralize, or in any way respond to, or assess the effects of pollutants; or
> (b) Claim or suit by or on behalf of a governmental authority for damages because of testing for, monitoring, cleaning up, removing, containing, treating, detoxifying or neutralizing, or in any way responding to, or assessing the effects of pollutants.Pollutants means any solid, liquid, gaseous, or thermal irritant or contaminant, including smoke, vapor, soot, fumes, acids, alkalis, chemicals and waste. Waste includes materials to be recycled, reconditioned or reclaimed.

While part (2) of the exclusion retains the "arising out of" language found in the absolute pollution exclusion, part (1) creates a "but for" test to be applied in determining whether the exclusion will apply: Would the injury have occurred "in whole or part but for the actual, alleged or threatened discharge, dispersal, seepage, migration, release or escape of pollutants at any time"? Although on its face the total pollution exclusion appears to weave a tighter net with which to prevent pollution claims from penetrating the realm of covered losses, it has been shown not be immune from some of the same types of concerns as courts have expressed regarding the application of the sudden and accidental and absolute pollution exclusions to particular scenarios.

Courts have held that the total pollution exclusion applied to bar coverage for damage caused by sedimentation runoff from construction sites,[85] sewage backup,[86] and carbon monoxide vapors.[87] Several courts, however, have held that the exclusion was ambiguous as applied to the facts of the claim at issue.[88] For example, the Tenth Circuit Court of Appeals concluded, applying Mississippi law, that the term "escape" found in the exclusion was ambiguous as to the issue of whether it applied to the expulsion of a container of ethyl parathion from a moving vehicle and reversed summary judgment in favor of the insurer on the issue of whether the exclusion barred coverage for losses resulting therefrom.[89] The court justified its conclusion by observing that many courts finding pollution exclusions ambiguous have done so in the context of nontypical fact patterns such as the one at bar. Further, the court commented that the lack of case law on the total pollution exclusion rendered a fact-intensive analysis particularly appropriate. In other words, the court just wanted to find coverage.

What Qualifies as a Pollutant?

Both the absolute and the total pollution exclusions define "pollutant" as the following:

> any solid, liquid, gaseous, or thermal irritant or contaminant, including smoke, vapor, soot, fumes, acids, alkalis, chemicals and waste. Waste includes materials to be recycled, reconditioned or reclaimed.

Despite the use of relatively concrete and specific terms, numerous disputes have arisen with regard to whether a particular substance falls within the definition. Forensics consultants can provide especially use-

[85]See *Pennsylvania Nat'l Mut. Cas. Ins. Co. v. Triangle Paving, Inc.*, No. CV-892-5-3-BR (E.D. N.C. Dec. 30, 1996), aff'd, 1997 U.S. App. LEXIS 19274 (4th Cir. 1997).

[86]See *Panda Management Co., Inc. v. Wausau Underwriters Ins. Co.*, 62 Cal. App. 4th 992, 73 Cal. Rptr. 2d 160 (Cal. Ct. App. 1998).

[87]See *Reliance Ins. Co. v. VE Corp.*, 1995 U.S. Dist. LEXIS 13872 (E.D. Pa. September 19, 1995).

[88]See, e.g., *Nautilus Ins. Co. v. Jabar*, 1999 U.S. App. LEXIS 20803 (1st Cir. August 30, 1999); *Red Panther Chemical Co. v. Insurance Co. of the State of Pennsylvania*, 43 F.3d 514 (10th Cir. 1994).

[89]See *Red Panther Chemical Co. v. Insurance Co. of the State of Pennsylvania*, 43 F.3d 514 (10th Cir. 1994).

ful advice when a coverage attorney is constructing an argument on this issue.

While courts have virtually universally held that substances such as benzene, chlorinated solvents, and ammonia are "pollutants" for the purposes of applying the exclusions, disputes regarding other substances have yielded splits in authority. For example, while many courts have held that asbestos qualifies as a "pollutant,"[90] others have reached the opposite conclusion.[91] Courts have also split regarding whether carbon monoxide should be viewed as a "pollutant."

The Supreme Court of Illinois held that the absolute pollution exclusion did not bar coverage for carbon monoxide poisoning caused by defective furnaces.[92] The court justified its conclusion by stating that the exclusion's "otherwise potentially limitless application" should be restricted to only those hazards traditionally associated with environmental pollution.[93] In a recent decision, a New Jersey appellate court cited similar concerns in holding that because the absolute pollution exclusion was "uncertain or ambiguous and susceptible to differing interpretations" as applied to a claim for personal injury caused by the indoor residential ingestion of lead paint chips, flakes, or dust, it could be interpreted as excluding coverage therefor.[94] Other courts, however, have held that the exclusions' definition of "pollutant" applies to substances that are not typically considered to be traditional environmental pollutants, such as *E. coli* bacteria[95] and slaughterhouse fumes.[96]

Care, Custody, and Control (Owned Property) Exclusion

The care, custody, and control exclusion, often referred to as the owned property exclusion, basically represents an attempt by insurers to prevent third-party liability policies from functioning as first-party policies. The typical owned property exclusion contains the following language:

[90]See, e.g., *Kosich v. Metropolitan Prop. and Cas. Ins. Co.,* 1995 N.Y. LEXIS 3599 (N.Y. App. Div. 1995); *Sunset-Vine Tower, Ltd. v. Committee & Indus. Ins. Co.,* No. C 738 874 (Cal. Super. 1993).

[91]See, e.g., *Flintkote Co. v. American Mut. Liab. Ins. Co.,* No. 808-594 (Cal. Super. 1993); *Owens-Corning Fiberglass Corp. v. Allstate Ins. Co.,* No. 90-2521 (Ohio Com. Pleas 1993).

[92]*American States Ins. Co. v. Koloms,* 177 Ill. 2d 473, 687 N.E.2d 72(1997).

[93]Id., at 489, 687 N.E. 2d at 79.

[94]See *Byrd v. Blumenreich,* No. A-3810-97T3 (N.J. App. Div. 1999).

[95]See *East Quincy Services Dist. v. Continental Ins. Co.,* 864 F. Supp. 976 (E.D. Cal. 1994).

[96]See *Zacky Farms v. Central Nat'l Ins. Co. of Omaha,* 92 F.3d 1195 (9th Cir. 1996).

This insurance does not apply to:

j. "Property damage" to:
   (1) Property you own, rent, or occupy;
   (2) Premises you sell, give away or abandon, if the "property damage" arises out of any part of those premises;
   (3) Property loaned to you;
   (4) Personal property in the care, custody or control of the insured;
   (5) That particular part of real property on which you or any contractors or subcontractors working directly or indirectly on your behalf are performing operations, if the "property damage" arises out of those operations. . . .

The owned property exclusion has been held to generally preclude coverage for costs incurred in cleaning up the insured's own property.[97] Several issues have arisen, however, relating to the application of this general principle.

Does the Exclusion Apply to Groundwater?

One issue that has frequently arisen with respect to the application of the owned property exclusion is whether the term "property" includes groundwater underneath the insured's site. Most courts addressing this issue have distinguished between soil and groundwater contamination, holding that the exclusion does not apply to the latter.[98] Explaining its holding in accord with this principle, one court noted that allegations of potential groundwater contamination may reasonably be read as including possible off-site contamination, thus not necessarily being limited to the insured's property.[99] Other courts have simply opined that because groundwater is not "owned property," the exclusion does preclude coverage for damage thereto.[100] There are isolated decisions, however, in which courts have questioned whether the exclusion could not apply to damage to groundwater, especially if the groundwater were found to be within the "care, custody, or control" of the insured.[101]

[97]See, e.g., *Shell Oil Co. v. Winterthur Swiss Ins. Co.,* 12 Cal. App. 4th 715, 15 Cal. Rptr. 2d 815 (Cal. Ct. App. 1993), rev. den. (Cal. May 13, 1995); *Summit Assocs., Inc. v. Liberty Mut. Fire Ins. Co.,* 229 N.J. Super. 56, 550 A.2d 1235 (App. Div. 1988).

[98]See, e.g., *LaSalle Nat'l Trust v. Schaffner,* 818 F. Supp. 1161 (N.D. Ill. 1993); *Strnad v. The North River Ins. Co.,* 292 N.J. Super. 476, 679 A.2d 166 (App. Div. 1996).

[99]See *Vann v. Travelers Cos.,* 39 Cal. App. 4th 1610, 46 Cal. Rptr. 2d 617 (Cal. Ct. App. 1995).

[100]See, e.g., *Strnad v. The North River Ins. Co.,* 292 N.J. Super. 476, 679 A.2d 166 (App. Div. 1996).

[101]See, e.g., *Shell Oil Co. v. Winterthur Swiss Ins. Co.,* 12 Cal. App. 4th 715, 15 Cal. Rptr. 2d 815 (Cal. Ct. App. 1993), rev. den. (Cal. May 13, 1995).

Does the Owned Property Exclusion Apply to Costs of Cleaning Up the Insured's Owned or Occupied Property for the Purpose of Preventing Off-Site Migration?

In situations where the insured endeavors to clean up its own property in order to prevent contamination located thereon from traveling to another's property, the question arises whether finding the owned property exclusion to bar coverage for these costs would not necessarily follow from the goal of preventing CGL coverage from serving as first-party property insurance. Some courts have thus held that an exception to the general rule exists with respect to such claims.[102] The rationale for this exception is that to the extent the insured has expended sums for the purposes of ensuring that another's property is not damaged, the purpose behind the owned property exclusion is not being circumvented. In New Jersey, the inapplicability of the exclusion in such cases is conditioned on the existence of actual damage to a third party's property, not just potential damage.[103] Other courts have refused to embrace this exception, some basing their position in part on a literal interpretation of CGL policy language providing that coverage is available only for damage "to which this insurance applies."[104] In light of the foregoing concerns expressed by the courts, a forensic expert's advice regarding whether particular costs were allocated to cleanup efforts of third-party property or property of the insured can prove crucial to an attorney's coverage analysis.

CLAIMS-MADE POLICIES

In the foregoing discussions, we have made reference mainly to CGL policies, most of which provide coverage on an occurrence basis. Some CGL and virtually all EIL policies only cover injury or damage from a claim that is made during the policy period. While many of the issues that are of concern in cases involving occurrence-based policies are identical or at least similar when a claims-made policy is involved, the latter type of policy has characteristics that merit an independent treatment.

A typical coverage provision found in claims-made policies contains the following language:

[102]See, e.g., *Shell Oil Co. v. Winterthur Swiss Ins. Co., supra; CPS Chem. Co. v. Continental Ins. Co.,* 222 N.J. Super. 175, 536 A.2d 311 (App. Div. 1988).

[103]See *State Dep't of Env'l Protection v. Signo Trading Int'l, Inc.,* 130 N.J.51, 612 A.2d 932 (1992).

[104]See, e.g., *Bausch & Lomb, Inc. v. Utica Mut. Ins. Co.,* 330 Md. 758, 625 A.2d 1021 (1993).

1. Environmental Impairment Liability and Claims Made Clause: To pay on behalf of the Insured all sums in excess of the deductible amount stated in the Declarations which the Insured shall become legally obligated to pay as damages as a result of CLAIMS FIRST MADE AGAINST THE INSURED DURING THE POLICY PERIOD for
   a. Personal injury;
   b. Property damage;
   c. Impairment or diminution of or other interference with any other environmental right or amenity protected by law; caused by environmental impairment in connection with the business of the Insured. . . .

Claims-made policies also typically define "claim" as "a demand received by the Insured for money or services, including the service of suit or institution of arbitration proceedings against the Insured." Courts have interpreted the term "claim" in the context of claims-made policies as contemplating the assertion of a legal right by a third party for damages caused by the insured's conduct.[105] It has also been concluded that a "claim" relates to "an assertion of legally cognizabledamage" and "must be a type of demand which can be defended, settled, and paid by the insurer."[106]

Another characteristic of claims-made policies that distinguishes them from occurrence-based policies is that they generally provide coverage for occurrences going back in time to before the policy's inception, limited only by a retroactive date. This should not be confused with coverage for claims prior to the policy's inception, coverage for which is not available under claims-made policies. Further, application for claims-made policies requires prospective policyholders to disclose the existence of claims against them with respect to the sites to be insured. Failure to disclose such claims can constitute misrepresentation in the application and can result in a denial of coverage for claims for damages resulting from activitiesat the site(s). Retention of a forensics expert to determine whether prior claims have been made and the nature thereof may be of value in evaluating coverage under a claims-made policy.

Unlike policies in which an "occurrence" triggers coverage, the insurer's obligations under a claims-made policy are generally triggered by the claim against the insured and, thereafter, the insured's notice of a claim

[105]See, e.g., *Atlas Underwriters, Ltd. v. Meredith-Burda, Inc.*, 231 Va. 255, 343 S.E.2d 65 (1986).

[106]See *Evanston Ins. Co. v. GAB Business Services, Inc.*, 132 A.2d 180, 521 N.Y,.S.2d 692 (1st Dept. 1987).

to the insurer. Failure to comply with the notice requirement can result in a finding of no coverage for the claim.[107] A type of claims-made policy known as a claims-made-and-reported policy conditions the availability of coverage on both a claim being made during the policy and notice being provided within the policy period. In contrast, occurrence-based policies allow notice to be provided after the policy has expired, contingent on compliance with requirements that notice be given "as soon as practicable," "as soon as possible," or as otherwise specified in the notice provision.

EIL policies generally contain the following definition of "environmental impairment":

> Environmental impairment, whenever used in this policy means any one or a combination of the following:
> a. Emission, discharge, dispersal, disposal, release, escape or seepage of smoke, vapors, soot, fumes, acids, alkalis, toxic chemicals, liquids or gases, waste materials or other irritants, contaminants or pollutants into or upon land, the atmosphere or any watercourse or body of water;
> b. the generation of odor, noises, vibrations, light, electricity, radiation, changes in temperature or any other sensory phenomena; arising out of or in the course of the Insured's operations, installations or premises, all as specifically described in the Declarations *provided (a) or (b) is not sudden and accidental* [emphasis added].

Note that the requirement that the pollution-causing incident not be sudden and accidental is the reverse of that found in a CGL policy's sudden and accidental exclusions. In contrast to CGL policies, EIL policies were designed to provide coverage for gradual pollution. A similarity to CGL insurance is the EIL policy's exclusion of coverage for "claims for or costs or expenses of or in connection with . . . conditions on the premises of the insured." Technical experts are thus often invaluable in attempts at determining whether pollution is onsite or offsite, and, if both, identifying the percentages of each.

SUMMARY

As the foregoing discussion has attempted to stress, the issues, questions, and problems encountered in many coverage matters involving environmental contamination cannot be analyzed effectively without obtaining the assistance of experts in the relevant fields. The reason for this is

[107]See, e.g., *Graman v. Continental Cas. Co.*, 87 Ill. App. 3d 896, 409 N.E.2d 387 (Ill. Ct. App. 1980).

simple: The principles and rules that courts formulate and apply do not exist in a conceptual vacuum. Rather, they are formed by and intended to shape our business relationships with respect to the realities of contemporary industrial society and the need to protect the environment while avoiding undue restrictions on commerce. The coverage attorney who understands these underlying concerns will also appreciate the value of forensic consultation.

The issues impacting coverage are many more and are more complex than we have discussed herein. We have, however, provided an introduction to coverage for environmental losses and a framework for the resolution of same.

Chapter 5

Site Histories: The Paper Trail How and Why to Follow It

Shelley Bookspan and Julie Corley

The central questions compelling an environmental forensics investigation involve sorting out who caused a contamination problem, when, how, and, among multiple potentially responsible parties (PRPs), how much of the problem is appropriately assigned to each. These are historical questions. Answering them may in the end involve the expertise of multiple professionals: soil scientists, hydrogeologists, forensic chemists, process engineers, and so on. The beginning of the answer, however, lies in the work of the site historian. Not only will the site history answer many of the questions compellingly through a trail of documentary evidence, but it will both narrow and structure the remaining questions. So, for example, a site history of a property with several industrial uses over the years may uncover the whereabouts of a former, now graded-over, liquid waste disposal sump. Through historic maps, the sump may be found on the current property, and, through historic annual reports submitted to a now-defunct water agency, some of the types of process wastes deposited there may be identified. With this information, the site scientists can develop a sampling and analytical plan to locate and identify residual chemicals from the historic operation and to compare them to the contamination problem at hand.

Sometimes, if scientific studies have preceded historical ones, the analytical data may defy reasonable interpretation. The site historian can assist even at this late stage, although with less effect on the efficiency of the scientists, who may have run a number of tests and models attempting a laboratory explanation of their findings. In one case in point, the owners of a former zinc smelter in Arkansas, along with their environmental consultants, called the historical research specialist with a particular request: Help prove the accuracy of their theory about the origin of lead contamination found on scattered lawns in a residential neighborhood northwest of the defunct facility. Although downwind, the lead findings were too random to fit any air dispersion model. The consultants had, therefore, ruled out deposition from the smelter's smokestack. In some cases, the elevated levels occurred on lawns several blocks apart, but not on intervening or even adjoining ones. This seemed to rule out an underlying old slag pile, as did the relative superficiality of the lead. The alternate theory for which the consultants and their clients wanted historical support was that the lead had actually been applied for some reason by the residents themselves. Since there was no known reason why residents would now be doing such a thing, the theory held that there had been a reason in the past, and it would be possible for the historians to uncover that.

In spite of the possible pitfalls of inductive reasoning, uncover it they did. Through a multiple-front investigation that involved identifying and locating past residents and neighborhood merchants while simultaneously unearthing documentation containing express details about historic lawn care products generally, and one lead-based product in particular, developed and marketed at a retail level for Arkansas residential growing conditions, they were able to find witnesses and primary source documentation to undergird what was otherwise just a theory. Significantly, also, they were able in their research to document the chemical composition of the suspected former lawn care products; the environmental science consultants used these latter findings to devise additional laboratory tests that further supported the conclusion that the smelter was not culpable.

It is perhaps more usual for an environmental historian's findings to precede the authorship of a whodunit contamination theory. In an example of a more typical circumstance, the same historian mentioned previously assisted a private industrial client whose historic metals processing facility was assumed to be the source of mercury and other heavy metals found in two functioning water wells that the municipal water district had to take out of service. Although, indeed, the facility had once been the major employer in the area, had operated for more than three decades, and had processed the offending metals, it was a mile or more distant from the affected wells. Numerous other industrial and light industrial businesses had had intervening facilities during those same decades, and many of them had handled the same metals. It was the historian's job to identify all such facilities that had operated within a mile radius of each affected well, to find evidence of their use, storage, and disposal of any of the types of metals affecting the wells, and to prepare individual narration packages for each. The client company used these digests of relevant findings to dissuade the municipal water district from pursuing it for response costs. In this case, it was unnecessary to ascertain exactly whodunit, but it was vital to implicate multiple suspects.

Then, again, frequently in a pollution liability dispute, whodunit and how much are exactly the issues. At times, the solution to such a dispute may elude certainty; confidence in conclusions based on circumstantial evidence, however cumulative, can depend almost entirely on the reliability of such evidence and the plausibility of the resulting narrative. This chapter will review the background of the development and uses of site history, outline site history techniques, and exemplify an ongoing pollution liability dispute involving the U.S. military and several of its former World War II bases. This dispute pits historical narrative against historical narrative.

SITE HISTORIES: WHAT THEY ARE

Before bore holes are drilled on a property, before samples are taken, analytical tests run, and results interpreted, a thorough site history will reveal where to sink those holes, where not to sink them, what chemicals to test for, and to what interpretive end. Conversely, a scientific investigation in the absence of solid historical site information can overlook distinguishing analytes, miss "hot" spots on a property, and occasionally exacerbate or even cause an environmental problem, if, say, a drill bit punctures a theretofore unknown underground storage tank. Although it is generally known in the environmental field that a site history is a necessary part of the homework done in advance of, or in iteration with, scientific site investigations, it is less well understood just how much there is to be known about a site through documents and their interpretation. The prevailing American Society for Testing materials (ASTM) Phase I environmental site assessment standards have helped earn history a permanent place in environmental practice, but they define historical work in a mostly nominal and formulaic way. The goals of standard setting may be met thereby, but environmental professionals engaged in in-depth site analysis are often untrained in historical methodology. Often, however, historical sources, when systematically collected and professionally interpreted, provide the basis for a comprehensive understanding of the types of land uses that have occurred on any given property over time. This includes any associated engineering processes, chemical usage, and waste disposal practices. It is almost always possible to develop a differential list of distinguishing site and operational features and of chemicals and/or waste or storage practices so that wastes on or under a site with a long history of multiple tenants can be tied to a specific time period or process. So it is that the most convincing environmental forensic analyses start with a good site history, that is, a history that really tells the relevant story of the site.

What is such a history? To understand the elements of a good site history, first let us explore what it is not. It is not a collection of documents, maps, or aerial photographs, although these may be some of the necessary sources or "raw materials." Nor is a good site history a recitation of the findings contained in each of these types of sources. It may be of value to know the content of a particular data source, but a history fits the relevant information where it belongs into a larger picture. Nor is a good history a timeline. Although a timeline of selected significant events can provide a road map by which to navigate the complexities of a site history, it is by its nature episodic in coverage; unexplained gaps may indicate either absence of data or absence of relevant data, and are, therefore, easily subject to misinterpretation. Likewise, and perhaps more important, thematic connections are unexamined, and perhaps even unrecognizable, when there is no apparent temporal relationship, and, therefore, the story of the site goes untold.

What, then, is a good site history? The following are necessary conditions:

- The findings are presented in narrative form. This is important because the transitional terms, conjunctions, and other verbal linkages the historian chooses to facilitate the narrative actually provide an interpretation of the data in the source materials, and show the significance of the data presented to the issues of concern at the site.
- The narrative is presented chronologically, intertwining pertinent themes as appropriate. Change (or stasis) over time is the organizing principle of history, and tracking it is the goal of a historical investigation. Even if there is no apparent causal relationship be-

tween sequential events at a site, their occurrence at a certain point in time, and not at another, will have consequential and explanatory value regardless of the site issues.

- Diverse sources, and types of sources, are used. Various types of records about sites have been generated at various points in time, by various (sometimes now extinct) agencies, regarding various and possibly changing aspects of site usage. All such sources are potentially useful and all are potentially flawed, that is, possibly inaccurate, possibly incomplete, possibly ambiguous. A good site history will synthesize the disparate materials, corroborate data, note and assess contradictory information, and mediate between/among inconsistencies. It will make no assumptions at all, but will explain all interpolations or extrapolations.
- Findings are reproducible. Like good science, a good history is reproducible. Other professionals with similar training conducting research into the same site questions would seek and review the same source materials. Other professionals reviewing the same materials would reconstruct the history of the site in the same, or similar, way, or would at least concur in the plausibility and defensibility, given the evidence, of the given reconstruction. In order to allow for this sort of reproduction to occur, the good site history contains clear notations as to the source of the information or finding cited, and to the source of that source, that is, the repository and file locators containing the original source.[1]

For the user, keeping these four criteria in mind will allow for a competent critical evaluation of the historical work product. For the practitioner, the next question is: What do I need to do to meet these criteria? The precise answer to that question will vary from site to site and from situation to situation. In all cases, however, the site historian, whether a professionally trained historian or another professional acting in that capacity for the instant matter, will proceed systematically, incorporating at least the following steps.[2]

[1]There are many good sources that explain the techniques of history practice. One straightforward one, with good reference value, is William Kelleher Storey, *Writing History*, Oxford University Press (1998).

[2]Adapted from Shelley Bookspan, "History Requited: Historians and Toxic Waste," in Mock, P., Ed., *History and Public Policy*, Malabar, FL: Krieger Publishing Company (1991), pp. 75-95. Another useful publication is Craig E. Colten and Diane Mulville-Friel, *Guidelines and Methods for Conducting Property Transfer Site Histories*, Springfield, IL: Illinois State Museum (1990).

Project Objectives

The site historian must first learn the reason for which the site history is being undertaken. In a pollution liability situation or in a transactional matter, the historical investigation will be part of a larger analytical plan. As with any type of investigation, it is necessary to start with an understanding of its goal or goals. From this, the historian can develop an initial work plan, and from then forward, the historian can revise that work plan as appropriate rather than wait until one strategy is exhausted. Perhaps, for example, a site is known to be contaminated with polychlorinated biphenyl (PCB) oils, and the historian is hired to identify who in the past released those oils onto the property, or who may have released them. To accomplish this, the historian will want to learn about the historical use of PCBs, when they came about, and in what circumstances they would have been introduced to a site in the first place. Since PCBs were not in common commercial or industrial use prior to the 1950s, information about the use of arsenic pesticides on site crops in the 1910s and dichlorodiphenyl trichloroethane (DDT) in the 1940s will not further the aims of the investigation. The historian, aware of the project's goal, will not include documentation of those agricultural uses predating PCB, no matter how interesting they may be. On the other hand, if a site sits atop a regional groundwater plume of mixed solvents and other organic chemical contaminants, the historian's goal may be to uncover all of the land uses associated with the subject property, without any particular time or use limitation. In such a case, not only would the information regarding historic pesticide use be important, but so would information about historic land features, such as ditches and wells, that can still underlie property and affect the subsurface environment.

A Historical Frame of Reference

To capture relevant documentation, the historian must think historically when developing a work plan. Remembering that the structure of today's society has evolved over the years is the first step toward thinking historically. While many important documents such as air emission permits, underground storage tank permits, RCRA permits, and the like reside in known current local, state, or federal agencies' files, it may well be that predecessor agencies existed and required submission of documents containing similar or otherwise relevant information. What now lies within the jurisdiction, say, of a state environmental agency may have previously been under the aegis of the state's fish warden. What now is subject to oversight by a state's solid waste management agency once probably came under local public works control. Economic poisons now controlled by the EPA may have once been the province of a state or county agricultural commissioner. Knowledge of the regulatory heritage of a subject site, chemical, or industry will provide the basis for a work plan that includes repositories, such as state, federal, university, or private archives, where defunct business or succeeded agency records may be located. Moreover, confidence in knowing the types of historic records once generated will enable more effective searches to be done of file indices and more convincing inquiries to be made of current agency personnel, many of whom are reluctant to conduct detailed historical work without substantial guidance. Thinking historically means thinking in the same terms that those who produced the desired records would have thought. It means imagining files so old that they are not findable by any computerized indexing system. It means calling such files by what may now be obsolete terminology, such as "sewer permit," rather than "NPDES," or 'bills of lading," instead of "manifests,' so that they may be located.

Thinking historically can also assist in obtaining all of the necessary identifiers of the subject site. This is important to ensure that all site files sought are, in fact, found. Not only do different agencies use different types of identifiers (e.g., street addresses, tax assessor's parcel numbers, section/township/range) by which to sort site files, but often historic designations may have changed as streets have been reconfigured, as property parcelization has occurred, or as localities have been annexed or incorporated. So it is that building permits for a subject site now known by, say, street address 121 Elm Street may also be found in the file of its old address, 101 Elm Street, or 121 Myrtle Street. Frequently, although file merging may be intended when a jurisdiction undergoes this type of change, it may never have, in fact, occurred.

One of the best ways to begin the process of thinking historically, at least when working on a site in the United States, is to obtain a series of historic U.S. Geological Survey topographic maps, and perhaps a series of historic aerial photographs and, if available, Sanborn maps as well. Local university libraries often house these types of cartographic sources. Because they have been produced in series, these graphic sources depict change over time so that using them will help the site historian understand when the site was raw land, by what time it was first in any use by humans, and when it began to look as it currently does. Moreover, these types of sources display the context within which the site's history developed, so that the influence of off-site land uses and the growth of relevant jurisdictional regulations may be better understood, and so that a work plan to find the resulting public records may be better designed.

What is key is that there is no one-size-fits-all site history work plan. The research goals and the site's historical context are the bases on which a problem- and site-specific work plan will be tailored.

PHASE I ENVIRONMENTAL SITE ASSESSMENTS: A DIFFERENT SORT OF HISTORY

Those familiar with the site history sections of Phase I environmental site assessments (ESAs) will recognize that there is little adherence therein to the good site history precepts discussed previously. In general, Phase I ESAs adhere instead to standards developed for transactional purposes, and the elements of work those standards outline represent a compromise between the goals of thoroughness and reliability on one hand and speed and cost containment on the other.

The American Society for Testing Materials, a longstanding trade association for the setting of professional engineering standards, issued its environmental site assessment guidelines in 1993, in response to a need to define the clause in CERCLA (the Comprehensive Environmental Response, Compensation and Liability Act, 1980) that outlined a defense against liability for environmental response costs for certain owners of contaminated property [1]. Under most conditions, the current property owner of a contaminated site, according to this federal statute, holds cleanup liability, although that property owner may, in turn, sue past or off-site contributors, as defined in four categories of potentially responsible parties, for recompense.[3] Under Section 107(B), however, CERCLA offered the following circumstance under which the current property owner can be considered an innocent landowner, and, therefore, without liability for cleanup of contamination:

> . . . [if] the real property on which the facility concerned is located was acquired by the defendant after the disposal or placement of the hazardous

[3] 42 U.S.C. 9601 et seq. Categories of potentially responsible parties are site owners or operators at the time of the disposal or release, generators of the waste contaminants; haulers to the site of the wastes, and arrangers for the disposal of the wastes at the site.

> substance on, in, or at the facility, and . . . : (i). At the time the defendant acquired the facility the defendant did not know and had no reason to know that any hazardous substance which is the subject of the release or threatened release was disposed of on, in, or at the facility.

The section then elaborated on the "no reason to know" standard, as follows:

> (B) To establish that the defendant had no reason to know, as provided in clause (i) . . . the defendant must have undertaken, at the time of acquisition, all appropriate inquiry into the previous ownership and uses of the property consistent with good commercial or customary practice in an effort to minimize liability.

This defense against environmental liability, known generally as the "innocent landowner defense," then, was one of two equally important reasons for the expansion of prepurchase due diligence investigations into queries that would expressly reveal information about past activities on a subject site that could result in residual contamination. The second reason was, of course, actually to identify and evaluate existing environmental problems for the purpose of making an informed purchase decision.

Section 107 left otherwise undefined what, in fact, was necessary to do in order to conduct an inquiry that a court or an agency would consider "appropriate," and how to document that such an inquiry actually occurred prior to purchase. The continuing language in this section, however, implied that the authors of the statute themselves did not know how to document the site's past activities, at least not those that occurred previous to those remaining on site or remaining in the memory of the current owner:

> . . . the court shall take into account any specialized knowledge or experience on the part of the defendant, the relationship of the purchase price to the value of the property if uncontaminated, commonly known or reasonably ascertainable information about the property, the obviousness of the presence or likely presence of contamination at the property, and the ability to detect such contamination by appropriate inspection.

Section 107 offered some measure of security to parties involved in property transfer, and it helped to spur the exponential growth of the environmental consulting business in the 1980s. However, it gave almost no guidance about what would constitute "reasonably ascertainable information about the property," or "appropriate inspection," and it remained for the new Phase I consultants and their clients to develop their own

Phase I methodologies. Most of those early Phase I consultants, however, were engineers or physical scientists, more accustomed to drilling and testing than to researching documents and conducting interviews or even to inspecting properties for evidence of past uses. Untrained in noninvasive techniques and unguided by published standards, these consultants produced Phase I reports widely variable in types of content, reliability, reproducibility, and overall quality.

It was perhaps lending institutions that led the way toward Phase I standardization. Faced, through CERCLA, with the possibility of acquiring a liability instead of an asset if forced to foreclose on a property found to be contaminated, lending institutions began to require Phase I environmental site assessments to be submitted as part of the loan package for any commercial property. Many of these institutions, moreover, required their customers to hire Phase I contractors from among a group of prequalified consultants, and required the consultants to adhere to a certain content and/or format for their studies and reports. This, of course, was an effort to achieve a predictable content and level of reliability. Eventually, there was enough experience among consultants and consumers that three basic elements of a Phase I came generally to be expected: (1) a site inspection conducted by some knowledgeable professional (often otherwise undefined), (2) a review of regulatory agency records for the subject site and for surrounding sites to identify any environmental problem known to such an agency to exist, and (3) a site history. A proposed amendment to CERCLA, the Innocent Landowner Defense Amendment of 1993, would have codified these elements and, furthermore, would have limited the site history to the 50 years just past and would have defined the site history research (i.e., the research regarding prior ownership and uses, rather than the research into known environmental hazards) as

> a review of . . . the following sources of information concerning the previous ownership and uses of the real property:
>
> - Recorded chain of title documents regarding the real property, including all deeds, easements, leases, restrictions, and covenants for a period of 50 years.
> - Aerial photographs which may reflect prior uses of the real property and which are reasonably obtainable through State or local government.[4]

[4]H.R. 570 (the Weldon Bill, or the Innocent Landowner Defense Amendment of 1993), 103rd Cong., 1st Sess., January 25, 1993.

This proposed law codifying these limited sources of site histories for environmental site assessment purposes motivated a group of professional historians, all active in the environmental consulting field, to respond to the bill's author, Congressman Curt Weldon from Pennsylvania. As well intended as the definition was, its concept of a site history was inadequate if, in fact, the intention of conducting such a study was to develop a usable understanding of the property's past. Some of the objections raised in the historians' letter are as follows:

> a fifty-year search is wholly inadequate for several reasons: (a) many sources of toxic wastes began operation prior to 1940, particularly companies working with persistent substances, such as toxic metals, (b) there are innumerable instances of businesses that closed before or during the Great Depression of the 1930s that left behind significant quantities of hazardous wastes and that would escape discovery in a fifty-year documentary search, (c) it takes relatively little additional time to review land uses back to initial development of a site, when this work is performed by professionals, (d) with each passing year, important records will be ignored as the 50-year envelope moves away from the historical time periods associated with completely unregulated use and disposal of hazardous materials, and (e) the succession of land uses, such as factories built on old water wells or irrigation ditches, is often what accounts for the presence of hazards. Love Canal, for example, was closed as a chemical dump in 1952. Using current wording, [soon], a buyer would not be required to determine prior land uses that created the nation's most notorious toxic dump.

Moreover, the historians objected to the identification of a single land use history source, aerial photographs, a longstanding tool of engineers and physical scientists. They suggested, "[a]fter 'aerial photographs,' add 'business directories and other applicable cartographic sources,' explaining that such sources often depict past land uses more accurately and at a more detailed scale than aerial photographs. . . . " They also recommended, finally, an additional paragraph, to encourage an analytical rather than formulaic approach to the site history: "In the absence or inadequacy of the previously listed sources, other pertinent and reasonably obtainable historical records that document past land uses . . . to allow environmental professionals to supplement the existing list of sources when there are local failures in the historical record."[5]

[5]Phil Scarpino, president, National Council on Public History, to Curt Weldon, n.d.

ASTM Standards

The Innocent Landowner Defense Amendment did not progress far toward law, nor did, or has, any proposed CERCLA amendment. Perhaps in this case, however, the promulgation of ASTM Standard E1527-93, in 1993, obviated, or seemed to obviate, the need to define "all appropriate inquiry" through the mechanism of legislation. Through the ASTM vehicle, environmental practitioners could guide themselves and set peer review standards.

The ASTM standards for a site history are encompassed in Section 7.3. That section starts with defining the objective of the site history, in keeping with the good practice of knowing the research goals before beginning. According to the ASTM, "[t]he objective of consulting historical sources is to develop a history of the previous uses or occupancies of the *property* and surrounding area in order to identify those uses or occupancies that are likely to have led to *recognized environmental conditions* in connection with the *property*." The italicization is in the original, and it indicates words or phrases that are defined in earlier sections of the standards. The "property," of course, is the property undergoing assessment, and "recognized environmental conditions" are "the presence or likely presence of any hazardous substances or petroleum products on a property under conditions that indicate an existing release, a past release, or a material threat of a release. . . ."[6] (Italics not reprinted here.)

The methods then described for achieving the objective of identifying uses suggesting a release or a threatened release fall short of meeting the criteria discussed previously for a site history. Section 7.3 indeed lists a useful array of "standard historical sources": aerial photographs, fire insurance maps, property tax files, recorded land title records, U.S. Geological Survey topographic maps, local street directories, building department records, zoning/land use records, and "other historical sources." Nonetheless, according to the standards, it is only necessary for the environmental professional to consult one such source that is "reasonably ascertainable," and for such intervals as the professional determines is necessary to "adequately establish a site history from the present back to 1940."[7] Following these standards, there is no need to use disparate sources to corroborate information, and there is therefore little likelihood of recognizing inconsistencies or inaccuracies. There also is little chance of telling the story of a site from such limited research, although a timeline of docu-

[6]ASTM, op. cit., Sections 3.2.27 and 3.3.28.
[7]Ibid., Sections 7.3.4 and 7.3.2.

mented uses may be possible. If one of these uses is typically associated with chemical wastes or underground fuel storage, then another, invasive phase of investigation is likely to be recommended. If no such use is found, then all that can be known with confidence is that a Phase I environmental site assessment meeting ASTM standards was done.

In spite of the limits of these Phase I site histories, the ASTM has successfully authored minimum standards for professional environmental practice and has provided a reference of sorts for those otherwise untrained who would conduct a historical investigation. Unfortunately, customers purchasing Phase I ESAs are unaware of the limitations inherent in the standards and are often unwilling to pay the additional costs associated with conducting a complete site history.

POTENTIALLY RESPONSIBLE PARTY SEARCHES: THE USE AND LIMITS OF HISTORY IN ENVIRONMENTAL FORENSICS

It is more common for parties involved in a pollution liability dispute to commission a complete site history than it is for parties in a real property or business purchase transaction. That is because, in the former situation, a contamination problem is known to exist, and the costs associated with its cleanup are substantial. While much of the final resolution depends on the success of legal arguments, such as what it means to be an "arranger," or on negotiations, the task of identifying parties whose activities contributed to the problem and where such parties or their successors are today is undertaken in a PRP search, or potentially responsible party search. A site history following the criteria for a good history outlined previously is an integral part of a PRP search.

A work plan for a PRP-related site history will take into account the nature of the contamination and the overall situation. PRP searches can be undertaken for single industrial or commercial sites in which there have been successive users. Or, they may be undertaken for landfill, recycling, or other such sites where materials generated by multiple parties off site have been transported to and commingled on a property. Or, they may be undertaken for a district, such as a groundwater basin or an above-ground waterway, which has had multiple, stationary contributors over time. Each of these types of situations demands a distinct historical approach. Furthermore, each of the approaches must also be subject to the availability of records relative to the pertinent time, place, and activities of concern. There really is no such thing as a standard PRP search.

Nor is there a standard type of documentation that the historian can seek, which, if found, will predictably persuade a party of its culpability. Property ownership is often indisputable, and, therefore, evidence of it can be crucial and relatively easily found. Assigning the time of release to that period of ownership, however, can be more difficult. Various documents and multiple witnesses all supporting the same claim are often necessary, and, as time passes, reliable witnesses to any given release become fewer and harder to locate. Of perhaps even greater importance is how few "smoking gun" documents exist to prove the theory of a given party's contribution. It may well be possible to find a public health inspector's report from 1942 about a smelly acid sludge pit. That report may refer to discussions with "oil company representatives," and it may describe the pit as 20 feet in diameter and 10 feet deep. However, that report is unlikely to give the exact location of the pit by map coordinates; it is unlikely to identify the oil company that was the source of the sludge; it is unlikely to reveal the fate of the sludge. So, it will be necessary to use that document as a link in a story told in conjunction with other circumstantial documents, such as a subsequent special-use permit application filed with the county, hearing transcripts, and water district minutes.

How convincing are such histories, based as they are on the piecing together of miscellaneous documents by someone who was not present as a witness to the events being told? They are often very convincing, especially to a jury that is likely to understand the language of a historian better than the language of a scientist. They can be less convincing, however, to an uncooperative PRP. The following is a case in point.

On numerous sites throughout the United States, private or municipal parties, usually current landowners, are facing liability for the cleanup of chlorinated hydrocarbon solvent contamination, particularly trichloroethylene (TCE) contamination, found in the groundwater underlying the site, and possibly in an extended down-gradient plume. What many of these sites have in common is that they were previously military bases, built expressly to function during World War II and decommissioned and sold for other uses subsequently. During its occupation of these sites, the military often used large quantities of chemicals, especially degreasing and cleaning solvents and fuels. Postwar industry, purchasing wartime structures and engaged in the manufacture of defense equipment, often occupied the identical facilities and also used large quantities of identical or similar chemicals. Within this continuity lies the core of a dispute over historical evidence that is at present taking place in several venues: Must there be site-specific and chemical-specific documentation of the military's responsibility or can circumstantial and general chemical-type documentation satisfy the complaining party's burden of proof?

Through Public Laws 99-190 and 99-949, passed in 1983, Congress required the Department of Defense to investigate the environmental condition of military bases. Formerly Used Defense Sites (FUDS) became part of this Defense Environmental Restoration Program (DERP). Through fiscal year 1996, Defense had spent over $11 billion on DERP, inclusive of FUDS, during which time the department identified 10,660 sites as needing no further environmental action [2]. Most of the environmental action the military has taken to close these sites has been the location and removal of old underground gasoline or oil storage tanks. Simultaneously, many private parties or local agencies have conducted environmental investigations designed to identify residual chemicals in soils and groundwater at (among others) sites that encompass all or portions of former World War II bases. Many of these investigations have uncovered groundwater contamination caused by TCE.[8] TCE is an industrial solvent, prized for its degreasing abilities, that grew in use and popularity throughout the war and the 1950s. The military has been reluctant to accept responsibility for TCE cleanup at World War II bases, even some where there has been no substantial subsequent industrial use, and its actual use at specific bases can be difficult and costly to document. The generality, scattered location, and poor (if any) indexing of many of the pertinent records have led private industrial parties hoping to persuade the military to participate in cleanup costs to hire professional historians to unearth the evidence of the military's actual chemical use and waste disposal practices at given bases. Such evidence includes direct documentation, circumstantial documentation such as federal specifications and technical manuals, oral testimony of witnesses, and the systematic construction of narratives.

In keeping with good history practice, the historians first come to understand the contextual history of their subject contaminant, in these cases TCE. Rare in commercial, industrial, or military use prior to World War II, TCE came to be in almost ubiquitous use during and after the war. TCE is a chlorinated hydrocarbon solvent whose origin may be traced to the re-

[8]Comprehensive Environmental Response, Compensation and Liability Act (1980), 42 U.S.C. 9607 et seq.; and Superfund Amendments and Reauthorization Act (1986).

search of early 19-century European scientists. The availability of chlorinated hydrocarbon solvents for extensive industrial applications did not occur until the early 20th century, however, when a process for liquifying chlorine was developed. This advance rendered chlorine transportable and the manufacture of chlorinated solvents commercially viable [3].

Coinciding with the availability of chlorinated solvents was the development of solvent degreasing of metals by the European metal-working industry in the 1920s. Then, as now, metal processing required that after the initial treatment, such as machining, the metal had to be cleaned of all impurities before painting or other final finishing processes could be started. Chlorinated hydrocarbon solvents, especially TCE, were superior to other washers in that they would dissolve oils and greases on metals, hence cleansing them more thoroughly. In addition, they did not harm the metals and they were nonflammable. Initially, metal parts were dipped or otherwise subjected to cold solvent bathing, but metal workers soon discovered that the most effective cleaning technique was the immersion of metal in the vapors rising from boiling solvent. The advantage of vapor degreasing, as this method was called, was that once the contaminants were lifted off the metal by the vapors and carried away in the condensate runoff, the metal did not come into contact with the solvent or the removed dirt. Thus, no residue remained on the metal, as was possible after immersion in liquid solvent baths. Vapor degreasing did not replace the immersion method in industry, as solvent bathing continued to be employed as a first treatment prior to vapor cleaning. The superior cleaning ability of the vapor degreasing process, however, caused it to be widely adopted in factories and in repair and maintenance of metal products [4].

TCE was first manufactured in the United States in the mid-1920s, and two companies, G.S. Blakeslee & Company and Detroit Rex Products Company, made the first commercially successful solvent vapor degreasers in the early 1930s. The manufacturing capacity of TCE, however, expanded 655% between 1939 and 1950, to accommodate World War II and postwar manufacturing growth as well as direct military TCE usage in cold cleaning, immersion cleaning, and chemical medium functions [5–9].

Although there is an abundance of evidence showing that the military used TCE at its wartime bases in numerous ways, the evidence primarily consists of technical orders, procedural manuals, and chemical statistics. Much of this evidence is supported, for any given wartime base, by unit histories showing that the types of activities for which the military prescribed TCE occurred there and, occasionally, by a living witnesss—now quite elderly—who can describe the use of solvents and their disposal to the bare ground. Arguing that for more than 25 years following World War II, industrial manufacturers continued to use TCE in significant quantities, and predominantly as a metal degreaser, the military has declined to participate in cleanup responsibility based on these types of documents. Instead, the military demands increasingly specific documentation of TCE shipments to subject bases, knowing that such shipping records may never have existed. Moreover, in spite of convincing evidence to the contrary, the military argues that a civilian agency, the War Production Board, controlled military access to TCE during the war (it did not), that TCE was scarce during the war (only temporarily, and then additional production capacity came about), that the military did not use degreasers (it did), and so on. Without eye witnesses or smoking gun documents, it has taken tremendous efforts of historical investigators, and reams of documentation, to gain military participation at any of these sites.

A specific such site is located in southern California. After finding a groundwater plume underlying the former base, the EPA sent a 104(e) request to the U.S. Army Corps of Engineers (COE) on March 19, 1993.[9] The COE responded to the EPA's request on May 13, 1993, and informed the regulatory agency that the COE was conducting a preliminary assessment on the site and would submit its findings in August.[10] The EPA received a copy of the site report on November 10, 1993, although the document was dated July 20, 1993.[11] The EPA did not consider the information contained in it sufficient to answer its questions, and asked for more information. The EPA sent one of its requests to the COE's Office of History, which responded with minimal information about the site, suggested that the EPA hire a contract historian regularly used by the COE, and wished the EPA civil inspector the "best of luck."[12]

In May 1995, after more than two years of trying to obtain information from the COE and frustrated at the Army's lack of response, the EPA met with the COE and discovered that the Army had placed a very low prior-

[9]David Jones, EPA, to Lewis Walker, COE, March 19, 1993. EPA Superfund file room, 2363-00068.

[10]Michael Fellows, COE, to Colette Kostelec, May 13, 1995. EPA Superfund file room, 2363-90169.

[11]According to a letter dated May 25, 1995, from Joseph Stejskal, Municipal Water District to Jerry Lewis, House of Representatives, the EPA received the INPR on November 10, 1993. EPA Superfund file room, 2363-90446.

[12]Charles Hendricks, COE, to Cliff Davis, EPA, January 11, 1994. EPA Superfund file room, 2363-00355.

ity on the site.[13] Having already spent approximately $4 million, and anticipating that cleanup costs would eventually rise to nearly $100 million, the EPA redoubled its efforts to obtain the COE's cooperation. Further contact between COE and EPA in October 1995 led the EPA to conclude that "the Army is spending most, if not all, its investigative effort preparing a technical rebuttal to our suggestion that the Army is responsible for the . . . contamination."[14]

The EPA and COE met on November 29, 1995, to discuss the status of the COE's responses to the EPA's information requests.[15] At the meeting, the COE informed the EPA that the COE contractor's final report would not be submitted to the COE until June 1996, and that the COE would then prepare a response to the EPA's 104(e) requests based on the information in that report. The EPA continued to exert pressure on the Army for responses to its information requests. On February 2, 1996, the COE submitted some documentation to the EPA, and claimed that the slow award and contracting process and security issues under the Privacy Act as reasons for not providing information. Moreover, the COE claimed that it was not required to do archive research for 104(e) response. The COE stated:

> It is clear that the extent of the groundwater contamination in this case is extensive. However, the Army's archive research has produced no evidence confirming that the Army used any of the contaminants of concern on the premises during its ownership of the property. Indeed, general documents indicate such items were in short supply during World War II. . . ."[16]

The COE continued to send documents and manuals to the EPA with no analysis or explanation of the relevance of the materials. The EPA found the attitude of the COE, as well as its responses to information requests, insufficient. By June 28, 1996, the EPA drafted an Administrative Order for Protective Measures pursuant to the authority vested in the EPA under RCRA.[17] Other parties involved with the case have since filed suit against the government, which continues to resist participation.

[13]EPA Internal Memorandum, Kevin Mayer to Tom Kremer, Carmen Gonzalez, and Cliff Davis, May 30, 1995. EPA Superfund file room, 2363-00392.

[14]EPA Memorandum, Kevin Mayer to Carmen Gonzales, October 26, 1995. EPA Superfund file room, 2363-00461.

[15]"Agenda, EPA and U.S. Army Corps of Engineers—Sacramento, November 29, 1995." EPA Superfund file room, 2363-00499.

[16]Raymond Fatz, COE, to Keith A. Takata, EPA, February 2, 1996. EPA Superfund file room, 2363-90566.

[17]Administrative Order, Keith Takata, EPA, to Raymond Fatz, COE, June 28, 1996. EPA Superfund file room, 2363-00515.

So it is that the site history may not be enough to solve the environmental whodunit problem to all parties' satisfaction, especially not when large sums of money are at stake at sites across the country. In the case of former military bases succeeded by industry using the same chemical compounds, perhaps only adjudication will resolve the adequacy of the site history. In other cases, however, the site history may provide the distinguishing key by which environmental scientists can unlock the solution, such as an additive to TCE that was used only during the war years and that can be found on special laboratory analysis. Documentation of such an additive may exist, and a motivated client may, in fact, hire a site historian to find it.

REFERENCES

1. "Standard Practice for Environmental Site Assessments: Phase 1 Environmental Site Assessment Process," American Society for Testing Materials, Standard E1527-93.

2. *1996 Annual Report to Congress*, Vol. 2, Defense Environmental Restoration Program, pp. 4, 7 (1997).

3. T.W. Chestnut, "The Market Responses to the Government Regulation of Chlorinated Solvents: A Policy Analysis," RAND Graduate School (October 1988).

4. "Dry Cleaning Fluids," *The Iron Age*, 64–65 (June 18, 1942).

5. "Trichloroethylene," *Chemical and Engineering News*, 31(3): 234–237 (January 19, 1953).

6. L.P. Litchfield, "Safe Operation of Solvent Degreasers," *The Iron Age*, 58–62 (March 1, 1946).

7. "Industrial Metal Degreasing," *Metal Industry*, 473–474 (December 1932).

8. D.H. Byers, "Chlorinated Solvents in Common Wartime Use," *Industrial Medicine*, 12(7): 440–443 (July 1943).

9. M. Morse and L. Goldberg, "Chlorinated Solvent Exposures," *Industrial Medicine*, 12(10): 706–713 (October 1943).

Chapter 6

The Forensic Application of Contaminant Transport Models

Tad W. Patzek

The use of contaminant transport models to predict the future distribution of specific contaminants in groundwater and, thus, help in the design and implementation of a corrective action is a fundamental component of the remedial process. This is not, however, the role of contaminant transport models when dealing with environmental forensics.

Environmental forensic questions that normally need to be answered are (1) what was the origin or source of contamination and (2) when did the contamination occur? This information is often found through historic documents or, on occasion, the obvious (such as a leaking underground storage tank [UST]). However, there are times when questions require "inverse modeling" either to confirm or support other information or to directly address the questions in lieu of any other alternative. Inverse modeling is much more difficult because the answer may be inaccurate or imprecise while, at the same time, it is expected to be more accurate and precise than that in the plume modeling for remedial applications.

Accurate and precise answers require one key ingredient: information. The information required to run inverse contaminant transport models is based on knowing both the characteristics of the contaminant release and the hydrologic characteristics of the site being modeled.

Information concerning the mass of the contaminant and the time period over which it would have been released (i.e., hours, days, weeks, etc.)

into the aquifer needs to be known but usually is not available. As a result, the modeler must make assumptions about the release based on related site information (i.e., site data or testimony), and accuracy is necessarily lost. Furthermore, the degree to which the site has been characterized highly influences the ability of the modeler to develop accurate modeling predictions. Thus, the quality and the amount of information available generally increase the degree of accuracy and precision, or confidence, in the model predictions.

The purpose of this chapter is to illustrate that the science and mathematics that are used to make contaminant transport modeling predictions *do* work. It is also shown that model predictions are not necessarily as precise as one might desire. Consequently, the ultimate reliability of model predictions is based on the forensics expert's skill at making an astute analysis of the site data and providing a critical evaluation of the parameters selected as model inputs.

THE CONTAMINANT MODELING PROCESS

Finding past contamination sources from present sparse measurements is an inherently difficult task and answers are always nonunique to some extent. For this reason, one must be very thorough and cautious in analyzing the available data and in extrapolating them away from the monitoring well locations.

The well-known steps [1, 2] of a protocol for site model formulation are as follows:

1. Define the purpose of a site model.
2. Define the model scope by inventorying available site data.
3. Form a conceptual site model by analyzing the site data.
4. Select a suitable mathematical model to describe contaminant transport.
5. Select a suitable numerical algorithm[1] to implement the mathematical model.
6. Verify that the numerical algorithm can do what it is asked to do.
7. Use the conceptual model as inputs to the computer model.
8. Calibrate the computer model by adjusting inputs.

[1]This means that one writes one's own computer code, or one gets somebody else's code.

9. Verify the computer model.
10. Present the results.

Define the Purpose

It is usually helpful to define in a couple of sentences what one wants to do.

Define the Scope

Before any mathematical modeling of contaminant fate is attempted, the following steps must be completed. The relevant *hard data* may consist of

1. Triangulated *X*, *Y* surface locations of groundwater monitoring wells
2. Measured levels of groundwater at one or more times
3. Aquifer thickness measured from geological logs
4. Periodically measured groundwater concentrations of contaminants
5. Measured concentrations of chloride and sulfate ions[2]
6. Well test data to determine the aquifer conductivity
7. Start and end dates for various facilities at the site
8. Volumes of waste streams

The relevant *soft data* may consist of

9. Geological logs, soil types, and layering
10. Approximate surface locations of some entry points of contaminants
11. Regional geology information

Form a Conceptual Model of the Site

Before attempting modeling of *any sort*, a thorough exploratory data analysis must be performed. This analysis forms the foundation of all future mathematical and numerical models; without it, the numerical models can serve only as GIGO (garbage in, garbage out) machines to deceive and misinform the nonexperts.

[2]Determination of chloride plumes is seldom done, but may be extremely useful because chloride is a natural tracer.

The necessary steps of the exploratory data analysis for a site might be:

1. Evaluate the groundwater flow direction and gradient.
2. Evaluate the aquifer porosity from geological logs.
3. Evaluate the aquifer thickness from geological logs.
4. Estimate the contaminated aquifer intervals from vertical samples.
5. Evaluate well tests to obtain the aquifer conductivity.
6. If possible, evaluate the distribution of aquifer porosity and conductivity.
7. Calculate the groundwater flow velocity and estimate the possible range of values.
8. Evaluate the spatial correlation of chloride ion concentrations (calculate semi-variograms).
9. Calibrate the groundwater velocity against growth of the nonadsorbing chloride plume.
10. Evaluate the spatial correlation of concentrations of contaminants (calculate semi-variograms).
11. Using these semi-variograms and an optimal interpolation method, plot contaminant plume shapes.
12. Given the aquifer mineralogy and porosity, and relying on the results of published laboratory and *in situ* measurements, evaluate one or more contaminant retardation coefficients.
13. Using the aquifer porosity, scale the remaining retardation coefficients.
14. From analogous aquifer tests and general scaling techniques, estimate the longitudinal and transverse dispersion of the chloride plume.
15. Evaluate the historical volumes of contamination discharge at the site.
16. Estimate aquifer locations of contaminant discharge.

Now that a conceptual model of contaminant flow has been fixed, it is time to choose the type of numerical model.

Select a Suitable Mathematical Model

The mathematical flow model appropriate in most cases is the standard convection[3]–dispersion–adsorption (CDA) model of solute transport. In

[3]Others say "advection."

other cases, a more sophisticated, three-dimensional compositional simulator must be used (see [3]).

Find a Suitable Numerical Algorithm

The sparseness and low quality of site data often warrant the use of a two-dimensional model of the aquifer with the average porosity, hydraulic conductivity, and the contaminant retardation factors. The area encompassed by the model may be large, thousands of feet in each direction, and there may be multiple sources of contamination scattered across the model area. To properly distinguish among the various contamination sources, the model area must be subdivided into smaller elements (numerical or analytic). Often with reasonably small elements, the model grid may consist of over one million elements. The requirement of such a fine grid resolution may preclude a priori the use of a commercial code, such as MODFLOW with a contaminant transport module.

Later in this chapter, we develop a simple semianalytic model and apply it to a well-described site at Borden, Ontario. Such a model handles dozens of sources and millions of analytic elements, and still works on a personal computer (PC). Because our model is very fast, it allows one to perform many what-if calculations in a short time. The forward plume transport model may also be linked with an optimal inverse-calculation engine [4]. Then the source locations, durations, and strengths that satisfy all available data may be determined automatically. Only after the simplified exploratory calculations are done, a more sophisticated computer code may be used if necessary.

Design a Computer Implementation of the Conceptual Model

Although the preceding remarks may seem technical, they demonstrate that a proper and fast contaminant transport model may determine one's ability to resolve the problem at hand. For example, one may have to use 1,000,000 elements and 10 sources, and still obtain the plume shapes in minutes. If one were to use MODFLOW/MT3D with 1,000,000 grid cells, one would have to wait for a *week* to get a single answer. This is why the choice of a proper numerical algorithm is so crucial. Choosing a wrong computer code may prevent one from analyzing the problem at hand thoroughly and from investigating all possible scenarios.

Calibrate the Computer Model

Now the parameters of the conceptual model must be adjusted so that the calculated plume concentrations agree with the concentrations measured in groundwater monitoring wells at a particular time. This calibration involves mostly adjusting the number, strength, and location of contamination sources. Each such adjustment may take several iterations and a *very* long time to accomplish when done by trial and error.

Why Is It Important?

This chapter covers all the steps necessary to validate a numerical code for single-phase solute transport in porous media. Mastering its content will be helpful in debugging a computer code you are about to use in your consulting work.

PROBLEM STATEMENT: CONVECTION–DISPERSION–ADSORPTION MODEL

We pose the flow problem in two dimensions because vertical sampling in groundwater monitoring wells is virtually nonexistent outside of research projects. In the literature, we can find the convection–dispersion–adsorption (CDA) equation for the one-phase flow of a single conservative aqueous solute in a homogeneous porous medium[4]:

$$\frac{\partial}{\partial t}\left(R(x, y, t)c\right) + V(x, y, t)\frac{\partial c}{\partial x} - \frac{\partial}{\partial x}\left(D_L(x, y, t)\frac{\partial c}{\partial x}\right)$$
$$\frac{\partial}{\partial y}\left(D_T(x, y, t)\frac{\partial c}{\partial y}\right) = q \tag{6.1}$$

where c is the vertically averaged solute concentration; φ is the rock porosity; q is the volumetric solute source or sink term; $V = U_x(x, y, t,)/\varphi(x, y)$ is the linear solute velocity and U_x is the superficial (Darcy) velocity of one-dimensional water flow aligned with the x axis; D_L and D_T are the longitudinal and transverse dispersion coefficients; and

$$R = 1 + \frac{\rho_b K_d(x, y, t)}{\varphi(x, y)} \tag{6.2}$$

[4]See [4–9].

is the retardation factor for a linearly, reversibly, and independently sorbing, but otherwise nonreacting solute at local thermodynamic equilibrium with the porous medium. Note that small changes in the bulk density of the medium, $\rho_b = (1\text{-}\varphi)\rho_{grain}$, and its porosity, are dwarfed by the large changes in K_d (cubic centimeters of aqueous contaminant per gram of solid), the equilibrium partition coefficient of the solute. Usually, it is assumed that R does not depend on time. However, it now appears [10, 11] that K_d can be a function of time for several organic solutes. For hydrophobic organic solutes, such as chlorinated hydrocarbons, the linear partition coefficient is further expressed [10–13] as

$$K_d = K_{oc} f_{oc} \tag{6.3}$$

where K_{oc} is the organic carbon partition coefficient and f_{oc} is the weight fraction of organic carbon in the porous medium material. The use of (6.3) reduces a several-order-of-magnitude variation of K_d to a two- to threefold variation of K_{oc}. The use of (6.3), however, can lead to erroneous estimates of sorption, since the partition coefficients apply to particular solute–sorbent systems and, more important, these equations may not apply when the organic carbon content is below 0.1% [13]. The total concentration of the solute in the porous medium is, of course, cR.

It is also often [5–8, 14–16] assumed that

$$\begin{aligned} D_L &= D^* + \alpha_L(x, y)V(x, y, t) \\ D_T &= D^* + \alpha_T(x, y)V(x, y, t) \end{aligned} \tag{6.4}$$

where D^* is the coefficient of molecular diffusion of the solute in water and α_L and α_T are the longitudinal and transverse dispersivities. In many flow situations, molecular diffusion, D^*, can be neglected in comparison with hydrodynamic dispersion.

To address the issue of validity of (6.1), we first assume that in all problems of interest there exists a well-defined mean flow velocity, V_0, as well as a mean retardation factor, R_0, and mean longitudinal and transverse dispersion coefficients, D_{L0} and D_{T0}. They can be obtained by averaging $R(x, y, t)$, $V(x, y, t)$, $D_L(x, y, t)$, and $D_T(x, y, t)$ over the relevant space and time. We admit that these averages depend on the spatial and temporal scales and that they must be updated iteratively as new data become available. By dropping the subscript *0* from the averages, (6.1) becomes

$$R\frac{\partial c}{\partial t} + V\frac{\partial c}{\partial x} - D_{\mathrm{L}}\frac{\partial^2 c}{\partial x^2} - D_{\mathrm{T}}\frac{\partial^2 c}{\partial y^2} = 0 \tag{6.5}$$

In (6.5) all of the transport coefficients are constant.

Instantaneous Point Source Solution of the CDA Equation

The instantaneous point source solution to (6.5) is very important to all other developments in this chapter. Because mistakes have been made in the literature,[5] this solution is derived in Sidebar 6.1:

$$c(x, y, t) = \frac{M}{4\pi\varphi Ht\sqrt{D_{\mathrm{L}}D_{\mathrm{T}}/R^2}} \exp\left[-\frac{(x - Vt/R)^2}{4D_{\mathrm{L}}t/R} - \frac{y^2}{4D_{\mathrm{T}}t/R}\right] \tag{6.6}$$

Here M is the injected solute mass and H is the aquifer thickness. This particularly simple solution *does* not approximate well plumes with long-lasting sources, such as a spill of separate-phase nonaqueous phase liquid (NAPL) or continuous seepage of aqueous contaminant solution. We shall derive different, more complicated solutions for some of those cases. However, the instantaneous point source solution can be used to validate (6.5) in well-controlled field experiments with plumes that are not too close to their short-lived sources. Because (6.6) is so simple, it allows one to pose several questions that can be answered analytically. Among such questions are: Given the current size of a plume and the measured concentration levels, can we find the plume age? The location of its instantaneous source? The strength of the source? Of course, to answer such questions, one must know the average properties of the aquifer transporting the plume.

To obtain the peak concentration of the plume as a function of time, we simply set the terms $x - Vt/R$ and y to 0. Thus, the maximum concentration at the center of mass (CM) of the contaminant plume is

$$c_{\max}(t) = \frac{M}{4\pi\varphi Ht\sqrt{D_{\mathrm{L}}D_{\mathrm{T}}/R^2}} \tag{6.7}$$

We shall assume that this concentration has been measured at time t and is known.

[5]For example, de Marsily [8] has missed the porosity in his Eq. (10.3.6), p. 270

Sidebar 6.1 Instantaneous Point Source Solution

We seek a solution to,

$$R\frac{\partial c}{\partial t} + V\frac{\partial c}{\partial x} - D_L\frac{\partial^2 c}{\partial x^2} - D_T\frac{\partial^2 c}{\partial y^2} = 0 \quad \text{(SB.1)}$$

subject to the following initial (IC) and boundary (BC) conditions:

$$\begin{aligned} &\text{IC:} \quad c(x, y, t = 0) = \delta(x)\delta(y) && \text{for all } x \text{ and } y \\ &\text{BC:} \quad \lim_{|x|\to\infty,\, |y|\to\infty} c(x, y, t) = 0 && \text{for } t > 0 \end{aligned} \quad \text{(SB.2)}$$

Hence, at $t = 0$, at the origin we have a point source of strength

$$\varphi H \int_{-\infty}^{+\infty}\int_{-\infty}^{+\infty} \frac{M}{\varphi H}\,\delta(x)\delta(y)\,dx\,dy = M$$

In order to eliminate the first derivative with respect to x, let us apply the following change of variables: $t' = t$, $x' = x - (V/R)t$, and $y' = y$. The inverse change of variables is characterized by $t = t'$, $x = x' + (V/R)t'$, and $y = y'$. Therefore,

$$\frac{\partial}{\partial t} = \frac{\partial}{\partial t'} + \frac{\partial}{\partial x'}\frac{\partial x'}{\partial t} = \frac{\partial}{\partial t'} - \frac{V}{R}\frac{\partial}{\partial x'} \qquad \frac{\partial}{\partial x} = \frac{\partial}{\partial x'} \qquad \frac{\partial}{\partial y'} = \frac{\partial}{\partial y}$$

Hence, in the new variables the equation takes on the form

$$R\frac{\partial c}{\partial t'} - D_L\frac{\partial^2 c}{\partial x'^2} - D_T\frac{\partial^2 c}{\partial y'^2} = 0$$

Further, put $x'' = \sqrt{R/D_L}\,x'$ and $y'' = \sqrt{R/D_L}\,y'$. Then

$$\frac{\partial c}{\partial t'} = \frac{\partial^2 c}{\partial x''^2} + \frac{\partial^2 c}{\partial y''^2} \quad \text{(SB.3)}$$

Since $\delta(\alpha x) = 1/\alpha\delta(x)$, one obtains the following initial condition for the function $c(x'', y'', t)$:

(continues)

(Sidebar 6.1, continued)

$$c(x'', y'', t = 0) = \frac{R}{\sqrt{D_L D_T}} \frac{M}{\varphi H} \delta(x'')\delta(y'') \tag{SB.4}$$

Clearly, the boundary conditions at ∞ remain the same in the new variables:

$$\lim_{|x''| \to \infty, |y''| \to \infty} c(x'', y'', t) = 0 \tag{SB.5}$$

The solution to the boundary value problem (SB.3)–(SB.5) is well known [17]:

$$c(t', x'', y'' = 0) = \frac{1}{4\pi t'} \frac{R}{\sqrt{D_L D_T}} \frac{M}{\varphi H} \exp\left(- \frac{x''^2 + y''^2}{4t'} \right)$$

In the original variables, one obtains the desired solution

$$c(t, x, y) = \frac{1}{4\pi t} \frac{R}{\sqrt{D_L D_T}} \frac{M}{\varphi H} \exp\left(- \frac{\frac{R}{D_L}\left(x - \frac{V}{R}t\right)^2 + \frac{R}{D_T}y^2}{4t} \right) \tag{SB.6}$$

We define the leading edge of the plume, $X_{le}(t)$, as the leading tip of a constant concentration contour. The concentration to be chosen, c_{min}, may correspond to twice the detection limit or to the drinking water standard:

$$c_{min}[X_{le}(t)] = c_{max}(t)\exp\left[- \frac{(X_{le} - X_{CM})^2}{4D_L t/R} \right] \tag{6.8}$$

Similarly, the trailing edge of the plume is defined as the trailing tip of the same concentration contour

$$c_{min}[X_{te}(t)] = c_{max}(t)\exp\left[- \frac{(X_{te} - X_{CM})^2}{4D_L t/R} \right] \tag{6.9}$$

where

$$X_{\rm CM} = \frac{Vt}{R} = \frac{1}{2}(X_{\rm le} + X_{\rm te}) \tag{6.10}$$

From (6.8) and (6.10), one readily obtains

$$c_{\min}[X_{\rm le}(t)] = c_{\max}(t)\exp\left[-\frac{L^2}{16D_{\rm L}t/R}\right] \tag{6.11}$$

where $L \equiv (X_{\rm le} - X_{\rm te})$ is the plume length. Hence, the plume age is

$$t = \frac{RL^2}{16D_{\rm L}\ln(c_{\max}/c_{\min})} \tag{6.12}$$

the current position of its center of mass is

$$X_{\rm CM} = \frac{VL^2}{16D_{\rm L}\ln(c_{\max}/c_{\min})} \tag{6.13}$$

and the released mass of contaminant is

$$M = c_{\max}(t)4\pi\varphi Ht\sqrt{\frac{D_{\rm L}D_{\rm T}}{R^2}} = \frac{\pi\varphi HC_{\max}L^2}{4\ln(c_{\max}/c_{\min})}\sqrt{\frac{D_{\rm T}}{D_{\rm L}}} \tag{6.14}$$

Thus, we have expressed the plume age, the current position of its center of mass, and the total mass of contaminant released into the aquifer through the measured extent $(X_{\rm le}, X_{\rm te})$ of a contour of constant concentration $(c_{\min})$, the measured peak concentration $(c_{\max})$, and the aquifer properties.

The same contour crosses any other coordinate, X_0 (e.g., a property boundary), at a different time, t_0, given implicitly by

$$t_0 = \frac{RL^2}{16D_{\rm L}}\,\frac{c_{\max}(t_0)/c_{\min}}{\ln(c_{\max}(t_0)/c_{\min})}\exp\left[-\frac{(X_0 - Vt_0/R)^2}{4D_Lt_0/R}\right] \tag{6.15}$$

From the preceding equations, one can estimate the relative error of any of the derived parameters as a function of the individual errors in the determination of the independent variables. For example, the relative error in the determination of the plume age is no larger than

$$\left|\frac{\delta t}{t}\right| \leq \left|\frac{\delta R}{R}\right| + \left|\frac{2\delta L}{L}\right| + \left|\frac{\delta D_{\mathrm{L}}}{D_{\mathrm{L}}}\right| + \frac{1}{\ln(c_{\max}/c_{\min})}\left(\left|\frac{\delta c_{\max}}{c_{\max}}\right| + \left|\frac{\delta c_{\min}}{c_{\min}}\right|\right) \quad (6.16)$$

Here δ stands for the individual errors.

SOLUTION OF CDA EQUATION FOR CONTINUOUS INJECTION OF CONTAMINANT

Transient Solution with a Single Source

We want to solve the CDA equation with constant average coefficients

$$R\frac{\partial c}{\partial t} + V\frac{\partial c}{\partial x} - D_{\mathrm{L}}\frac{\partial^2 c}{\partial x^2} - D_{\mathrm{T}}\frac{\partial^2 c}{\partial y^2} = \frac{\delta(x)\delta(y)c_0(t)Q(t)}{\varphi H} \quad (6.17)$$

subject to the boundary and initial conditions (SB.2) in Sidebar 6.1. Here $\delta(x)$ and $\delta(y)$ are Dirac delta functions.

Thus, initially, there is no solute in the aquifer. At the origin ($x=0, y=0$), there is continuous injection of solute. Injection starts at $t=0$, and continues with a volumetric flow rate $Q(t)$ and solute concentration $c_0(t)$. Note that the total mass of solute in the porous medium at time t is

$$\varphi H \int_{-\infty}^{+\infty}\int_{-\infty}^{+\infty}\int_{0}^{t} \frac{\delta(x)\delta(y)c_0(\tau)Q(\tau)}{\varphi H}\, dx\, dy\, d\tau$$

$$= M(t) = \int_0^t c_0(\tau)Q(\tau)\, d\tau \qquad \forall t > 0 \quad (6.18)$$

We assume that $Q(t)$ is so small that it does not disturb the parallel water flow in the x direction. Often, $Q(t)$ is assumed to result in infiltration with a velocity equal to the local water flow velocity $V(X = 0, y = 0, t)$. Per unit time, an instantaneous mass slug of solute $dM(t)/H$ is injected per unit thickness of aquifer. The contribution of an instantaneous slug at time τ to the total solute concentration at time t is given by (6.6):

$$c_\tau(x, y, t) = \frac{c_0(\tau)Q(\tau)}{4\pi\varphi H(t-\tau)\sqrt{D_{\mathrm{L}}D_{\mathrm{T}}/R^2}} \times \exp\left[-\frac{(x - V(t-\tau)/R)^2}{4D_{\mathrm{L}}(t-\tau)/R} - \frac{y^2}{4D_{\mathrm{T}}(t-\tau)/R}\right] \quad (6.19)$$

The summation of all such slugs over the entire time interval [0, t] leads to the following convolution equation for the solute concentration:

$$\begin{aligned} c(x, y, t) &= \int_0^t c_\tau(x, y, t)\, d\tau \\ &= \frac{1}{4\pi\varphi H \sqrt{D_L D_T / R^2}} \\ &\quad \times \int_0^t \frac{c_0(\tau) Q(\tau)}{t - \tau} \exp\left[\frac{(x - V(t-\tau)/R)^2}{4D_L(t-\tau)/R} - \frac{y^2}{4D_T(t-\tau)/R} \right] d\tau \end{aligned} \tag{6.20}$$

Steady-State Solution with a Single Source

At steady state ($t \to \infty$) and constant injection rate and concentration, we easily obtain [4]:

$$\begin{aligned} c(x, y, t \to \infty) &= \frac{c_0 Q}{4\pi\varphi H \sqrt{D_L D_T / R^2}} \\ &\quad \times \exp\left[\frac{xV}{2D_L}\right] K_0 \left[\sqrt{\frac{V^2}{4D_L R} \left(\frac{x^2}{D_L/R} + \frac{y^2}{D_T/R} \right)} \right] \end{aligned} \tag{6. 21}$$

where K_0 is a modified Bessel function of the second kind and zero order. This steady-state solution can be used to determine the shapes of very long-lasting plumes.

Transient Solution with Multiple Sources

Note that for N point sources, each located at (x_i, y_i), and injecting at $Q_i(t)$ and $c_{0i}(t)$, the right-hand side of (6.17) is modified as

$$\sum_{i=1}^{N} \frac{\delta(x_i)\delta(y_i) c_{0i}(t) Q_i(t)}{\varphi H} \qquad \text{leading to}$$

$$M(t) = \sum_{i=1}^{N} \int_0^t c_{0i}(\tau) Q_i(\tau)\, d\tau \qquad \forall t > 0 \tag{6.22}$$

The contribution of each source to the total solute concentration is then

$$c_i(x, y, t) = \frac{1}{4\pi\varphi H\sqrt{D_L D_T/R^2}} \times \int_0^t \frac{c_{0i}(\tau)Q_i(\tau)}{t-\tau} \exp\left[-\frac{(x - x_i - V(t-\tau)/R)^2}{4D_L(t-\tau)/R} - \frac{(y-y_i)^2}{4D_T(t-\tau)/R}\right] d\tau \qquad i = 1, 2, \ldots, N \tag{6.23}$$

and the total solute concentration from all N sources is obtained by linear superposition, because the governing equation (6.17) with the source (6.22) are linear:

$$c(x, y, t) = \sum_{i=1}^{N} c_i(x, y, t) \tag{6.24}$$

Steady-State Solution with Multiple Sources

In particular, the steady-state solution for N sources, each injecting at a different, but constant, concentration and rate is

$$c(x, y, t \to \infty) = \frac{1}{4\pi\varphi H\sqrt{D_L D_T/R^2}} \times \sum_{i=1}^{N} c_{0i}Q_i \exp\left[\frac{(x-x_i)V}{2D_L}\right] \times K_0\left[\sqrt{\frac{V^2}{4D_L R}\left(\frac{(x-x_i)^2}{D_L/R} + \frac{(y-y_i)^2}{D_T/R}\right)}\right] \tag{6. 25}$$

This solution is useful in estimating the locations and strengths of several long-lasting aqueous contaminant sources. In field units, the solute concentration is in parts per million, the volumetric injection rate is in cubic feet per day, x and y are in feet, the linear solute velocity is in feet per day, and the dispersion coefficients are in square feet per day.

SOLUTION OF CDA EQUATION FOR A SLUG SOURCE OF CONTAMINANT

Transient Solution with a Single Slug Source

The solution for a slug of any finite duration begins with the instantaneous point source solution (6.19). We denote the total log duration time as T and the mass injection rate of the slug as $\dot{m}(\tau) = c_0(\tau)Q(\tau)$. Hence, the solute mass injected over an infinitesimal time interval $d\tau$ is simply $\dot{m}\, d\tau$. Note that T is sometimes referred to as the "release time" or the "injection time." For times less than T, the contribution of each slug source is given by (6.20). This equation is further simplified through a similarity transformation

$$\xi = \frac{V^2(t - \tau)}{4D_L R} \tag{6.26}$$

which combines distance and time. If the mass injection rate is constant, then after some algebra

$$c(x, y, t \leq T) = \frac{\dot{m}R}{4\pi\varphi H\sqrt{D_L D_T}} \times \exp\left(\frac{Vx}{2D_L}\right)\int_0^{V^2t/4D_LR}\frac{1}{\xi}\exp\left(-\xi - \frac{b^2}{4\xi}\right)d\xi \tag{6. 27}$$

where

$$b^2 = \frac{Vx}{4D_L^2}\left(x^2 + y^2\frac{D_L}{D_T}\right) = \frac{V^2}{4D_L R}\left(\frac{x^2}{D_L/R} + \frac{y^2}{D_T/R}\right) \tag{6.28}$$

is an affine transformation that stretches the coordinate perpendicular to the flow by the ratio of $\sqrt{D_L D_T}$. Note that if $t \to \infty$ and $T \to \infty$ simultaneously, (6.27) and (6.28) reduce to (6.21) as the integral converges to $K_0(b)$.

Our next concern is to obtain an equation describing plumes in which a slug point source existed for a time interval T, but the elapsed time, t, is greater than T. Put $\xi = V^2(t - \tau)/4D_L R$. For $t > T$, the upper limit of integration in (6.20) becomes T, and in terms of the new variable χ the limits of integration are given by

$$\xi = \begin{cases} \dfrac{V^2(t - T)}{4D_L R} & \text{if } \tau = T \\[2ex] \dfrac{V^2 t}{4D_L R} & \text{if } \tau = 0 \end{cases} \tag{6.29}$$

Hence,

$$c(x, y, t) = \frac{\dot{m}R}{4\pi\varphi H\sqrt{D_L D_T}} \exp\left(\frac{Vx}{2D_L}\right) \int_{V^2(t-T)/4D_L R}^{V^2 t/4D_L R} \times \frac{1}{\xi} \exp\left[-\xi - \frac{b^2}{4\xi}\right] d\xi \qquad (6.30)$$

Transient Solution with Multiple Slug Sources

Now we assume that there are N point sources, each injecting at a different rate $\dot{m}_i$, starting at a different time $T_{\text{start}, i}$, and ending at a different time $T_{\text{end}, i}$. Then the right-hand side of (6.17) is modified as

$$r_{i,\text{inj}}(x_i, y_i, t) = \begin{cases} \dfrac{\delta(x_i)\delta(y_i)c_{0i}(t)Q_i(t)}{\varphi H} & T_{\text{start}, i} \le t \le T_{\text{end}, i} \qquad i = 1, 2, \ldots, N \\ 0 \quad \text{otherwise} \end{cases} \quad \text{or}$$

$$\varphi H \int_{-\infty}^{+\infty}\int_{-\infty}^{+\infty}\int_{T_{\text{start}, i}}^{\min(t, T_{\text{end}, i})} r_{i,\text{inj}}(x, y, \tau)\, dx\, dy\, d\tau = M_i(t) = \int_{T_{\text{start}, i}}^{\min(t, T_{\text{end}, i})} c_{0i}(\tau)Q_i(\tau)\, d\tau \qquad \forall t > T_{\text{start}, i} \qquad i = 1, 2, \ldots, N \qquad (6.31)$$

For a constant concentration and rate source, the solute concentration is then given by

$$c(x, y, t) = \frac{R}{4\pi\varphi H\sqrt{D_L D_T}} \sum_{i=1}^{N} c_{0i}Q_i \exp\left[\frac{(x - y_i)V}{2D_L}\right] \times \begin{cases} \displaystyle\int_{\theta_{l, i}}^{V^2(t - T_{\text{start}, i})/4D_L R} \frac{1}{\xi}\exp\left(-\xi - \frac{b_i^2}{4\xi}\right) d\xi & \text{when } t > T_{\text{start}, i} \\ 0 & \text{otherwise} \end{cases} \qquad (6.32)$$

where

$$b_i^2 = \frac{V^2}{4D_L^2}\left((x - x_i)^2 + (y - y_i)^2 \frac{D_L}{D_T}\right) = \frac{V^2}{4D_L R}\left(\frac{(x - x_i)^2}{D_L/R} + \frac{(y - y_i)^2}{D_T/R}\right)$$

$$\theta_{l, i} = \begin{cases} 0 & \text{when } T_{\text{start}, i} \le t \le T_{\text{end}, i} \\ \dfrac{V^2(t - T_{\text{end}, i})}{4D_L R} & \text{when } t > T_{\text{end}, i} \end{cases} \qquad (6.33)$$

NUMERICAL INTEGRATION IN STIFF PROBLEMS

The integral

$$\int_c^a \frac{1}{\xi} \exp\left[-\xi - \frac{b^2}{4\xi}\right] d\xi \tag{6. 34}$$

has no singularities if $b \neq 0$. The integrand is a smooth function on $\{\xi > 0\}$. Its derivatives of an arbitrary order are also smooth functions with zero limit value at $\xi = 0$. However, numerical computation of the integral may be difficult if $c = 0$ or if c is very close to 0. Then in a neighborhood of the lower limit of integration, the values of the integrand are calculated from a fraction with a very small numerator and denominator. A small absolute computational error is inevitable. Under these circumstances, however, a small error in the numerator may produce a large error due to the division by a small number. Thus, the problem is how to calculate the values of the integrand when ξ approaches $c = 0$.

Integration by Parts

The easiest way to handle this problem is to integrate by parts:

$$\int_c^a \frac{1}{\xi} \exp\left[-\xi - \frac{b^2}{4\xi}\right] d\xi = \frac{1}{b^2} \int_c^a \xi \exp(-\xi) \frac{b_2}{\xi^2} \exp\left[-\frac{b^2}{4\xi}\right] d\xi$$

$$= \frac{4}{b^2} \int_c^a \xi \exp(-\xi) d\left(\exp\left[-\frac{b^2}{4\xi}\right]\right)$$

$$= \frac{4a}{b^2} \exp\left(-a - \frac{b^2}{4a}\right)$$

$$\times\left(1 - \frac{c}{a} \exp\left(a - c - b^2 \frac{c-a}{4ac}\right)\right)$$

$$- \frac{4}{b^2} \int_c^a \exp\left(-\xi - \frac{b^2}{4\xi}\right)(1 - \xi)\, d\xi$$

When $c \to 0$, the exponential term involving c in the denominator vanishes. The integrand in the last expression just goes to 0 as $\xi \to 0$ and does not include a division by a small number except in the exponent. This allows one to avoid large errors and makes numerical integration easier and more accurate. Other ways of increasing the accuracy of numerical integration are described in [4]. In some commercial codes, the difficulties with numerical integration may lead to negative calculated concentrations.

MODEL IMPLEMENTATION

The semianalytical model with multiple slug sources at different locations and of different durations has been implemented in MATLAB 5.2 [18], a general computing environment for engineers and scientists. The results of calculations are displayed immediately, together with the well locations and surface images and maps. MATLAB offers very robust computation algorithms, helpful in solving stiff numerical problems, such as (6.32). A 20-point Gauss integration method is used for high numerical accuracy. An efficient model implementation takes 40 seconds of central processing unit (CPU) time on a 333-MHz Pentium Pro Dell Workstation 400 for two plumes, each 4000 ft long and each resolved with a 5-ft increment along the flow direction and a 2.5-ft increment perpendicular to the flow direction. For each plume, this implementation involves the highly vectorized calculation of only half of the plume shape on a local mesh sized for that plume. This is followed by a reflection of the plume shape along its longitudinal axis of symmetry and mapping of the local plume mesh onto the global mesh that covers the entire area of interest. The example calculation described previously is equivalent to a finite-element solution with uniform aquifer properties and with 840,000 nodal points. An equivalent finite-element numerical solution would take many hours or days of CPU time per time step.

MODEL LIMITATIONS

According to (6.4), both dispersivities α_L and α_T are time independent and become constants when averaged over all relevant scales of spatial heterogeneity. However, are they time independent or "Fickian" in practical flow situations? In other words, is the CDA equation (6.1) applicable [11, 16, 19–25] to situations of practical interest? In most cases, (6.1) is valid, but there is no unequivocally positive or negative answer, simply because of the paucity of field data. Here we shall use a well-documented field study to illustrate the difficulties in applying (6.1) to large aqueous contaminant plumes.

THE BORDEN SITE FIELD EXPERIMENT

Consider a large-scale field experiment at the Borden site, Ontario, Canada [10, 11, 21, 22]. The experiment was conducted in the unconfined sand aquifer underlying an inactive sand quarry. The quarry is located about 350 m north of a municipal landfill operated between 1970 and 1976. The leachate plume from this landfill has been studied extensively [26–28]. The aquifer extends about 9 m beneath the horizontal quarry floor and is underlain by a thick, silty clay deposit. The aquifer is composed of clean, well-sorted, fine- to medium-grained sand. The median grain sizes for a set of 846 samples taken from 11 undisturbed cores range from 0.070 to 0.69 mm. Clay size fractions are very low, with 739 of the samples having no measurable clay fraction, and only 8 samples showing clay fractions greater than 15% by weight. Although the aquifer is quite homogeneous, the cores reveal distinct horizontal and parallel bedding features, although some cross- and convolute bedding is observed. The mean aquifer porosity is 0.33 ± 0.017 and its bulk density is 1.81 ± 0.045 g/cm^3. The average water level is about 1.0 m below the quarry floor and seasonal water table fluctuation is about 1.0 m. The horizontal hydraulic gradient is 0.0035 to 0.0054 and the best estimate of a yearly average horizontal gradient is 0.0043. Multiple tests have yielded hydraulic conductivities varying from 5×10^{-5} to 1×10^{-4} cm/s, with a mean of 7×10^{-5} cm/s. Lognormality cannot be rejected as the distribution of the conductivity values. An exponential spatial correlation function has been proposed to describe the covariance structure of the conductivity field. An isotropic correlation length of 2.8 m reasonably describes the horizontal structure, while a correlation length of 0.1 m is estimated for the vertical. The various estimates of average linear velocity of groundwater are listed in Table 6.1.

On August 23, 1982, approximately 12 m^3 of solution was injected into the aquifer over the vertical interval of 2 to 3.6 m below ground (BG). There were nine aqueous solute injectors. Each injected over 14.75 h into an aquifer volume that was considered sufficiently large to avoid flow disturbance. Hence, the solute injection can be treated as an almost instantaneous, but spatially distributed source. Given sufficient time from the injection, the plume may be approximated as generated by the point pulse and described by (6.6). The average composition of the injected solution is shown in Table 6.2.

The monitoring system consisted of a dense network of multilevel sampling devices spaced every 1 to 4 m horizontally and 0.2 to 0.3 m vertically.

Table 6.1 Comparison of Methods for Estimating Average Linear Groundwater Velocity [22]

	Estimates of Hydraulic Conductivity $\times 10^5 (m/s)$		
Method	*Range*	*Mean*	*V (m/d)*
Slug tests	5–10	7	0.078[a]
Core sample analyses			
Grain size analyses 11 cores (site 2)	0.03–76	7.1	0.079[a]
Permeameter analyses 2 cores (UW-1, UW-2)	0.1–15	6.7	0.076[a]
32-core set (UW-3)	0.1–15	7.2	0.081[a]
Natural gradient tracer test			0.091

[a] Calculated using estimated hydraulic conductivity and assuming the mean hydraulic gradient and aquifer porosity are 0.0043 and 0.33, respectively.

Over 19,900 water samples have been collected and analyzed at 20 different times spanning 1038 days since the injection.

Monitoring of the chloride and bromide concentrations allowed the estimation of mean velocity and trajectory of the center of mass (CM) of the migrating plume. Both tracers have been observed to travel along an almost straight line at a mean velocity of 0.091 m/day. This measured tracer velocity is larger than those calculated from the hydraulic conductivities, hydraulic gradient, and the aquifer porosity (0.076–0.081 m/day). The reported chloride concentrations have been averaged over the vertical interval from 1.5 to 7.5 m BG, while those for carbon tetrachloride have been averaged between 1.5 and 6 m BG.

Dispersion at Borden

Freyberg [21] calculated numerically the first moments of the injected bromide and chloride concentrations (solute masses) and their second moments about the CM (spatial covariance tensor $\boldsymbol{\sigma}$. From the second moments and the tracer velocity, one can calculate the dispersion coefficient tensor $\boldsymbol{D}$ and the apparent dispersivities $\boldsymbol{\alpha}$; cf. (6.4). Under rather restrictive assumptions of a steady, homogeneous pore water velocity, an arbitrary pulse input of solute, and validity of the classic CDA equation (6.5), the dispersion coefficient tensor is the time derivative of the covariance tensor [29]:

olor plate 1 Longitudinal and transverse dispersivities for the romide plume at Borden, calculated from Eq. (6.36) as chord opes. Note that both dispersivities are not constant and do not em to reach an infinite time "Fickian" value after three years of ow, contrary to an assertion by Neuman [29].

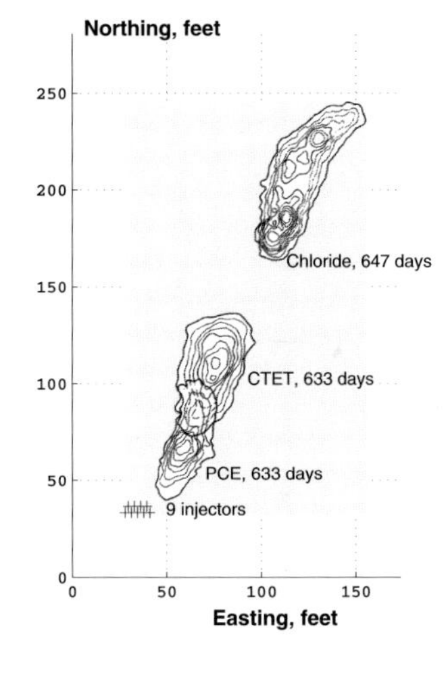

lor plate 2 Figure 5 in Roberts [6] imaged and reproduced in e coordinate system used in the present forward model of the rden site contaminant plumes. The vertically averaged chloride ncentrations are plotted in 5-ppm increments, starting from the ıme outline at 5 ppm. The vertically averaged PCE and CTET ncentrations are plotted in 0.1-ppb increments starting from e plume outlines, each at 0.1 ppb.

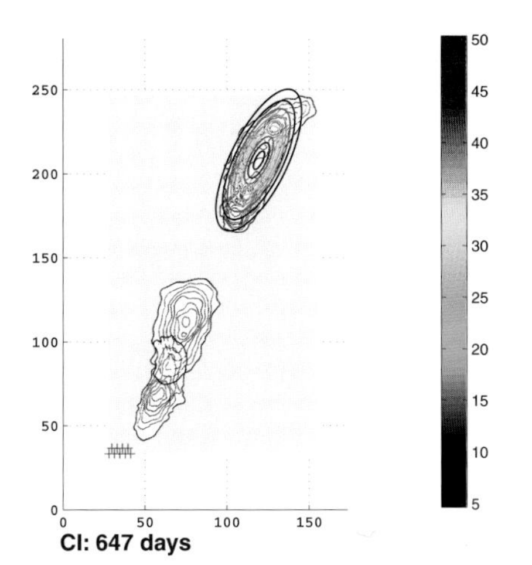

lor plate 3 The results of forward modeling of the Borden loride plume at 647 days. Our forward semi-analytic model uses published aquifer parameters [5, 6, 19, 20]. The groundwater ocity is based on the natural gradient tracer injection test. The me location and extent match very well the field data. The calated chloride contours are plotted at 5-ppm increments from the ter contour of 5 ppm. The plotted concentration contours agree ly well with the vertically averaged chloride contours shown in ure 5 in Roberts [6] and reproduced in Color plate 2. Note the culated peak concentration falls exactly in the middle of the bicated peak chloride concentrations observed at the Borden site. us, the forward model matches well the average transport propies of the aquifer.

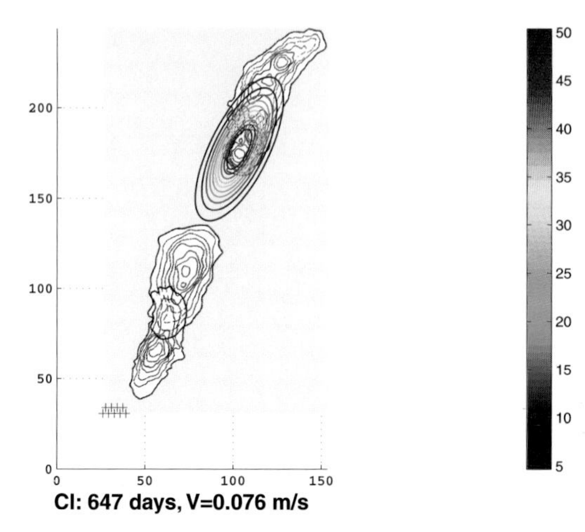

Color plate 4 The results of forward modeling of the Borden chloride plume at 647 days. The groundwater velocity in this calculation is based on the permeameter measurements in two cores. The forward semi-analytic model uses the published aquifer parameters [5, 6, 19, 20]. The chloride contours are plotted at 5-ppm increments from the outer contour of 5 ppm. Now the calculated plume is significantly retarded relative to the field data; however, the calculated plume center overlays the delayed peak chloride concentrations observed at the Borden site.

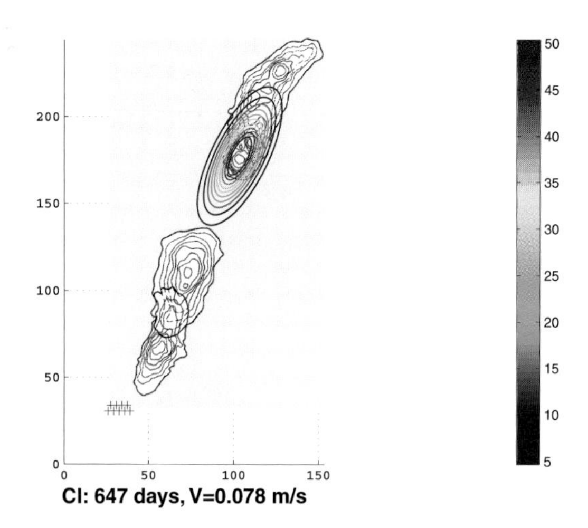

Color plate 5 The results of forward modeling of the Borden chloride plume at 647 days. The groundwater velocity is based on the slug tests. The forward semi-analytic model uses the published aquifer parameters [5, 6, 19, 20]. The calculated chloride contours are plotted at 5-ppm increments from the outer contour of 5 ppm. The calculated plume is retarded relative to the field observations and its center overlays the second, delayed peak chloride concentrations observed at the Borden site.

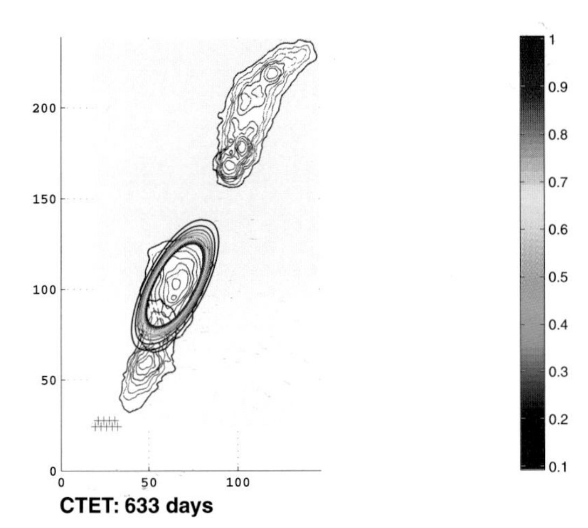

Color plate 6 The results of forward modeling of the Borden carbon tetrachloride (CTET) plume at 633 days. The semi-analytic model uses the published aquifer parameters [5, 6, 19, 20]. The plume extent matches very well the field data. The CTET contours are plotted at 0.1-ppb increments from the outer contour of 0.1 ppb. The plotted contours correspond to the vertically averaged CTET contours shown in Figure 5 in Roberts [6] and reproduced Color plate 2. The concentration gradient in the calculated plume is significantly steeper than that observed in the field.

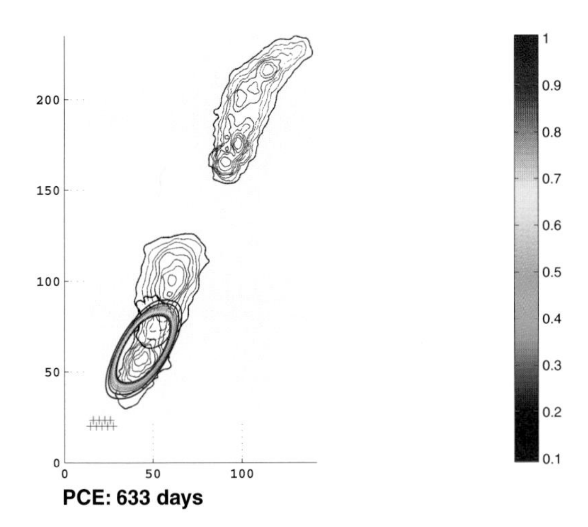

Color plate 7 The results of forward modeling of the PCE plume at 633 days. The forward model uses the published aquifer parameters [5, 6, 19, 20]. The plume extent matches well the field data, although the observed plume shape seems to be shifted east by a few feet near the sources. The calculated PCE contours are plotted at 0.1-ppb increments from the outer contour of 0.1 ppb. The calculated contours correspond to the vertically averaged PCE contours shown in Figure 5 in Roberts [6] and reproduced in Color plate 2. The concentration gradient in the calculated plume is significantly steeper that that observed in the field.

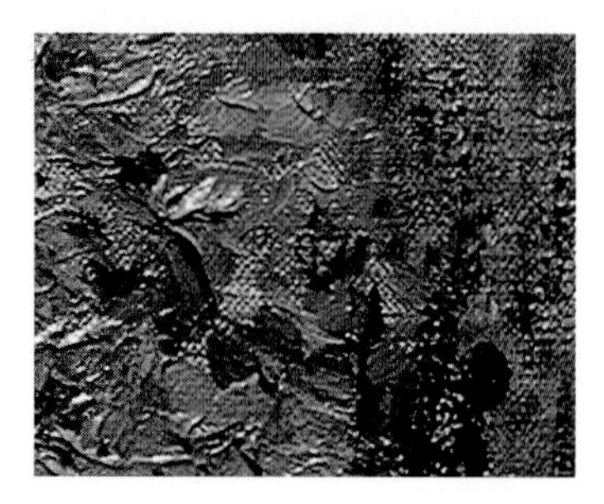

Color plate 8 Average characterization of a contaminated site may be compared to brush strokes.

Color plate 9 Average characterization of a contaminated site may also be compared to picture fragments.

Color plate 10 Out of the sparse brushstrokes and picture fragments, counsel insist on obtaining a rich picture of reality and coloring it their way. (Gustave Caillebotte, The Yellow Fields at Gennevilliers, 1884, oil on canvas. From the Fondation Corboud, on permanent loan to the Wallraf-Richartz-Museum, Cologne.)

Table 6.2 Solutes Injected at the Borden Site.

Solute	*Average Concentration (ppb)*	*Mass Injected (g)*
Chloride (Cl^-)	892,000	10,700
Bromide (Br^-)	324,000	3,870
CTET	32	0.37
PCE	31	0.36
Bromoform	32	0.38
1,2-dichlorobenzene	332	4.0
Hexachloroethane	20	0.23

$$\boldsymbol{D} = \frac{1}{2}\frac{d\boldsymbol{\sigma}}{dt} \tag{6.35}$$

Further, by assuming (see, e.g., [6]) that $D = D^* \boldsymbol{I} + \boldsymbol{\alpha}\,|\boldsymbol{V}|$,

$$\boldsymbol{\alpha} = \frac{1}{2\,|\boldsymbol{V}|}\frac{d\boldsymbol{\sigma}}{dt} \tag{6.36}$$

With spatially nonuniform pore water velocity, the dispersion tensor defined in (6.35), as well as the dispersivities (6.36), are no longer constant material properties, at least for a perfectly stratified aquifer [20, 30]. Hence, the CDA models of real plumes may require sequential calibration [30]. In this approach, the prior solutions to the CDA equation with constant, spatially uniform parameters are recalibrated after each observation session, and new scale-dependent transport coefficients of the CDA equation are obtained.

In the Borden tracer plume, the dispersivities have not been constant [21]. Color plate 1 shows these dispersivities, as well as their least squares fit to the constant dispersion tensor model.

Retardation Factors at Borden

Laboratory experiments were conducted by Curtis [10] to determine whether the field retardation of bromoform ($CHBr_3$, "BROM"), carbon tetrachloride (CCl_4, "CTET"), tetrachloroethylene ($Cl_2C{=}CCl_2$, "PCE"), 1,2-dicholorobenzene ($C_6H_4Cl_2$, "DCB"), and hexachloroethane (C_2Cl_6, HCE), observed at the Borden site, could be explained by the linear, reversible, equilibrium adsorption model. The five halogenated organic solutes, which have octanol–water partition coefficients ranging from 200 to 4000, were the same as those used in the field study. The sorbent, a medium-grained sand containing 0.02% organic carbon, was excavated 11.5 m from the sampling well field at the Borden site. Sorption isotherms were linear in the aqueous concentration range from 1 to 50 parts per billion (ppb) and could be described by a single distribution coefficient K_d, (6.3).

The experimentally determined K_d exceeded those predicted by the hydrophobic sorption model that accounts only for partitioning into organic matter [13], by factors ranging from 3.4 for hexachloroethane to 3.9 for tetrachloroethylene. Retardation factors inferred from the laboratory-determined distribution coefficients were always lower, but fell within the range estimated from spatial sampling data in the field experiment. The respective retardation factors are listed in Table 6.3.

The spatial or synoptic retardation factors have been evaluated as ratios of the plume and groundwater velocities

$$R_i^{(S)} = \frac{V_{Cl}}{V_i} \tag{6.37}$$

These ratios were calculated at six different times. All retardation factors kept increasing with time. The retardation factor for PCE increased (the plume decelerated) more than twofold in less than two years, while the chloride plume decelerated by about 10% over the same period. The temporal or time series retardation factors have been calculated from break-

Table 6.3 Comparison of Retardation Factors [10]

Contaminant	*Predicted*[a]	*Batch Experiments*	*Temporal Field Data*[b]	*Spatial Field Data*[b]
CTET	1.2	1.9±0.1	1.6–1.8	1.8–2.5
BROM	1.2	2.0±0.2	1.5–1.8	1.9–2.8
PCE	1.3	3.6±0.3	2.7–3.9	2.7–5.9
DCB	2.3	6.9±0.7	1.8–3.7	3.9–9.0
HCE	2.3	5.4±0.5	4.0	5.1–7.9

[a] Calculated from the regression [13] with $f_{oc} = 0.02\%$.

[b] From Tables 1 and 5 in [11].

through times of the individual contaminants at the sampling wells. Of course, both estimates illustrate the inherent duality of the chromatographic behavior in convection-sorption systems: Solutes with different sorption affinities to the solid tend to separate in both space and time.

INVERSE CALCULATIONS WITH THE INSTANTANEOUS POINT SOURCE SOLUTION

Verification of (6.6) to (6.14)

Equations (6.6) to (6.14) can be verified by inverting the Borden chloride plume contours, plotted in Figure 8 in [22], into the plume origin, age, and released mass of the chloride tracer. The parameters used to characterize both the aquifer and the chloride plume are listed in Tables 6.4 and 6.5, and the calculation results are given in Tables 6.6 and 6.7. The reported chloride and bromide concentrations have been averaged over a 6-m vertical interval (from 1.5 to 7.5 m BG).

As we can see, there is fair-to-good agreement between the field data and the simplistic inverse calculations. It is obvious, however, that the Bor-

Table 6.4 Plume and Aquifer Parameters

M (kg)	10.7
H (m)	6
α_L (m)	0.36
α_T (m)	0.039
V (m/day)	0.091
R	1

Table 6.5 Characterization of Chloride Plume

Quantity	*462 days*	*647 days*
c_{max} (ppm)	65	45
c_{min} (ppm)	5	5
$X_{le} - X_{te}$ (m)	27	28

Table 6.6 Agreement with Field Data at 462 Days

Quantity	*Calculated*	*Real*	*Percentage of Error*
Plume age (d)	542	462	17
CM location (m)	50	42	20
M (kg)	28.7	10.7	168
M (kg) (H=2.5 m)[a]	11.9	10.7	12
c_{max} (ppm)[b]	86	65	32

[a] The chloride plume is known to have spanned a vertical interval of approximately 2.5 m.

[b] The peak concentration occurs at a mathematical point and it cannot be measured in the field. Instead, concentrations close to one standard deviation from the peak may be measured. Consequently, one contour increment of 5 ppm has been added to the largest contoured concentration in the field.

den plume was very well characterized and its sources were short lived and limited to a tiny volume of the aquifer. In remediation practice, on the other hand, characterization of contaminant plumes and aquifers is usually poor and incomplete; the plume sources are spatially distributed, vary with time, and last long. Still, this simple model may be useful in some practical situations.

Table 6.7 Agreement with Field Data at 647 Days

Quantity	*Calculated*	*Real*	*Percentage of Error*
Plume age (d)	681	647	5
CM location (m)	63	59	7
M (kg)	24.9	10.7	133
M (kg) (H=2.5 m)[a]	10.4	10.7	−3
c_{max} (ppm)[b]	62	45	38

[a] The chloride plume is known to have spanned a vertical interval of approximately 2.5 m.

[b] The peak concentration occurs at a mathematical point and it cannot be measured in the field. Instead, concentrations close to one standard deviation from the peak may be measured. Consequently, one contour increment of 5 ppm has been added to the largest contoured concentration in the field.

FORWARD CALCULATION OF THE BORDEN PLUMES WITH THE DISTRIBUTED SLUG SOURCES

Now we shall apply to the Borden site the more sophisticated semianalytical model described previously. The aquifer parameters and the retardation factors have been set to the values published in the four Borden papers and listed in Tables 6.2 and 6.3. The calculation results are shown in color plates 2 to 7.

It seems that with a well-characterized aquifer and well-characterized, but not too large plumes, the two-dimensional semianalytic model performs quite admirably. Unfortunately, in most practical field cases, the system description is incomplete at best and erroneous at worst. The effects of subtle systematic errors in the estimation of the aquifer transport velocity are shown in color plates 4 and 5. Such errors may cause incorrect estimates of the plume age and the source locations. No degree of numerical-model sophistication can hide the poor aquifer characterization.

SUMMARY

As we have demonstrated using the Borden site, the two-dimensional CDA equation with constant parameters models adequately the aqueous solute plumes generated by multiple, spatially distributed slug sources in a well-characterized homogeneous and uniform aquifer. The importance of chloride or another nonadsorbing ion as an aquifer calibration tracer cannot be overemphasized. The effect of small systematic errors in the determination of the average aquifer transport velocity can be very significant. The effect of the retardation factor values can be equally important. Therefore, even with a high level of accurate characterization, one may expect systematic errors in the estimation of plume age and masses of released solutes.

In contrast, the level of characterization of sites that are the subject of litigation is nowhere close to that at the Borden site. Most of the characterization effort is concentrated on the periodic acquisition of vertically averaged concentrations of solutes of concern. These samples are usually acquired in a few biased locations around the suspected contamination sources or property boundaries. Characterization of the aquifer transport properties (i.e., conductivity, dispersivity, porosity, layering, retardation factors, etc.), is usually poor or nonexistent.

When the poor aquifer characterization becomes the foundation for the very detailed questions posed in a litigation, the result is confusion, competing incomplete interpretations, and a general inability to resolve the problem to the satisfaction of counsel.

When the input data are poor and incomplete, the resulting simulations may be lacking in quality. The latter observation leads to an inevitable, albeit somewhat self-serving, conclusion that a knowledgeable and honest expert is indispensable in untangling the web of confusion, misconceptions, and false expectations by the litigants. The expert's role is very difficult and may be compared to the reconstruction of a painting from brush strokes (e.g., aqueous contamination samples; see color plate 8) and fragments (e.g., geological logs; see color plate 9).

Out of these partial and incomplete input data, the expert is expected to paint a full, detailed, and beautiful picture that will enchant both litigating parties (see, e.g., color plate 10). Not everyone has such formidable painting skills!

REFERENCES

1. M. P. Anderson and W. W. Woessner, *Applied Groundwater Modeling Simulation of Flow and Advective Transport*, San Diego: Academic Press (1992).

2. P. B. Bedient, H. S. Rifai, and C. J. Newell, *Ground Water Contamination—Transport and Remediation*, Englewood Cliffs, NJ: Prentice Hall (1992).

3. A. E. Adenekan, T. W. Patzek, and K. Pruess, "Modeling of Multiphase Transport of Multicomponent Organic Contaminants and Heat in the Subsurface: Numerical Model Formulation," *Water Resources Research*, 29(11): 3727–3740 (1993).

4. T. W. Patzek and D. B. Silin, *E241—Mathematical and Numerical Methods in Earth Sciences*, Berkeley: University of California class notes (1998).

5. S. G. Taylor, "Dispersion of Soluble Matter in Solvent Flowing Slowly Through a Tube," *Proceedings of the Royal Society of London*, 219 (Ser. A): 186–203 (1953).

6. J. Bear, *Hydraulics of Groundwater*, New York: McGraw-Hill (1979).

7. G. Sposito, V. K. Gupta, and R. N. Bhattacharya, "Foundational Theories of Solute Transport in Porous Media: A Critical Review," *Advances in Water Resources*, 2: 5969 (1979).

8. G. De Marsily, *Quantitative Hydrogeology*, San Diego: Academic Press (1986).

9. S. Y. Chu, and G. Sposito, "A Derivation of the Macroscopic Solute Transport Equation for Homogeneous, Saturated, Porous Media," *Water Resources Research*, 16(3): 542–546 (1980).

10. G. P. Curtis, P. V. Roberts, and M. Reinhard, "A Natural Gradient Experiment on Solute Transport in a Sand Aquifer, 4. Sorption of Organic Solutes and Its Influence on Mobility," *Water Resources Research*, 22(11): 2059–2067 (1986).

11. P. V. Roberts, M. N. Goltz, and D. M. Mackay, "A Natural Gradient Experiment on Solute Transport in a Sand Aquifer, 3. Retardation Estimates and Mass Balances for Organic Solutes," *Water Resources Research*, 22(11): 2047–2058 (1986).

12. S. W. Karickhoff, D. S. Brown, and T. A. Scott, "Sorption of Hydrophobic Pollutants on Natural Sediments," *Water Resources Research*, 13(3): 241–248 (1979).

13. R. P. Schwartzenbach and J. Westall, "Transport of Nonpolar Organic Compounds from Surface Water to Groundwater. Laboratory Sorption Studies," *Environmental Science and Technology*, 15(11): 1360–1367 (1981).

14. A. L. Scheidegger, *The Physics of Flow Through Porous Media*, Toronto: University of Toronto Press (1960).

15. D. R. F. Havelman and R. R. Rumer, "Longitudinal and Lateral Dispersion in an Isotropic Porous Medium," *Fluid Mechanics*, 16: 385–394 (1963).

16. G. Sposito, W. A. Jury, and V. K. Gupta, "Fundamental Problems in the Stochastic Convection–Dispersion Model of Solute Transport in Aquifers and Field Soils," *Water Resources Research*, 22(1): 77–88 (1986).

17. J. C. Carslaw, and J. C. Jaeger, *Conduction of Heat in Solids*, Oxford: Clarendon Press (1959).

18. MATLAB—The Language of Technical Computing, Natick, MA: MathWorks (1998).

19. F. W. Schwartz, "Macroscopic Dispersion in Porous Media: The Controlling Factors," *Water Resources Research*, 13(4): 743–752 (1977).

20. G. Matheron and G. De Marsily, "Is Transport in Porous Media Always Diffusive? A Counterexample," *Water Resources Research*, 16(5): 901–917 (1980).

21. D. L. Freyberg, "A Natural Gradient Experiment on Solute Transport in a Sand Aquifer, 2. Spatial Moments and the Advection and Dispersion of Nonreactive Tracers," *Water Resources Research*, 22(11): 2031–2046 (1986).

22. D. M. Mackay, D. L. Freyberg, and P.V. Roberts, "A Natural Gradient Experiment on Solute Transport in a Sand Aquifer, 1. Approach and Overview of Plume Movement," *Water Resources Research*, 22(13): 2017–2029 (1986).

23. J. J. Boggs, S. C. Young, and L. M. Beard, "Field Study of Dispersion in a Heterogeneous Aquifer, 1. Overview and Site Description," *Water Resources Research*, 28(12): 3281–3291 (1992).

24. E. E. Adams, and L. W. Gelhar, "Field Study of Dispersion in a Heterogeneous Aquifer, 2. Spatial Moments Analysis," *Water Resources Research* 28(12): 3293–3307 (1992).

25. K. R. Rehfeldt, J. M. Boggs, and L. W. Gelhar, "Field Study of Dispersion in a Heterogeneous Aquifer, 3. Geostatistical Analysis of Hydraulic Conductivity," *Water Resources Research* 28(12): 3309–3324 (1992).

26. B. C. E. Egboka, J. A. Cherry, R. N. Farvolden, and E. O. Frind, "Migration of Contaminants in Groundwater at a Landfill,'" *Journal of Hydrology*, 63: 51–80 (1983).

27. D. S. MacFarlane, J. A. Cherry, R. W. Gilham, and E. A. Sudicky, "Migration of Contaminants in Ground Water at a Landfill: A Case Study, 1. Groundwater Flow and Plume Delineation," *Journal of Hydrology*, 63: 1–29 (1983).

28. R. V. Nicholson, J. A. Cherry, and E. J. Reardon, "Migration of Contaminants in Ground Water at a Landfill: A Case Study, 6. Hydrogeochemistry," *Journal of Hydrology*, 63: 131–176 (1983).

29. R. Aris, "On the Dispersion of a Solute in a Fluid Flowing Through a Tube," *Proceedings of the Royal Society of London,*, 235: 67–78 (1956).

30. E. A. Sudicky, J. A. Cherry, and E. O. Frind, "Migration of Contaminants in Groundwater at a Landfill: A Case Study, 4. A Natural Gradient Dispersion Test," *Journal of Hydrology*, 63: 81–108 (1983).

31. S. P. Neuman, "Universal Scaling of Hydraulic Conductivities and Dispersivities in Geologic Media," *Water Resources Research*, 26(8): 1749–1758 (1990).

Chapter 7

Chemical Fingerprinting

James E. Bruya

Chemical fingerprinting is widely used in many different disciplines and fields. It has been used to identify the origin or source of bomb parts and fire accelerants [1] and drugs and other contraband [2], and it has been used to track the path of chemical pollution [3]. It has been used to provide information on the climatological conditions of the past [4]. Chemical fingerprinting has been used by archeologists to document ancient trading routes [5]. It has been used to identify species of plants and animals [6]. Chemical fingerprinting also has been used to support the current theories regarding life on Mars and the extinction of many species of dinosaurs (on Earth and not Mars) by an asteroid [7].

In most environmental cases, site investigations and/or remedial investigations will have been completed prior to retaining a forensic expert. As part of the investigative process, routine Environmental Protection Agency (EPA) protocols are utilized to chemically characterize air, soil, and groundwater pollution. If the collected analytical data need to be evaluated in greater detail (i.e., to determine their value in litigation) or if additional studies need to be conducted in order to answer specific litigation issues, an environmental forensic expert in the area of chemical fingerprinting may be retained.

Within the field of environmental forensics, chemical fingerprinting is used to answer some of the who, what, when, where, and why questions that may exist with respect to chemical pollution. For example:

- If two parties contributed similar compounds to a pollution event, what methods of sampling and analysis can be used to determine their respective percentage contribution?
- If pollution is suspected at a given property, what methods of sampling and analysis are needed to accurately characterize the pollution and how is this information used? (This is usually undertaken during the site investigation phase.)
- In insurance litigation, the question as to when a specific pollution event first occurred often needs to be addressed. This usually requires methods that can be used to determine the age of a specific pollutant or mixture of pollutants.
- In some cases, the source of a specific pollutant or mixture of pollutants may be unknown; however, there may be one or more adjacent sources. Thus, chemical fingerprinting can be employed to identify the source of the pollution.

Any or all of these situations may occur on a single site. Thus, the forensic expert must provide guidance to clients in the ways in which chemical fingerprinting can be used. In general, once a litigation issue has been defined and the existing data analyzed, the steps involved in the fingerprinting process include (1) the generation of chemical data, which is then followed by (2) a data processing or evaluation step. The general outcome is the creation of a result or a story that addresses the litigation issue(s).

When conducting or reviewing a chemical fingerprinting process, each step should be evaluated to determine if the assumptions made or conclusions arrived at are reasonable and supported by the known facts. The review can take many different approaches. With respect to the data generation step, attention can be focused on compliance with particular testing procedures. It can also address data usability issues and the error or bias associated with a particular measurement. With respect to the data processing or data evaluation step, review may center on the multiple assumptions that are inherent in these activities.

In the remainder of this chapter, various chemical fingerprinting procedures will be discussed. The first section will address the testing procedures used to generate forensic data. Emphasis will be placed on developing the reasons why these tests are used and their potential bias. Such information can be used to ensure the applicability of a particular fingerprinting approach. The second section will focus on a few of the ways in which data are processed or evaluated.

CHEMICAL FINGERPRINTING TEST METHODS

When selecting the test or tests used for chemical fingerprinting purposes, there are two issues that should be considered. The first deals with the actual test selected to generate the fingerprinting data. The test used must be reliable and reproducible and must be based on known scientific principles. The common tests that are used for chemical fingerprinting purposes have been discussed by a number of authors [8–11]. Second, there must be some reason why the test is used; that is, there must be a basis for using the test results for fingerprinting purposes. When discussing the common tests used for fingerprinting purposes, we will try to provide the reasons why such tests might be used, as well as the bias or limitations associated with each test.

Boiling Range Determinations

The boiling point of a chemical compound is a fundamental chemical property that imparts particular physical properties on a chemical or chemical product. For example, water boils at 212°F or 100°C and nitrogen boils at −78°C. Boiling water can cook food and boiling nitrogen will freeze it. This is important because the boiling point of a chemical can help identify an unknown material, as well as provide information on its physical properties and potential uses. Where chemicals are mixed together, the resulting mixture will typically not have a single boiling point, but rather a boiling range. The boiling range will typically span the temperature range from the boiling point of the lowest boiling compound to the boiling point of the highest boiling compound in the mixture.

Low boiling compounds readily evaporate and can quickly change from a liquid to a vapor. This is an important property of cleaning agents, solvents, and gasoline. On the other hand, high boiling compounds do not readily vaporize. They tend to be used in products that must remain relatively constant in their physical properties. High boiling compounds are used in oils and lubricants or as additives such as softening agents to plastics. Medium boiling compounds can be used for products where rapid volatilization is not needed or desired, but where some volatilization is necessary. Products like fuel oil, diesel fuel, and slow-drying solvents such as those used in ink contain medium boiling compounds.

Bulk commodity products such as gasoline and diesel fuel are produced throughout the world by hundreds of different manufacturers. They can be used in vehicles that are new, as well as some that are very old. The ability to have such an expansive production base that can be used in both new and old technology lies in the specifications used to produce these products. These bulk commodities were originally produced to supply an invention with an economical source of energy. The commodity products were generally replacements for existing fuels. Gasoline engines were first developed using alcohols as fuels, while diesel engines used vegetable oils.

The design of gasoline and diesel engines dictated many of the physical properties needed for a fuel. For example, a gasoline engine required a fuel that readily vaporized in the combustion chamber where it was ignited using a spark plug. The combustion chamber was partially sealed by using rings around the piston that slid along the walls of the cylinders. Because of this partial seal, unburned high boiling compounds from the combustion chamber could collect in the underlying lubricating oil where they could change the viscosity of the oil. To eliminate this problem, high boiling compounds were excluded from the fuel used in gasoline engines. These and other requirements of the inventions were eventually developed into production specifications. Tables 7.1 and 7.2 show the production specifications for gasoline and diesel fuel since the early 1900s. Of particular interest for fingerprinting purposes is the fact that the boiling ranges or distillation requirements have remained fairly constant over time. This means that the boiling range information from an unknown commodity fuel can be compared to these production specifications for identification purposes.

For both commodity and noncommodity products, the general boiling range can be documented through material safety data (MSD) sheets. The composition and boiling range of specialty chemicals like paint, cleaner, and cutting fluid can change from supplier to supplier. These products can also show significant changes over time as product formulations are changed to lower costs and to meet consumer or regulatory demands. All of these changes can be found in the MSD sheets.

There are several ways to measure or establish a boiling point or boiling range. The classical approach, using a thermometer to measure the temperature of a boiling liquid, is rarely used in forensic investigations due to the large quantity of product needed. Instead, a gas chromatograph (GC) is used for measuring the boiling point or range of a chemical product. The use of gas chromatography to establish the boiling range of petroleum is the basis for ASTM Standard D2887—Simulated Distillation. With the proper choice of GC operating conditions, the elution or retention time of a given compound is directly related to its boiling point. This relationship is used to convert the typical time scale of a GC trace to a temperature scale. An example of this process is shown in Figure 7.1. Here, a series of normal alkanes whose boiling points are known are used to establish the temperature scale.

When performing boiling range determinations, direct conversion to a temperature scale is not often carried out. Instead, the boiling range is described using the identities of particular compounds. For example, rather than describe a boiling range as 345 to 736°F (174 to 391°C), it can be reported as boiling from n-C_{10} to n-C_{24} (boiling points 345 and 736°F, respectively). By using the specific compound identifications rather than the actual temperatures, one conveys the boiling range information, as well as the fact that a GC was used for the boiling range determination.

Figures 7.2 to 7.15 show the GC traces from the analysis of a variety of materials. All analyses were conducted such that the temperature scale shown in Figure 7.1 can be used. When reviewing such GC traces, it

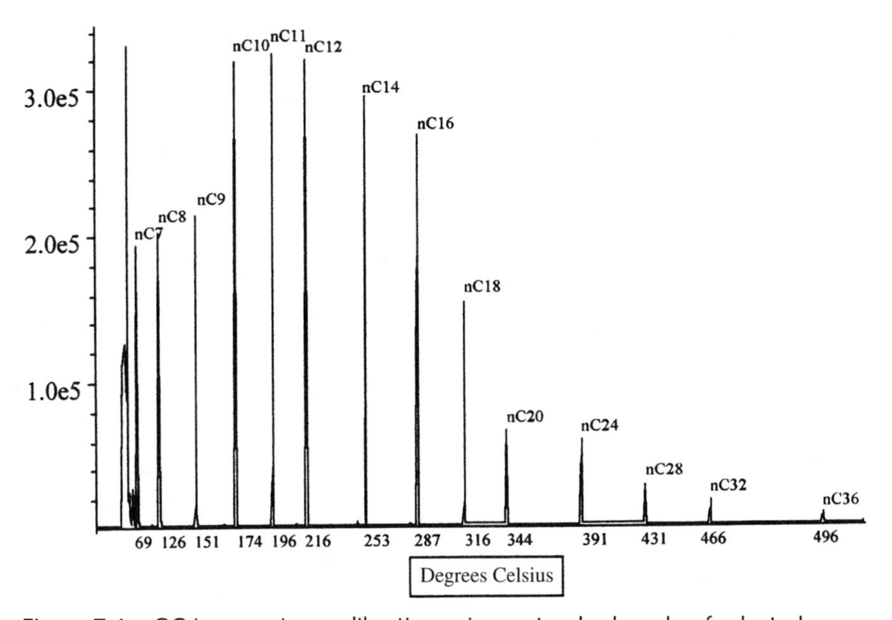

Figure 7.1 GC temperature calibration using a standard made of selected n-alkanes.

Table 7.1 Gasoline Production Specifications

Test	*Federal Limits 1919*	*Federal Limits 1929*	*ASTM-D-439 1940*	*ASTM-D-4814 1989 Unleaded*	*ASTM-D-4814 1995 Unleaded*
Distillation (°F)					
Initial BP	140 max				
10% recovered		176 max 122 min	140–167 max depending on season and grade of gasoline	158–122 max depending on volatility class	158–122 max depending on distillation (volatility) class
20% recovered	221 max				
50% recovered	284 max	284 max	284–257 max depending on grade of gasoline	170 min 250–230 max depending on volatility class	170 min 250–230 max depending on volatility class
90% recovered	374 max	200 max	392–356 max depending on grade of gasoline	374–365 max depending on volatility class	374–365 max depending on volatility class
End point	437 max	437 max		437 max	437 max
Amount recovered after distillation (% vol)		95% min			
Distillation residue (% vol)			2% max	2% max	2% max

Copper strip corrosion	Passes	No. 1 max	No. 1 max
Vapor pressure (psi)	9.5–13.5 depending on season	9.0–15 depending on volatility class	7.8–15 depending on volatility class
Octane (average of research and motor octane ratings)	70 min for regular and 77 min for premium grade	85–90 (recommended)	Report
Existent Gum (mg gum per 100 mL gasoline)	7mg/mL max	5 mg/100 mL max	5 mg/100 mL max
Lead (grams lead per gallon)		0.05 max unleaded 4.2 max leaded[a]	0.05 max
Sulfur (% mass)		0.10 max	0.10 max
Oxidation Stability (min)		240 max	240 min
Vapor Lock protection (vapor/liquid ratio)		20 max	20 max
Water tolerance		Seasonal	Seasonal

[a] EPA regulations limited the lead concentration in leaded gasoline to no more than 0.1 g/gal averaged per calendar quarter for each refinery.

Table 7.2 Diesel Fuel Production Specifications

Test	*International Petroleum Commission 1912*	*ASTM-D-975 1953 Diesel #2*	*ASTM-D-975 1981 Diesel #2*	*ASTM-D-975 1994 Diesel #2 Low Sulfur*
Specific gravity (API gravity)	0.860–0.895			
Flash point (°C)	60 min	52 min	52 min	52 min
Distillation (°C)				
90% recovered	350 max	357 max	282 min 338 max	282 min 338 max
Coke or ash (g/100 mL)	0.5 max			
Ash content (% weight)	Negligible	0.02 max	0.01 max	0.01 max
Acidity	Free of acid			

Water and sediment (% vol)	Clear	0.10	0.05% max	0.05% max
Heating value (BTU per pound)	18,000 min			
Viscosity (° Engler)	2.50 max			
Carbon residue (%)		0.35	0.35 max	
Copper strip corrosion		No. 3	No. 3	No. 3
Cloud point			per agreement	per agreement
Cetane Number		40 min	40 min	40 min
Aromaticity (%)				35% max
Sulfur (% weight)		1.0	0.50%	0.05% max
Viscosity Kinematic (mm^2/S @ 40°C)		1.8 min 5.8 max	1.9 min 4.1 max	1.9 min 4.1 max

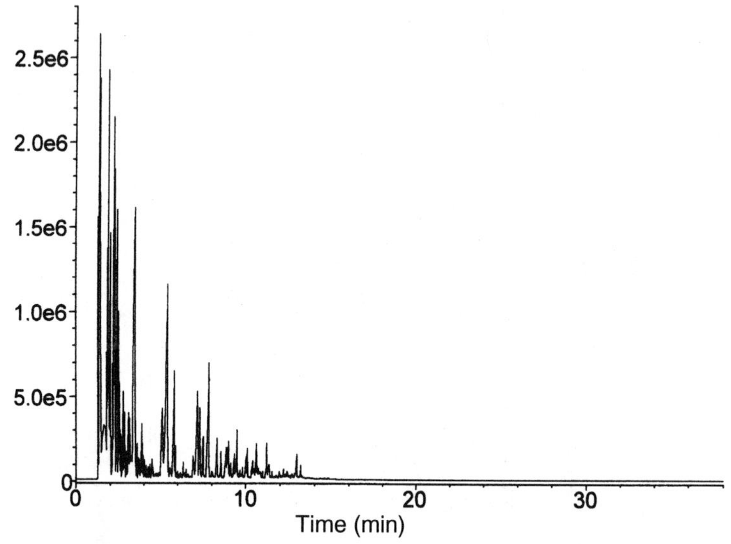

Figure 7.2 Gasoline—Example 1.

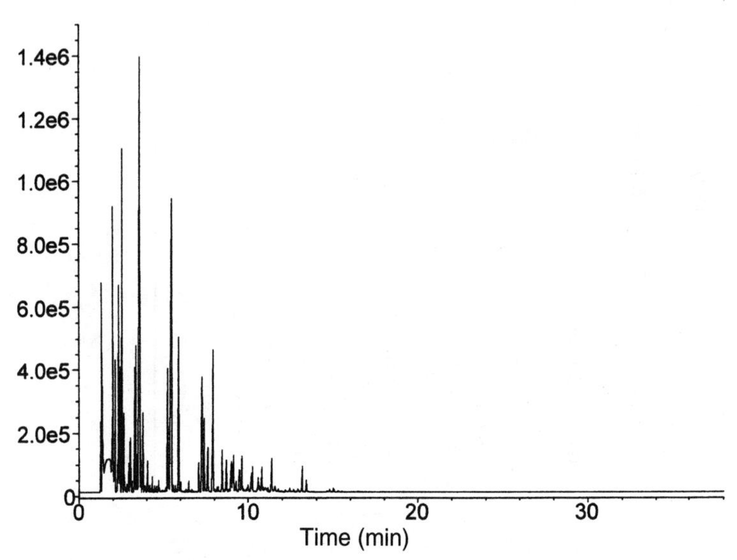

Figure 7.3 Gasoline—Example 2.

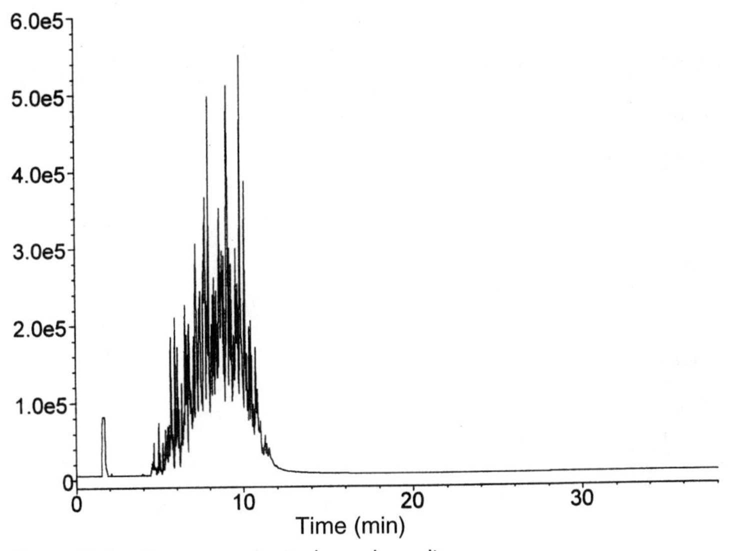

Figure 7.4 Degreaser (petroleum based).

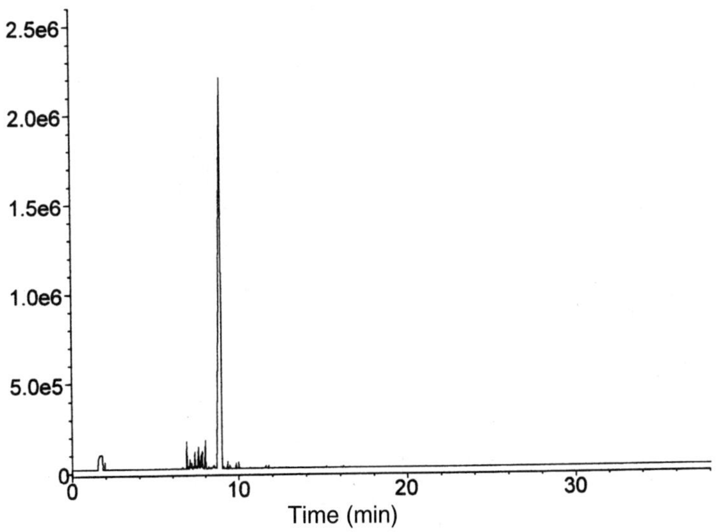

Figure 7.5 Degreaser (citrus oil based).

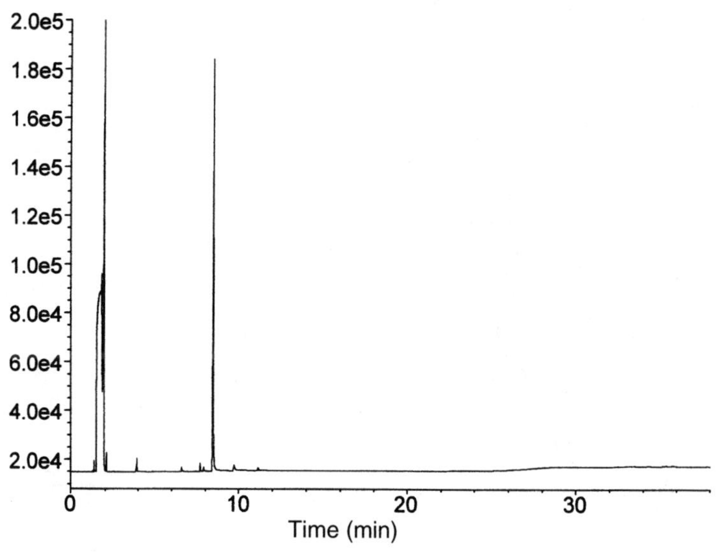

Figure 7.6 Orange juice.

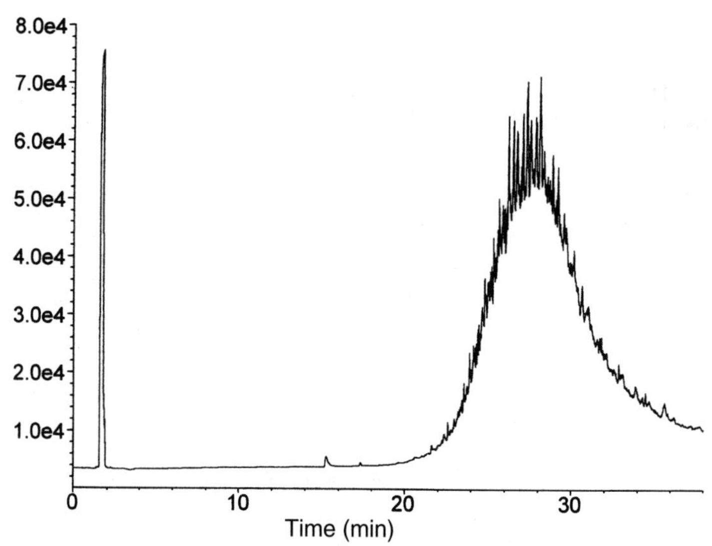

Figure 7.7 Petroleum-based hydraulic fluid.

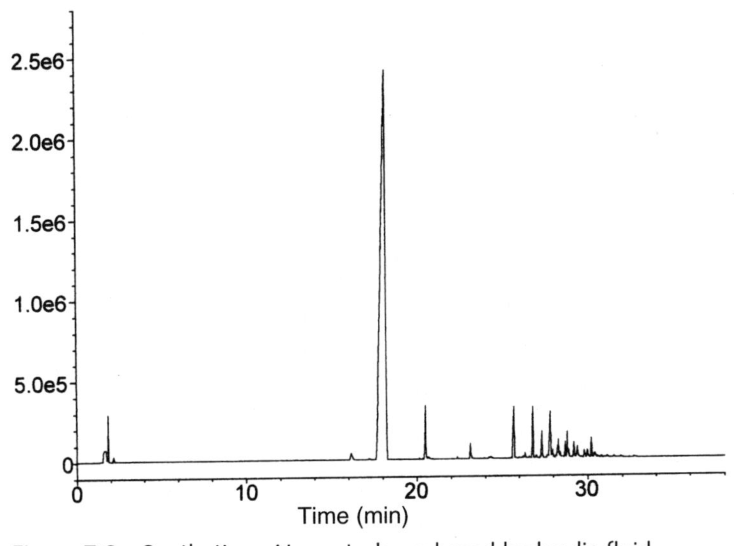

Figure 7.8 Synthetic or Nonpetroleum based hydraulic fluid.

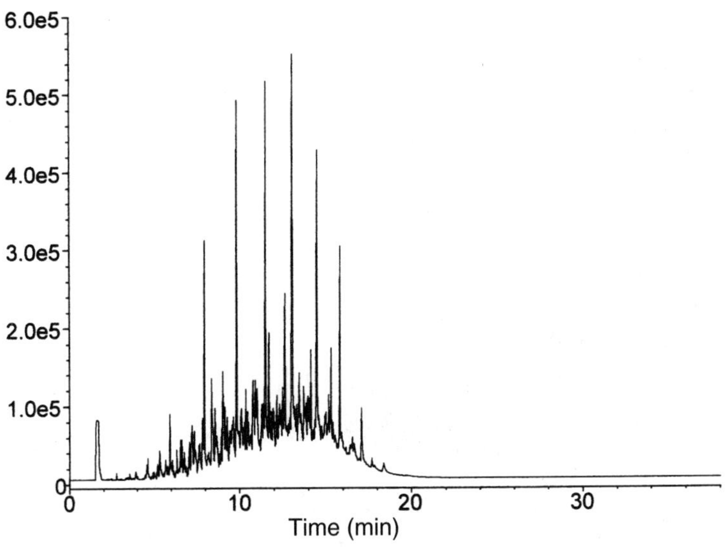

Figure 7.9 Kerosene.

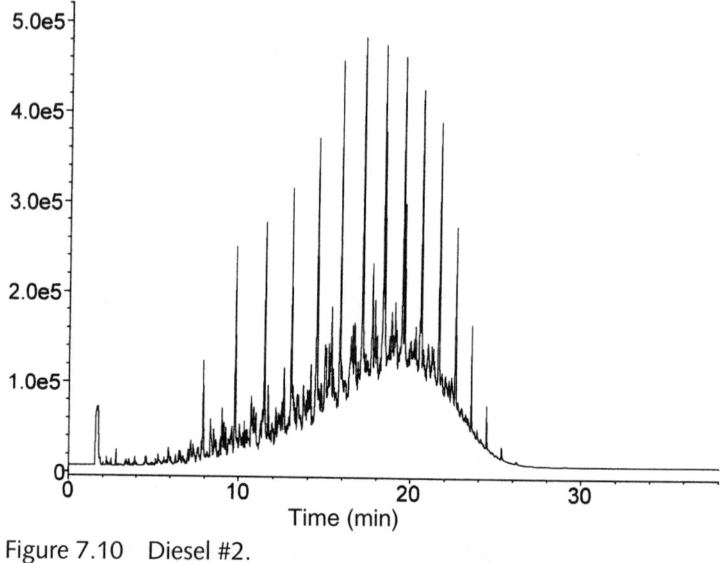

Figure 7.10 Diesel #2.

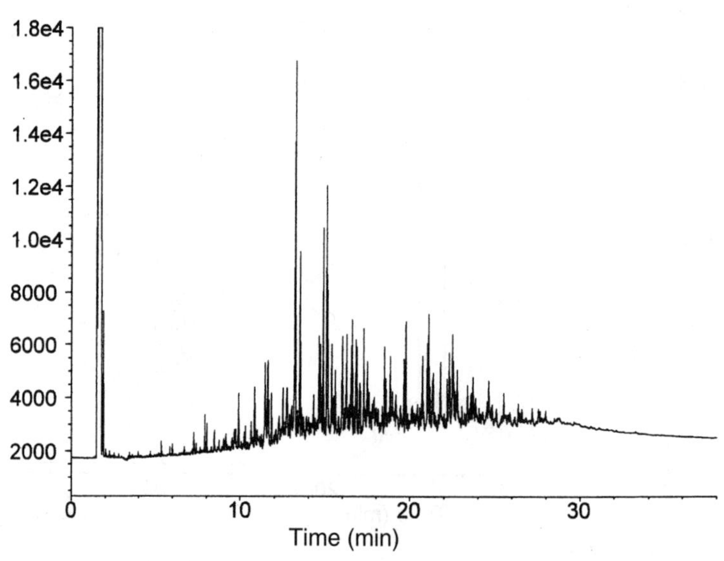

Figure 7.11 Bunker C—Example 1.

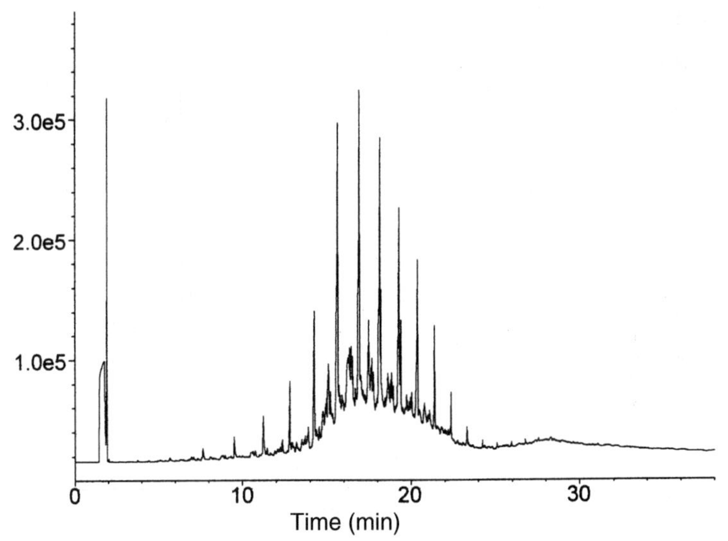

Figure 7.12 Bunker C—Example 2.

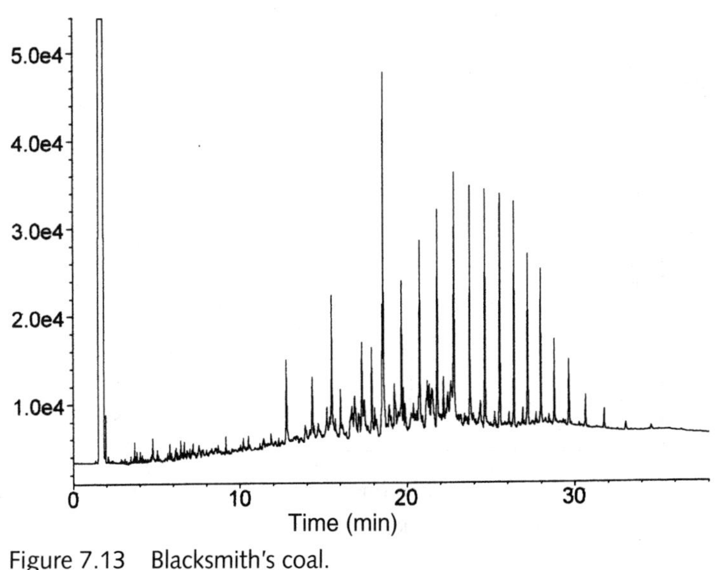

Figure 7.13 Blacksmith's coal.

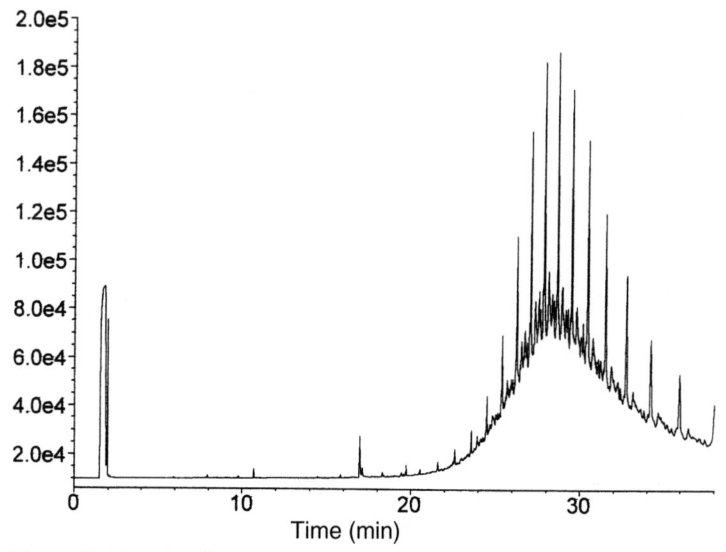

Figure 7.14 Vasoline.

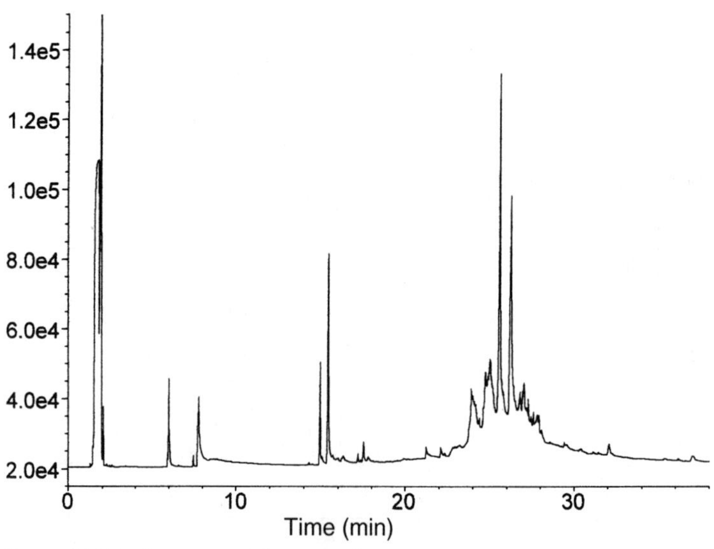

Figure 7.15 Hops used to make beer.

is important to know that the first large peak seen at the left-hand side of each GC trace is usually due to a solvent. In the case of Figures 7.1 to 7.15, carbon disulfide was used as a solvent. Because of the high sensitivity of these instruments, the solvent was used to dilute the sample.

Different GC operating conditions will cause the eluting time or temperature scale to vary. It is, therefore, necessary to establish an individual temperature scale for each GC analysis that is conducted. This can be done by inclusion of a GC analysis of a known material such as diesel fuel.

In looking at the GC traces, it is important to note that the relative pattern of peaks varies from one product to another and occasionally between two samples of the same product. The pattern of peaks provides some information on the composition of the products, and the location of the peaks on the x axis provides information on the boiling range. Although the composition of petroleum products can change, their boiling range will stay relatively constant. It is important to remember that commodity petroleum products are manufactured according to production specifications. These products are typically not manufactured to match a particular chemical composition or to match a particular GC trace. A GC trace of a petroleum product is a snapshot of what one particular batch of petroleum can look like.

Specific Compound Identifications

The presence or absence of specific chemical compounds can also be used for chemical fingerprinting purposes. The presence of specific chemicals can be used to show differences and similarities between two or more samples or between a sample and product specific information such as that provided on an MSD sheet. For example, the presence of specific compounds like methyl cellosolve (found in a number of cleaning products) or one of the organic lead compounds (e.g., tetraethyl lead, tetramethyl lead, trimethylethyl lead, dimethyldiethyl lead, or methyltriethyl lead that make up leaded gasoline) can sometimes be used to identify specific chemical products.

To detect and measure individual chemical compounds, one can use a variety of analytical methods. The most common fingerprinting technique uses gas chromatography. Fingerprinting analyses can use a wide variety of other analytical techniques, including high-pressure liquid chromatography, thin-layer chromatography, and visible spectroscopy to name a few.

The choice of one or more of these techniques depends on the target analyte, the type of matrix analyzed, and the goal of the fingerprinting analysis.

Analysis for metal-containing materials is also useful for fingerprinting purposes. Metal and metallic species are common constituents of a multitude of industrial materials and are also introduced into some products as wear metals. Paints, oils, and soaps can all contain some form of metals. Fingerprinting analyses for metals can involve the use of atomic emission and atomic absorption spectrometry.

There are a multitude of other fingerprinting methods that are available for use for chemical fingerprinting purposes. One that is of particular interest is the use of isotopic differences to match or compare samples. Recent analytical advancements have made such evaluations relatively inexpensive, which should greatly increase their use for fingerprinting purposes. The other techniques are too numerous to mention and a detailed discussion of their use is beyond the scope of this chapter.

Gas Chromatographic Analyses

Gas chromatographic analyses are some of the most common analyses conducted by chemical testing laboratories [12]. This technique uses a gas chromatograph. It is commonly used for the detection and measurement of individual chemical compounds. Many of the EPA and state environmental testing methods are based on gas chromatographic analyses. Because of the widespread use of gas chromatography, its use for fingerprinting analyses is relatively straightforward. In addition, the data from a GC can be computerized, digitized, and made available for further processing. The computerization of the GC data can greatly simplify their processing and evaluation.

The ability of a GC to identify a particular chemical compound is based on the time at which a compound will appear or elute on a GC trace. The retention or elution time of a particular compound is established through the analysis of standards containing the compound or compounds of interest. Because there can be some slight variations in the retention time from one analysis to another, a retention time window spanning the expected retention time for each compound is established. The window should be wide enough so that a compound will always appear within the retention time window. If the window is too narrow, a compound may appear outside its retention time window and its presence may not be detected or reported. When a compound is present but goes unreported, the absence of any finding is said to be a false negative.

One limitation of a GC analysis is that more than one compound can elute within the retention time window of a particular compound. This is an inherent problem with all chromatographic analyses and chemists have designed several ways to minimize this potential error. First, the chemist must consider the possibility that two compounds may appear as a single peak. This situation is called coelution. Second, the chemist must consider the possibility that two or more compounds may appear as separate peaks, all of which are present in the retention time window of the compound of interest. Both of these situations can give rise to a situation where a false-positive response is generated by a GC analysis.

The first classical solution to establish whether a false positive exists is to identify all of the compounds that may be present in a sample. The retention time windows for all of the compounds are determined and any overlapping retention time windows will be clearly identified. If any overlap exists, modification of the GC operating conditions may be possible, which could eliminate any false positive issue. Where it is impossible to establish the identity and retention time windows of all of the potential constituents in a sample, this approach cannot be used.

A second approach to address the issue of false positives is to confirm a GC identification by looking for the presence of other compounds known to coexist with the compound of interest. This approach is sometimes called pattern recognition. It is applicable where there exists a group of compounds that are known to coexist. The identification of one compound can be confirmed by the presence of the other members of the group. This approach is based on the assumption that it is unlikely that the detection of all members within a group would be due to false positives. Where the relative proportion of each member in the group is known, a comparison of the relative proportions of each member can provide added confidence to any identifications made based on pattern recognition. This approach works well when the target analytes are major constituents of a sample.

Pattern recognition is often used in the environmental testing field to assist in the identification of benzene. Benzene, as well as toluene, ethylbenzene, and the xylenes, are well-known constituents of gasoline. Analysis of a sample may show the presence of a peak that falls within the retention time window established for benzene. Because of the potential that gasoline contains other compounds that could also fall within the same retention time window, the data can be examined to see if other gasoline constituents such as toluene, ethylbenzene, and the xylenes are also present. If so, their presence lends some support that the sample also contains benzene.

The most common approach to reduce the potential for false positives is to use a GC fitted with selective detectors. Some of these detectors, such as an electron capture detector (ECD), Hall conductivity detector (HCD), and photoionization detector (PID), respond to certain compounds and not to others. The ECD and HCD are particularly useful for the detection of halogenated compounds (chlorinated solvents and lead scavengers, ethylene dibromide, and ethylene dichloride). They are also useful for detecting organic lead compounds that were added in the past to leaded gasoline.

The PID uses ionizing radiation to selectively detect compounds. Compounds will display different ionizing energies and there are several selections that one can choose among for use with this detector. Table 7.3 shows the ionization energies (ionization potential) for selected compounds. By the proper choice of the ionization energy used in a PID detector, one can selectively detect certain compounds. This detector is particularly useful for the detection of benzene, toluene, ethylbenzene, and the xylenes.

The advantage with the use of selective detectors is that they can be so selective at times that all of the potential compounds that are detected can be identified. This allows one to establish retention time windows for all of the potential compounds and ensure that false positives cannot occur.

A mass spectrometer (MS) is another example of a selective detector that can be used with a GC. It is different from other GC detectors in that it is a separate analytical instrument that can be used without a GC to identify compounds. The MS detector ionizes compounds with sufficient energy to break them apart. Molecules break apart in a predictable and reproducible fashion. The various pieces or fragments produced and their relative abundance are compared to a library of standards to identify the compounds that are present. The mass spectra of several compounds are shown in Figures 7.16 to 7.21. Use of a mass spectrometer as a GC detector is particularly helpful in identifying a single compound in a complex chemical mixture.

The most common GC detector is the flame ionization detector (FID). This detector responds to ions that are generated when a compound is burned. Because of this, the FID is a nonselective detector. The nonselective nature of this detector means that it can be used for a very wide range of compounds. The FID is also useful for showing that a compound of in-

Table 7.3 Ionization Potential (IP) for Selected Organic Compounds

Analyte	*IP (eV)*
Alkanes	
Methane	12.98
Ethane	11.65
Propane	11.07
n-Butane	10.63
i-Butane	10.57
n-Pentane	10.35
n-Hexane	10.18
n-Heptane	10.08
Isooctane	9.86
Aromatics	
Benzene	9.245
Toluene	8.82
Ethyl Benzene	8.76
n-Propyl benzene	8.72
i-Propyl benzene	8.69
n-Butyl benzene	8.69
o-Xylene	8.56
m-Xylene	8.56
p-Xylene	8.445
Naphthalene	8.12
1-Methyl naphthalene	7.69
2-Methyl naphthalene	7.955
Oxygenates	
Methyl alcohol	10.85
Ethanol	10.48
i-Propyl ether	9.20

terest is not present. Because of the nonselective nature of the FID, the absence of a peak for a compound of interest is usually strong evidence that the compound is not present. However, when a peak is detected, there is the potential for a false-positive identification.

The response of a detector can also be used to establish the amount or quantity of a compound present in a sample. When using a GC to provide quantitative results, standards of known quantity or concentration are analyzed and the detector responses recorded. The data are then evaluated to determine if a predictable relationship exists between the quantity of a

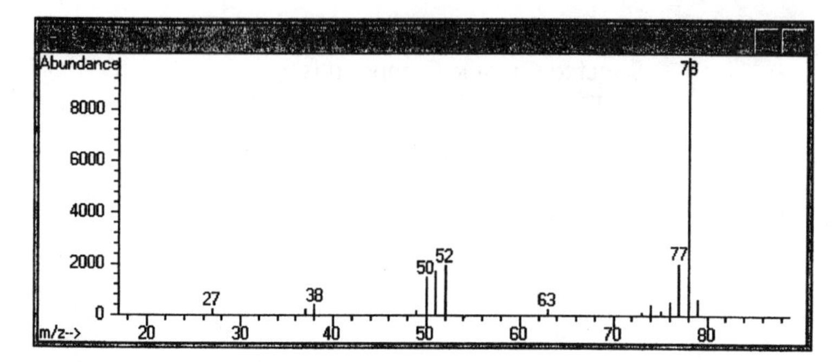

Figure 7.16 Mass spectra of benzene.

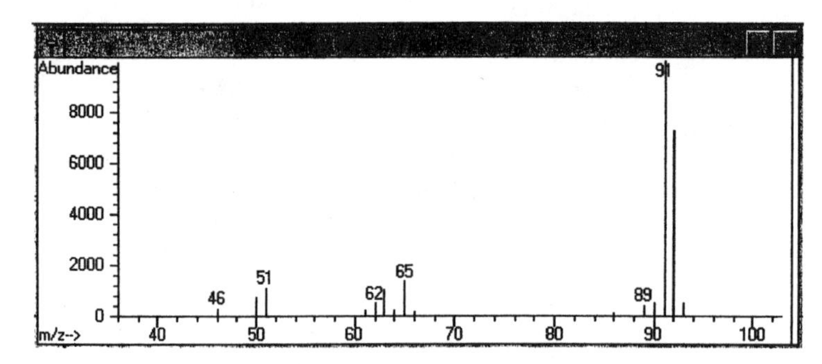

Figure 7.17 Mass spectra of toluene.

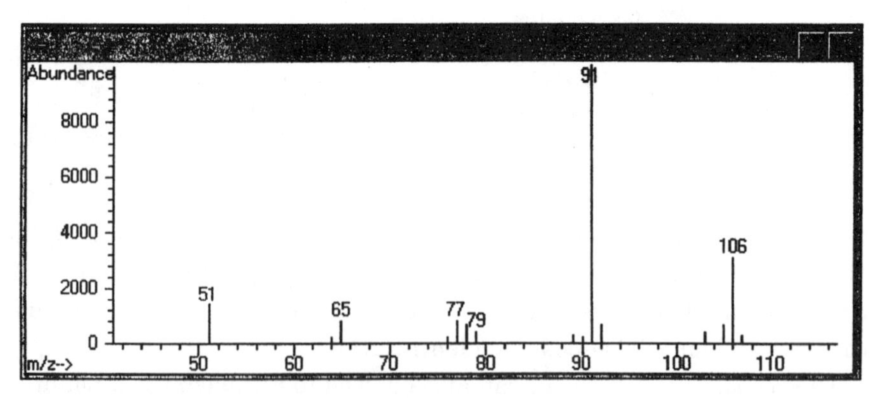

Figure 7.18 Mass spectra of ethylbenzene.

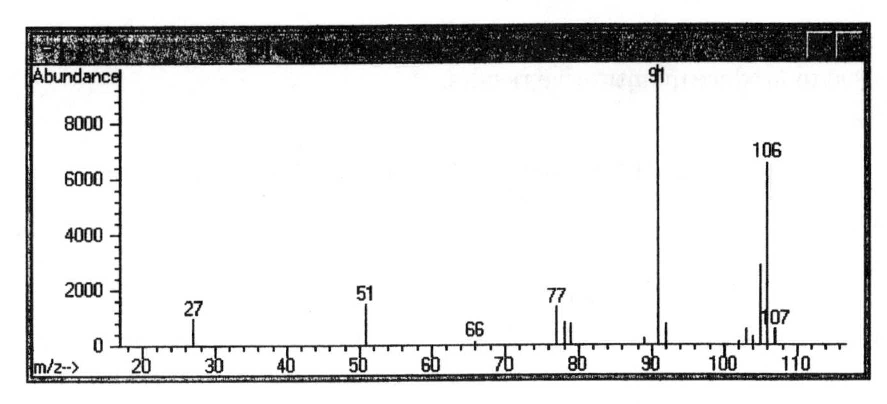

Figure 7.19 Mass spectra of *m*-xylene.

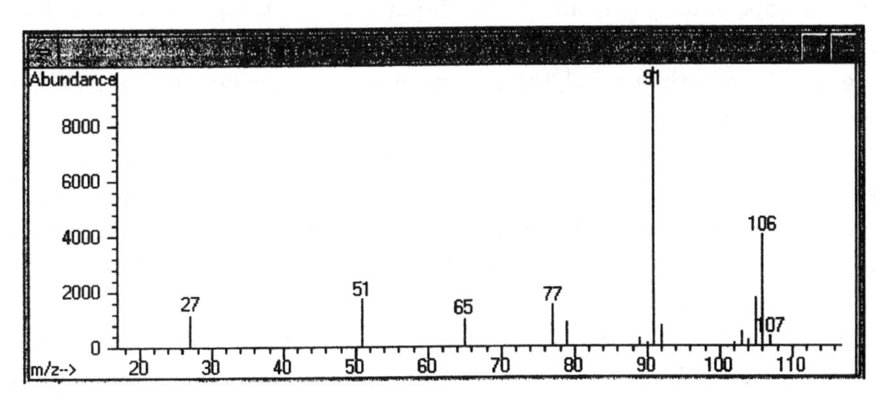

Figure 7.20 Mass spectra of *o*-xylene.

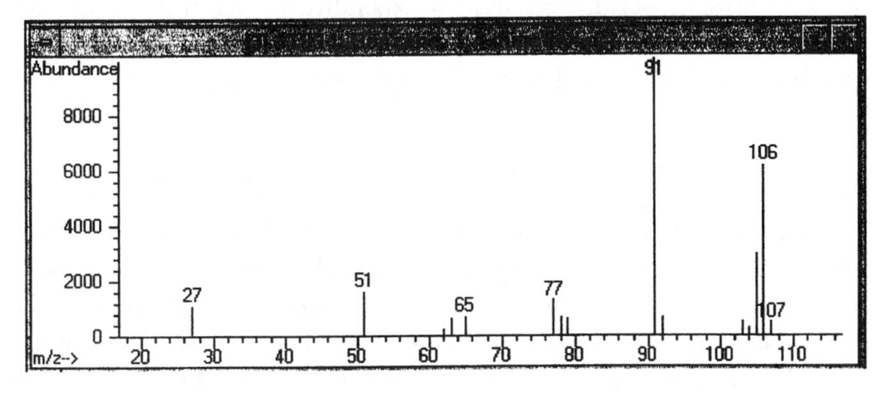

Figure 7.21 Mass spectra of *p*-xylene.

compound and its detector response. If a correlation does exist, it can be used to produce quantitative results.

High-Pressure Liquid Chromatographic Analyses

High-pressure liquid chromatography is an analytical technique similar to gas chromatography. It uses a high-pressure liquid chromatograph (HPLC) to separate a mixture into individual components that appear as peaks on a chromatogram. It differs from gas chromatography in one very important way. An HPLC is a liquid separation system, whereas a GC is a gas separation system. Because of this, the high temperatures needed by a GC to convert compounds into a gas phase are not required with the HPLC. The HPLC can be operated at room temperature or even below room temperature. This allows for the analysis of temperature-sensitive compounds, as well as nonvolatile compounds that cannot be measured using a GC. As with the GC, the retention time of a compound is used for identification purposes.

A wide variety of detectors can be used with an HPLC. Some, such as a refractometer, are nonspecific and have a high potential for coelution and false positives. Other HPLC detectors, such MS or ultraviolet (UV)/fluorescence, are standalone analytical detectors and can be used to provide highly reliable data.

An HPLC is particularly well suited for the analysis of thermally unstable compounds such as explosives and some insecticides [13, 14]. It is also one of the few analytical instruments that can be used for the analysis of nonvolatile materials such as polymers and other high-molecular-weight compounds. The need to design or fine-tune an HPLC to fit a particular analysis generally requires the expenditure of a significant amount of time before a suitable analysis can be established. This has likely limited the use of HPLC for chemical fingerprinting purposes.

Thin-Layer Chromatographic Analyses

Thin-layer chromatographic analyses are simplified versions of high pressure liquid chromatographic analyses. It uses thin-layer chromatography (TLC) to separate a mixture into various fractions or components. TLC is a relatively low-tech analytical method that was one of the first chromatography-based analytical techniques developed. TLC has been widely used for the analysis of a multitude of chemical compounds for over 50 years. It has been used for the analysis of dyes, pigments, food additives, explosives, natural products, pharmaceuticals, vitamins, lipids, sugars, and a multitude of other chemicals [15].

TLC is particularly useful in the analysis of dyes [16]. Dyes are used in inks and in a variety of foodstuffs, plastics, oils, and fuels. Dyes are often added to impart a particular color to a material for marketing or tracking purposes. Historically, dyes were required additives to leaded gasoline as far back as 1926 [17]. They were also used to distinguish among various grades of fuel within a refinery and fuel distribution system. Currently, dyes are used to distinguish high sulfur fuels from low sulfur fuels. TLC can also be useful for screening samples for particular chemical compounds. It can be used for the analysis of a variety of chemical compounds, including PCB, explosives, petroleum fuels, lubricating oils, asphalt, residual petroleum products, and coal tar [18].

One serious limitation of TLC is its potential for false positives. Due to the low cost and ease of operation, this technique is excellent for screening samples for possible contamination. Where a positive screening result is obtained, additional and more costly analyses can be conducted to establish the presence or absence of a chemical of concern. One of the perceived limitations of TLC is its lack of digital or analog outputs. The detector system generally relies on the human eye. Rather than generate a chromatogram consisting of a series of peaks, TLC produces a series of spots on a plate that must generally be interpreted visually. Because of its simplicity, TLC is easy for nonchemists to understand and interpret. As such, it can be included in a simple demonstration.

Analyses by Atomic Spectrometry

Atomic emission spectrometry (AES) and atomic absorption spectrometry (AAS) are analytical techniques commonly used to characterize inorganic compounds [19]. In these techniques, the sample is dissociated into atoms or ions through excitation, usually carried out by heating the sample. The excited species are then measured using a spectroscopic apparatus that monitors the response at different wavelengths. This is the same procedure used by astrophysicists to detect elements on distant planets and in far-away galaxies. These techniques may be used to measure compounds that emit or absorb light at ultraviolet, visible, and near-infrared wavelengths, making them very versatile.

The most common excitation source in current use is an inductively coupled plasma (ICP). The ICP-AES excites the atoms or ions by using an argon plasma, a cloud of argon gas that is isolated from the surrounding air and heated using radiofrequency energy. It is well suited to the determination of inorganic compounds over wide concentration ranges. However, because ICP-AES is based on measurement of electromagnetic spectra, spectral interferences are common. In addition to spectrometers, other detectors can be coupled with ICP such as a mass spectrometer or HPLC. The combination of these techniques can be used, in many cases, to avoid the spectral interferences common in ICP-AES.

The sample types that can be analyzed using AES and AAS include blood, paint, grease, petroleum, semiconductors, antifreeze, and metal alloys. As an example, ICP-AES and similar techniques have been used to determine the relative abundance of vanadium and nickel in crude oils. The data from this type of testing can be used to identify the geologic origin of the crude oils. Similarly, characterization of other compounds such as organometallics and wear metals can aid in the delineation of the toxicity and source of materials, as well as, in some cases, the approximate date of manufacture.

Isotopic Determinations

The detection and measurement of individual elemental isotopes is of current interest for chemical fingerprinting puposes. Developments in analytical instrumentation have resulted in the widespread use of ICP/MS, isotope ratio mass spectrometer detectors for GC analyses (GCIRMS), and new atomic emission detectors for GC analyses. The availability of these instruments has lowered the costs associated with the collection of isotopic data and has fostered their consideration for fingerprinting purposes.

Isotopic determinations look at differences in the atomic or elemental composition of a material. Isotopes occur when an element contains varying numbers of neutrons, one of the three main particles (protons, electrons, and neutrons) of an element. Isotopes are useful for fingerprinting purposes because the ratios of isotopes of some elements can vary from one location to another. In addition, certain chemical processes can preferentially select one isotope over another. The result is that every material has an isotopic signature. Sometimes, differences in the isotopic compositions can be measured and used to distinguish between a number of potential sources.

Within the realm of isotopes, there are stable isotopes and radioactive isotopes. Stable isotopes are stable forms of an element that will not spontaneously change over time. Radioactive isotopes are forms of an element that will spontaneously break apart to form a different isotope or one or more different elements. For example, carbon-14 will spontaneously decompose into nitrogen-14. Each radioactive isotope has a characteristic rate at which it decays, called its half life. The half-life for carbon-14 is 5730 years. This means that half of the carbon-14 present in a sample today will be lost or converted to nitrogen-14 in 5730 years. Half of the carbon-14 remaining after 5730 years will be lost in the following 5730 years and this will presumably continue until the end of time.

One of the most widely known forms of chemical fingerprinting using isotopes is carbon-14 age determinations used for archeological purposes. The amount of carbon-14 in the atmosphere is relatively constant. It is created through the bombardment of nitrogen-14 with cosmic rays in the upper atmosphere. This carbon is oxidized to form carbon dioxide, which is distributed throughout the atmosphere within 2 to 3 years. At the same time that the carbon-14 is being produced, it is converted, through its radioactive decay process, back to nitrogen-14. The result is that there is a relatively constant amount of carbon-14 in our atmosphere.

The carbon dioxide containing carbon-14 can be fixed or converted into other forms of carbon. Photosynthesis will fix the radioactive form of carbon from the atmosphere into plant material. This carbon is the basis of natural chemicals and fibers. If eaten by animals, it will be distributed throughout the carbon-based chemicals that make up the animal. Because the carbon is now fixed, it cannot be replenished by incorporation of new carbon-14-containing carbon dioxide. This means that the amount of carbon-14 present in these materials will decrease at the rate dictated by its half-life. By knowing the change in the amount of carbon-14 over time, the half-life can be used to calculate the time required for such a change to occur.

Analysis of stable isotopes is much more common for chemical fingerprinting purposes than radioactive analysis. There are many more stable isotopes than there are radioactive isotopes. In addition, the analyses used for stable isotopes can also be used with radioactive isotopes. The ratio of isotopes present in a sample can come from a number of sources. First, the ratio of isotopes can vary based on the location from which a material is obtained. Different ore bodies can have different isotope ratios based on their origins. Changes in the ratio of isotopes can also occur during various chemical processes. Heavier isotopes, due to their greater mass, are sometimes slower to react in chemical reactions. When this happens, the resulting chemical products will be enriched in the lighter isotopes and the remaining unreacted material will be enriched in the heavier isotopes. The degree to which the relative proportion of the isotopes are changed is a function of the temperature at which the reaction is carried out. Colder temperatures will increase the isotopic discrimination in the final product. If the temperature is high enough, no isotopic discrimination may occur.

Isotopic discrimination can occur in a variety of natural situations. The fixation of carbon dioxide through photosynthesis is an example of a chemical process that can enrich the final products in one isotope over another. When carbon is fixed by plants to form sugars, cellulose, and other organic chemicals, the air temperature will impact the ratio of carbon isotopes in the plant chemicals. Cool temperatures will result in the fixation of a higher proportion of light carbon-12 over the heavier carbon-13 and carbon-14. When these natural chemicals are converted into crude oil and then into different commercial products, the isotopic signature imparted during the carbon fixation process can be carried along. The results are differences in the isotopic composition of products based on the temperature at which the carbon was fixed from the atmosphere [20–22]. Isotopic differences have been used to distinguish separate origins of crude oil, as well as products made from these materials. Isotopic differences exist with other elements, and investigation into the isotopic ratios of other elements such as hydrogen, oxygen, nitrogen, and chlorine can be useful for many organic compounds [23].

A large number of metals also show variations in their isotopic composition. These variations have been widely used to distinguish among various sources of environmental contamination in the past [24]. A large amount of work has been conducted on lead and fairly definitive answers have been obtained using this approach [25]. Various ore bodies have been shown to have unique isotopic composition and the products made from these ores can be tracked back to a particular source.

The detection and measurement of individual isotopes can be accomplished with the use of a mass spectrometer. The mass spectrometer can be coupled with a GC or ICP to provide additional discrimination for characterization purposes. Because standards of pure isotopes are difficult, if not impossible to obtain, isotopic analyses are generally conducted using a standard that has a fixed ratio of isotopes. The data reported are sometimes reported as differences from the isotopic ratio of the standard used.

FINGERPRINTING APPLICATIONS

There are an infinite number of ways that chemical analyses can be used for fingerprinting purposes. In general, there are two technical approaches that can be used. One approach is to identify an unknown material. Here, some knowledge is needed as to what materials were potentially present and their composition or expected chemical properties. The second approach is to simply compare or match two or more samples. No information as to the composition of a material is needed in this latter approach.

Identification of an Unknown

There are many different sources of information that can be used to identify unknown materials. MSD sheets, production specifications, and historical information can all provide information as to the historical composition of materials. Several good references for petroleum products are available [26–28]. Using this information, tests can be run on an unknown material to determine its properties. The results may allow a direct identification to be made.

A boiling range determination is a common fingerprinting technique. The boiling range of an unknown material can be compared to the historical information available concerning production specifications, as well as examples of products that may be part of a laboratory's library of chemical substances. Using this information, an identification can often be made. A check can sometimes be performed using pattern recognition to identify the type of refining or production processes that may have been used to make a particular product.

Identifications can also be established based on the presence of individual chemical compounds. For many industrial chemicals, the individual chemicals present are known. The chlorinated solvent trichloroethylene contains trichloroethylene and perc contains tetrachloroethylene. Citrus oil–based cleaning solvents contain limonene. Common antifreeze can contain ethylene or propylene glycol.

With petroleum products, their composition is less well known. There are a number of additives that have been added to petroleum products and analysis for their presence can be conducted as part of a fingerprinting process. GC analysis for organic lead compounds is of particular use when dealing with gasoline contamination. Historically, a variety of organic lead compounds have been added to gasoline. These include tetramethyl lead, trimethylethyl lead, dimethyldiethyl lead, methyltriethyl lead, and tetraethyl lead (TEL). TEL was first used in gasoline in the early 1920s [29]. Beginning in the early 1960s, other organic lead compounds came into use [30]. These new organic lead compounds, generally added as mixtures, improved the octane boost over that provided by the use of TEL alone.

A number of mixtures were available and the particular mixture used was based on the manufacturing processes employed by a refinery. The various organic lead species can be readily detected using a GC/ECD analysis. Occasionally, levels are sufficiently high to use a GC/MS for the analysis of these compounds. Lead can also be determined using AES and AAS. When these latter methods are used, interference from inorganic or natural lead may be of concern. The potential interference from inorganic lead can be reduced by suitable sample processing or careful design of a sampling plan. The presence of organic lead can be used to clearly identify the presence of leaded gasoline.

Another additive to petroleum products was (methylcyclopentadienyl)-manganese tricarbonyl, or MMT. This compound was commercially available in 1958 and was often added to gasoline along with organic lead. For a short period of time, MMT was added to unleaded gasoline sold in the United States. In 1978, the EPA banned the use of MMT in unleaded gasoline. This did not stop its continued use in leaded gasoline nor its use in unleaded gasoline outside the United States. Analysis for MMT can be carried out using GC/ECD, GC/MS, AES, and AAS. Because MMT was used in both leaded and unleaded gasoline, it cannot be used to distinguish between leaded and unleaded gasoline.

Oxygenates are gasoline fuel additives that can be used for chemical fingerprinting purposes [31, 32]. Oxygenates were first used in gasoline in the 1920s when methanol and ethanol were added to boost the octane rating of gasoline stocks. They continued to be used as both octane boosters and icing inhibitors but their distribution was limited within the United States. In the 1970s, ethanol and methanol became more widely used due to the rapid rise in crude-oil prices. The passage of the Clean Air Act Amendments greatly expanded the use of the oxygenates in the 1990s. GC/MS is particularly suited for the analysis of oxygenates, especially when gasoline is present in the sample matrix.

Identification of unknown materials can sometimes be accomplished by using detailed chemical composition data that are sometimes available. Tables 7.4 to 7.10 provide compositional information from the analysis of petroleum products [33]. This type of information can provide the types of compounds present in a particular product, as well as an estimate of its compositional variation. When using this type of data, one must be concerned with the potential presence of false positives in the reference data.

Matching Samples

The matching of samples is often the most direct method of fingerprinting. There are several ways in which one can match samples when conducting chemical fingerprinting. One may have a sample of the suspect material. Here, the fingerprinting goal is to determine if one or more samples match the source material. This is often accomplished by establishing the detailed composition of all products. A second approach is to simply group samples that have similar chemical compositions.

Changes in the chemical refining operations used to make a product can result in changes in the composition of the final product [34, 35]. This compositional information can sometimes be used to distinguish between different batches of the same product, especially where the different batches were manufactured using different processes. There are several analyses that can be used to obtain detailed compositional information. A PIANO (paraffins, isoparaffins, aromatics, naphthenes, and olefins) or PONA (paraffins, olefins, naphthenes, and aromatics) analysis is particularly suited for the analysis of gasoline. This analysis can be performed using a nonselective flame ionization detector (FID) or an MS detector [36, 37]. The low boiling nature of the gasoline allows one to establish GC conditions to minimize the potential for coelution.

Compositional information for materials that boil higher than gasoline is generally obtained by using a GC/MS analysis. Selected ions or fragments of molecules that are characteristic of particular types or classes of compounds can be monitored to distinguish between different sources of materials [38–40].

For chemical fingerprinting purposes, carbon-14 dating can be used to distinguish products produced from renewable resources from those made from crude oil or coal. The carbon compounds from renewable resources will typically be carbon-14 active. The carbon compounds that make up crude oil or coal are often older than 100,000 years, which is the current limit of radioactive carbon dating. This means that the carbon from crude oil or coal will be carbon-14 dead. When measuring the carbon-14 activity of a sample, the absence of activity will indicate that crude-oil or coal-based carbon is present, whereas high activity will indicate the presence

Table 7.4 Detailed Composition of Crude Oil

Compound	*Number of Carbons*	*EC*	*Weight Percent*	*Reference*
Straight chain alkanes				
n-Hexane	6	6	0.7–1.8	API, 1993[a]
n-Heptane	7	7	0.8–2.3	API, 1993
n-Octane	8	8	0.9–1.9	API, 1993
n-Nonane	9	9	0.6–1.9	API, 1993
n-Decane	10	10	1.8	API, 1993
n-Undecane	11	11	1.7	API, 1993
n-Dodecane	12	12	1.7	API, 1993
Branched Chain Alkanes				
2,2-Dimethylbutane	6	5.37	0.04	API, 1993
2,3-Dimethylbutane	6	5.68	0.04–0.14	API, 1993
2-Methylpentane	6	5.72	0.3–0.4	API, 1993
3-Methylpentane	6	5.85	0.3–0.4	API, 1993
3-Ethylpentane	7		0.05	API, 1993
2,4-Dimethylpentane	7	6.31	0.05	API, 1993
2,3-Dimethylpentane	7	6.69	0.1–0.6	API, 1993
2,2,4-Trimethyl-pentane	8	6.89	0.004	API, 1993
2,3,3-Trimethyl-pentane	8	7.58	0.006	API, 1993
2,3,4-Trimethyl-pentane	8	7.55	0.005	API, 1993
2-Methyl-3-Ethyl-pentane	8	7.66	0.04	API, 1993
2-Methylhexane	7	6.68	0.7	API, 1993
3-Methylhexane	7	6.76	0.19–0.5	API, 1993
2,2-Dimethylhexane	8	7.25	0.01–0.1	API, 1993
2,3-Dimethylhexane	8	7.65	0.06–0.16	API, 1993
2,4-Dimethylhexane	8	7.38	0.06	API, 1993
2,5-Dimethylhexane	8	7.36	0.06	API, 1993
3,3-Dimethylhexane	8	7.45	0.03	API, 1993
2,3-Dimethylheptane	9	8.64	0.05	API, 1993
2,6-Dimethylheptane	9	8.47	0.05–0.25	API, 1993
2-Methyloctane	9		0.4	API, 1993
3-Methyloctane	9	8.78	0.1–0.4	API, 1993
4-Mehtyloctane	9	8.71	0.1	API, 1993
Cycloalkanes				
Cyclopentane	5	5.66	0.05	API, 1993
Methylcyclopentane	6	6.27	0.3–0.9	API, 1993

1,1-Dimethylcyclopentane	7	6.72	0.06–0.2	API, 1993
1-*trans*-2-Dimethylcyclopentane	7	6.87	0.15–0.5	API, 1993
1-*cis*-3-Dimethylcyclopentane	7	6.82	0.2	API, 1993
1-*trans*-3-Dimethylcyclopentane	7	6.85	0.2–0.9	API, 1993
1,1,2-Trimethylcyclopentane	8	7.67	0.06	API, 1993
1,1,3-Trimethylcyclopentane	8	7.25	0.3	API, 1993
1-*trans*-2-*cis*-3-Trimethylcyclopentane	8	7.51	0.3–0.4	API, 1993
1-*trans*-2-*cis*-4-Trimethylcyclopentane	8		0.2	API, 1993
1-*trans*-2-Dimethylcyclohexane	8	7.94	0.3	API, 1993
Ethylcyclohexane	8	8.38	0.2	API, 1993
Cyclohexane	6	6.59	0.7	API, 1993
1-*trans*-2-*trans*-Trimethylcyclohexane	9		0.2	API, 1993
Alkyl benzenes				
Benzene	6	6.5	0.04–0.4	API, 1993
Toluene	7	7.58	0.09–2.5	API, 1993
Ethylbenzene	8	8.5	0.09–0.31	API, 1993
o-Xylene	8	8.81	0.03–0.68	API, 1993
m-Xylene	8	8.6	0.08–2.0	API, 1993
p-Xylene	8	8.61	0.09–0.68	API, 1993
1-Methyl-4-ethylbenzene	9	9.57	0.03–0.13	API, 1993
1-Methyl-2-ethylbenzene	9	9.71	0.01–0.09	API, 1993
1-Methyl-3-ethylbenzene	9	9.55	0.04–0.4	API, 1993
1,2,3-Trimethylbenzene	9	10.06	0.1	API, 1993
1,2,4-Trimethylbenzene	9	9.84	0.13–0.69	API, 1993
1,3,5-Trimethylbenzene	9	9.62	0.05–0.18	API, 1993

(continues)

Table 7.4 (*Continued*)

Compound	*Number of Carbons*	*EC*	*Weight Percent*	*Reference*
1,2,3,4-Tetramethyl-benzene	10	11.57	0.2	API, 1993
Biphenyl	12	14.26	0.006–0.4	API, 1993
Naphtheno benzenes				
Indan	9	10.27	0.07	API, 1993
Tetralin (tetrahydro-naphthalene)	10	11.7	0.03	API, 1993
5-Methylthtrohydro-naphthalene	11		0.08	API, 1993
6-Methylthtrohydro-naphthalene	11		0.09	API, 1993
Fluorene	13	16.55	0.003–0.06	API, 1993
Alkyl / naphthalenes				
Naphthalene	10	11.69	0.02–0.09	API, 1993
Polynuclear aromatics				
Phenanthrene	14	19.36	0.003–0.05	API, 1993

[a] API, 1993 Petroleum Product Surveys, American Petroleum Institute, Washington, DC.

Source: J. B. Gustafson, J. G. Tell, and D. Orem, *Selection of Representative TPH Fractions Based on Fate and Transport Considerations—Total Petroleum Hydrocarbon Criteria Working Group Series*, Vol. 3, Amherst, MA: Amherst Scientific Publishers, Appendix A. With permission.

Table 7.5 Detailed Composition of Gasoline

Compound	*Number of Carbons*	*EC*	*Weight Percent*	*Reference*
Straight chain alkanes				
Propane	3	3	0.01–0.014	LUFT, 1988[a]
n-Butane	4	4	3.93–4.70	LUFT, 1988
n-Pentane	5	5	5.75–10.92	LUFT, 1988
n-Hexane	6	6	0.24–3.50	LUFT, 1988
n-Heptane	7	7	0.31–1.96	LUFT, 1988
n-Octane	8	8	0.36–1.43	LUFT, 1988
n-Nonane	9	9	0.07–0.83	LUFT, 1988
n-Decane	10	10	0.04–0.50	LUFT, 1988
n-Undecane	11	11	0.05–0.22	LUFT, 1988
n-Dodecane	12	12	0.04–0.09	LUFT, 1988
Branched chain alkanes				
Isobutane	4	3.67	0.12–0.37	LUFT, 1988
2,2-Dimethylbutane	6	5.37	0.17–0.84	LUFT, 1988
2,3-Dimethylbutane	6	5.68	0.59–1.55	LUFT, 1988
2,2,3-Trimethylbutane	7	6.36	0.01–0.04	LUFT, 1988
Neopentane	5	4.32	0.02–0.05	LUFT, 1988
Isopentane	5	4.75	6.07–10.17	LUFT, 1988
2-Methlypentane	6	5.72	2.91–3.85	LUFT, 1988
3-Methlypentane	6	5.85	2.4 (vol)	LUFT, 1988
2,4-Dimethylpentane	7	6.31	0.23–1.71	LUFT, 1988
2,3-Dimethylpentane	7	6.69	0.32–4.17	LUFT, 1988
3,3-Dimethylpentane	7	6.55	0.02–0.03	LUFT, 1988
2,2,3-Trimethylpentane	8	7.37	0.09–0.23	LUFT, 1988
2,2,4-Trimethylpentane	8	6.89	0.32–4.58	LUFT, 1988
2,3,3-Trimethylpentane	8	7.58	0.05–2.28	LUFT, 1988
2,3,4-Trimethylpentane	8	7.55	0.11–2.80	LUFT, 1988
2,4-Dimethyl-3-ethylpentane	9		0.03–0.07	LUFT, 1988
2-Methylhexane	7	6.68	0.36–1.48	LUFT, 1988
3-Methylhaxane	7	6.76	0.30–1.77	LUFT, 1988
2,4-Dimethylhexane	8	7.38	0.34–0.82	LUFT, 1988
2,5-Dimethylhexane	8	7.36	0.24–0.52	LUFT, 1988
3,4-Dimethylhexane	8	7.74	0.16–0.37	LUFT, 1988
3-Ethylhexane	8	7.79	0.01	LUFT, 1988
2-Methyl-3-ethylhexane	9		0.04–0.13	LUFT, 1988
2,2,4-Trimethylhexane	9	7.93	0.11–0.18	LUFT, 1988
2,2,5-Trimethylhexane	9	7.87	0.17–5.89	LUFT, 1988
2,3,3-Trimethylhexane	9		0.05–0.12	LUFT, 1988

(continues)

Table 7.5 (*Continued*)

Compound	*Number of Carbons*	*EC*	*Weight Percent*	*Reference*
2,3,5-Trimethylhexane	9	8.24	0.05–1.09	LUFT, 1988
2,4,4-Trimethylhexane	9	8.07	0.02–0.16	LUFT, 1988
2-Methylheptane	8	7.71	0.48–1.05	LUFT, 1988
3-Methylheptane	8	7.78	0.63–1.54	LUFT, 1988
4-Methylheptane	8	7.72	0.22–0.52	LUFT, 1988
2,2-Dimethylheptane	9	8.28	0.01–0.08	LUFT, 1988
2,3-Dimethylheptane	9	8.64	0.13–0.51	LUFT, 1988
2,6-Dimethylheptane	9	8.47	0.07–0.23	LUFT, 1988
3,3-Dimethylheptane	9	8.42	0.01–0.08	LUFT, 1988
3,4-Dimethylheptane	9	8.62	0.07–0.33	LUFT, 1988
2,2,4-Trimethylheptane	10		0.12–1.70	LUFT, 1988
3,3,5-Trimethylheptane	10		0.02–0.06	LUFT, 1988
3-Ethylheptane	9	8.77	0.02–0.16	LUFT, 1988
2-Methyloctane	9		0.14–0.62	LUFT, 1988
3-Methyloctane	9	8.78	0.34–0.85	LUFT, 1988
4-Methyloctane	9	8.71	0.11–0.55	LUFT, 1988
2,6-Dimethyloctane	10	9.32	0.06–0.12	LUFT, 1988
2-Methylnonane	10	9.72	0.06–0.41	LUFT, 1988
3-Methylononane	10	9.78	0.06–0.32	LUFT, 1988
4-Methylnonane	10		0.04–0.26	LUFT, 1988
Cycloalkanes				
Cyclopentane	5	5.66	0.19–0.58	LUFT, 1988
Methylcyclopentane	6	6.27	Not quantified	LUFT, 1988
1-Methyl-*cis*-2-ethylcyclopentane	8		0.06–0.11	LUFT, 1988
1-Methyl-*trans*-3-ethylcyclopentane	8		0.06–0.12	LUFT, 1988
1-*cis*-2-Dimethylcyclopentane	7	7.21	0.07–0.13	LUFT, 1988
1-*trans*-2-Dimethylcyclopentane	7	6.87	0.06–0.20	LUFT, 1988
1,1,2-Trimethylcyclopentane	8	7.67	0.06–0.11	LUFT, 1988
1-*trans*-2-*cis*-3-Trimethylcyclopentane	8	7.51	0.01–0.25	LUFT, 1988
1-*trans*-2-*cis*-4-Trimethylcyclopentane	8	0.03	0.03–0.16	LUFT, 1988
Ethylcyclopentane	7	7.34	0.14–0.21	LUFT, 1988
n-Propylcyclopentane	8	7.1	0.01–0.06	LUFT, 1988

Isopropylcyclopentane	8		0.01–0.02	LUFT, 1988
1-*trans*-3-Dimethylcyclohexane	8	7.99	0.05–0.12	LUFT, 1988
Ethylcyclohexane	8	8.38	0.17–0.42	LUFT, 1988
Cyclohexane	6	6.59	0.08	API, 1993
Straight chained alkenes				
cis-2-Butene	4	4.25	0.13–0.17	LUFT, 1988
trans-2-Butene	4	4.1	0.16–0.20	LUFT, 1988
Pentene-1	5	4.89	0.33–0.45	LUFT, 1988
cis-2-Pentene	5	5.16	0.43–0.67	LUFT, 1988
trans-2-Pentene	5	5.08	0.52–0.90	LUFT, 1988
cis-2-Hexene	6	6.14	0.15–0.24	LUFT, 1988
trans-2-Hexene	6	6.05	0.18–0.36	LUFT, 1988
cis-3-Hexene	6	6.03	0.11–0.13	LUFT, 1988
trans-3-Hexene	6	6.02	0.12–0.15	LUFT, 1988
cis-3-Heptene	7	7.01	0.14–0.174	LUFT, 1988
trans-2-Heptene	7	7.05	0.06–0.10	LUFT, 1988
Branched chain alkenes				
2-Methyl-1-butene	5	4.96	0.22–0.66	LUFT, 1988
3-Methyl-1-butene	5	4.57	0.08–0.12	LUFT, 1988
2-Methyl-2-butene	5	5.21	0.96–1.28	LUFT, 1988
2,3-Dimethyl-1-butene	6	5.7	0.08–0.10	LUFT, 1988
2-Methyl-1-pentene	6	5.89	0.20–0.22	LUFT, 1988
2,3-Dimethyl-1-pentene	7		0.01–0.02	LUFT, 1988
2,4-Dimethyl-1-pentene	7	6.48	0.02–0.03	LUFT, 1988
4,4-Dimethyl-1-pentene	7		0.60 (vol)	LUFT, 1988
2-Methyl-2-pentene	6	6.07	0.27–0.32	LUFT, 1988
3-Methyl-*cis*-2-pentene	6	6.11	0.35–0.45	LUFT, 1988
3-Methyl-*trans*-2-pentene	6	6.22	0.32–0.44	LUFT, 1988
4-Methyl-*cis*-2-pentene	6	5.69	0.04–0.05	LUFT, 1988
4-Methyl-*trans*-2-pentene	6	5.73	0.08–0.30	LUFT, 1988
4,4-Dimethyl-*cis*-2-pentene	7	6.47	0.02	LUFT, 1988
4,4-Dimethyl-*trans*-2-pentene	7	6.23	Not quantified	LUFT, 1988
3-Ethyl-2-pentene	7	7.07	0.03–0.04	LUFT, 1988
Cycloalkenes				
Cyclopentene	5	5.55	0.12–0.18	LUFT, 1988
3-Methylcyclopentene	6	6.1	0.03–0.08	LUFT, 1988
Cyclohexene	6	6.74	0.03	LUFT, 1988
Alkyl benzenes				
Benzene	6	6.5	0.12–3.50	LUFT, 1988

(continues)

Table 7.5 (*Continued*)

Compound	*Number of Carbons*	*EC*	*Weight Percent*	*Reference*
Toulene	7	7.58	2.73–21.80	LUFT, 1988
Ethylbenzene	8	8.5	0.36–2.86	LUFT, 1988
o-Xylene	8	8.81	0.68–2.86	LUFT, 1988
m-Xylene	8	8.6	1.77–3.87	LUFT, 1988
p-Xylene	8	8.61	0.77–1.58	LUFT, 1988
1-Methyl-4-ethylbenzene	9	9.57	0.18–1.00	LUFT, 1988
1-Methyl-2-ethylbenzene	9	9.71	0.19–0.56	LUFT, 1988
1-Methyl-3-ethylbenzene	9	9.55	0.31–2.86	LUFT, 1988
1-Methyl-2-*n*-propylbenzene	10		0.01–0.17	LUFT, 1988
1-Methyl-3-*n*-propylbenzene	10		0.08–0.56	LUFT, 1988
1-Methyl-2-isopropylbenzene	10		0.01–0.12	LUFT, 1988
1-Methyl-3-*t*-butylbenzene	11		0.03–0.11	LUFT, 1988
1-Methyl-4-*t*-butylbenzene	11	10.92	0.04–0.13	LUFT, 1988
1,2-Dimethyl-3-ethylbenzene	10	10.93	0.02–0.19	LUFT, 1988
1,2-Dimethyl-4-ethylbenzene	10	10.75	0.50–0.73	LUFT, 1988
1,3-Dimethyl-2-ethylbenzene	10	10.81	0.21–0.59	LUFT, 1988
1,3-Dimethyl-4-ethylbenzene	10	10.75	0.03–0.44	LUFT, 1988
1,3-Dimethyl-5-ethylbenzene	10	10.51	0.11–0.42	LUFT, 1988
1,3-Dimethyl-5-*t*-butylbenzene	12	0.02	0.02–0.16	LUFT, 1988
1,4-Dimethyl-2-ethylbenzene	10	10.68	0.05–0.36	LUFT, 1988
1,2,3-Trimethylbenzene	9	10.06	0.21–0.48	LUFT, 1988
1,2,4-Trimehtylbenzene	9	9.84	0.66–3.30	LUFT, 1988
1,3,5-Trimethylbenzene	9	9.62	0.13–1.15	LUFT, 1988
1,2,3,4-Tetramethylbenzene	10	11.57	0.02–0.19	LUFT, 1988
1,2,3,5-Tetramethylbenzene	10	11.09	0.14–1.06	LUFT, 1988
1,2,4,5-Tetramethylbenzene	10	11.05	0.05–0.67	LUFT, 1988
1,2-Diethylbenzene	10	10.52	0.57	LUFT, 1988
1,3-Diethylbenzene	10	10.4	0.05–0.38	LUFT, 1988
n-Propylbenzene	9	9.47	0.08–0.72	LUFT, 1988
Isopropylbenzene	9	9.13	<10.01–0.23	LUFT, 1988
n-Butylbenzene	10	10.5	0.04–0.44	LUFT, 1988
Isobutylbenzene	10	9.96	0.01–0.08	LUFT, 1988
sec-Butylbenzene	10	9.98	0.01–0.13	LUFT, 1988
t-Butylbenzene	10	9.84	0.12	LUFT, 1988
n-Pentylbenzene	11	11.49	0.01–0.14	LUFT, 1988
Isopentylbenzene	11	0.07	0.07–0.17	LUFT, 1988

Naphtheno benzenes				
Indan	9	10.27	0.25–0.34	LUFT, 1988
1-Methylindan	10		0.04–0.17	LUFT, 1988
2-Methylindan	10	11.39	0.02–0.10	LUFT, 1988
4-Methylindan	10	11.33	0.01–0.16	LUFT, 1988
5-Methylindan	10	11.28	0.09–0.30	LUFT, 1988
Tetrailin (tetrahydro-naphthalene)	10	11.7	0.01–0.14	LUFT, 1988
Alkyl naphthalenes				
Naphthalene	10	11.69	0.09–0.49	LUFT, 1988
Polynuclear Aromatics				
Pyrene	16	20.8	Not quantified	LUFT, 1988
Benz[*a*]anthracene	18	26.37	Not quantified	LUFT, 1988
Benz[*a*]pyrene	20	31.34	0.19–2.8 mg/kg	LUFT, 1988
Benz[*e*]pyrene	20	31.17	Not quantified	LUFT, 1988
Benz[*ghi*]perylene	22	34.01	Not quantified	LUFT, 1988

[a] LUFT, 1988 Leaking Underground Fuel Tank Manual: Guidelines for Site Assessment, Cleanup, and Underground Storage Tank Closure, State of California Leaking Underground Fuel Tank Force, May 1988.

Source: J.B. Gustafson, J.G. Tell, and D. Orem, *Selection of Representative TPH Fractions Based on Fate and Transport Considerations—Total Petroleum Hydrocarbon Criteria Working Group Series*, Vol. 3, Amherst, MA: Amherst Scientific Publishers, Appendix A. With permission.

Table 7.6 Detailed Composition of Diesel Fuel

Compound	*Number of Carbons*	*EC*	*Weight Percent*	*Reference*
Straight chain alkanes				
n-Octane	8	8	0.1	BP, 1996[a]
^{n}-Nonane	9	9	0.19–0.49	BP, 1996
n-Decane	10	10	0.28–1.2	BP, 1996
n-Undecane	11	11	0.57–2.3	BP, 1996
n-Dodecane	12	12	1.0–2.5	BP, 1996
n-Tridecane	13	13	1.5–2.8	BP, 1996
n-Tetradecane	14	14	0.61–2.7	BP, 1996
n-Pentadecane	15	15	1.9–3.1	BP, 1996
n-Hexadecane	16	16	1.5–2.8	BP, 1996
n-Heptadecane	17	17	1.4–2.9	BP, 1996
n-Octadecane	18	18	1.2–2.0	BP, 1996
n-Nonadecane	19	19	0.7–1.5	BP, 1996
n-Eicosane	20	20	0.4–1.0	BP, 1996
n-Heneicosane	21	21	0.26–0.83	BP, 1996
n-Docosane	22	22	0.14–0.44	BP, 1996
n-Tetracosane	24	24	0.35	BP, 1996
Branched chain alkanes				
3-Methylundecane	12		0.09–0.28	BP, 1996
2-Methyldodecane	13		0.15–0.52	BP, 1996
3-Methyltridecane	14		0.13–0.30	BP, 1996
2-Methyltetradecane	15		0.34–0.63	BP, 1996
Alkyl benzenes				
Benzene	6	6.5	0.003–0.10	BP, 1996
Toluene	7	7.58	0.007–0.70	BP, 1996
Ethylbenzene	8	8.5	0.007–0.20	BP, 1996
o-Xylene	8	8.81	.001–0.085	BP, 1996
m-Xylene	8	8.6	0.018–0.512	BP, 1996
p-Xylene	8	8.61	0.018–0.512	BP, 1996
Styrene	9	8.83	<.002	BP, 1996
1-Methyl-4-isopropyl-benzene	10	10.13	0.003–0.026	BP, 1996
1,3,5-Trimethylbenzene	9	9.62	0.09–0.24	BP, 1996
n-Propylbenzene	9	9.47	0.03–0.048	BP, 1996
Isopropylbenzene	9	9.13	<0.01	BP, 1996
n-Butylbenzene	10	10.5	0.031–0.046	BP, 1996
Biphenyl	12		0.01–0.12	BP, 1996

Naphtheno benzenes				
Fluorene	13	16.55	0.034–0.15	BP, 1996
Fluoranthene	16	21.85	0.0000007–0.02	BP, 1996
Benz[*b*]fluoranthene	20	30.14	0.0000003–0.000194	BP, 1996
Benz[*k*]fluoranthene	20	30.14	0.0000003–0.000195	BP, 1996
Inden[1,2,3-*cd*]pyrene	22	35.01	0.000001–0.000097	BP, 1996
Alkyl naphthalenes				
Naphthalene	10	11.69	.01–0.80	BP, 1996
1-Methylnaphthalene	11	12.99	0.001–0.81	BP, 1996
2-Methylnaphthalene	11	12.84	0.001–1.49	BP, 1996
1,3-Dimethylnaphthalene	12	14.77	0.55–1.28	BP, 1996
1,4-Dimethylnaphthalene	12	14.6	0.110–0.23	BP, 1996
1,5-Dimethylnaphthalene	12	13.87	0.16–0.36	BP, 1996
Polynuclear aromatics				
Anthracene	14	19.43	0.000003–0.02	BP, 1996
2-Methylanthracene	15	20.73	0.000015–0.018	BP, 1996
Phenanthrene	14	19.36	0.000027–0.30	BP, 1996
1-Methylphenanthrene	15	20.73	0.000011–0.024	BP, 1996
2-Methylphenanthrene	15		0.014–0.18	BP, 1996
3-Methylphenanthrene	15		0.000013–0.011	BP, 1996
4 & 9-Methylphenanthrene	15		0.00001–0.034	BP, 1996
Pyrene	16	20.8	0.000018–0.015	BP, 1996
1-Methylpyrene	17		0.000002–0.00137	BP, 1996
2-Methylpyrene	17		0.0000037–0.00106	BP, 1996
Benz[*a*]anthracene	18	26.37	0.0000021–0.00067	BP, 1996
Chrysene	18	27.41	0.000045	BP, 1996
Triphenylene	18	26.61	0.00033	BP, 1996

(*continues*)

Table 7.6 (*Continued*)

Compound	*Number of Carbons*	*EC*	*Weight Percent*	*Reference*
Cyclopenta[*cd*]pyrene	18		0.000002–0.0000365	BP, 1996
1-Methyl-7-isopropyl-phenanthrene	18		0.0000015–0.00399	BP, 1996
3-Methylchrysene	19		<0.001	BP, 1996
6-Methylchrysene	19		<0.0005	BP, 1996
Benz[*a*]pyrene	20	31.34	0.000005–0.00084	BP, 1996
Benz[*e*]pyrene	20	31.17	0.0000054–0.000240	BP, 1996
Perylene	20	31.34	<0.0001	BP, 1996
Benz[*ghi*]perylene	22	34.01	0.0000009–0.00004	BP, 1996
picene	22		0.0000004–0.000083	BP, 1996

[a] BP, 1996 Summary tables of laboratory analysis for diesel and fuel oil #2, personal communication from B. Albertson, Friedman & Bruya, Inc., Seattle, WA, developed for British Petroleum.

Source: J. B. Gustafson, J. G. Tell, and D. Orem, *Selection of Representative TPH Fractions Based on Fate and Transport Considerations—Total Petroleum Hydrocarbon Criteria Working Group Series*, Vol. 3, Amherst, MA: Amherst Scientific Publishers, Appendix A. With permission.

Table 7.7 Detailed Composition of Fuel Oil #2

Compound	*Number of Carbons*	*EC*	*Weight Percent*	*Reference*
Straight chain alkanes				
n-Octane	8	8	0.1	BP, 1996[a]
n-Nonane	9	9	0.20–0.30	BP, 1996
n-Decane	10	10	0.5	BP, 1996
n-Undecane	11	11	0.80–0.90	BP, 1996
n-Dodecane	12	12	0.84–1.20	BP, 1996
n-Tridecane	13	13	0.96–2.00	BP, 1996
n-Tetradecane	14	14	1.03–2.50	BP, 1996
n-Pentadecane	15	15	1.13–3.20	BP, 1996
n-Hexadecane	16	16	1.05–3.30	BP, 1996
n-Heptadecane	17	17	0.65–3.60	BP, 1996
n-Octadecane	18	18	0.55–2.50	BP, 1996
n-Nonadecane	19	19	0.33–1.30	BP, 1996
n-Eicosane	20	20	0.18–0.60	BP, 1996
n-Heneicosane	21	21	0.09–0.40	BP, 1996
n-Docosane	22	22	0.1	BP, 1996
Alkyl benzenes				
Benzene	6	6.5	<0.125	BP, 1996
Toluene	7	7.58	0.025–0.110	BP, 1996
Ethylbenzene	8	8.5	0.028–0.04	BP, 1996
Biphenyl	12		0006–0.009	BP, 1996
Naphtheno benzenes				
Acenaphthene	12	15.5	0.013–0.022	BP, 1996
Acenaphthylene	12	15.06	0.006	BP, 1996
Fluorene	13	16.55	0.004–0.045	BP, 1996
Fluoranthene	16	21.85	0.000047–0.00037	BP, 1996
2,3-Benzofluorene	17	23.83	<0.0024	BP, 1996
Benzo[a]fluorene	17		<0.0006	BP, 1996
Benzo[ghi]fluoranthene	18		<0.0024	BP, 1996
Benz[b]fluoranthene	20	30.14	<0.0024	BP, 1996
Benz[k]fluoranthene	20	30.14	<0.00006	BP, 1996
Indeno[1,2,3-cd]pyrene	22	35.01	<0.0012	BP, 1996
Alkyl naphthalenes				
Naphthalene	10	11.69	0.009–0.40	BP, 1996
1-Methylnaphthalene	11	12.99	0.29–0.48	BP, 1996
2-Methylnaphthalene	11	12.84	0.36–1.00	BP, 1996
1,4-Dimethylnaphthalene	12	14.6	0.043–0.045	BP, 1996

(*continues*)

Table 7.7 (*Continued*)

Compound	*Number of Carbons*	*EC*	*Weight Percent*	*Reference*
Polynuclear aromatics				
Anthracene	14	19.43	0.00010–0.011	BP, 1996
2-Methylanthracene	15	20.73	0.009–0.017	BP, 1996
9,10-Dimethylanthracene	16		0.002–0.006	BP, 1996
Phenanthrene	14	19.36	0.009–0.170	BP, 1996
1-Methylphenanthrene	15	20.73	0.017	BP, 1996
2-Methylphenanthrene	15		0.768	BP, 1996
Pyrene	16	20.8	0.00–0.012	BP, 1996
Benz[a]anthracene	18	26.37	0.000002–0.00012	BP, 1996
Chrysene	18	27.41	0.000037–0.00039	BP, 1996
Triphenylene	18	26.61	0.00002–0.00014	BP, 1996
Benzo[b]chrysene	19		<0.0036	BP, 1996
Benz[a]pyrene	20	31.34	0.000001–0.000060	BP, 1996
Benz[e]pyrene	20	31.17	0.0000020–0.000010	BP, 1996
Benzo[ghi]pyrene	20	31.17	0.0000010–0.0000070	BP, 1996
Perylene	20	31.34	<0.0024	BP, 1996
3-Methylcholanthrene	21		<0.00006	BP, 1996
Benz[ghi]perylene	22	34.01	0.0000057	BP, 1996
Picene	22		<0.00012	BP, 1996
Coronene	24	34.01	<0.000024	BP, 1996

[a]BP, 1996 Summary tables of laboratory analysis for diesel and fuel oil #2, personal communication from B. Albertson, Friedman & Bruya, Inc., Seattle, WA, developed for British Petroleum.

Source: J. B. Gustafson, J. G. Tell, and D. Orem, *Selection of Representative TPH Fractions Based on Fate and Transport Considerations—Total Petroleum Hydrocarbon Criteria Working Group Series*, Vol. 3, Amherst MA: Amherst Scientific Publishers, Appendix A. With permission.

Table 7.8 Detailed Composition of JP-4

Compound	*Number of Carbons*	*EC*	*Weight Percent*	*Reference*
Straight chain alkanes				
n-Butane	4	4	0.12	API, 1993[a]
n-Pentane	5	5	1.06	API, 1993
n-Hexane	6	6	2.21	API, 1993
n-Heptane	7	7	3.67	API, 1993
n-Octane	8	8	3.8	API, 1993
n-Nonane	9	9	2.25	API, 1993
n-Decane	10	10	2.16	API, 1993
n-Undecane	11	11	2.32	API, 1993
n-Dodecane	12	12	2	API, 1993
n-Tridecane	13	13	1.52	API, 1993
n-Tetradecane	14	14	0.73	API, 1993
Branched chain alkanes				
Isobutane	4	3.67	0.66	API, 1993
2,2-Dimethylbutane	6	5.37	0.1	API, 1993
2,2,3,3-Tetramethylbutane	8	7.3	0.24	API, 1993
2-Methylpentane	6	5.72	1.28	API, 1993
3-Methylpentane	6	5.85	0.89	API, 1993
2,2-Dimethylpentane	7	6.25	0.25	API, 1993
2-Methylhexane	7	6.68	2.35	API, 1993
3-Methylhexane	7	6.76	1.97	API, 1993
2,2-Dimethylhexane	8	7.25	0.71	API, 1993
2,4-Dimethylhexane	8	7.38	0.58	API, 1993
2,5-Dimethylhexane	8	7.36	0.37	API, 1993
3,3-Dimethylhexane	8	7.45	0.26	API, 1993
2-Methylheptane	8	7.71	2.7	API, 1993
3-Methylheptane	8	7.78	3.04	API, 1993
4-Methylheptane	8	7.72	0.92	API, 1993
2,4-Dimethylheptane	9	8.34	0.43	API, 1993
2,5-Dimethylheptane	9	8.47	0.52	API, 1993
4-Ethylheptane	9	8.69	0.18	API, 1993
2-Methyloctane	9		0.88	API, 1993
3-Methyloctane	9	8.78	0.79	API, 1993
4-Methyloctane	9	8.71	0.86	API, 1993
2-Methylundecane	12		0.64	API, 1993
2,6-Dimethylundecane	13		0.71	API, 1993

(*continues*)

Table 7.8 (*Continued*)

Compound	*Number of Carbons*	*EC*	*Weight Percent*	*Reference*
Cycloalkanes				
Methylcyclopentane	6	6.27	1.16	API, 1993
1-*cis*-2-Dimethylcyclopentane	7	7.21	0.54	API, 1993
1-*cis*-3-Dimethylcyclopentane	7	6.82	0.34	API, 1993
1-*trans*-3-Dimethylcyclopentane	7	6.85	0.36	API, 1993
Ethylcyclopentane	7	7.34	0.26	API, 1993
1-*cis*-3-Dimethylcyclohexane	8	7.75	0.42	API, 1993
Cyclohexane	6	6.59	1.24	API, 1993
Methylcyclohexane	7	7.22	2.27	API, 1993
1-Methyl-2-ethylcyclohexane	9		0.39	API, 1993
1-Methyl-3-ethylcyclohexane	9		0.17	API, 1993
1,3,5-Trimethylcyclohexane	9		0.99	API, 1993
1,1,3-Trimethylcyclohexane	9	8.45	0.48	API, 1993
n-Butylcyclohexane	10		0.7	API, 1993
Alkyl benzenes				
Benzene	6	6.5	0.5	API, 1993
Toluene	7	7.58	1.33	API, 1993
Ethylbenzene	8	8.5	0.37	API, 1993
o-Xylene	8	8.81	1.01	API, 1993
m-Xylene	8	8.6	0.96	API, 1993
p-Xylene	8	8.61	0.35	API, 1993
1-Methyl-4-ethylbenzene	9	9.57	0.43	API, 1993
1-Methyl-2-ethylbenzene	9	9.71	0.23	API, 1993
1-Methyl-3-ethylbenzene	9	9.55	0.49	API, 1993
1-Methyl-2-isopropylbenzene	10		0.29	API, 1993
1,2-Dimethyl-4-ethylbenzene	10	10.75	0.77	API, 1993
1,3-Dimethyl-5-ethylbenzene	10	10.51	0.61	API, 1993
1,4-Dimethyl-2-ethylbenzene	10	10.68	0.7	API, 1993
1,2,4-Trimethylbenzene	9	9.84	1.01	API, 1993
1,3,5-Trimethylbenzene	9	9.62	0.42	API, 1993
1,3-Diethylbenzene	10	10.4	0.46	API, 1993
n-Propylbenzene	9	9.47	0.71	API, 1993
Isopropylbenzene	9	9.13	0.3	API, 1993

Alkyl naphthalenes				
Naphthalene	10	11.69	0.5	API, 1993
1-Methylnaphthalene	11	12.99	0.78	API, 1993
2-Methylnaphthalene	11	12.84	0.56	API, 1993
2,6-Dimethylnaphthalene	12	14.6	0.25	API, 1993

[a] API, 1993 Petroleum Product Surveys, American Petroleum Institute, Washington, DC.

Source: J. B. Gustafson, J. G. Tell, and D. Orem, *Selection of Representative TPH Fractions Based on Fate and Transport Considerations—Total Petroleum Hydrocarbon Criteria Working Group Series*, Vol. 3, Amherst, MA: Amherst Scientific Publishers, Appendix A. With permission.

Table 7.9 Detailed Composition of JP-5

Compound	*Number of Carbons*	*EC*	*Weight Percent*	*Reference*
Straight chain alkanes				
n-Octane	8	8	0.12	API,1993[a]
n-Nonane	9	9	0.38	API, 1993
n-Decane	10	10	1.79	API, 1993
n-Undecane	11	11	3.95	API, 1993
n-Dodecane	12	12	3.94	API, 1993
n-Tridecane	13	13	3.45	API, 1993
n-Tetradecane	14	14	2.72	API, 1993
n-Pentadecane	15	15	1.67	API, 1993
n-Hexadecane	16	16	1.07	API, 1993
n-Heptadecane	17	17	0.12	API, 1993
Branched chain alkanes				
2,4,6-Trimethylheptane	10		0.07	API, 1993
3-Methyloctane	9	8.78	0.07	API, 1993
4-Methyldecane	11		0.78	API, 1993
2-Methyldecane	11		0.61	API, 1993
2,6-Dimethyldecane	12		0.72	API, 1993
2-Methylundecane	12		1.39	API, 1993
2,6-Dimethylundecane	13		2	API, 1993

(continues)

Table 7.9 (*Continued*)

Compound	*Number of Carbons*	*EC*	*Weight Percent*	*Reference*
Cycloalkanes				
1-Methyl-4-ethylcyclohexane	9		0.48	API, 1993
1,3,5-Trimethylcyclohexane	9		0.09	API, 1993
1,1,3-Trimethylcyclohexane	9	8.45	0.05	API, 1993
n-Butylcyclohexane	10		0.9	API, 1993
Heptylcyclohexane	13		0.99	API, 1993
Straight chain alkenes				
Tridecene	13		0.45	API, 1993
Alkyl benzenes				
o-Xylene	8	8.81	0.09	API, 1993
m-Xylene	8	8.6	0.13	API, 1993
1,2,4-Trimethylbenzene	9	9.84	0.37	API, 1993
1,3-Diethylbenzene	10	10.4	0.61	API, 1993
1,4-Diethylbenzene	10	10.46	0.77	API, 1993
1,2,4-Triethylbenzene	12	12.29	0.72	API, 1993
1-*t*-Butyl-3,4,5-trimethyl-benzene	13		0.24	API, 1993
n-Heptylbenzene	13		0.27	API, 1993
n-Octylbenzene	14		0.78	API, 1993
Biphenyl	12		0.7	API, 1993
Phenylcyclohexane	12		0.82	API, 1993
Alkyl naphthalenes				
Naphthalene	10	11.69	0.57	API, 1993
1-Methylnaphthalene	11	12.99	1.44	API, 1993
2-Methylnaphthalene	11	12.84	1.38	API, 1993
2,3-Dimethylnaphthalene	12	15	0.46	API, 1993
2,6-Dimethylnaphthalene	12	14.6	1.12	API, 1993
1-Ethylnaphthalene	12	14.41	0.32	API, 1993

[a] API, 1993 Petroleum Product Surveys, American Petroleum Institute, Washington, DC.

Source: J. B. Gustafson, J. G. Tell, and D. Orem, *Selection of Representative TPH Fractions Based on Fate and Transport Considerations—Total Petroleum Hydrocarbon Criteria Working Group Series*, Vol. 3, Amherst, MA: Amherst Scientific Publishers, Appendix A. With permission.

Table 7.10 Detailed Composition of JP-8

Compound	*Number of Carbons*	*EC*	*Weight Percent*	*Reference*
Straight chain alkanes				
n-Heptane	7	7	0.03	API, 1993[a]
n-Octane	8	8	0.9	API, 1993
n-Nonane	9	9	0.31	API, 1993
n-Decane	10	10	1.31	API, 1993
n-Undecane	11	11	4.13	API, 1993
n-Dodecane	12	12	4.72	API, 1993
n-Tridecane	13	13	4.43	API, 1993
n-Tetradecane	14	14	2.99	API, 1993
n-Pentadecane	15	15	1.61	API, 1993
n-Hexadecane	16	16	0.45	API, 1993
n-Heptadecane	17	17	0.08	API, 1993
n-Octadecane	18	18	0.02	API, 1993
Branched chain alkanes				
2,4,6-Trimethylheptane	10		0.07	API, 1993
3-Methyloctane	9	8.78	0.04	API, 1993
2-Methyldecane	11		0.41	API, 1993
2,6-Dimethyldecane	12		0.66	API, 1993
2-Methylundecane	12		1.16	API, 1993
2,6-Dimethylundecane	13		2.06	API, 1993
Cycloalkanes				
1-Methyl-4-ethylcyclohexane	9		0.1	API, 1993
1,3,5-Trimethylcyclohexane	9		0.06	API, 1993
1,1,3-Trimehtylcyclohexane	9	8.45	0.06	API, 1993
n-Butylcyclohexane	10		0.74	API, 1993
n-Propylcyclohexane	9		0.14	API, 1993
Hexylcyclohexane	12		0.93	API, 1993
Heptylcyclohexane	13		1	API, 1993
Straight chain alkenes				
Tridecane	14		0.73	API, 1993
Alkyl benzenes				
o-Xylene	8	8.81	0.06	API, 1993
m-Xylene	8	8.6	0.06	API, 1993
1-Methyl-2-isopropylbenzene	10		0.56	API, 1993
1,3-Dimethyl-5-ethylbenzene	10	10.51	0.62	API, 1993
1,2,4-Trimethylbenzene	9	9.84	0.27	API, 1993

(continues)

Table 7.10 (*Continued*)

Compound	*Number of Carbons*	*EC*	*Weight Percent*	*Reference*
1,2,4-Triethylbenzene	12	12.29	0.99	API, 1993
1,3,5-Triethylbenzene	12	12.1	0.6	API, 1993
n-Heptylbenzene	13		0.25	API, 1993
n-Octylbenzene	14		0.61	API, 1993
Biphenyl	12		0.63	API, 1993
Phenylcyclohexane	12		0.87	API, 1993
Alkyl naphthalenes				
Naphthalene	10	11.69	1.14	API, 1993
1-Methylnaphthalene	11	12.99	1.84	API, 1993
2-Methylnaphthalene	11	12.84	1.46	API, 1993
2,3-Dimethylnaphthalene	12	15	0.36	API, 1993
2,6-Dimethylnaphthalene	12	14.6	1.34	API, 1993
1-Ethylnaphthalene	12	14.41	0.33	API, 1993

[a]API, 1993 Petroleum Product Surveys, American Petroleum institute, Washington, DC.

Source: J. B. Gustafson, J. G. Tell, and D. Orem, *Selection of Representative TPH Fractions Based on Fate and Transport Considerations—Total Petroleum Hydrocarbon Criteria Working Group Series*, Vol. 3, Amherst MA: Amherst Scientific Publishers, Appendix A. With permission.

of carbon from renewable resources. Intermediate activity will indicate the incorporation of new carbon into old carbon, possibly through biological processes.

The use of isotopic discrimination has been generally limited to the analysis of carbon or low-molecular-weight compounds. The recent development of ICP/MS analytical systems has greatly expanded the ability to measure isotopes of heavy elements. This method can be used to distinguish metals refined from different ore bodies. A significant amount of this type of work has been conducted with lead due to its toxicological impact on humans [41–43]. This work documents the ability to use the ratio of lead isotopes to identify the source of lead in the environment. Isotopic analysis of lead has been reported to be a useful tool for fingerprinting gasoline [44].

REVIEW OF CHEMICAL FINGERPRINTING EVALUATIONS

A chemical fingerprinting evaluation is basically a story that uses chemical data as part of its story line. When developing a fingerprinting story, it is important to show relevance to the reader or intended audience and to try to maintain their interest in the subject. Chemistry and chemical data are generally very abstract, often taking the form of a table of numbers, a chart, or a graph. Because of the highly abstract nature of chemical data, there is often a challenge to establish relevance between the subject and the intended audience and to maintain their interest.

Relevance can be established by linking chemical data to facts that are pertinent to the story. Story facts can include historical information or something that an individual can imagine touching. These reference points may be a particular person, a newspaper article, a place on a map, a building, or a street. Things like a monitoring well or geographic information system (GIS) coordinate may be somewhat foreign to an individual and may require additional time or extra effort to process. If this occurs too often, there is the potential that the individual will stop making the extra effort to process data, will simply lose interest in the story, or will simply fall behind and lose the story line.

Relevance and interest can be established by linking chemical concepts to everyday situations. Biodegradation processes can be discussed in terms of the eating habits of children. Molecules that are long, skinny, and easy to grab and hold onto are like hot dogs and carrots, which are easy for a small child to eat. Molecules that are round and bulky are like big juicy hamburgers, which are more difficult for a child to hold and eat. Boiling water can be used to establish relevance with a distillation process and the smell of fresh baked bread can be used when discussing evaporation or the movement of organic vapors.

When presenting a fingerprinting story, it is important to understand how individuals process information. Typically, individuals will process data in small packets. Once the information is containerized or understood, then additional information can be processed. This learning process is not linear, nor does it proceed at a constant rate. Information needs to be provided and then time allowed for its assimilation. New information can then be provided only after the previous information has been processed. In addition, review or some type of repetition of the information is usually required. The learning process can be looked at as a series of loops or circles that work from the beginning of the story to its end. It is not linear in time; it skips around, and it does not proceed at a constant rate. It is important to incorporate the requirements of learning into a chemical fingerprinting story.

The chemical fingerprinting story has parts. The beginning of the story is an introduction into the foundation or fundamental ideas or concepts on which the fingerprinting relies. Next, comes a presentation of the findings. To this point, all of the information provided can exist in isolation. The middle of the story establishes relationships or links between the previously isolated information. The final part of the story is the conclusion or summary.

In written form, a chemical fingerprinting evaluation can be readily applied to a learning process. Each paragraph or section can be read, thought about, and reread until the information is assimilated. The relationships can be evaluated, thought about, and checked. If necessary, each and every part of the story can be reread and reprocessed.

A verbal presentation of a chemical fingerprinting evaluation is very different from a written evaluation. A verbal presentation has an added element of time and is restricted by the fact that time is unidirectional. Time forces the story into a linear presentation and establishes a rate at which the story is told. It also prevents the intended audience from reprocessing information when needed. Verbal presentations should compartmentalize information, allow time for assimilation, as well as provide a review of the information. It is easy for technical professionals to forget that their own instruction often occurred over several months, or even years, where periods of instruction (daily lectures) were followed by extended assimilation times. This type of a segmented learning process should be replicated in a verbal presentation of a chemical story.

Review of Fingerprinting Evaluations

The review of a chemical fingerprinting evaluation can be a very complex and difficult task. There are many points that can require examination. The main point to understand is that the fingerprinting evaluation is the story. The story line should contain the bases or technical foundations that are used. It should also contain the reasoning used in reaching any conclusions. The second key point to consider is the reliability of the technical data. Review of this portion of the fingerprinting evaluation is often relatively straightforward. The issues are those that chemists have dealt with over many years, ever since they started to detect and measure chemicals that are, for the most part, invisible. The third key point to consider is whether there is more than one story that can be told.

Main Story Line Review

When reviewing a story line, one should carefully evaluate the technical foundations and arguments used. There is generally some technical basis for the main story line. Even if one disagrees with the technical basis for a fingerprinting evaluation, it is usually difficult to confront such disagreements head-on. Direct confrontation can work well within a technical field where those involved are well versed in the technical jargon used. However, such an approach may appear to be nitpicking to those outside the technical field. An indirect approach that can show the incongruity of a technical assumption with facts known to the intended audience can prove effective.

For example, there is an age-dating model whose authors claim can be used to date the time of a diesel fuel release to within plus or minus 2 years [45]. This article was peer reviewed and published in a widely recognized scientific journal. It uses the ratio of two compounds, normal heptadecane (n-C_{17}) and pristane, to establish the age of a diesel fuel release. One compound, n-C_{17}, is known to be susceptible to degradation, while the other compound, pristane, is recalcitrant to degradation. The model claims that the amount of n-C_{17} will decrease much faster than the amount of pristane. This is what one would expect based on differences in the rates of degradation of these two compounds. The model further claims that by knowing the relative amounts of these two compounds, one can calculate the age of a release.

There are several technical aspects of this age dating model that are the subject of disagreements. First, the model required the use of averaged data from a site rather than the use of data from an individual sample. Here, the technical issue is whether one can use the data from a single sample to determine the age of a diesel fuel release or whether one must use averaged data. Second, the model could not be used with other compounds known to essentially mimic the behavior of n-C_{17} and pristane in the environment. Since the model could not be used with compounds expected to have similar properties, the question of repeatability is raised. Due to the fact that the model successfully underwent peer-review acceptance prior to its publication, anyone arguing against the technical aspects of the model may appear to be isolated and a nonconformist.

The validity of a technical argument can sometimes be brought into question by other approaches. Occasionally, the technical argument can be used to evaluate a data set where the answer is known to the intended audience. The result from the technical argument can then be checked against the expected result to provide a sort of gut check. If the two results do not agree, questions with regards to the technical argument will exist.

With respect to the age-dating model, the authors provided information on the variability of the n-C_{17} and pristane ratio for fresh or undegraded diesel fuel samples. All of these samples had been recently procured and had not been subjected to conditions that would have allowed for degradation. Each would be expected to give an age of 0 years if analyzed by the model. Based on the claims of the model, they should be no more than 2 years old. Actual use of the model on the data from the undegraded samples gave results that indicated ages as old as 16 years, clearly outside the error range claimed by the model. This evaluation raises questions with regards to the specific age dates provided by the model.

Perhaps the most difficult situation when reviewing fingerprinting evaluations occurs when the evaluation is based on faulty logic. Arguments about logic in abstract fields such as chemistry can be overwhelming to nontechnical individuals. One can either see the fault in the logic or not. For example, the following text was used to support a chemical fingerprinting evaluation:

> The EPA began regulatory restrictions on lead in 1975, with reductions in lead to 0.5 g/gal by 1979 and 0.1 g/gal in 1985. By 1991, leaded gasoline production was 3% of gasoline produced. Nearly all of this gasoline was produced for aviation or nonroad uses. Nearly no leaded gasoline was produced for retail sale by 1991. The indications of TEL indicate the gasoline present was from 1985 or earlier.

Dissection of this argument reveals that there are three facts, two statements, and one conclusion provided. The facts provided are: (1) EPA restricted the level of lead in gasoline to 0.5 g/gallon in 1979; (2) EPA restricted the level of lead in gasoline to 0.1 g/gallon in 1985; and (3) by 1991, leaded gasoline production was 3% of the total volume of gasoline produced. These facts can be checked and evaluated to determine their significance. As an example, the EPA lead restrictions were based on the average level of lead present in gasoline manufactured during a 3-month period. The actual level of lead in any two batches of gasoline would vary. Because the lead restrictions referenced were based on an average, they provide no information as to the actual level of lead that may have been present in any single batch of leaded gasoline delivered to the site. Aside from this added information, which may be able to show that the facts provided are somewhat shallow, there is little with which to disagree with regards to the facts provided.

Two statements are also provided with respect to the amount of leaded gasoline that was available for use in cars and trucks. These statements are simply generalizations that are also difficult to disagree with.

The major point of contention is with the conclusion provided. The conclusion appears to be that leaded gasoline could not have been used at the site after 1985. This appears to contradict both the facts and the statements provided by the fingerprinting author. The facts identify that leaded gasoline was available, and presumably available for use at this site, after 1985. The statements provided also support the position that leaded gasoline was potentially available for use at the site after 1985. Nothing has been provided that can support the position that leaded gasoline could not have been delivered to the site after 1985.

Unfortunately, faulty logic is difficult to address from a technical point of view. Either faulty logic is readily apparent to an individual or it is not. Where faulty logic is incorporated within stories designed to be convoluted and purposely confusing, addressing faulty logic on a technical level can be very difficult. Unfortunately, the use of faulty logic seems to be becoming more common.

Technical Data Review

Review of the technical data developed for a fingerprinting evaluation is usually straightforward. Because of the complexity involved in the collection and analysis of environmental samples, there are many potential areas that may be considered. There is the potential impact that the sampling, shipment, and storage may have on the data generated. There are the errors and bias inherent in the testing protocols that were used. A complete technical review would be very time consuming with little assurance that any worthwhile information would be uncovered.

Rather than provide complete reviews, technical reviews can focus on those areas likely to provide useful information. Since most chemical fingerprinting evaluations are based on the presence or absence of specific chemicals, a useful technical review can focus on the presence of false positives and false negatives. Both are possible regardless of the analytical method used, and an in-depth review of these issues can be important. The widespread use of split samples is usually sufficient to identify those situations where false positives or false negatives exist. At other times, a detailed review of the raw analytical data generated by the reporting laboratory may be necessary.

Where quantitative results are used in a fingerprinting evaluation, the areas typically requiring review increase. Once again, the issue of false positives and false negatives exists. In addition, the levels of the chemicals present can be important to examine. If standard testing protocols were followed, there may be little benefit of a detailed review of the actual levels reported. Instead, one might look at how the data are used to determine if the data will support the conclusions provided.

In most fingerprinting evaluations, large data sets are often generated. Their size is suited for processing using statistical approaches such as principal component analysis. This can be very useful to identify trends or subsets of data within large databases. For fingerprinting purposes, it is important to understand the reasons why matches were or were not made. The issues of concern are the variability in the composition of the samples analyzed and the variability in the numerical data provided.

Identifying the source of the variability in the composition of a sample is an important consideration when one is attempting to match two or more samples. Two samples may differ for several reasons. First, they may have come from separate sources, each of which shows a difference in its chemical composition. In addition to source variation, differences can also come from variations in the degradation of a single material over a given area. For example, chemical changes will occur when a chemical product is released into the environment. The chemicals present can evaporate, dissolve into water, or simply react or change via some chemical process such as biodegradation. A problem exists when these processes occur at varying rates at different locations within an area. When a difference in the rate of degradation occurs, a variation will be seen among samples of formerly identical material. In addition, variation can occur in the laboratory where one set of samples is analyzed under slightly different conditions than another set of samples.

When using data comparison techniques to match samples, it is important to carefully evaluate the source of the variation seen. Variability is rarely evaluated in fingerprinting evaluations. Such evaluations can be easy to conduct, but may be time consuming if the data sets are large. The absolute variability in the analytical data can be estimated and the expected range in the absolute values can then be subjected to analysis.

Alternative Solutions

It can be very important to consider the possibilities for multiple solutions to a fingerprinting evaluation from the onset of a fingerprinting process. To best use a fingerprinting process, it should be adopted or modified to fit a scientific evaluation. In a scientific evaluation, facts are reviewed and various hypotheses are proposed to explain the facts as they are known. Scientific tests are then designed to distinguish between the various hypotheses. The data are evaluated to determine if one or more of the hypotheses can be ruled out. Should additional facts become known, additional hypotheses may be proposed, possibly requiring further testing. This process can continue until there is one or possibly several probable hypotheses remaining.

This process of elimination fits well with a chemical fingerprinting evaluation. By identifying the likely hypotheses at the beginning of the investigation, one can better select and design a fingerprinting investigation that will reach a definite conclusion or provide the client with useful information. For example, take a typical situation where chemical fingerprinting is used to try to identify the source of contamination given two or more potential sources. Often, a sample is collected and submitted without any information to maintain "impartiality." The result can be a technical dissertation that is beyond the comprehension of most individuals. If the process is designed along a scientific investigation, the potential sources of the contamination are identified. The tests necessary to distinguish between the various possibilities can be identified and discussed. The test results can then be reported and shown how they either support or rule out the various hypotheses. And the conclusion should be clearly understood. The use of a scientific approach generally fits well within a story line.

The ability to understand the scope of a problem before conducting chemical tests also has the advantage that tests can be designed to help the client understand the significance of various chemical findings or tests. Chemical contamination is a very powerful and scary concept for most individuals. It invokes many fears due to the unknown or invisible nature of the potential threat. A chemical fingerprinting evaluation can help ex-

Table 7.11 TPH Testing Results from the Analysis of Natural Materials Analyses Performed Using GC/FID

	TPH as Gasoline (ppm)	*TPH as Diesel (ppm)*
Spinach	<10	60
Carrots	<10	10
Orange juice	300	<10
Cedar tree	1400	2200
Pine tree	450	400
Dandelion	<10	140
Daisy	40	40
Moss	<10	<10

plain or put into perspective the chemical results and allow the intended audience to better evaluate a technical issue.

Several years ago, there was a dispute where one of the issues centered around the reported presence of total petroleum hydrocarbons (TPHs). One side claimed that the soils from the site contained unacceptable levels of chemical contamination and justified the relocation of a public utility. The pertinent state health and environmental agencies had adopted a neutral position.

TPH test results are occasionally of concern due to the nonselective nature of the analytical method used. These will detect and measure many compounds, not just petroleum hydrocarbons. In an attempt to educate those individuals who had reviewed the TPH test results, samples of native materials were collected and analyzed. These included mint leaves, sage brush, and cow manure. All of these native materials showed levels of "contamination" that exceeded those that were present in the soils at the site. The presence of TPH in materials that were familiar to the intended audience, hopefully, allowed them to place the TPH results into situations that they were familiar with and understood. A table of TPH results from the analysis of other natural materials is provided in Table 7.11.

Chemical fingerprinting evaluations can be designed into interesting detective stories. By explaining the basic concepts and associations using experiences encountered in our daily lives, an interesting and relevant story can be developed. Once chemistry becomes understandable and commonplace, it can be used to make reasonable and rational decisions.

REFERENCES

1. M. L. Fulz and J. D. DeHann, "Gas Chromatography in Arson and Explosives Analysis," *Gas Chromatography in Forensic Science* (1992); also I. Tebbett and E. Horwood, pp. 109–163 (1992).

2. L. A. Kaine et al., "Use of Ion Chromatography for the Verification of Drug Authenticity," *Journal of Chromatography*, 671: 303–308 (1994).

3. I. E. Gabriel and D. T. Patten, "Establishing a Standard Sonoran Reference Plant and Its Application in Monitoring Industrial and Urban Pollution Throughout the Sonoran Desert," *Environmental Monitoring and Assessment*, 36: 27–43 (1995).

4. A. Nissenbaum and D. Yakir, "Stable Isotope Composition of Amber," *Amber, Resinite, and Fossil Resins*; also K. B. Anderson and J. C. Crelling, Eds., *ACS Symposium Series*, 617: 32–42 (1995).

5. I. E. Gabriel and D. T. Patten, "Establishing a Standard Sonoran Reference Plant and Its Application in Monitoring Industrial and Urban Pollution Throughout the Sonoran Desert," *Environmental Monitoring and Assessment*, 36: 27–43 (1995).

6. B. K. Lavine and D. A. Carlson, "Chemical Fingerprinting of Africanized Honeybees by Gas Chromatography/Pattern Recognition Techniques," *Microchemical Journal*, 42: 121–125 (1990).

7. L. W. Alvarez et al., "Extraterrestrial Cause for the Cretaceous-Tertiary Extinction," *Science*, 208: 1095–1108 (1980).

8. R. D. Morrison, *Environmental Forensics, Principles and Applications*, New York: CRC Press (1999).

9. T. J. Bois II and B. J. Luther, *Groundwater and Soil Contamination*, New York: John Wiley & Sons (1996).

10. L. G. Bruce and G. W. Schmidt, "Hydrocarbon Fingerprinting for Application in Forensic Geology: Review with Case Studies," *AAPG Bulletin*, 78: 1692–1710 (November 1994).

11. D. T. McCarthy et al., "The Use of Hydrocarbon Analyses for Environmental Assessment and Remediation," *Journal of Soil Contamination* 1(3): 197–216 (1992).

12. R. L. Grob, *Modern Practice of Gas Chromatography*, New York: John Wiley & Sons (1995).

13. J. Yinon, and S. Zitrin, *The Analysis of Explosives*, Elmsford, NY: Pergamon Press (1981).

14. EPA SW-846 Method 8318 (September 1994).

15. E. Stahl, *Thin-Layer Chromatography*, New York: Springer-Verlag (1967).

16. G. Davies, *Forensic Science*, ACS Symposium Series, Vol. 13, Washington, DC: American Chemical Society (1975).

17. L. M. Gibbs, "Gasoline Additives—When and Why," *SAE Transactions*, Vol. 99, Paper 902104 (1990).

18. J. E. Bruya and A. J. Friedman, "On-Site Analysis of Petroleum Hydrocarbons Using Thin Layer Chromatography," *Hydrocarbon Contaminated Soils, Volume 1: Remediation Technologies, Regulatory Considerations*, Lewis Publishers (1991).

19. A. Montaser and D. W. Golightly, *Inductively Coupled Plasmas in Analytical Atomic Spectrometry*, New York: VCH Publishers (1992).

20. M. Schoell, "Recent Advances in Petroleum Isotope Geochemistry," *Organic Geochemistry* (1984).

21. P. D. Jenden and I. R. Kaplan, "Origin of Natural Gas in Sacramento Basin, California," *73 AAPG Bulletin*, 431–453 (1989).

22. K. H. Hayes et al., "Compound-Specific Isotopic Analyses: A Novel Tool for Reconstruction of Ancient Biogeochemical Processes," *Organic Geochemistry*, 16: 1115–1128 (1990).

23. N. Tanaka and D. Rye, "Chlorine in the Stratosphere," *Nature*, 353:707 (1991).

24. A. Montaser and D. W. Golightly, *Inductively Coupled Plasmas in Analytical Atomic Spectrometry*, New York: VCH Publishers (1992).

25. M. B. Rabinowitz and G. W. Wetherill, "Identifying Sources of Lead Contamination by Stable Isotope Techniques," *Environmental Science and Technology*, 6(8): 705–709 (1972).

26. V. B. Guthrie, Petroleum Products Handbook, New York: McGraw-Hill (1960).

27. D. Klamann, "Lubricants and Related Products," *Verlag Chemie* (1984).

28. E. L. Marshall and K. Owen, "Motor Gasoline," *The Royal Society of Chemistry* (1995).

29. L. M. Gibbs, "How Gasoline Has Changed," *SAE Transactions*, Section 4, Paper 932828 (1993).

30. L. M. Gibbs, "Gasoline Additives—When and Why," *SAE Transactions*, Vol. 99, Paper 902104 (1990).

31. K. Owen and T. Coley, *Automotive Fuels Reference Book*, Warrendale, PA: Society of Automotive Engineers (1995).

32. *Intraagency Assessment of Oxygenated Fuels*, National Science and Technology Council (1997).

33. J. G. Gustafson et al., *Selection of Representative TPH Fractions Based on Fate and Transport Considerations—Total Petroleum Hydrocarbon Criteria Working Group Series,* Vol. 3, Amherst, MA: Amherst Scientific Publishers, (1997).

34. V. B. Guthrie, *Petroleum Products Handbook*, York, PA: Maple Press Company (1960).

35. J. I. Thornton et al., "The Implications of Refining Operations to the Characterization and Analysis of Arson Accelerants," *Arson Analysis Newsletter*, 1–16 (May 1979) and 1–16 (August, 1979).

36. D. C. Mann, "Comparison of Automotive Gasolines Using Capillary Gas Chromatography II: Limitations of Automotive Gasoline Comparisons in Casework," *Journal of Forensic Science*, 32(3): 616–628 (1987).

37. "Composition of Canadian Summer and Winter Gasolines 1993," CPPI Report No. 94–5, Ottawa: Canadian Petroleum Products Institute (June 1994).

38. I. R. Kaplan and Y. Galprin, "How to Recognize a Hydrocarbon Fuel in the Environment and Estimate Its Age of Release," T. J. Boise II and B. J. Luther, Eds., in *Groundwater and Soil Contamination*, New York: John Wiley & Sons (1996).

39. P. D. Boehm et al., "Application of Petroleum Hydrocarbon Chemical Fingerprinting and Allocation Techniques After the Exxon Valdez Oil Spill," *Marine Pollution Bulletin*, 599–613 (1997).

40. Z. Wang and M. Fengas, "Developments in the Analysis of Petroleum Hydrocarbons in Oils, Petroleum Products and Oil-Spill-Related Environmental Samples by Gas Chromatography," *Journal of Chromatography*, 774: 51–78 (1997).

41. M. B. Rabinowitz and G. W. Wetherill, "Identifying Sources of Lead Contamination by Stable Isotope Techniques," *Environmental Science and Technology*, 6: 705–709 (1972).

42. H. Shirahata et al., "Chromological Variations in Concentrations and Isotopic Compositions of Anthropogenic Atmospheric Lead in Sediments of a Remote Subalpine Pond," *Geochimica et Cosmochimica Acta*, 44: 149–162 (1980).

43. K. J. R. Rossman et al., "Isotopic Evidence to Account for Changes in the Concentration of Lead in Greenland Snow Between 1960 and 1988," *Geochimica et Cosmochimica Acta*, 58: 3265–69 (1994).

44. R. W. Hurst et al., "The Lead Fingerprints of Gasoline Contamination," *Environmental Science and Technology*, 30(7): 304A-307A (1996).

45. L. B. Christensen and T. H. Larson, "Method for Determining the Age of Diesel Oil Spills in the Soil," *Ground Water*, 142–149 (Fall 1993).

Chapter 8

Forensic Applications of Environmental Health Sciences

Paul C. Chrostowski and Sarah A. Foster

The field of environmental health sciences is multidisciplinary in nature. It relies on the application of different scientific disciplines to examine and evaluate the impacts of chemicals on human health, welfare, and the environment. These disciplines include the basic sciences such as chemistry, biology, toxicology, epidemiology, and physiology, as well as the applied sciences such as engineering, industrial hygiene, medicine, statistics, and risk assessment. Each of these sciences contributes to our understanding of environmental health, and each of these sciences has developed along its own evolutionary path.

A variety of scientific disciplines, encompassed within the field of environmental health sciences, are well suited to forensic applications involving environmental toxicants. These scientific disciplines include toxicology, environmental toxicology, and epidemiology. Toxicology is often considered to be the branch of basic science that deals with poisons. Toxicology is both an experimental and observational science that focuses on the disposition and impacts of agents (chemical, physical such as radiation, or biological) on an organism. Experimental data gained from laboratory bioassays and *in vitro* experiments provide much of the foundation for toxicological knowledge. Environmental toxicology is concerned with agents that are of significance to humans or other organisms following environmental exposure. Pesticides, metals, radiation, solvents, and plants are all examples of common environmental toxicants. Epidemiological studies are primarily observational rather than experimental in nature. The foundations of epidemiological knowledge are studies that examine the distribution of disease frequency in human populations, often by comparing the presence of an effect in a group that has been exposed to a toxicant to the presence in a control group that has not been exposed. Important information concerning the toxicity of agents is also derived from clinical medical or veterinary practice case studies or surgical reports. For example, a physician may report on the conditions and outcome of a suicide attempt using an environmental toxicant. Worker health and safety have been important since the time of the Roman Empire. Systematic industrial hygiene studies involving workplace monitoring, biomonitoring, or occupational health have been significant contributors to the environmental health sciences for over 100 years. Consideration of these disciplines is important in forensic studies, in particular because environmental health sciences did not become an identifiable discipline of its own until the late 1970s.

Environmental Health Sciences in Forensic Investigations

Toxicology, one of the important scientific disciplines in forensic environmental health science, is probably the oldest of the forensic sciences. Toxicology substantially predates the practice of the environmental health sciences. Taylor's 1848 treatise [1] on poisons was directed to the practicing forensic toxicologist who was required to give expert testimony in cases of accidental or intentional poisoning. The most common application of forensic toxicology involves the identification of agents that can serve as causative agents in inflicting death or injury on humans or other organisms or in causing property damage. Included among the many other applications of forensic toxicology are regulatory toxicology and urinalysis to test for illicit drug use. The practice of forensic toxicology is also inextricably linked to that of analytical chemistry, often requiring the analysis of chemicals that may exert adverse effects on living organisms. Obviously, if analytical technology does not exist to identify or quantify a potential toxicant, it would not be possible to demonstrate causation.

In many legal cases involving applications of forensic environmental health science, it is important to be able to determine when there was sufficient knowledge concerning an agent to be able to determine that it could pose a threat to human health, welfare, or the environment under appropriate conditions of exposure. Cases ranging from insurance claims to personal injury (toxic tort) to natural resources damage often involve a determination of expected or intended damage. In essence, these determinations answer questions regarding a potentially liable party's knowledge that the introduction of a material in the environment was intended or expected to cause harm.

The Scientific Method in the Environmental Health Sciences

There are many facets to finding answers to questions regarding the potential adverse effects of a substance in the environment; however, one important aspect is a historical probe of knowledge. Because science is evolutionary by nature, the investigator should have a firm understanding of the history of science and the scientific method as it pertains to environmental health science before embarking on an investigation for a particular substance. Further, historical actions should be judged not in the light of current knowledge, but rather by the state of knowledge that prevailed at the time when the actions occurred.

Basic science is characterized by the application of the scientific method. Although much has been and can be written about the scientific method, only a brief summary is necessary for the development of this chapter. Prior knowledge is used to formulate hypotheses that are tested by experiment and observation. A tested hypothesis becomes a theory. If the basis for a theory is replicated and becomes generally accepted, it becomes a scientific law. Taken together, hypotheses, theories, and laws constitute a scientific body of knowledge. There need not be consensus among all scientists regarding components of this body. Rather, general acceptance and application reflect the content of the body of knowledge at a point in time. The continuing dialogue regarding scientific proof of Darwin's evolutionary theories is an example of a theory that has yet to become law; however, these theories have enjoyed general acceptance throughout the scientific community for some time.

Applied science uses the results of basic science to meet other goals. Thus, the physician relies on knowledge of physiology to heal the sick, and an engineer relies on knowledge of physics to build a bridge. In themselves, these are not scientific endeavors; however, they would not be possible without reliance on a body of scientific knowledge. Knowledge of and reliance on basic scientific principles by applied scientists is a strong indicator that a scientific body of knowledge has gained general acceptance.

Toxicology, Dose–Response Relationship, and Epidemiology

Toxicology is a science of antiquity. Early civilizations often classified chemicals taken into the body as "food" or "poison." There was consensus among early Greek natural philosophers (the equivalent of modern scientists), for example, that hemlock was a poison that would cause death if ingested in sufficient amounts. Much of the science of toxicology in the Middle Ages and the Renaissance was concerned with the classification of poisons into various categories. In the 16th century, the controversial Swiss natural philosopher Paracelsus[1] stated that all substances were poisons and that the right dose differentiated the poison from the remedy. This statement has come to be known as the "dose–response relationship."

The dose–response concept did not enjoy immediate acceptance in the field of toxicology. Toxicology was not viewed as a quantitative science until the early 19th century when Orfila published his influential treatise on poisons. Despite Orfila's advocacy of quantitative methods, his work still focused on the categorization of poisons. Toxicology in the mid-19th century explicitly rejected the dose–response relationship. Taylor's 1848 book on poisons [1] criticized the dose–response relationship as arbitrary and adopted the following definition, "A poison is a substance which, when taken internally, is capable of destroying life without acting mechanically on the system." This definition persisted into the 20th century. For example, in a 1926 text [2] for pharmacists and physicians, a poison was defined as "a substance the administration of which is injurious to health." There was some crude recognition of the dose–response relationship at this point as this text further excludes substances that are toxic at doses greater than 50 g from the definition.

Following World War II, there was a rapid development both in quantitative chemical methods and in medicine that fostered development of the concept of the dose–response relationship. During the early 1950s, the prevailing concept in toxicology was that of hazard rather than a dose–

[1] The secular deification of Theophrastus von Hohenheim, who used the alias Paracelsus, by some risk assessors is a relatively recent phenomenon. Far from being a paragon of scientific thought, Paracelsus's work is fairly typical of medicine in the 16th century. His writings are replete with references to cosmology, astrology, and secret remedies. He particularly disdained the sciences of anatomy and physiology, preferring to rely on inductive reasoning. A contemporary rumor alleging that Paracelsus was forced to leave Basel because he poisoned a patient has never been substantiated, but may indicate that the distinction between poison and pharmaceutical dose was not clear in his mind [3].

response relationship. This concept was not limited to esoteric discussions among scientists, but was a key element of national public policy. In the 1950s, the U.S. House of Representatives held hearings and debates on the issue of chemicals used as food additives. This activity started in 1950 and culminated in 1958 with passage of the Delaney Amendment to the Food, Drug, and Cosmetic Act (21 USC 348(c)(3)(A) and 376(b)(5)(B)), which stated that no chemical shown to cause cancer in humans or animals could be added to cosmetics or food. Thus, the critical early question regarding the state of knowledge relates to whether a particular substance would be regarded as a poison, or from the Delaney perspective, as a carcinogen.

In the mid-1970s, the U.S. Food and Drug Administration formalized the modern practice of chemical risk assessment. This practice postulated the existence of a toxicological threshold. A toxicological threshold is defined as that point on the dose–response continuum where the presence of an effect can be distinguished from the absence of an effect. If the dose of the substance was below the threshold, the probability of an adverse health effect was considered to be negligible or even zero. Doses above the threshold were assumed to result in a health effect. In 1983, the practice of risk assessment was codified by an influential work of the National Research Council. This document advocated a clear distinction between the concept of hazard (characterization as a poison) and the dose–response relationship.[2] In 1986, the EPA published its *Superfund Public Health Evaluation Manual*, which presented detailed instructions for conducting risk assessments for chemicals in the environment.

Epidemiology is also an old science. Although Hippocrates expressed the idea that disease may be connected with a person's environment almost 2400 years ago, it was not until the mid-20th century that epidemiology became an identifiable and named discipline. Early evidence of epidemiological science can be traced, however, to the mid-17th century when John Graunt analyzed various features of birth and death data in London. The roots of modern epidemiology did not appear until the early 19th century when William Farr examined vital data to understand problems of public health in England and Wales, and the mid-19th century when John Snow used principles of epidemiology to demonstrate that cholera could be spread by fecal contamination of drinking water. Although epidemiology

[2] It should be noted that this is still controversial, although accepted by many scientists. Environmental activists and many European regulatory agencies have rejected use of a dose–response relationship in environmental health in favor of using the hazard concept.

is primarily observational in nature, some early experimental data also exist, including experiments involving the use of fresh fruit to treat scurvy and experiments with the cowpox vaccination in the late 18th century. Over the course of the Industrial Revolution, epidemiology played an increasingly important role in identifying diseases associated with exposures to toxicants in the workplace.

Questions regarding the aquatic toxicity of industrial wastes that were discharged to water became prominent in the second decade of the 20th century. During the decades that followed, it became routine to perform aquatic toxicity testing on whole wastes and individual components of the wastes. The standard chemical industry position in the mid-1950s was given by the Monsanto Chemical Company. Monsanto noted that when a new product is manufactured, it behooved the manufacturer to determine what effects the waste products could have on the aquatic environment. Specifically, Monsanto advocated developing both human and aquatic life toxicity data for use in water pollution control. In 1957, Monsanto published the results of aquatic toxicity testing on several of its own wastes.

Methods for Investigating Individual Substances

The best way to start an investigation of the historical state of knowledge of an individual substance's toxicity is often to state concisely what is currently known about that substance. This allows one to identify important issues regarding the substance's potential effects on human health, welfare, and the environment and to differentiate between current and historical state of knowledge. When there is controversy concerning the ability of a chemical to cause an adverse effect, it is important that the summary of current knowledge reflects an objective unbiased viewpoint. There are indubitably experts who will testify that one molecule of almost any chemical can cause an adverse effect as well as experts who will testify that all chemicals are basically harmless. The forensic environmental health scientist should determine the mainstream scientific position on a chemical, which is likely to be somewhere between these two extremes. Points of controversy, scientific uncertainties, and data gaps should be clearly identified in presenting the results of the investigation.

There are many sources of contemporary information on toxicology and environmental impacts. Regulatory and public health literature available from the Agency for Toxic Substances and Disease Registry (ATSDR), the

U.S. Environmental Protection Agency (EPA), the International Agency for Research on Cancer (IARC), and the World Health Organization (WHO) is often a good starting place for information. Contemporary monographs such as Chang's *Toxicology of Metals* [4] or Rom's *Environmental and Occupational Medicine* [5] are also good sources for overview statements regarding the current status of knowledge for the toxicity of a particular substance. The investigator should, however, always critically review secondary sources and obtain the original literature when necessary to resolve or understand ongoing scientific controversies. Once the current information has been compiled, the investigator will have a good idea of what to look for in the historical literature. For example, a chemical substance such as asbestos may be viewed as a carcinogen. Thus, the investigator should focus historical searching on dates and facts that answer questions about when that knowledge became generally known, or known to a specific segment of the population.

Literature searching in contemporary science has become a computerized, often routine, endeavor. Although this is often acceptable for current knowledge, it soon breaks down regarding historical knowledge. Most historical scientific literature searching must be performed manually with access to a good history of science library. One notable exception to this statement is the database HISTLINE maintained by the National Library of Medicine's History of Medicine Division. This database contains references to secondary literature dealing with the history of medicine and related fields published since 1964, and is accessible on the Internet (http://igm.nlm.nih.gov/).

The key to searching historical literature is the development of a literature "tree" for a particular topic. Citations available in contemporary sources may be viewed as the trunk of the tree. Each one of them can lead off into "branches" of older published literature, which, in turn, can lead to "leaves" of original publications. In addition to the systematic literature tree approach, it is often helpful to do some supplemental random reading in older significant secondary sources that many forensic environmental health scientists will already have on their bookshelves. For example, the works of Paracelsus, Orfila, Taylor, and Hamilton can often reveal sources that are not found in systematic literature searches.

Prior to the current era, industrial hygiene literature often was the source of documentation of the adverse effects on humans and the environment. Before the environmental health sciences became recognized as a separate discipline, industrial hygienists often had responsibility for all aspects of industrial environmental control [6]. Thus, industrial hygiene should not be overlooked as a source of information, even though the focus of the investigation may not be on the industrial environment. As we will show through case studies, similar statements can be made about the study of adverse side effects from the use of a chemical as a therapeutic agent. Many substances that are now considered to be environmental pollutants were originally developed or applied in medicine, for example, chloroform as an anesthetic or coal tar for the treatment of skin diseases. The medical literature often contains reports of adverse effects noted by physicians or veterinarians then using these materials and should not be overlooked.

Once historical literature has been obtained, it must be critically reviewed. As discussed previously, it is important to judge the quality of historical scientific research by the standards that prevailed at the time it was conducted. For example, *Daphnia magna* became important as an aquatic test organism in the 1940s; however, one would not judge the reliability of a 1940s' bioassay using the current American Society for Testing and Materials (ASTM) standard for performing such tests. The critical review will help to separate mainstream thinking from scientific extremism, in addition to culling out material that clearly could not be judged to be reliable by any scientific standard.

The historical state of knowledge can then be tracked by preparing a formalized chronology of information based on material presented in the remaining secondary and primary literature sources. Various databases that can sort data by date of publication and information presented can be used to assist in this process. Ultimately, the investigator can determine benchmark dates for particular adverse effects.

CASE STUDIES

Two case studies will be presented to illustrate the information gathered by applying the forensic environmental health science methods discussed in this chapter. The case studies involve investigations of human health, environmental health, and welfare (i.e., resource impacts) associated with a naturally occurring substance, the metal chromium, and a family of anthropogenic substances, halogenated organic solvents. As discussed previously, the case studies open with discussions of the contemporary state of knowledge and proceed to a historical investigation.

Chromium

Current Knowledge Regarding Chromium [7]

Chromium was discovered in the late 18th century. Chromium salts became widely used as dyes and tanning agents and as ingredients in the chrome plating of metals. The most significant chrome salts are oxyanions where chromium is in a 3+ (trivalent) or 6+ (hexavalent) valence state.[3] The chemical, physical, and biological properties of chromium depend on both the valence state and the associated cation. Currently, chromium is regulated in the environment and in the workplace. These regulations range from the application of water quality standards to outright bans on the use of hexavalent chromium in cooling towers.

Human Health Effects of Chromium [4,8]

Most salts of trivalent chromium are generally considered to be less toxic to humans than salts of hexavalent chromium. The EPA, ATSDR, and the New Jersey Department of Environmental Protection have summarized the current regulatory toxicology of chromium. When administered by ingestion, chromium(VI) is associated with gastrointestinal effects and local irritation. When administration is by inhalation, chromium(VI) causes lung cancer in addition to local effects on the upper respiratory tract, including nasal perforation. Dermal exposure to chromium(VI) may cause contact sensitization in addition to contact dermatitis.

Environmental Impacts of Chromium [9,10]

Chromium is hazardous to fish, invertebrates, plants, and wildlife. Both chromium(III) and chromium(VI) are acutely toxic to aquatic life at low concentrations. Although chromium(III) is generally less acutely toxic than chromium(VI), the difference for aquatic life is not as striking as it is for human health effects. Chromium(III) and chromium(VI) are also chronically toxic to aquatic life. There is little, if any, difference between the quantitative chronic toxicity of the two forms of chromium to aquatic life. The aquatic toxicity of chromium salts is heavily dependent on water chemistry factors such as pH and hardness.

The toxicity of chromium has also been studied with respect to birds,

[3] Hexavalent chromium, chromium(VI), Cr(VI), and Cr+6 are synonyms. Trivalent chromium, chromium(III), Cr(III), and Cr+3 are also synonyms.

terrestrial wildlife, domestic animals, invertebrates, and plants both in the laboratory and in the field. A variety of both acute and chronic effects have been found, including mortality, teratogenicity, developmental effects, reproductive effects, and growth inhibition. Generally chromium(VI) is more toxic than chromium(III), although there are some exceptions.

Human Welfare (Resource) Impacts of Chromium

There are two primary areas in which chromium impacts human welfare (i.e., resources). Chromium salts have a toxic or inhibitory action on naturally occurring microorganisms that are used in sewage treatment plants to degrade sewage. The introduction of chromium into sanitary or combined sewers has resulted in the dysfunction of sewage treatment plants and the discharge of untreated sewage to surface waters. Chromium salts can also have an impact on groundwater, rendering it unfit to be used as a water source. This effect is largely limited to soluble forms of chromium, especially chromium(VI) compounds.[4] The land disposal of chromium(VI) may result in groundwater becoming colored or exceeding drinking water or groundwater standards through the infiltration of chromium into groundwater from the surface. Chromium(VI) is visible in water at concentrations less than 10 parts per million (ppm).

Historical Knowledge of Chromium

Human Health Chromium was discovered in 1797 by Vauquelin. In the first quarter of the 19th century, Christison classified it as a poison. The first reports of occupational illness (skin lesions) were made in 1826/1827. By 1833, the chromium industry was able to distinguish the toxicity of the various chromium chemical species. The first report of nasal perforation in workers in the United States was in 1852. The last half of the 19th century was marked by many discoveries in chromium science, including associations between chromium exposure and asthma, hypotension, and adenocarcinoma, as well as characteristic lesions of the nose and skin.

In 1900, the United Kingdom instituted the first worker protection regulations that dealt specifically with workers in chromate and dichromate manufacturing plants. The rules included personal protection devices, en-

[4] Most Cr(VI) compounds are highly soluble in water and most Cr(III) compounds are slightly soluble. Although there are some exceptions, as a general rule, soluble chromium may be assumed to be Cr(VI) and insoluble chromium to be Cr(III). Cr(VI) salts are generally yellow or orange in color and Cr(III) salts are generally blue or green in color.

gineering controls, medical monitoring, and personal hygiene. A detection limit for chromium in air as low as 0.74 mg/m^3 was reported in 1902. Between 1900 and 1930, there were numerous reports in the literature of skin problems and nasal perforation associated with exposure to chromium. The U.S. Bureau of Labor Statistics recognized chromium workers as being employed in a hazardous occupation in 1922. "Chromic acid diabetes" was recognized in 1902; chromium-induced nephritis was recognized in 1922; and dermal sensitization was recognized in 1931. It was recognized at this point that women were more sensitive to the effects of chromium than men.

Concerns of chromium toxicity were not confined to worker populations. Chromium was used therapeutically in the late 19th and early 20th centuries and there are several reports of side effects, including death, from dermal or ocular absorption of topical chromium. In the 1930s, there was a substantial amount of concern about exposure of the general public to chromium leaching into milk and other foodstuffs in contact with chrome steels or chrome-plated metals. At this time, the analytical detection limit for chromium in water was 100 parts per billion (ppb).

An explosion of information regarding chromium toxicity marked the 1930s. Papers were published on inhalation bioavailability, pharmacokinetics, and acute toxicity. More important, the first publication linking exposure of workers to chromium and lung cancer appeared in 1935. The concern regarding lung cancer continued through the next decade. In 1948, the first modern occupational epidemiologic study of cancer in chromium workers was published. Elevated incidences were reported for cancers of the lung, digestive tract, oral region, nose, pharynx, and all sites. It was further reported that nasal perforation could not reliably be used as an index of exposure in predicting cancer incidence. At this point, the air analytical detection limit was less than 0.01 mg/m^3. Also, in the 1940s, a maximum allowable concentration for chromium in the workplace of 0.1 mg/m^3 was published.

The number of studies that were published associating airborne exposure to chromium with lung cancer distinguishes the decade of the 1950s. One of the most important studies was published by the U.S. Public Health Service in 1953. This study examined 897 workers from chromate plants. These investigators found that 56.7% of the workers had perforated nasal septa. These investigators further reported an incidence rate for bronchiogenic carcinoma of 1115 per 100,000 compared to 20.8 per 100,000 in the control group. When considering cases where a chromate etiology was supported by histological findings, the rate in chromate workers was 892 per 100,000 compared to 18.6 per 100,000 in the general population.

The concern about chromium worker lung cancer was sufficient to prompt the Ohio Department of Health to conduct a study of chromium in the outdoor air around a manufacturing facility in 1949. These investigators found elevated levels of chromium in the air around the plant out to 10,000 feet from the source. The authors did not draw any public health conclusions from these findings; however, this is an example where knowledge of workplace problems alerted investigators to possible general environmental concerns.

Environmental Impacts The first reports regarding the aquatic toxicity of chromium appeared for fish in 1937 and for invertebrates in 1940 [10]. During the 1940s, there were further investigations of aquatic toxicity. Toxicity thresholds of less than 0.3 ppm sodium dichromate to *Daphnia magna* and 20 ppm for potassium dichromate and chromate in rainbow trout were established. In 1949, the toxicity of chromium to several species of algae was established. A substantial amount of aquatic toxicity research was conducted and published in the 1950s and 1960s. By the mid-1960s, the qualitative or quantitative toxicity of chromium to *Daphnia magna*, *Microregna* spp. (protozoa), *Gammarus pulex, Polycelis nigra*, *Macrocystis pyrifera* (a kelp species), several species of trout, several species of salmon, goldfish, largemouth bass, sea urchin eggs, American oyster, bluegills, sunfish, young eels, sticklebacks, minnows, and *Gambusia affinis* (mosquitofish) had all been reported. Specialized studies were also undertaken in the early 1960s to investigate the pharmacokinetics of chromium in fish, the effect on early life stages, and the comparative toxicity of chromium(III) and chromium(VI).

Welfare Impacts Reports regarding groundwater contamination with chromium first appeared in 1942. By the end of the 1940s, reports had been published concerning chromium contamination of groundwater in New York, California, and Michigan. A report published in 1941 noted that objectionable conditions were produced when wastes containing chromium were discharged to sewers or streams. In 1942, the U.S. Public Health Service (PHS) published drinking water standards that stated, "salts of barium, hexavalent chromium, heavy metal glucosides, or other substances with deleterious physiological effects shall not be allowed in the water supply system." In 1946, the PHS recommended a water quality standard of 0.05 ppm for hexavalent chromium. In 1946 and again in 1948, the International Joint Commission referred to chromium wastes as "highly toxic wastes" in the context of controlling water pollution. In 1952, a review of water quality standards conducted for the California State Water Pollution Board reviewed the potential impacts of chromium in water on its use as a potable source, for livestock watering, and to support aquatic life. The PHS continued its drinking water standard of 0.05 ppm in 1962. In 1963, the California State Water Quality Control Board recommended that total chromium (trivalent and hexavalent) be regulated to 0.05 ppm for water to be used as a domestic supply, 5.0 ppm for water to be used for stock and wildlife watering, 1.0 ppm for the protection of fish, and 0.05 ppm for the protection of other aquatic life. The color threshold for hexavalent chromium in water has been reported between 1.5 and 10 ppm; thus, water with the characteristic color of hexavalent chromium salts would likely exceed water quality guidelines and standards that have been in place since 1946 and continue to this date.

The first reports of the deleterious impacts of chromium on sewage treatment plants appeared in 1939. These reports continued into the 1940s, including specialized research into the effects of chromium on activated sludge and nitrifying bacteria. Contrary to the situations with humans and most aquatic life, chromium(III) was thought to be more toxic to sewage treatment microorganisms than chromium(VI).

Halogenated Organic Solvents

Current Knowledge of Halogenated Organic Solvents [8,11]

Halogenated organic solvents (HOSs) represent a class of chemicals that includes chlorinated aliphatics (including monochloro carbon compounds, alkanes, alkenes, and alkynes), alicyclics, and aromatics. There are numerous members of this class; however, only a few of these will be addressed in this chapter for illustrative purposes.[5] TCE and PCE[6] are the most significant members of this class from an environmental and industrial standpoint. In the natural environment, these materials may transform

[5] In addition to trichloroethylene, tetrachloroethylene, and vinyl chloride, the family of HOSs includes chloroform, carbon tetrachloride, methylene chloride, 1,1,1-trichloroethane (111-TCA) (also known as methyl chloroform), chlorobenzene, among others.

[6] Trichloroethylene, trichloroethene, and TCE are synonyms. Tetrachloroethylene, tetrachloroethene, perchloroethylene, and PCE are synonyms. Vinyl chloride, chloroethene, and VC are synonyms.

into another member of this class, vinyl chloride (VC), which is substantially more persistent and toxic than the parent compounds. HOSs are regulated in the environment and workplace by a variety of programs.

Human Health Impacts of HOSs

The primary human health effects of TCE and PCE exposure are central nervous system (CNS) dysfunction and liver damage. CNS effects include depression, dizziness, headache, vertigo, and behavioral effects. Liver effects include steatosis, centrilobular necrosis, and hepatic failure. Other adverse effects have been noted on mucous membranes, eyes, skin, kidneys, and lungs. TCE and PCE have been found to cause cancer in laboratory animals under certain circumstances; however, their human carcinogenicity is inconclusive. VC, on the other hand, is a known human carcinogen that has been associated with a signature disease, hepatic angiosarcoma, in addition to noncancerous liver damage.

Environmental Impact of HOSs

TCE and PCE are acutely and chronically toxic to plants, invertebrates, and most forms of aquatic life. Generally speaking, PCE is more toxic than TCE. Depending on dose, chronic effects include narcosis and liver dysfunction in fish.

Welfare Impacts of HOSs

HOSs are both volatile and mobile in soils and the subsurface environment. Because of their properties, their volume of production, and historical disposal practices, they represent the most significant source of groundwater contamination in the United States. Although the extent of the problem has not been fully defined, over 10% of U.S. groundwater supplies may contain detectable levels of HOSs (excluding trihalomethanes that are introduced through water purification). Reductive dechlorination of PCE and TCE results in the formation of VC in the subsurface. Under appropriate conditions, TCE and PCE may decompose to yield hydrogen chloride and/or phosgene. HOSs also participate in large-scale atmospheric pollution phenomena by having a deleterious impact on the atmospheric ozone layer.

HOSs exert a biological oxygen demand when discharged to surface waters. They have relatively low odor thresholds and can impart taste and odor to potable water supplies, rendering them unfit for use.

Historical Knowledge of HOSs

TCE was first synthesized by Fisher in 1864. It was first commercially produced in Europe in 1908 and in the United States in 1925. Production of TCE was about 12 million pounds per year in 1935, climbing to a peak of 610.8 million pounds per year in 1970, after which it declined dramatically. PCE was first synthesized by Faraday in 1821. By 1940, production of PCE in the United States had reached 12 million pounds per year. PCE production also peaked in 1970. The primary uses of PCE and TCE were for degreasing metals, dry cleaning clothes, and other applications involving dissolution of hydrocarbons and other forms of grease.

Human Health Effects The first academic report of the mammalian toxicity of chlorinated solvents was Behr's dissertation, published in 1903. This was followed by studies reported in the dissertations of Quadflieg (1908), Franz (1909), Herrmann (1910), and Knoblauch (1910). The first industrial reports of the toxicity of TCE occurred in 1915 when an acute toxic syndrome was noted. In 1925, TCE was proposed for use as an antihelmintic; however, reservations were expressed related to its potential for hepatotoxicity by analogy to carbon tetrachloride.

Most aspects of the toxicology of TCE were established in the 1930s. In 1932, the first extensive medical study of industrial health effects from TCE was published. In 1937, von Oettingen reviewed the industrial toxicology of HOSs. In addition to a discussion of TCE and PCE, he also described the toxicity of 1,1,1-trichloroethane, ethylene dichloride, and vinyl chloride. Adverse effects to the central nervous system, gastrointestinal system, and circulatory system were observed. Trichloroacetic acid (TCA) was identified as a metabolite of TCE in 1939.

Due to its narcotic action, TCE was introduced as a medical anesthetic in 1935,[7] which allowed investigation of its toxicity on a diverse population. For example, in 1944, cases of cranial nerve injuries, including cranial nerve palsy associated with the use of TCE in anesthesia, were published. The acute toxicity of TCE in the industrial environment is well known. In the United Kingdom, for example, there were 414 cases of TCE gassing reported between 1941 and 1961; 33 of these cases were fatal. The pathology of these deaths was undiagnosed in many cases. The death of a worker from TCE exposure in a dry-cleaning establishment was reported in 1951. In 1955, a case of death from massive liver necrosis associated

[7] Some authorities give this date as 1941.

with occupational exposure to TCE was reported. The cardiovascular toxicity of TCE was also reported in 1955. TCE has been recognized as a substance with the potential for intentional abuse. A case of fatal addiction to TCE was reported in 1963. Cases of hepatorenal toxicity associated with intentional self-administration of TCE occurred in 1968 and 1969.

Carpenter and co-workers reported the chronic toxicity of PCE to laboratory animals in 1937. These investigators found a lowest observed adverse effect level of 230 ppm in rats by inhalation. The most sensitive target organ was the kidney. Despite some controversy in the 1940s over the toxicity of PCE, the maximum allowable air concentration for the workplace was reduced from 200 ppm to 100 ppm in 1947. Also in 1947, a case of death due to the administration of PCE as an antihelmintic was reported.

The chlorinated ethenes, including TCE and PCE, have long been known to be chemically unstable under a variety of conditions both in the workplace and in the environment. In 1937, von Oettingen noted that TCE could decompose to form phosgene, dichloroacetylene, hexachlorobenzene, chlorine, and/or hydrochloric acid when heated. Also in 1937, McNally reported a case of occupational illness due to phosgene resulting from the decomposition of TCE. Phosgene is a highly toxic substance that was used as a chemical warfare agent in World War I. As noted previously, it was known in the 1930s that TCE could be biologically transformed into TCA. The adverse effects associated with exposure to dichloroacetylene and/or phosgene in TCE were reported in 1960.

The question of the human carcinogenicity of the HOSs has been much debated. Experiments with rats reported in 1971 clearly showed that VC had the ability to induce a variety of cancers.[8] In the mid-1970s, VC was identified as an etiologic agent for hepatic angiosarcoma in humans. This cancer is sufficiently rare and the association with inhaled VC exposure sufficiently strong that it is often thought of as a signature disease. The association is sufficiently strong that researchers have devoted a considerable effort to determining if other chlorinated ethenes are human carcinogens. In 1975, the National Cancer Institute issued a widely disseminated memorandum of alert concerning the possible carcinogenicity of TCE. The International Agency for Research on Cancer (IARC) first considered the case of TCE in 1976. The data at this point were highly equiv-

[8] Prior to 1970, although not associated with cancer, VC was not thought to be harmless. Early reports identified liver damage, narcosis, and acro-osteolysis as effects of VC exposure.

ocal. Based on additional evidence, the IARC reconsidered TCE in 1979. At that time, the IARC concluded that there was limited evidence that TCE was carcinogenic in mice. The IARC made a similar conclusion regarding PCE in 1979. In 1978, the National Institute for Occupational Safety and Health (NIOSH) recommended that PCE be handled in the workplace as if it were a human carcinogen.

Environmental Impacts Many HOSs, including TCE and PCE, were formerly used as insecticides.[9] Since insecticides are designed to have environmental toxicity, the potential for adverse effects to the environment is not surprising.

The toxicity of TCE and PCE to insects was first reported in 1954. The first reports of the toxicity of TCE to fish and daphnia were published in 1956. The toxicity of ethylene dichloride to fish was reported in 1957. The phytotoxicity of TCE (to beans, corn, cotton, cucumber, and tomato) was also reported in 1956. The acute and chronic aquatic toxicity and bioaccumulation potential of the HOSs for a wide variety of species was fully established by the mid- to late 1970s.

Resource Impacts HOSs are among the most prevalent of toxic groundwater contaminants. Contamination of groundwater with chemicals has been recognized as a problem in the United States since the first decade of the 20th century. Concerns over this problem expressed themselves as industrial standard practices. For example, in 1948, the Manufacturing Chemists Association (MCA) recommended disposal of TCE and PCE by evaporation. Despite these precautions, in 1949 Lyne and McLachlan reported two cases of TCE contamination of groundwater, relying on a detection limit of 1 ppm. At a 1961 symposium on groundwater contamination held by the U.S. Public Health Service, it was noted that TCE contamination of groundwater had occurred within the last 3 years. PCE was reported as a groundwater contaminant in 1974.

Biochemical oxygen demand (BOD) is a technique for measuring aggregate organic constituents in water and wastewater. The effect of BOD is to depress oxygen in natural waters, which, in turn, may have an adverse impact on aquatic life or other beneficial uses. The concept of BOD was originally developed in the United Kingdom in the 19th century and

[9] Methyl bromide, a member of this family, is currently used as an insecticide (fumigant), although this use is controversial.

technically refined to close to its present form in the 1930s in the United States. In the early 1950s, states and other jurisdictions (e.g., water compacts and commissions) began using BOD as an environmental standard for organic chemicals. BOD limitations were reviewed in the 1963 Water Quality Criteria published by the California State Water Quality Control Board. The HOSs all exert a BOD when discharged into water.

The first drinking water standard regarding HOSs was published by the PHS in 1962. In this standard, the PHS limited the carbon chloroform extract (CCE) of potential drinking water to 200 mg/L. CCE is an analytical parameter, developed by Ettinger in 1960, for measuring organic toxic chemicals in water. TCE and PCE are included in the CCE parameter. The California State Water Quality Control Board designated TCE, and TCA, a metabolite of TCE, as potential pollutants in 1961. TCE and PCE were designated as priority pollutants under the Clean Water Act in 1976, and the EPA proposed ambient water quality criteria for TCE and PCE in 1979.

The 1970s were a period of increasing knowledge about the extent of groundwater pollution by HOSs. By the end of this decade, the nature and extent of TCE, PCE, and VC contamination of groundwater throughout the United States was well defined. By 1980, it was additionally known that the highly toxic VC could be formed from biological degradation of other HOSs.

As was the case with chromium, HOSs were found to have adverse impacts on the operation of sewage treatment plants through their toxicity to microorganisms. This was first reported in 1944. Anaerobic digestion was particularly impacted with additional occurrences being reported in 1963, 1965, and 1969.

The HOSs are mostly odoriferous compounds with low odor thresholds. For example, TCE has an odor threshold of 10 mg/L in water. Water with TCE above this level will not be potable because of its organoleptic qualities. The odor thresholds of the HOSs have been known since the early 1960s.

HOSs have been implicated in large-scale regional atmospheric pollution problems. TCE was classified as a photochemically reactive smog constituent in 1966. The HOSs, especially methyl chloroform, have also been regulated on the basis of their impact on the atmospheric ozone layer.

SUMMARY

Forensic environmental health science can be used to answer questions regarding knowledge that the introduction of a material in the environment could pose a threat to human health, welfare, or the environment.

Cases ranging from insurance claims to personal injury (toxic tort) to natural resources damage often involve a determination of expected or intended damage. In essence, these determinations answer questions regarding a potentially liable party's knowledge that the introduction of a material in the environment was intended or expected to cause harm.

A variety of scientific disciplines, encompassed within the field of environmental health sciences, are well suited to forensic applications. These include toxicology, environmental toxicology, epidemiology, industrial hygiene, and medicine, among others. Toxicology, for example, is probably the oldest of the environmental health sciences, dating back to antiquity. The roots of modern epidemiology appeared in the early 19th century with examinations of vital data. Industrial hygiene studies involving workplace monitoring, biomonitoring, and occupational health evaluations have been significant contributors to the environmental health sciences for over 100 years [12].

In the application of forensic environmental science, historical actions are judged not in light of current knowledge, but rather by the state of knowledge prevailing at the time when the actions occurred. State of knowledge can be determined from a chronological evaluation of data from both the basic and the applied sciences and consideration of whether certain scientific bodies of knowledge had gained general acceptance during specified periods of time. The identification of the current state of knowledge about individual substances, such as environmental toxicants, is a useful starting point in forensic studies. Historical state of knowledge can then be tracked by obtaining and critically analyzing secondary and primary literature sources published contemporaneously and prior to the time periods of interest.

The case studies show that information may be found in the scientific literature that documents the history of chemicals that are as dissimilar in nature as a metal and a synthetic organic chemical. The case studies also show the importance of industrial hygiene and clinical medicine as sources of information and demonstrate the important interplay among the various scientific disciplines comprising the modern practice of environmental health sciences.

REFERENCES

1. Taylor, A. S., *Poisons in Relation to Medical Jurisprudence and Medicine,* Philadelphia: Lea & Blanchard (1848).

2. Sollman, T., *A Manual of Pharmacology and Its Applications to Therapeutics and Toxicology,* Philadelphia: W.B. Saunders Company (1926).

3. Jacobi, J., *Paracelsus: Selected Writings*, Princeton, NJ: Princeton University Press (1995).

4. Chang, L. W., *Toxicology of Metals*, Boca Raton, FL: CRC Lewis Publishers (1996).

5. Rom, W. N., *Environmental and Occupational Medicine*," Boston: Little, Brown and Company (1992).

6. *"The Industrial Environment—Its Evaluation and Control*, National Institute for Occupational Safety and Health (NIOSH), Washington, DC: U.S. Government Printing Office (1973).

7. Krebs, R. W., *The History and Use of Our Earth's Chemical Elements*, Westport, CT: Greenwood Press (1998).

8. Klaassen, C. D., *Casarett & Doull's Toxicology*, New York: McGraw-Hill (1996).

9. Dugan, P. R., *Biochemical Ecology of Water Pollution*, New York: Plenum/ Rosetta (1972).

10. Fisher, L. M., "Pollution Kills Fish," *Scientific American*, 160: 144–146 (1938).

11. Browning, E., *Toxicology of Industrial Organic Solvents*, Medical Research Council Industrial Health Research Board, London: Her Majesty's Stationary Office (1953).

12. Nelson, N., "Lectures in Toxicology—An Introduction," *Archives of Environmental Health*, 16: 492 (1968).

Chapter 9

Risk Assessments

Susan L. Mearns

Risk assessments are used to achieve many different goals in environmental engineering and environmental forensics. Risk assessments are used in environmental engineering to evaluate the impacts of known residual concentrations of chemicals in soil, surface water, sediments, and groundwater to human health and the environment. Risk assessments are also used to derive health-based cleanup goals used in remediation, such that the resultant remedial action has target levels or "action levels" of contaminants determined to be safe to human health and therefore safe to be left in place after cleanup. Risk assessments are used to assess the potential human health risks due to exposure to a chemical release or accident and to communicate human health risk to the public. Finally, risk assessments are used in litigation to determine whether human health has been impacted and to determine the potential extent of the impact.

Risk assessments as defined previously appear to be very straightforward. However, risk assessments are as much an art as they are a science. They are a tool that should be applied by a toxicologist familiar with their nuances and the project's objective.

DATA COLLECTION

The first step in performing a valid risk assessment is the data collection. Data collected at a site can be used to satisfy many project objectives, including risk assessments. The collection of data should be coordinated among the various experts involved in the project so as to facilitate collection of data that is useful to all disciplines. The risk assessor (and indeed all experts) is interested in collecting data that will be of sufficient quality to withstand the rigors of validation, including reproducibility, statistical analysis, and modeling.

Usually, data are collected from source or chemically impacted areas, in addition to background or nonimpacted areas, and submitted for analysis of suites of chemicals associated with the chemical processes or waste streams from the facility. The laboratory selected to provide the analytical services must meet regulatory agency criteria regarding holding times for the media prior to analysis and analytical equipment criteria regarding analyses requested. Detection limits (i.e., the ability of the equipment to detect concentrations of the chemicals) must be consistent with regulatory agency guidelines; otherwise, the data will not be usable for the risk assessment.

The anticipated future use of the site will affect the depths (soil and groundwater) at which samples are collected for use in the risk assessment. Usually, risk assessors are concerned with the potential exposure to the residual concentrations of the "chemicals of concern" by individuals who may come into contact with the chemicals (i.e., receptors), as well as with the effects of the residual chemicals on the environment.

In order to select the chemicals of concern that will be evaluated in the risk assessment, all the data collected from past investigations and assessments should be examined to determine whether the data meet risk assessment data quality objectives, are representative of the conditions at the site, adequately characterize the current site conditions, and were analyzed by a qualified laboratory that used equipment capable of detecting low concentrations of chemicals. At this time, data gaps should be identified and steps taken to fill the data gaps.

DATA GAPS

A data gap exists when a site has not been adequately characterized; that is, there is insufficient data regarding the areal extent of the site as well as site contaminants, and/or there is insufficient data describing historic operations, past uses, and waste disposal areas, thereby failing to adequately define the vertical and/or horizontal extent of these areas. Data gaps can also exist when incorrect analyses are performed by a laboratory, for example, when a laboratory used one method for analysis of volatile organic compounds, such as U.S. Environmental Protection Agency (EPA) Method 8020, instead of EPA Method 8260 (a more rigorous method), and/or when an incomplete analytical list was submitted (e.g., chromium was used in manufacturing in the trivalent and hexavalent forms, therefore it is essential that the data contain analytical results for both trivalent and hexavalent chromium when operations such as metal plating, anodizing processing, bushing works, or metal refurbishing occurred or when paint was manufactured on the property).

Identified data gaps can be filled by taking additional samples at the site and submitting them for the analyses needed. Obviously, this data will represent current site conditions and will not be representative of past conditions associated with historic operations. However, as the risk assessor is using the data generated to determine whether residual concentrations of chemicals of concern at the site pose an adverse impact to human health and the environment, having data representative of current conditions typically best suits the purposes of the risk assessor. If the collection and analyses of additional samples is not possible, data gaps can be filled by extrapolation.

Extrapolation of data generated from past investigations can be accomplished through the use of models, such as fate and transport models, chemical degradation models, and geostatistical models. However, if modeling is used, the assumptions made in each step of the application of the model should be justified, keeping in mind that the assumptions will be challenged by the opposition's experts. Whenever possible, regulatory agency–approved values for parameters and variables should be employed in the models. Standard of practice recognized within the industry also should be followed, in anticipation of being challenged by opposing counsel.

EXPOSURE

In addition to understanding past site practices, including the location of historic operations, waste disposal practices, and personnel, it is equally as important to understand the intended future use of the site. The intended future use will determine the receptors selected to be evaluated in the risk characterization portion of the risk assessment. It is entirely possible to have a site with residual concentrations of chemicals of concern present and no risk. Risk is determined by evaluating the potential effect of the exposure on the receptors. Therefore, if there is *no exposure* to the residual concentrations, there can be *no risk*.

Exposure is evaluated by understanding the intended future use of the property, the receptors that may be potentially exposed to the residual concentrations of the chemicals of concern, the exposure pathways by which these receptors may be exposed, and the routes of exposure by which the chemicals may potentially adversely impact the receptors.

RECEPTORS

As an example, if the intended future use of the site is for a day care facility, senior citizen residence, hospital, school, or residence, the risk assessor will structure the risk assessment differently than if the intended future use is for an industrial building. Because the receptors that may potentially be impacted will be different, the exposure pathways will be different and the routes of exposure may be different. The receptors selected for evaluation in the risk assessment are clearly dependent on the intended future use. If the intended future use is unknown, or if the intent of the risk assessment is to be as health protective as possible, then the receptors selected for evaluation will be children and/or adults[1] who presumably will inhabit the property for 50 to 70 years. Otherwise, the receptors selected for evaluation will be specific to those expected to occupy the site. For example, senior citizens would be selected as receptors to be evaluated if the future property use is expected to be a hospital or senior citizen residence; industrial workers would be selected as receptors if the future property use is expected to be an industrial facility.

EXPOSURE PATHWAYS

The exposure pathways, or exposure scenarios, are specific to the future use of the property and the receptors selected for evaluation. If the intended future use of the property is as an industrial facility, commercial center, or retail shopping complex, then the receptors selected most likely will be adult workers, expected to be on the property for 40 hours per week, 50 weeks per year for 25 years. The exposure pathways by which these receptors could potentially be exposed will be dependent on the types of chemicals of concern detected at the site, the media in which these chemicals of concern were detected (soil or groundwater), the chemicals' phase (vapor, solid, or liquid), and the depth at which the chemicals were de-

[1]This group can be considered "default receptors" or a "most protected group."

tected. For example, if we are dealing with a probable human carcinogen, such as tetrachloroethylene (PCE), that has been detected in the vapor phase through a soil gas survey, at depths less than 5 feet below ground surface (BGS), and at concentrations that exceed regulatory agency–approved screening criteria, then the most likely exposure pathway to PCE is through diffusion of PCE into the buildings through cracks in the buildings' footings and foundations. The most likely route of exposure to PCE, then, using this example, would be the route of inhalation; that is, the workers would potentially be exposed to concentrations of PCE in the air inside the building (that had diffused upward from the ground into the building) through inhalation. Therefore, this risk assessment would evaluate the effects of PCE on workers exposed to PCE through inhalation for 40 hours a week, 50 weeks a year for 25 years.

ROUTES OF EXPOSURE

The routes of exposure whereby a receptor may be potentially exposed to residual concentrations of chemicals of concern associated with the past use of the property include ingestion, inhalation, and dermal absorption. The ingestion route of exposure includes ingestion of soil to which chemicals are adsorbed and ingestion of garden produce grown in soil that contains residual concentrations of the chemicals of concern. The inhalation route of exposure includes inhalation of volatile organic compounds in air, inhalation of dust generated during activities conducted on the property to which chemicals of concern are entrained, and inhalation of the vapor phase of chemicals of concern in steam generated while showering. The dermal absorption route of exposure includes skin absorption of chemicals of concern through contact with contaminated soil, sediments, surface water, or groundwater.

TOXICITY

The toxicity of the chemicals of concern to be evaluated in the risk assessment is an important factor. Toxicity determines whether the chemicals are assessed as carcinogens or noncarcinogens. The risk assessor should use the regulatory agency–approved cancer potency slope factors[2]

[2]Potency slope factors have been used as a basis for regulatory actions or standards, and are derived by extrapolating animal laboratory data to humans.

when evaluating carcinogens and reference doses when evaluating non-carcinogens. Occasionally, regulatory agency-approved slope factors and reference doses are not available for the chemicals of concern being evaluated. The toxicologist should then use the slope factor or reference dose of the most similar chemical for the chemical of concern being evaluated. This is another area where the opposition will request documentation and, if slope factors or reference doses of similar chemicals were used, will more than likely question the rationale of the choices made.

More often than not, the chemicals of concern consist of a short list of chemicals among which are one or two that appear to be "driving" the risk. In other words, even though there may be a plethora of data for the site, usually the chemicals of concern are a small subset of all the data representing chemicals or compounds used at the site in the past. Within that small subset of data, one or two (or a few at most) chemicals or compounds are typically present that have the greatest residual concentrations, the greatest number of detected concentrations, or are so toxic that if the risk due to exposure to these chemicals is evaluated and addressed such that the potential impacts to human health due to this exposure are ameliorated, the risk due to exposure to the remaining chemicals of concern is often considered negligible. These chemicals or compounds within the chemicals of concern are, therefore, "driving" the risk the site poses to human health.

However, even though there are "drivers" of risk, the risk assessor should evaluate the risk due to exposure to all the chemicals of concern, as the risk assessment evaluates cumulative, excess risk over the receptor's lifetime due to exposure to the chemicals of concern.

The chemicals of concern are selected from the body of data representative of the site by comparing the detected chemicals to (1) a list of chemicals known to be used, stored, handled, and/or disposed on the site; (2) background concentrations of the chemicals detected on the site; (3) chemicals known to be human nutrients or minerals; and (4) the frequency with which the chemicals were detected at the site. The risk assessor is relying on the quality of the data in order to select the list of chemicals of concern; therefore, the data should conform to the standard-of-practice data quality objectives. The selection of the chemicals of concern is another area where the opposition will challenge the selection process, as well as question the validity of the rationale used to decide which chemicals to evaluate further in the risk assessment. Make no mistake, that is what the risk assessor is doing, winnowing a list of many chemicals to a list of few, subjected to further evaluation in the risk assessment process. The process of selecting chemicals of concern, therefore, needs to be based on sound decision making and standard of practice, as it will be challenged by the opposition.

CALCULATION OF RISK

The next step in the risk assessment process, once the data have been collected and the models, receptors, exposure pathways, and routes of exposure selected, is the calculation of the risk values. Calculations to assess risk are provided in guidance documents available from regulatory agencies. Most usually, the EPA and state environmental agencies have guidance material that provides not only the equations to be used, but also default values to be used for the variables within each equation. The risk assessor requires knowledge of the future use of the property, in addition to the receptors expected to be on the property, in order to select the best default values. For example, when evaluating the dermal absorption route of exposure, the amount of exposed skin of the receptor being evaluated should be known, or assumed. The risk assessor should select the default values based on known information as much as possible. Any assumptions the risk assessor makes will more than likely be challenged by the opposition's experts.

RISK COMMUNICATION

After the risk assessment has been completed, the results may need to be disseminated to the public and/or the community, especially if the risk assessment was conducted to evaluate the health risks due to a release of chemicals from a facility. If the results are to be communicated to the public, the best time to initiate communications is before the risk assessment is started. Actually, the best time to communicate to the public and/or community is before the need for a risk assessment. The ability of the risk assessor to communicate to the public and gain the public's trust in the results is often dependent on the relationship that was established by the property owner prior to the release. In other words, the importance of having a good-neighbor policy cannot be overstressed.

The risk to human health due to exposure to the chemicals of concern is perceived differently by the public than the scientific community. The public experiences several emotions after a release, not the least of which is anger. The intensity of this anger is directly proportional to the sense of outrage the public has toward the facility that had the release. The anger and outrage will be directed at the risk assessor whenever the risk assessor attempts to communicate the results of the risk assessment to the public, especially if the public was not invited to participate in the risk assessment process (i.e., especially when the public was not made a stakeholder in the process).

The initial meeting with an outraged public can be an outpouring of anger directed at you, the risk assessor. The best strategy is to be in the position where the facility can admit that mistakes were made (i.e., accept responsibility for the release) apologize for the impacts of these mistakes to the public (i.e., say they are sorry), present a solution to the immediate situation, and pledge to ensure that the(se) mistake(s) will not happen again.

After the public's anger has been acknowledged, and the facility has begun to take action along with pledging to ensure that these mistakes will not result in the same situation in the future, the risk assessment process is ready to begin.

Inviting participants from the community, in addition to regulators, to form a risk assessment committee will facilitate acceptance of the results by both the public and the regulatory agency. The risk assessment committee should choose the exposure pathway, receptors, and routes of exposure to be evaluated in the risk assessment, in addition to the values to be used in the models and the equations used in the risk calculations. The objective is to achieve consensus among the stakeholders of what the risk assessment can provide to the community and how to conduct the risk assessment such that the results, regardless of what they may show, are acceptable to the community, the regulators, and the site owner.

Communication of risk to the public, in the manner outlined before, is much easier than if the risk assessor conducts the risk assessment without input from the public and then expects the public to accept the conclusions reached. There is no reason why the public should accept the conclusions of a risk assessment when they are not actively involved in the process of assessing the risks to their health.

LITIGATION

The role of risk assessments in litigation is as varied as how risk assessments can be used, that is, to derive health-based cleanup goals, to determine the impacts of known residual concentrations of chemicals of concern in media to human health and the environment, to determine the health risks due to exposure to a chemical release or accident, and to communicate the risk to human health to the public.

Risk assessments can be used in litigation to evaluate the cost/benefit of purchasing a property where residual concentrations of chemicals of concern have been identified. The risk assessment would then be used in this instance to derive health-based cleanup goals to determine safe levels of chemicals to leave behind and to identify areas of concern that would require some degree of remedial action. The cost of ameliorating the health impact of the areas of concern could then be determined.

For example, in this first case history the site was the location of a paint manufacturer for some 40 years. The current property owner purchased the property from the paint manufacturer in the 1970s and did not practice environmental due diligence prior to the transaction. The current property owner then attempted to sell the property; however, the prospective purchaser did perform due diligence and discovered the past use of the property and also ascertained that no site investigation or remedial work had been conducted. The prospective purchaser walked away from the impending sale of the property and the current owner initiated a lawsuit against the original owner, the paint manufacturer. A risk assessment determined safe levels of residual concentrations of chemicals of concern that could be left in place and would not pose a risk to human health or the environment. These health-based cleanup goals derived from the risk assessment were used to delineate areas of concern, or "hot spots," that had residual concentrations of chemicals of concern that exceeded the health-based cleanup goals. The costs of ameliorating the potential impacts of the localized "hot spots" were determined to be acceptable to the paint manufacturer, to the extent that the paint manufacturer purchased the property from the current owner, performed limited remediation at the "hot spots," leased the property, and looked forward to the prospect of selling the property, at a profit, at some future date.

Risk assessments can be used in litigation to resolve conflict regarding the most efficient manner to ameliorate the impacts of residual concentrations of chemicals of concern to human health. For example, in this second case history the property had been used for the manufacture of airplanes for 75 years. During the past 10 years, the property was investigated and remediated under the auspices of the regulatory agency. The chemicals of concern included volatile organic compounds (VOCs) in soil and groundwater. The best available technology for remediation of VOCs from soil and groundwater was unable to remove all the VOCs from a zone of 35 to 55 feet below ground surface. A developer purchased the property with the intent of constructing an office/retail sales complex. The potential impact of the VOCs diffusing from the soil into the buildings' footings and foundations through cracks and potentially impacting the health of workers was the source of conflict among the property owner, the developer, and the city within which the property was located. To resolve the conflict, a risk assessment was conducted to evaluate the potential health impacts of the residual concentrations of the VOCs in soil diffusing through cracks and entering the buildings. Health-based cleanup goals were derived and current data compared to these cleanup goals to delineate the areas of concern. A risk assessment was conducted to determine whether the office and retail sales workers would be exposed to residual concentrations of VOCs that would have a deleterious impact on their health. Based on the results of the risk assessment, the developer had several options to consider and negotiate with the property owner: (1) restructuring the development plan, such that buildings would not be constructed over the areas of concern; (2) installing vapor barriers underneath the building footprints that extended on top of the areas; or (3) conducting limited remediation to remove the soil containing the residual concentrations of VOCs.

Risk assessments can be employed in litigation to show the harmful effects of the chemicals of concern to human health and the environment, to the extent that either human health or the environment has been irreparably damaged.[3] For example, in this third case history the receptor is a freshwater lake that has been irreparably damaged. The freshwater lake is located downstream from several dairy farm operations. During the annual heavy rainfall events, which usually occur in late spring, liquids from the dairy farm waste lagoons overflow into a nearby stream and subsequently discharge into the lake. The dairy farm waste, being rich in phosphates, causes a nutrient imbalance in the lake, creating an ideal situation for an algal bloom. The algae consume all the phosphates in the water column, die, and settle to the bottom of the lake. The bacteria in the lake sediments feed on the dead and decaying algae and consume the dissolved oxygen in the water column while breaking down the algae into

[3]In this case, "irreparable damage" is defined as damage that affects the individual's quality of life (e.g., cancer, although treatable if detected early enough, affects the individual's quality of life). Included in this definition is damage to an organ such that the organ's function is harmed beyond repair.

macronutrients. The lack of oxygen in the water column creates anoxic conditions, at depth, in the lake during the summer. The end result is a fish kill. Thousands of freshwater fish suffocate and float to shore exacerbating the anoxic conditions as bacteria now start decomposing the fish, depleting the dissolved oxygen still further. Additionally, the odor and sight of thousands of dead fish floating to shore over several days hamper recreational use of the lake.

A risk assessment was used to evaluate the effects of the chemical of concern, namely, phosphates, on the lake. The results unequivocally determined that it was the phosphate loading that was causing the anoxic conditions, resulting in an annual fish kill. Moreover, the phosphates having settled into the sediments of the lake are reintroduced during fall and spring turnover, creating additional algal blooms and anoxic conditions. The conclusion of the risk assessment was that, as a result of this "cycle," the freshwater lake has been irreparably damaged due to the extended phosphate loading.

Conversely, risk assessments can be used to show that there is no exposure to the residual concentrations of chemicals of concern; therefore, there is no risk to human health. For example, in this final case history the defendant alleged exposure to human sewage from a ruptured sewer pipe that ran along her property line via an easement. The plaintiffs with the responsibility for the maintenance of the sewer pipe corrected the problem as soon as they were notified. However, the plaintiff alleged that she was exposed to human waste while gardening along her property line and while cleaning up after the break in the sewer line. The plaintiff alleged health impacts due to exposure to *Escherichia coli* and nickel. The health impacts alleged by the plaintiff included a bacterial lung infection and dermatitis, respectively. The onset of health impacts were alleged 4 months after repairs were made to the broken sewer pipe. In this instance, exposure to *E. coli* possibly could have resulted in a gastrointestinal problem that should have manifested itself within hours of exposure via ingestion. There have not been any documented cases where exposure to nickel in soils caused dermatitis. Moreover, as the defendant's cleanup efforts and gardening efforts were restricted to the surface, she was not exposed to the sewage that leaked from the ruptured sewer pipe at a depth of 14 inches below ground surface. A risk assessment was used to show that there was no exposure to the alleged chemicals of concern and, therefore, there could be no health impacts attributable to the sewer leak.

These case histories illustrate a few instances where risk assessments have been used in environmental forensics. The role of risk assessments in environmental forensics is to evaluate the efficacy of the claim of injury regardless of whether the injurious claim is to property, an environmental resource, or human health.

Chapter 10

Why Use Visuals

John Muir Whitney and Mark S. Greenberg

A PICTURE IS WORTH . . .

Environmental forensic testimony is, by virtue of its very subject matter, complex and often misunderstood by laypersons and experts alike. Nowhere else, for reasons explained in this chapter, does the adage *A picture is worth a thousand words* ring more true. As with many other types of litigation, it is your job, as the forensic expert witness, to communicate your ideas clearly and succinctly. Visuals help fulfill that mandate perfectly.

As the basic science of environmental forensics becomes more complex, so too does litigation in the field. It is vitally important for testimony to begin by conveying a fundamental understanding of the subject. If jurors feel that they have been left at the gate, your best opportunity to connect with them may be lost.

Begin by teaching the basic science before explaining case-specific facts and opinions, bearing in mind that the average juror often has no more than an eighth-grade education. Because of the pervasiveness of television and video in the lives of most Americans today, jurors expect to receive information visually and oftentimes will not be inclined to work to discern important facts solely by listening to oral testimony. Jurors are also keenly aware of the use of visuals during trial proceedings and have come to expect them, due in part to the effective use of visual presentations during the highly publicized 1995 criminal trial of O. J. Simpson.

Also noteworthy is the fact that most jurors take their role seriously. After all, they give up their time away from family and work and expect a well-prepared attorney to present a well-thought-out case. An ill-conceived presentation on your part will only annoy a judge or jury and could prejudicially color their opinion of you, the facts, and ultimately the party you represent.

DO YOU SEE WHAT I SEE?

Even before you set foot in a courtroom, visuals can help organize and clarify the case, both for you—the expert—and for the attorney with whom you are working. By visualizing the message you hope to communicate, you should derive several key benefits:

- Identification of the strengths and weaknesses of the case
- Recognition of the areas that need further explanation
- Refinement of the issues to their most basic, relevant components

It is at this stage, when you are satisfied that you are telling the story you want to tell, that the attorneys can incorporate your key points into their presentations to set the appropriate stage for your testimony.

The strategic use of visuals increases retention and helps focus judge and jury on your key message. Try this exercise: Gather together a small group of people and test one of your messages on them. Next ask them each individually to repeat what they remember. You will be astonished at the disparity in responses. Try it again, this time using a visual aid during your explanation. The difference in the rate of retention will be astounding.

Psychologists who study human behavior claim that we generally remember 15% of what we hear versus 85% of what we see, especially when shown the image(s) repeatedly. To see this point illustrated, all one need do is consider television advertising. The use of a visual aid in testimony will increase the likelihood that jurors remember similar details—that they all will see, understand, and remember the same story—*your* story.

Juries like stories to be neatly packaged. It is best to have an introduction, background statements, main points, and a conclusion. It is imperative that you wrap it up at the end—bend the judge or jury to *your* conclusions, rather than leaving them to draw their own based on their personal biases.

As any educator will agree, visuals stimulate learning. And jurors enjoy learning things. They have been observed literally moving to the edge of their seats when a new, visually compelling exhibit is presented and explained.

TIME IS OF THE ESSENCE

Generally speaking, you have a limited amount of time to tell your story. You will probably have to boil down to an hour a set of facts and opinions that may have taken weeks, months, or even years to develop. Conventional wisdom and experience tell us that jurors tune out after 30 or 40 minutes of oral testimony. The use of visuals breaks up spoken testimony and acts as an indicator that something interesting or important is about to happen.

Visuals also help link you and the jury. Use visual aids to explain issues that are beyond the normal scope of the jurors' lives. Show high-concept issues that are not tangible and explain scientific matters quickly and efficiently. Never appear to "speak down" to your audience.

Visuals lend increased impact and staying power, because they can be viewed for a sustained length of time and can provide an excellent framework for prolonged or difficult testimony.

ONE FROM COLUMN A AND ONE FROM COLUMN B

Two basic types of visuals are used in the presentation of expert testimony—thematic and tutorial exhibits. *Thematic exhibits* are used to present the themes or issues to be discussed in the case, your take on the issues, and why you believe your conclusions to be correct (or why your opposition is incorrect in its theories). Many times, trials are won or lost based on decidedly nonfactual, emotional issues. Telling the story through pictures gives a jury an emotional base from which to evaluate later testimony and evidence. Several examples of both thematic and tutorial exhibits are given in Figures 10.1 to 10.9.

Tutorial exhibits help teach or explain critical and complex, yet relevant concepts so that the lay jury has a frame of reference for understanding your facts and opinions. They need to form these basic understandings to later draw conclusions (hopefully yours) based on the evidence they will hear. Beginning from a common set of ideas will enable them to reach similar conclusions, with one another and with you.

Tutorials within an expert's testimony might typically begin with explanations of the basic science about to be brought into play. Flowcharts, comparative exhibits, and even simple photographs help a jury make strong associations throughout the testimony to provide a benchmark against which the opposition's testimony will be weighed.

Finally, almost every case can benefit from a basic timeline of key events. Providing continuity is essential. Again, your key goal is not to have jurors arrive at the same conclusion as you have—it is to explain your conclusion and have them agree with you. The distinction is critical.

In essence, your testimony should follow the simple conventions of public speaking: Tell your audience what you are going to talk about, say it, and wrap up by summarizing what you just said.

VARIETY IS THE SPICE OF LIFE

When factoring in the advancements in courtroom presentation technologies (discussed in the following section), there are a wide variety of media types available to the expert witness. It is possible to tell your story through words and pictures in such a way as to keep the jury interested, informed, entertained, *and* persuaded.

Begin by choosing the most appropriate media for your presentation. Many factors contribute to the selection of the appropriate medium or combination of media.

First, determine the level of detail the exhibit requires. Consider how different options affect image visibility within the presentation environment. What medium works best? You should be comfortable with the chosen format. Consider new technology such as video or computerized presentation graphics; it is becoming easier to use, as well as being highly cost effective.

Does your presentation require a dynamic (moving) or a static presentation? Animations and videos are dynamic presentations and offer higher impact. Graphics are generally static, although they can be interactive with the addition or removal of magnetic or Velcro® elements.

Which options are best suited to the budget for visuals? Costs vary from medium to medium and within category, depending on the desired effect and the level of complexity.

What are the venue preferences and limitations? Certain courtroom configurations can be limiting, and some judges are extremely particular about the types of equipment (if any) they allow in their courtrooms and have limits on how extensive and/or intrusive the equipment can be. Some courtrooms have systems in place that can be used or augmented with equipment you supply.

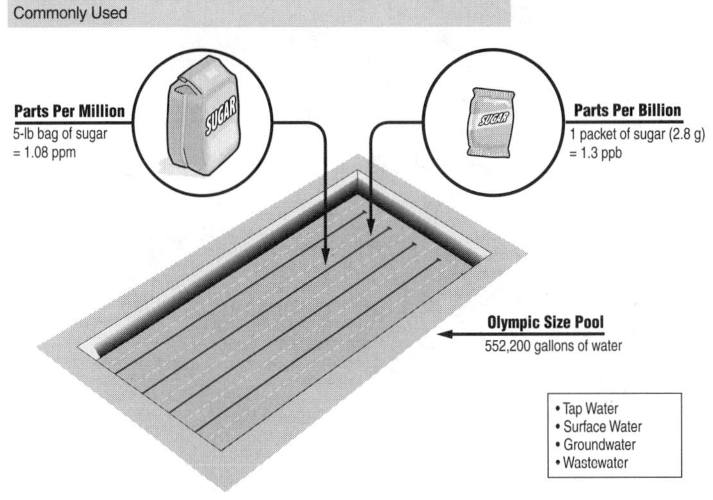

Figure 10.1 A tutorial exhibit that demonstrates the concept of concentration (the amount of a chemical dissolved in water) and the scientific units used to express concentration (i.e., how small is small) using familiar items. This exhibit also shows the common "types" of water in which chemicals are dissolved.

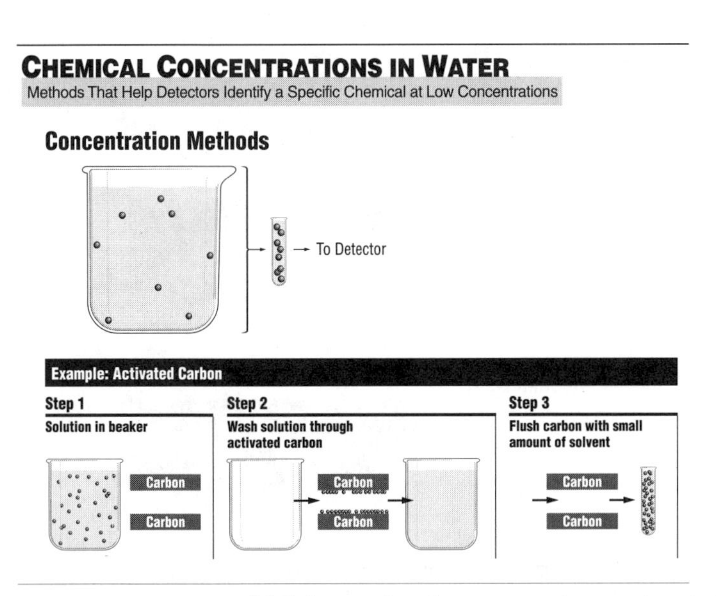

Figure 10.2 A tutorial exhibit used to demonstrate how a chemical is concentrated so that it may be more readily detected. This exhibit was designed to minimize the chemistry presented to a jury.

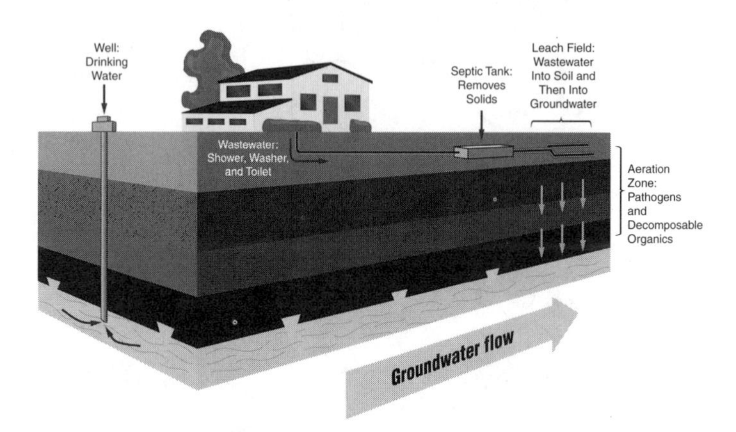

Figure 10.3 A tutorial exhibit designed to show a simple septic/ leach field system and its influence on groundwater quality. This exhibit could then be further used to discuss the biologic decomposition of household organic waste and the commonsense location of a drinking water well.

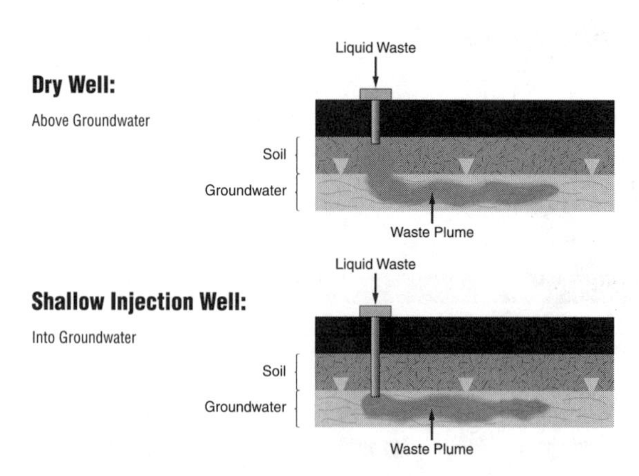

Figure 10.4 A tutorial exhibit designed to clarify technical jargon. In this example, a "dry well" (implying no water) contaminates groundwater no differently than an "injection well." In other words, the name may change but the results are identical.

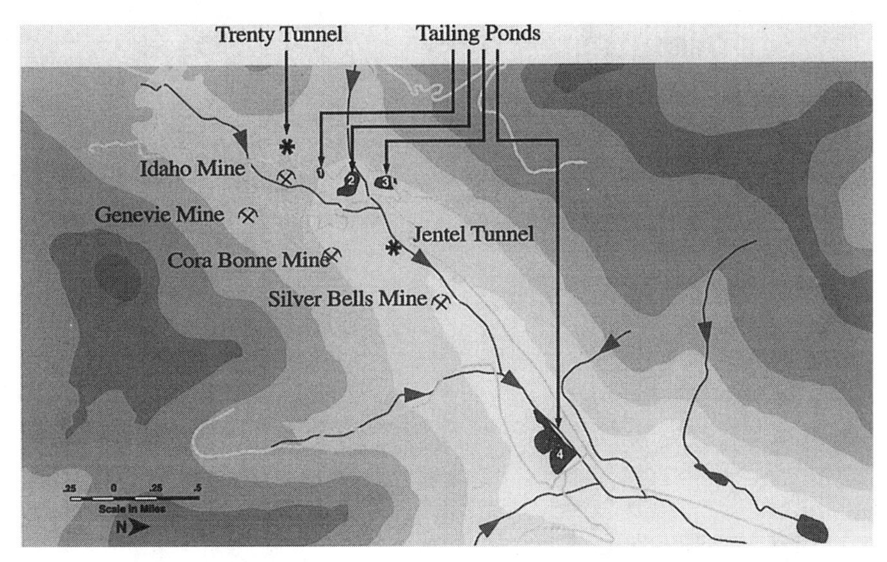

(A)

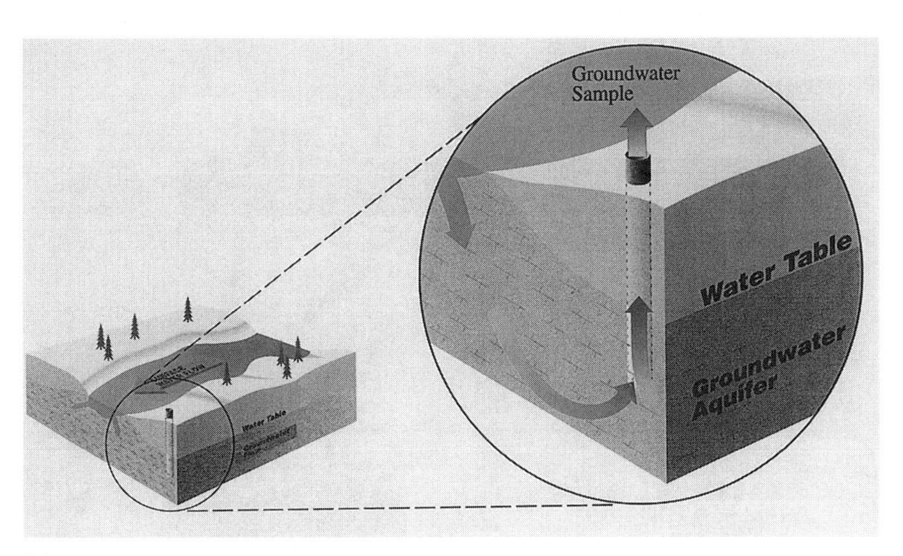

(B)

Figure 10.5 A two-part tutorial exhibit designed to show a sequential relationship. In this example, the relationship between the location of various mines (and, thus, acid/metal discharges) to the surface water drainage and flow direction is shown in Exhibit A. Exhibit B then illustrates the relationship of surface water to groundwater and groundwater sampling.

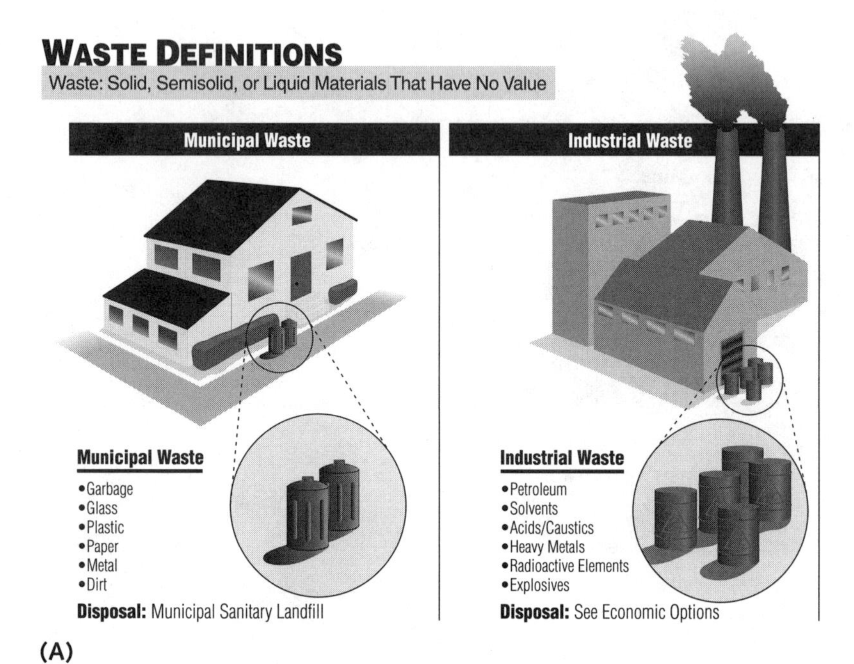

(A)

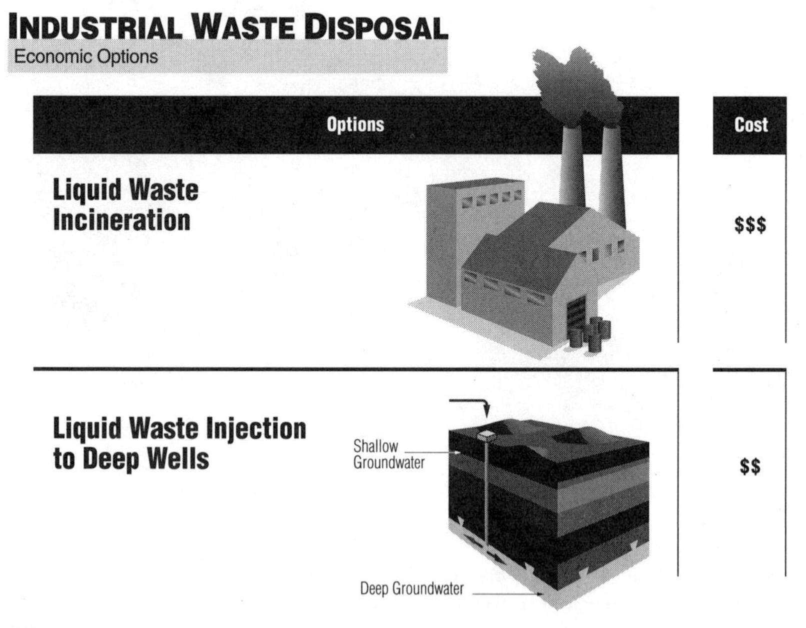

(B)

Figure 10.6 A four-part thematic exhibit developed to introduce the differences between municipal waste and industrial waste (Exhibit A), the methods of industrial waste disposal and their relative cost (Exhibits B and C), and the

Industrial Waste Disposal

Economic Options

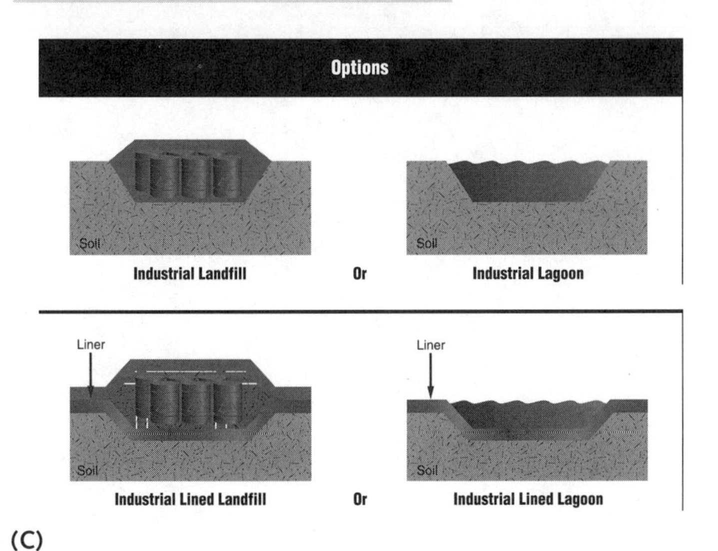

(C)

Typical Landfill

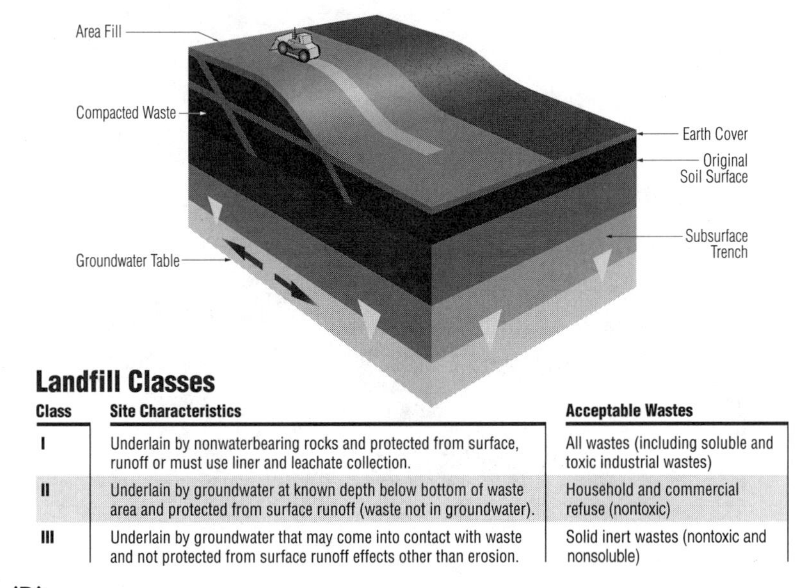

Landfill Classes

Class	Site Characteristics	Acceptable Wastes
I	Underlain by nonwaterbearing rocks and protected from surface, runoff or must use liner and leachate collection.	All wastes (including soluble and toxic industrial wastes)
II	Underlain by groundwater at known depth below bottom of waste area and protected from surface runoff (waste not in groundwater).	Household and commercial refuse (nontoxic)
III	Underlain by groundwater that may come into contact with waste and not protected from surface runoff effects other than erosion.	Solid inert wastes (nontoxic and nonsoluble)

(D)

use of landfills for both municipal and industrial waste disposal (Exhibit D). Each graphic was designed to present the least amount of information necessary in order for the expert to discuss and expand on each concept.

(A)

(B)

Figure 10.7 A two-part thematic exhibit (Exhibits A and B) that illustrates the use of aerial photographs to clearly demonstrate changes of land use and waste disposal practices.

INCINERATION

Industry State-of-the-Art Documents

Document	Year
"Incineration Solves a Waste Disposal Problem"	1948
"Incineration of Wastes from Large Pharmaceutical Establishments"	1954
"Catalytic Oxidation of Aqueous Wastes"	1957
"Multipurpose Incineration: Monsanto Recommends Its Wider Use for Pollution Abatement"	1957
"Waste Solvent Incineration: Successful at Upjohn Co., Kalamazoo, Michigan"	1957
"The Construction and Operation of an Incinerator for Chemical Plant Wastes"	1957
"New Incineration Facilities at Dow, Midland"	1959
"Process Waste Burner Destroys Liquid Organic Chemical Wastes Safely"	1964

Figure 10.8 Example of a simple thematic timeline for presentation to a jury.

Groundwater and Waste Disposal Practice

State-of-the-Art Documents	Year
The cholera epidemic of London, England was traced to groundwater contamination of the Broad Street Pump from a neighboring septic tank	1854
The theory of groundwater movement was defined by "Darcy's Law"	1856
A US Geological Survey report on the movement of groundwater was published	1902
The pollution of underground waters as the result of waste disposal practices was published	1911
An American Water Works Association report on the pollution of water supplies by industrial wastes	1923
A US Geological Survey report on the occurrence of groundwater in the United States	
Pollution Hazards of Ground Water Supplies	1931
An American Water Works Association report on impounded garbage contamination of groundwater eights months after disposal	1932
Tracing the Travel and Changes in Composition of Underground Pollution	1932
Industrial Pollution of Ground Waters	1935
An American Water Works Association report indicates that the closer the water table is to a source of contamination, the more likely that groundwater will be contaminated	1936
An American Water Works Association report discusses chemical and biological groundwater contamination thousands of feet from disposal pits	1940
Basic Factors Affecting the Pollution of Sub-Surface Water	1941
A text on Industrial Waste Treatment Practice reports that phenol waste is toxic to aquatic life at levels above 10 ppm	1942
An American Water Works Association report on tracing groundwater pollution using tracer dye	1943
An American Water Works Association report on groundwater contamination by refinery and industrial waste with attempted cleanup	1945
Formulating Legislation to Protect Ground Water from Pollution	1947
An American Water Works Association report lists nine cases of industries polluting groundwater (one by dichlorophenol)	1947
An American Water Works Association report that warns once a groundwater reservoir is polluted the contamination may persist almost indefinitely	1948
Reported groundwater contamination by trichloroethylene	1949
American Society of Civil Engineers Handbook on hydrology published	
A Sewage and Industrial Waste article on industrial waste lagoons that can contaminate groundwater if constructed in porous soils	1950
An American Water Works Association article on groundwater pollution from industrial wastes that do not biologically degrade; a long-term problem has been phenols and a charcoal-iron production plant	1952

(A)

Groundwater and Waste Disposal Practice

State-of-the-Art Documents

1953

An American Water Works Association article reports on groundwater contamination from industrial sources such as phenols, petroleum and cleaning fluids and that the "proper time to control underground pollution is before it occurs"

Sewage and Industrial Waste articles on industrial waste to land and their potential for groundwater contamination; soluble organic chemicals show relatively little change in groundwater

1954

An American Water Works Association article that reports that chemical movement through soil into groundwater can be expected with percolating liquids

Purdue Industrial Waste Conference reports that once injury to groundwater occurs, it may be expected to persist for a long time

Industrial and Engineering Chemistry article on groundwater contamination from land disposal of liquid wastes and that prior to any disposal the soil porosity, proximity to groundwater, and waste composition should be known; groundwater contamination is "difficult and expensive to correct"

1955

Sewage and Industrial Waste article on groundwater contamination from industrial sources and that soil filtration does not appreciably reduce the concentration of may chemical compounds

1955

The Impermeabilization of the Lagoon at the International Paper Co.

1956

A Sewage and Industrial Waste article recommends that in areas of groundwater use, lagoons should have impermeable bottoms

1956

World Health Organization issues a report on pollution of groundwater

1957

An American Water Works Association article on groundwater contamination; impoundments and lagoons were responsible for groundwater contamination in 20 states

1958

Purdue Industrial Waste Conference report that recommends surface impoundments for oil-refinery wastes be located in areas of impervious geological formation and very little groundwater

1959

An American Water Works Association article indicates that the cleanup of contaminated groundwater is far more difficult and time-consuming than pollution removal in surface water

American Society of Civil Engineers publication on Sanitary Landfills that recommends site selection based on characteristics considering underground water supplies and the dangers of pollution

1960

American Water Works Association articles on groundwater contamination stress that ponding of liquids are a hazard to groundwater as well as septic systems and cesspools

1960

The Federal Housing Administration publishes a report on the status of knowledge on groundwater contaminants

(B)

Figure 10.9 Example of a detailed thematic timeline (Exhibits A–C) used as a handout in a mediation. Figures 10.8 and 10.9 illustrate the difference in the level of detail presented in an exhibit, depending on the audience.

(*continues*)

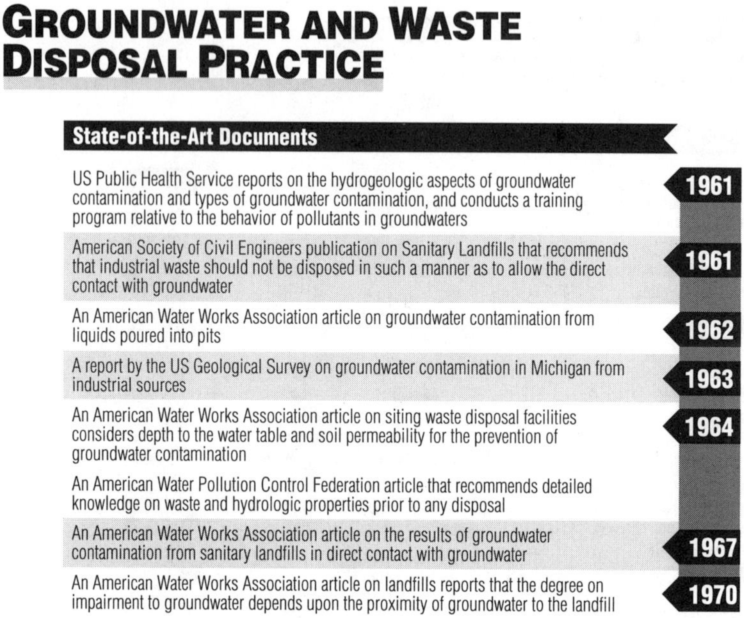

GROUNDWATER AND WASTE DISPOSAL PRACTICE

State-of-the-Art Documents

US Public Health Service reports on the hydrogeologic aspects of groundwater contamination and types of groundwater contamination, and conducts a training program relative to the behavior of pollutants in groundwaters	1961
American Society of Civil Engineers publication on Sanitary Landfills that recommends that industrial waste should not be disposed in such a manner as to allow the direct contact with groundwater	1961
An American Water Works Association article on groundwater contamination from liquids poured into pits	1962
A report by the US Geological Survey on groundwater contamination in Michigan from industrial sources	1963
An American Water Works Association article on siting waste disposal facilities considers depth to the water table and soil permeability for the prevention of groundwater contamination	1964
An American Water Pollution Control Federation article that recommends detailed knowledge on waste and hydrologic properties prior to any disposal	
An American Water Works Association article on the results of groundwater contamination from sanitary landfills in direct contact with groundwater	1967
An American Water Works Association article on landfills reports that the degree on impairment to groundwater depends upon the proximity of groundwater to the landfill	1970

(C)

Figure 10.9 *(continued)*

Specific Media Types

We begin with the basics: *photo/document enlargements.* Key points can easily be made when you present letters, reports, depositions, or other documents that plainly state the information you wish to convey. Depending on the courtroom presentation system, these documents can have pertinent passages enlarged, highlighted, or otherwise called out, and the resulting image can be printed out, shared with the jury, entered as evidence, or simply discarded. Suppose you have a letter from the president of Company A in which he clearly directs his subordinates to spare no time in disposing of some chemical waste products. "I don't care if you pull up to the riverbank and let it flow," he writes. A copy of that letter, now in your possession, supports your contention that the source of the river contamination was, in fact, illegal dumping. You pull up a copy of that letter, highlight the sentence, read it aloud for the jurors, and confidently move on to your next point.

Photographs also tell a story in seconds. Refer back to the previous example. You have just exposed the careless attitude of Company A's president by exhibiting his letter. Add to the blow by showing an aerial photograph taken days after the letter's date, depicting the residue of a chemical spill along the riverbank; zoom in to show tire tracks leading from the river's edge and the spill. Circumstantial, yes, but taking you in the right direction.

Next, we have *illustrative charts and diagrams*—those specially made graphics for which you have hired a professional litigation graphics studio to create. Working with you and a draft of your expert report (or other supporting documentation), graphic artists can design exhibits to convey specific information (see the previous discussion on thematic and tutorial exhibits).

Videotape presentations, including depositions, demonstrations, site inspections, and day-in-the-life videos, enhance testimony and directly address the jury's television predisposition.

Rounding out the list are *two- and three-dimensional animation.* Animation is often used when illustrating functions, such as the inner workings of a mechanism, or in time-lapse illustrations such as water moving through an aquifer or mine runoff passing through various sloughs and tailing ponds and into groundwater. Although such presentations are not cheap to produce and require time in planning and execution, they can be invaluable when communicating hard-to-understand concepts to lay jurors.

A comparison of both the pros and the cons of each media type is shown in Table 10.1.

TWO TIN CANS AND A STRING

Presentation methods vary with time, budget, and rules of conduct set by the presiding jurist. Even within those constraints, however, a little showmanship can go a long way.

Large-format printed visuals can be produced as static (only to be looked at) boards or they can be prepared for interactive use, using clear or colored overlays or magnetic or Velcro attachments. An interactive visual aid gives the jury something to focus on as they follow the progression of your testimony. It also enables you to control the pace of the presentation and to choose when to reveal specific information to the jury (and the rest of the court) for the most dramatic impact. Something as simple as a laminated exhibit board that allows the use of erasable or permanent marking pens can be very effective. (Although it is not wise to use this method if you have trouble drawing a circle or if your handwriting leaves something to be desired!)

Table 10.1 Comparison of Media Types

Tools	*Options*	*Pros*	*Cons*
Printed graphic boards	Magnetic or Velcro pieces Acetate overlays Matte or gloss varnishes (enables use of marking pens)	Flexible, interactive Can remain displayed in the courtroom indefinitely	Too many boards or boards too large in size can be bulky and unwieldy
Animation	Two- or three-dimensional representations or event re-creations Many levels of complexity and cost	Dynamic Strong impact and recall	Limited on-screen viewing time
Video	Instructional or educational Depositions	Dynamic Real-life quality	Limited on-screen viewing time

Exhibits printed on letter-sized paper or transparency film can be projected from the desktop using a device called an Elmo Visualizer™, which operates much like a traditional overhead projector. Pages are placed on a flatbed directly under a video camera, which sends the image to television monitors placed strategically in the courtroom.

Whether thematic or tutorial, printed or electronic, visuals help you connect with the jury and communicate your ideas clearly and succinctly, which is, of course, your chief aim in expert environmental forensic testimony. When presented in the appropriate media, visuals help organize and clarify the case, stimulate learning and increase retention, break up spoken testimony, and indicate that something significant is being said. Together with a little showmanship, an exhibit lends increased impact and staying power to spoken testimony. In the context of the courtroom, a thoughtfully prepared and meaningfully presented picture is indeed worth a thousand words.

Chapter 11

Evidence Issues: Getting Expert Opinions Past the Judicial Gatekeeper and Into Evidence

Timothy A. Colvig

The ultimate goal of environmental forensics in litigation is to produce a winning expert opinion that carries the day at trial or convinces other parties to settle. An opinion not allowed in evidence will do neither. Judges frequently bar experts from testifying for a variety of reasons. When that happens, the following very unhappy events often occur:

- The client loses the case.
- The client has spent thousands of dollars in expert and attorneys' fees toward an opinion that was never given.
- The client becomes extremely angry with the judge, attorney, and expert.
- The client cannot sue the judge, but may consider suing the attorney, and perhaps the expert, for malpractice.
- The attorney will likely blame the expert for not being expert enough, and the expert will likely blame the attorney for failing to educate the expert about the important evidentiary issues.
- Both the attorney and the expert suffer tarnished professional reputations.

On the other side, the winning party's attorney and expert have impressed their client with their acumen in causing the other expert to be excluded from evidence and enhanced their professional reputations. For these reasons, any environmental practitioner, whether attorney or expert, must address and revisit evidence admissibility issues early and often in the environmental forensic process. This applies both to making sure the expert will be allowed to testify and to taking the necessary actions to enhance the possibility of keeping the opposing expert from testifying.

Expert testimony is indispensable in resolving environmental disputes. Whether an issue arises under CERCLA, the Clean Water Act, toxic tort theory, or an insurance coverage dispute, an expert opinion is inevitably required for a party to present a credible claim. Expert witnesses present crucial testimony and evidence on these (and many other) subjects at trials, but even informal claims presentations and mediations generally require an expert opinion of some kind to persuade opposing parties to settle. An expert is of little assistance if not allowed to testify, but even a shaky expert who is allowed to testify will likely reduce the value of a claim, or the chance of winning the case. Thus, in order to be effective, an expert must be expert enough to provide forceful testimony. Equally important, the environmental practitioner must be able to foresee any potential obstacles to the admissibility of expert testimony and evidence, and make every effort to ensure that the court will allow the admission of a proposed expert's opinion.

The guiding case for anyone seeking the admission of expert testimony is *Daubert v. Merrell Dow Pharmaceuticals.*[1] In *Daubert*, the U.S. Supreme Court set the standard for determining the admissibility of expert testimony. *Daubert* requires federal court judges to act as judicial "gatekeepers" by barring expert testimony unless it is both relevant and reliable. The Supreme Court has further clarified the *Daubert* decision in two later decisions, *General Electric Company v. Joiner*[2] and *Kumho Tire Company, LTD., v. Carmichael,*[3] both of which confirm the importance and challenge of selecting and preparing experts who will meet a court's approval.

As in any other area of the law, practitioners in the environmental arena are required to demonstrate that the expert testimony they intend to introduce is relevant, reliable, and otherwise admissible. In this chapter, we will initially discuss the Supreme Court's expert admissibility requirements, based on its holdings in the *Daubert, Joiner*, and *Kumho* cases mentioned previously. Second, we will discuss clues of how courts may react to

[1]509 U.S. 579 (1993).
[2]522 U.S. 136 (1997).
[3]526 U.S. 137 (1999), 119 S. Ct. 1167.

expert testimony in environmental litigation, by reviewing several environmental cases decided after *Daubert*. Third, we will briefly discuss the logistics of bringing a *Daubert* motion in federal court or an equivalent motion in state court. Fourth, we will include a list of suggestions that environmental practitioners should consider when seeking to introduce or challenge expert testimony pursuant to *Daubert*. Finally, we will discuss additional evidentiary hurdles faced by environmental law practitioners in introducing expert testimony and evidence, even when an expert's testimony has passed the court's "gatekeeping" scrutiny, including laying the foundation for evidence, demonstrating its authenticity, and overcoming potential hearsay objections.

Obviously, there is no way to guarantee that a court will allow the testimony of any particular expert or the admission of a given piece of evidence. Nevertheless, there are certainly steps that can be taken to better the odds. With that in mind, this chapter is designed to provide information on the kinds of issues that judges evaluate when determining the admissibility of proffered expert testimony and other evidence in environmental cases.

HISTORY OF ADMISSIBILITY OF EXPERT TESTIMONY IN FEDERAL COURT

The initial focus of anyone seeking to introduce expert testimony should be whether a proposed expert is qualified and capable of meeting admissibility requirements, as established by the Supreme Court in the 1990s in the *Daubert*, *Joiner*, and *Kumho* cases. While the standards of admissibility for experts have evolved over time, the Court's recent decision in *Kumho* again confirmed and clarified that judges must act as "gatekeepers" when determining the admissibility of expert testimony, and must make certain that proffered testimony is both relevant and reliable. The history and current standard for determining the admissibility of expert testimony is described next.

FRYE "GENERAL ACCEPTANCE" TEST

The controversy concerning which standard a court should apply in evaluating expert testimony began almost 80 years ago in *Frye v. United States*.[4] *Frye* dealt with the admissibility of a lie detector test. In *Frye*, the

[4] 293 F. 1013 (D.C. Cir. 1923).

defendant was subjected to a scientific test designed to determine his innocence or guilt based on the fluctuation of his blood pressure when asked a series of questions related to the crime for which he stood accused. The defendant objected to the test and its results based on the novelty of the testing technique. In stating the rule, the *Frye* Court reasoned that

> [j]ust when a scientific principle or discovery crosses the line between the experimental and demonstrable stages is difficult to define. Somewhere in this twilight zone the evidential force of the principle must be recognized, and while courts will go a long way in admitting expert testimony deduced from a well-recognized scientific principle or discovery, the thing for which the deduction is made must be sufficiently established to have gained *general acceptance* in the particular field in which it belongs. [emphasis added].[5]

In light of the new rule, the *Frye* Court held that the blood pressure test at issue had not yet gained such standing and scientific recognition to justify admitting the expert testimony at hand. The *Frye* "general acceptance" test was applied by federal courts for more than 50 years, and was applied exclusively to expert testimony based on new scientific techniques. The *Frye* test was also adopted and applied by many state courts, some of which still apply the *Frye* test today.[6]

FEDERAL RULES OF EVIDENCE

The *Frye* test stood alone until 1975 when Congress enacted the Federal Rules of Evidence. The Federal Rules of Evidence (FRE) seemed to create a new standard for courts to evaluate the admissibility of expert testimony. FRE 104(a) placed the power of determining the qualifications of a witness in the hands of the district court:

> Preliminary questions concerning the qualifications of a person to be a witness, the existence of a privilege, or the admissibility of evidence shall be determined by the court, subject to the provision of subdivision (b) [pertaining to conditional admissions]. In making its determination it is not bound by the rules of evidence except those with respect to privilege.

In addition, the newly created FRE 702 seemed to set out new criteria for courts to employ when evaluating the admissibility of expert testimony:

[5] Id., at 1014.
[6] State law is discussed *infra*.

> If scientific, technical, or otherwise specialized knowledge will assist the trier of fact to understand the evidence or to determine a fact in issue, a witness qualified as an expert by knowledge, skill, experience, training, or education may testify thereto in the form of an opinion or otherwise.

DAUBERT RELEVANCE AND RELIABILITY

The coexistence of the *Frye* "general acceptance" test and FRE 702 created widespread confusion and division among the courts as to which test was applicable. The Supreme Court finally attempted to clarify the uncertainty in *Daubert v. Merrell Dow Pharmaceuticals*. In *Daubert*, plaintiffs were attempting to introduce expert testimony that they sustained birth defects because their mothers had ingested the drug Bendectin during pregnancy. The Supreme Court held that FRE 702 superseded *Frye*, and that the general acceptance test was not a precondition to the admissibility of scientific expert testimony under the Federal Rules of Evidence. Instead, the Court held that the trial court judge must act as a "gatekeeper" in determining the admissibility of expert testimony to ensure that the testimony rests on a reliable foundation and is relevant to the issue to be determined.[7]

Thus, under *Daubert*, the trial court judge must consider two separate issues regarding the proposed testimony:

- *Relevance* to the issues in the case
- *Reliability* of the basis for the testimony

To determine *relevance*, the trial court is required to ensure that expert testimony "will assist the trier of fact to understand or determine a fact in issue."[8] In other words, the testimony must fit the facts in issue. The *Daubert* Court articulated a number of different factors that a trial court may consider in determining *reliability*, including:

- whether the methodology can be and has been tested;
- whether the methodology has been subject to peer review or publication;
- whether the error rates are known;

[7] 509 U.S. 579.
[8] Id., at 591.

- the existence of standards controlling its operation; and
- whether the theory has achieved general acceptance in the relevant scientific community.[9, 10]

The Court noted that these factors are not a definitive list. For example, the Court acknowledged that peer review or publication is not always a dispositive consideration. "[I]t does not always correlate with reliability" and "[s]ome propositions, moreover, are too particular, too new, or of too limited interest to be published."[11] The facts of each particular case must be considered in reaching the admissibility determination. The Court also emphasized that even if weak evidence is found to be admissible by the trial court, the court and the parties still have the availability of "vigorous cross-examination, presentation of contrary evidence, and careful instruction on the burden of proof" to address perceived deficiencies.[12]

Thus, under *Daubert*, the expert must address two questions. First, is the science being used based on reliable methodologies? In other words, is it consistent with the scientific method in which a hypothesis is stated, tested, repeated, and subject to peer review? Or is it based on bad or junk science resting on hunches and baseless opinions? Second, does the science forming the basis for the opinion fit the particular situation?

GENERAL ELECTRIC COMPANY V. JOINER

As is often the case, the *Daubert* decision explained some issues, but greatly clouded others. The Supreme Court clarified some of them in *General Electric Company v. Joiner*. In *Joiner*, the plaintiff alleged that workplace exposure to PCBs and their derivatives "promoted" his small-cell lung cancer. The plaintiff was an admitted smoker. The district court granted summary judgment for the defendant, finding that there was no

[9]On remand, the Ninth Circuit Court of Appeals articulated another important consideration in determining admissibility of expert testimony. "Whether the experts are proposing to testify about matters growing naturally and directly out of research they have conducted independent of the litigation, or whether they have developed their opinions expressly for purposes of testifying. That an expert testifies for money does not necessarily cast doubt on the reliability of his testimony, as few experts appear in court merely as an eleemosynary gesture. But in determining whether proposed expert testimony amounts to good science, we may not ignore the fact that a scientist's normal work place is the lab or the field, not the courtroom or the lawyer's office." *Daubert v. Merrell Dow Pharmaceuticals, Inc.*, 43 F.3d 1311, 1317 (9th Cir. 1995).

[10]509 U.S. at 593–594.

[11]Id.

[12]Id., at 596.

causative link between the PCB exposure and the plaintiff's illness. The district court also found that the testimony of plaintiff's expert was based on either subjective belief or unsupported speculation. On appeal, the Eleventh Circuit Court of Appeals reversed, stating that the Federal Rules of Evidence relating to expert testimony favored admissibility, and therefore reviewing courts must apply a "particularly stringent standard of review to the trial judge's exclusion of expert testimony."[13] The Supreme Court then reversed the Eleventh Circuit Court, holding that appellate courts must apply a more deferential "abuse of discretion" standard when reviewing a trial court's decision to admit or exclude expert testimony.[14] In reviewing the admissibility of the expert testimony at issue under the more lenient abuse of discretion standard, the Court noted that the expert's causation opinion was based on studies involving the application of massive doses of PCBs to infant mice. The Court observed that these studies were factually too different from the plaintiff's case, and therefore did not provide the proper foundation for the expert's testimony. Moreover, although the expert relied on several other epidemiological studies, none of those studies could demonstrate a link between PCB exposure and the plaintiff's illness.

The Supreme Court also disagreed with the Eleventh Circuit's holding that, pursuant to *Daubert*, it was legal error for the trial court to focus on the expert's conclusions. According to the *Joiner* Court:

> Conclusions and methodology are not entirely distinct from one another . . . nothing in either *Daubert* or the Federal Rules of Evidence requires a district court to admit opinion evidence that is connected to existing data only by the *ipse dixit*[15] of the expert.[16]

The *Joiner* Court held that, even though *Daubert* requires courts to focus solely on "principles and methodology used, and not on the conclusions that they generate," they are not required to admit opinion evidence where there is too great an analytical gap between the data and the opinion proffered.[17]

[13]*Joiner*, 78 F.3d 524, 529 (1996).

[14]Under an abuse of discretion standard, the reviewing court gives great deference to the trial court judge's decision, and will reverse it only if the decision is manifestly erroneous (Id. at 141) or against all reason.

[15]*Ipse dixit* means "he himself said it; a bare assertion resting on the authority of an individual," *Blacks Law Dictionary*, 5th ed. (1979)—or in the common vernacular, it means "because I said so."

[16]*Joiner*, 522 U.S. at 146.

[17]Id.

KUMHO TIRE COMPANY V. CARMICHAEL

Another area left uncertain after *Daubert* was whether its holding applied to *non*scientific expert testimony. Because *Daubert* involved scientific expert evidence, the *Daubert* Court noted that its decision did not address "technical, or other specialized knowledge" also described in FRE 702, thus leaving that issue to be decided another day. As a result, another rift developed in the courts over whether trial courts were required to act as gatekeepers of nonscientific expert testimony. Some circuits held that *Daubert* applied to all expert testimony, while others held it only applied to expert scientific testimony.

In *Kumho*, the Court held that *Daubert* applies to all expert testimony. *Kumho* involved an automobile tire blowout, which caused a crash killing one and causing multiple injuries to others. The plaintiffs contended that the tire was defectively manufactured, basing their claim on studies conducted by their tire failure engineer. The defendants objected to the expert testimony based on the requirements of FRE 702 and the *Daubert* factors. After accepting the case for review, the Supreme Court held that *Daubert's* gatekeeping obligation, which requires an inquiry into both relevance and reliability, applies not only to scientific testimony, but to *all expert testimony*.[18]

The *Kumho* Court also confirmed that in evaluating the engineering testimony at issue, the trial judge could consider the *Daubert* factors to the extent they were relevant, but the application of such factors should depend on the nature of the issue, the expert's area of expertise, and the subject matter of the testimony. The *Daubert* factors were meant to be helpful and not definitive; thus, the gatekeeping responsibility must be tied to the facts of each particular case. Indeed, the trial judge may go beyond the *Daubert* factors in assessing relevance and reliability. The *Kumho* Court confirmed that, regardless of whether an expert's testimony is based on professional studies or personal experience, the court must ensure that the expert employ in the courtroom "the same level of intellectual rigor that characterizes the practice of an expert in the relevant field."[19]

In a concurring opinion, Justice Scalia reiterated the importance of the trial court's gatekeeper role. Accordingly, the discretion of courts to evaluate expert reliability "is not to perform the function inadequately. Rather, it is discretion to choose among reasonable means of excluding expertise

[18] *Kumho*, 526 U.S. 137; 119 S. Ct. at 1174.
[19] Id., at 1176.

that is fausse and science that is junky. Though . . . the *Daubert* factors are not holy writ, in a particular case the failure to apply one or another of them may be unreasonable, and hence an abuse of discretion."[20] Thus, although reviewing courts will address a trial judge's ruling on expert testimony under the deferential abuse of discretion standard, such rulings will be reversed if they unreasonably do not consider applicable *Daubert* factors.

EXPERT ENVIRONMENTAL TESTIMONY UNDER *DAUBERT, JOINER*, AND *KUMHO*

The *Daubert*, *Joiner*, and *Kumho* decisions have required most courts to significantly alter the way in which they evaluate expert testimony. One way for environmental practitioners to predict the way a court will react to disputed expert evidence is to review the decisions of numerous courts that have already applied the *Daubert* standards to scientific expert testimony introduced in environmental disputes. Accordingly, these decisions provide environmental practitioners with a strong indication of the preparation necessary to ensure that expert testimony is admitted into evidence. Although *Daubert* expressly states that the trial judge's evaluation of expert testimony is intended to be flexible and tailored to the facts of each individual case, the following cases provide some indication of how *Daubert* has been and will be applied in the environmental field in courts across the country.

AN EXPERT SHOULD RELY ON MORE THAN "EXPERIENCE" IN REACHING A CONCLUSION

When an expert depends solely on "experience" in reaching a conclusion, a court will usually exclude the testimony for failing to meet the reliability requirement set out in *Daubert*. The Court's holding in *Freeport-McMoran Resource Partners v. B-B Paint Corporation*,[21] provides a clear example of what *not* to do when seeking to admit expert testimony.

Freeport was a case decided by the District Court for the Eastern District of Michigan in which the plaintiff sought contribution for environmental cleanup costs under CERCLA. Several defendants moved to ex-

[20]Id., at 1179.
[21]1999 U.S. Dist. LEXIS 11009 (E.D. Mich. July 16, 1999).

clude the plaintiff's expert on grounds that his testimony failed the *Daubert* tests. The *Freeport* judge sided with the defendants, finding that the expert's proposed testimony was not reliable and in contravention of FRE 702.[22]

The proposed expert testified that he did not follow any published professional standards in reaching his opinions, used no peer-reviewed standards in reaching his conclusions, performed no scientific tests, and did not evaluate any margins of error in reaching his findings. The proposed expert admitted that he relied solely on his "experience" in concluding that residual contaminating solids remained in the various defendants' waste.[23] Moreover, he made no findings on the alleged quantity of hazardous substances for which each of the defendants was responsible, nor did he provide any proof that any particular defendant's hazardous waste was ever sent to the waste site that was the subject of the litigation.[24]

In line with the proposed expert's own testimony, the court found that the proposed expert's findings were not based on a technique or theory that could be or ever had been tested. Likewise, the court found that the proposed expert's theory and technique had not been subjected to peer review or publication. Because the expert's technique had never been tested, there was no known potential rate of error, nor any standards controlling evaluation of the expert's technique. Finally, the court found that the plaintiff had not demonstrated that the expert's proposed testimony enjoyed general acceptance within a relevant scientific community.[25]

Because the expert was unable to substantiate his conclusions other than through his own experience, the court found that the proposed expert's testimony did not meet the *Daubert* requirement of scientific reliability. The court held that the expert "utterly failed" to point to any evidence showing that he had followed scientific methodology,[26] noting that the proposed expert failed to even consider other research, such as learned treatises, policy statements of professional associations, published articles in reputable scientific journals, or the like. In sum, the court could point to absolutely no evidence to support admission of the proposed expert's conclusions.[27]

[22]Id., at *35.
[23]Id., at *30.
[24]Id., at *24–*25.
[25]Id., at *27.
[26]Id., at *30.
[27]Id.

SCIENTIFIC METHODOLOGY MUST BE FOLLOWED

In *Dombrowski v. Gould Electronics, Inc.*,[28] the plaintiffs resided near a battery-crushing and lead-processing plant, and sought an order requiring the defendant owners of the plant to establish a medical monitoring program for individuals exposed to emissions from the defendant's plant. In order to prove their entitlement to a medical monitoring program, the plaintiffs planned on introducing expert testimony of a purported expert in "toxic lead poisoning." Among other things, the proposed expert planned to use technology known as "KXRF bone lead testing" in the monitoring program. The defendants sought to exclude the proposed expert's testimony on grounds that the testimony would fail to meet the standards required by FRE 702, 703, 104, and *Daubert*.[29]

In response, the court set a special hearing to consider the *Daubert* issues. It found that the KXRF methodology, including the instrument proposed to be used in the methodology, did not meet "the standards of reliability, viability, and general acceptance, as outlined in cases such as *Daubert* and other cases, and writings, which guide a court in making this 'gatekeepers' determination as to whether or not such evidence should be presented to a jury."[30] The court's holding was based on the following factors:

- Even though the KXRF method could demonstrate levels of lead within bone structure, the test provided no set standard for determining whether a given level of lead exposure was harmful. In other words, it would be impossible to determine the significance of a particular reading.
- Tests showed that test results varied significantly based on the nature of the instrument, on who was using the instrument, and on how it was calibrated before its use.
- There was extensive evidence and testimony that the KXRF procedure had not gained widespread acceptance in the scientific community, except for experimental and research purposes. The court found there was little evidence that the technique was useful in the clinical sense as was its proposed use in the pending case.[31]

[28] 31 F. Supp. 2d 436 (M.D. Pa. 1998).
[29] Id., at 438.
[30] Id., at 443.
[31] Id., at 442.

The *Dombrowski* court also found that the KXRF technique likewise would not meet the requirements of FRE 403, which provides that "although relevant, evidence may be excluded if its probative value is substantially outweighed by the danger of unfair prejudice, confusion of the issues, or misleading the jury, or by considerations of undue delay, waste of time, or needless presentation of cumulative evidence."[32] Here, the court found that the presentation of evidence obtained through the use of the KXRF technique would most likely "create more confusion, would be misleading, and would possibly lead to an inappropriate decision."[33]

COURTS MAY OR MAY NOT RELY ON THE SPECIFIC FACTORS DESCRIBED IN *DAUBERT*

Although the *Daubert* factors are neither mandatory nor exclusive, many courts heavily rely on them in determining the admissibility of expert testimony. *Curtis v. M&S Petroleum*[34] was an action in the Fifth Circuit brought by refinery workers against the refinery's owner for exposure to excessive amounts of benzene. The plaintiffs sought to introduce expert testimony to prove that their health problems were caused by exposure to the benzene at the refinery.[35]

At the very outset, the court recited the four nonexclusive factors set out in *Daubert* intended to aid in determining whether the expert's methodology was reliable. The court found that "in accord with the [*Daubert*] principles set forth above, [the expert] provided generous support for his general causation theory that exposure to excessive levels of benzene will cause harm such as Plaintiffs experienced."[36]

The Third Circuit in *Paoli Railroad PCB Litigation v. Southeastern Pennsylvania Transportation Authority*[37] relied on the *Daubert* factors to determine admissibility, but also relied on additional factors set out in another Third Circuit case, *U.S. v. Downing*.[38] The additional factors included "the degree to which the expert testifying is qualified, the relationship of a technique to 'more established modes of scientific analysis,' and the 'non-

[32]FRE 403 is discussed *infra*.
[33]Id., at 443.
[34]174 F.3d 661 (5th Cir. 1999).
[35]Id., at 664.
[36]Id., at 669.
[37]35 F.3d 717 (3rd Cir. 1994).
[38]753 F.2d 1224 at 1238.

judicial uses to which the scientific technique are put.'"[39] The *Paoli* court continued by stating that "we now make clear that a district court should take into account all of the factors listed by either *Daubert* or *Downing* as well as any others that are relevant."[40]

THE *DAUBERT* INQUIRY IS FLEXIBLE AND TIED TO THE SPECIFIC FACTS OF EACH PARTICULAR CASE

An expert's reliance on an accepted and tested methodology may be admissible in one case but not another. In *Heller v. Shaw Industries, Inc.*,[41] the plaintiff brought suit against a carpet manufacturer alleging that the fumes generated by the carpet installed in her home caused respiratory illnesses. The plaintiff's expert applied a differential diagnosis theory in evaluating the plaintiff's condition. The proposed expert ruled out possible causes to the plaintiff's symptoms other than the carpet fumes, and studied the temporal relationship between the plaintiff's symptoms and the installation of the carpet. The proposed expert concluded that the fumes emitted from the carpet caused the plaintiff's symptoms.[42]

The *Heller* court recognized that "differential diagnosis consists of a testable hypothesis, has been peer reviewed, contains standards for controlling its operation, is generally accepted, and is used outside of the judicial context."[43] The court noted that "assuming the [proposed expert] conducted a thorough differential diagnosis and had thereby ruled out other possible causes of [the plaintiff's] illnesses, and assuming the he had relied on a valid and strong temporal relationship between the installation of the carpet and Heller's problems, we do not believe that this would be an insufficiently valid methodology."[44]

Although the *Heller* court determined that the proposed expert's *methodology* was valid, it also determined that the expert's *conclusions* did not reliably flow from his data and methodology. The proposed expert failed to demonstrate that the number of particulates emitted by the carpet caused illness in humans; nor was the expert able to demonstrate that the onset of the plaintiff's illness corresponded with the carpet's installation, the

[39] *Paoli*, 35 F.3d at 742, citing to *Downing*, 753 F.2d at 1238–1239.
[40] Id., at 742.
[41] 167 F.3d 146 (3rd Cir. 1999).
[42] Id., at 153.
[43] Id., at 154–155.
[44] Id., at 154.

time when the fumes would be the strongest. Accordingly, the proposed expert's testimony was excluded.[45]

The Fourth Circuit reached a different result in *Westberry v. Gislaved Gummi, AB,*[46] where it admitted expert findings based on a differential diagnosis analysis. In *Westberry*, the plaintiff brought suit against defendant Gislaved for occupational exposure to talc. Gislaved manufactured rubber gaskets for placement in window frames and coated them with a thick layer of talc before shipping them to clients. Gislaved did not warn its clients of the health risks associated with exposure to talc.[47]

Plaintiff James Westberry worked in a plant that used Gislaved's gaskets in manufacturing windows. He was responsible for removing the gaskets from their boxes and cutting them to size. In the process of doing his job, Westberry inhaled large amounts of airborne talc. He did not wear any protective clothing or protective breathing devices while working. As a result, Westberry claimed that he suffered unremitting sinus problems, to the point that he underwent sinus surgery in order to alleviate his sinus pain. Westberry brought suit against Gislaved under a strict liability theory, claiming that Gislaved failed to warn him that breathing airborne talc was dangerous, and that failure to warn caused his preexisting sinus condition to become aggravated.[48]

Westberry introduced the testimony of his treating physician, who used a differential diagnosis theory in determining that the exposure to talc had caused Westberry's problems. The defendants asserted that the expert's testimony should have been thrown out because he conducted no epidemiological studies, published and consulted no peer-reviewed published studies, conducted and consulted no animal studies, and presented no laboratory data to support a conclusion that talc inhalation had caused Westberry's malady. Moreover, the treating physician took no tissue samples to show that talc was found in Westberry's sinuses, nor did he demonstrate that talc at any level was the cause of sinus disease in anyone.[49]

The court disagreed with the defendants, recognizing that the differential diagnosis technique "has widespread acceptance in the medical community, has been subject to peer review, and does not frequently lead to in-

[45]Id., at 159.
[46]178 F.3d 257 (4th Cir. 1999).
[47]Id., at 260.
[48]Id.
[49]Id., at 262.

correct results."[50] The physician's reliance on the relationship between the time that Westberry's symptoms arose and his exposure to the talc was appropriate in light of the particular facts of the case.[51]

The *Westberry* court recognized that, under *Daubert*, its inquiry was to be a "flexible one" and that it should focus on the expert's principles and methodology, not on the expert's conclusions. The court recognized that it had broad latitude to consider whatever factors bearing on validity it found to be useful, and that the particular factors to be considered would depend on the facts of the case and the unique circumstances of the proffered expert testimony. Further, the offered expert testimony is irrefutable or correct with absolute certainty in order to be admissible. The validity of an expert's findings may be established through "vigorous cross examination, presentation of contrary evidence, and careful instruction on the burden of proof."[52]

OPINIONS SHOULD NOT BE DEVELOPED SOLELY FOR PURPOSES OF TESTIFYING

As the Ninth Circuit Court of Appeals observed in *Daubert* on remand, the court in *Mancuso v. Consolidated Electric Company of New York* noted that "[o]ne of the abuses, at which *Daubert* and its sequelae are aimed . . . [is] the hiring of reputable scientists, impressively credentialed, to testify for a fee to propositions that they have not arrived at doing their professional work, rather than being paid to give an opinion helpful to one side in a lawsuit."[53]

Mancuso v. Consolidated Electric Company of New York was brought under the Clean Water Act and other theories arising out of personal injuries suffered from alleged exposure to PCBs. The plaintiffs owned and operated a boat marina and claimed that Consolidated Electric released high levels of PCBs into the water surrounding the marina, causing them to suffer a variety of illnesses. Consolidated Electric made a summary judgment motion pursuant to *Daubert* seeking to bar the plaintiffs' designated expert witnesses from testifying. The court had previously excluded the

[50]Id.

[51]Id., at 265.

[52]Id., at 261, citing to *Daubert*, 509 U.S. at 596.

[53]*Mancuso v. Consolidated Edison Company of New York*, 1999 U.S. Dist. LEXIS 11681, at *62 (S.D. N.Y. July 30, 1999). See also *Daubert v. Merrell Dow Pharmaceuticals, Inc.*, 43 F.3d at 1317.

testimony of the plaintiffs' first expert and greatly limited the testimony of their second expert. Despite having disallowed expert testimony pursuant to *Daubert*, the court had exercised its broad discretion and allowed the plaintiffs additional time to locate alternative experts qualified to present evidence in support of their case.[54]

At the outset of its opinion, the court reviewed and confirmed its reason for disallowing testimony of the plaintiff's first expert and limiting that of the second. The court noted that the plaintiffs' first expert was not a toxicologist and had no training in environmental medicine or toxins. In addition, the court found the expert

- had no idea as to the level of PCB contamination of soil or water which would constitute a health hazard to persons exposed thereto—what is known as a "dose/response" relation;
- had no information as to the concentrations of PCBs in the soil or water at [the] marina;
- did not take blood samples of the plaintiffs to ascertain even the presence, much less the level, of PCBs in their system;
- gave each of the plaintiffs a single, rather brief physical examination, relying entirely upon their medical histories and subjective complaints and what his visual inspection revealed, without objective tests of any kind;
- rendered a written report in which he expressed that every one of their many and diverse symptoms of illness was caused by exposure to PCBs, a conclusion which was essentially ipse dixit, although ostensibly supported by citation or scientific literature not readily available to the Court, without attaching copies of the relevant portions of such publications or even quoting from them."[55]

The court also found that although the first expert was able to show that PCBs generally caused certain illnesses in humans, the expert did not and could not demonstrate that the toxins allegedly released by the defendant specifically caused the plaintiffs' illnesses. The expert did not consider whether the plaintiffs' illnesses resulted from other causes, and therefore could not "make the required differential diagnosis."[56] "He thus totally ig-

[54]*Mancuso*, 1999 U.S. Dist. LEXIS 11681, at *9–*10.
[55]Id., at *7–*8.
[56]Id., at *8–*9.

nored the methodology prescribed by both the World Health Organization (WHO) and the National Academy of Sciences (NAS) for determining whether a person has been adversely affected by a toxin."[57]

The court had also limited the testimony of the plaintiffs' second expert, who offered testimony that one of the plaintiffs suffered learning disabilities due to exposure to PCBs. The court disallowed the testimony because the expert was not a medical doctor, nor an expert on the effects of exposure to PCBs.[58]

In response to the court's invitation to find other qualified experts to support their claims, the plaintiffs selected an alternate expert. Unfortunately, that expert similarly failed to pass *Daubert* muster in order to testify that the PCBs allegedly released by Consolidated Electric were causing their ailments. Although the alternate expert stated that she was a licensed medical doctor in the state of New York, it was revealed that she never did her required medical internship and therefore was never licensed to practic—and thus never treated or worked with patients. In addition, she had no formal training regarding the effects of PCBs and testified that when plaintiffs contacted her she "did not know anything about [PCBs] and . . . thought it would be a nice thing to learn."[59] Her knowledge of PCBs was derived from reviewing medical literature during the three months after being selected.[60]

The court found again that this proposed expert was similarly not qualified. Not only did she know nothing about PCBs before being retained, but also she admitted that she had no knowledge of the background level of PCBs in human blood, even testifying that she did not believe that the average person had any PCBs in his or her blood. At a later deposition, the proposed expert had developed an understanding of background levels of PCBs in humans, yet her understanding was "wildly inaccurate."[61]

Again citing the methodology employed by WHO and NAS, the court found that each step of the plaintiffs' expert's methodology was flawed. The court detailed the expert's methodology and contrasted it with the methodology she should have employed. In the end, the court found that even if the plaintiffs had unlimited time to prove that they had suffered any in-

[57]Id., at *9.
[58]Id.
[59]Id., at *11.
[60]Id., at *11–*12.
[61]Id., at *22

jury, they would not be able to prove that they had suffered any injuries caused by exposure to PCBs.[62]

In its final analysis, the court remarked that the plaintiffs' expert's

> mantle of academic credentials and her dazzling, even if inappropriate, incantation of medical jargon . . . has the troubling potential of misleading a jury of laypersons naturally inclined to be sympathetic to a likable family punished by cruel circumstance and to give them economic relief at the expense of a deep-pocket defendant. That is precisely why the Supreme Court in *Daubert* and in the recent *Kumho* . . . has imposed on the trial court a "gatekeeper" responsibility. . . .[63]

A REVIEWING COURT MUST APPLY AN ABUSE OF DISCRETION STANDARD

In *Moore v. Ashland Chemical Inc.,*[64] the Fifth Circuit Court of Appeals determined that, pursuant to *Daubert* and *Joiner*, the trial judge did not abuse her discretion in excluding expert testimony that the plaintiff's exposure to toluene was the cause of his pulmonary illness. The *Moore* court cited *Joiner*, recognizing that its standard of review was merely for an abuse of discretion on the part of the trial judge. Also, as in *Joiner*, the court determined that there was too great of an analytical gap between the proposed expert's opinion on causation and the data used in support of that opinion. Some of the factors the Fifth Circuit relied on in making its determination are as follows:

- Although the proposed expert testified that he relied on his training and experience and his examination and test results in making his diagnosis, he provided no evidence to the court on how these items were helpful in reaching his diagnosis of the plaintiff.
- The proposed expert testified that he had never before treated a patient who had been exposed to toluene.
- The expert claimed that he relied on another expert's article in reaching his diagnosis. While courts have found that experts may rely on other experts' studies under *Daubert*, the authors of the study relied on in this case made it clear that their conclusions

[62] Id., at 63.
[63] Id., at 61.
[64] 151 F.3d 269 (5th Cir. 1998).

were speculative. Similarly, the study relied on by the proposed expert was based on significantly different facts.

- The other materials and evidence relied on by the expert did not in any way support his conclusions. For example, the expert relied on the temporal relationship between the time of exposure and the time of the injury, yet provided no evidence to support a causal link between the two. The court found that "in the absence of an established scientific connection between exposure and illness, or [other] compelling circumstances . . . the temporal connection between exposure to chemicals and an onset of symptoms standing alone, is entitled to little weight in determining causation."
- The proposed expert had no accurate data on the level of the plaintiff's exposure to toluene, if any, and therefore his theory that any exposure to toluene caused his pulmonary illness was unsupportable.
- The plaintiff's personal habits and other medical history were properly considered by the court, including the fact that plaintiff had smoked for 20 years, had just recovered from pneumonia before his exposure, and had suffered from asthma during childhood.[65]

EXPERT CONCLUSIONS BASED SOLELY ON SPECULATION AND POSSIBILITY ARE INSUFFICIENT

One issue that has posed significant problems to many practitioners is the blurred line between the court's ability to challenge an expert's *methodology* as opposed to an expert's *conclusions*. While the *Daubert* Court held that the court's focus should be on an expert's methodologies rather than his or her conclusions, the *Joiner* Court and others have found that "[c]onclusions and methodology are not entirely distinct from one another . . . [thus] [a] court may conclude that there is simply too great an analytical gap between the data and the opinion proffered."[66] In practice, courts have approached this issue differently.

In *Kalamazoo River Study Group v. Rockwell International Corp.*,[67] for example, the Sixth Circuit Court of Appeals found that "[t]he weighing of

[65] Id., at 279.
[66] 522 U.S. at 146.
[67] 171 F.3d 1065 (6th Cir. 1999).

evidence, credibility determinations, and the drawing of legitimate inferences from the facts are jury functions, and not those of the judge."[68] However, it also ruled that an expert's "facts" are not without challenge. Thus, a court is required to look beyond the conclusions of an expert and determine if an expert's testimony is based on a reliable foundation. In *Kalamazoo*, plaintiff Kalamazoo River Study Group (KRSG) sought contribution under CERCLA for costs from defendant Benteler Industries incurred in responding to the release of PCBs into Michigan's Kalamazoo River. To prevail, KRSG needed to prove that the defendant's release of PCBs caused the response costs incurred in cleaning up the hazardous waste.[69]

In a motion for summary judgment, Benteler admitted that its facility was contaminated with PCBs when it bought the site and that it had released more PCBs during its ownership. However, it disputed the theory that the hazardous waste migrated from its property to the river as alleged by the plaintiff. Benteler offered evidence that PCB contamination was limited to an area near its facility and could not have migrated to the Kalamazoo River. Benteler also offered the testimony of a hydrogeologist who backed the same theory and provided evidence that the PCBs could not have been carried to the river.[70]

In opposition to Benteler's motion, KRSG introduced its own expert's testimony criticizing Benteler's expert's findings. KRSG's expert offered additional evidence, though no affirmative evidence demonstrating that Benteler's contaminants were carried to the Kalamazoo River.[71]

While KRSG asserted that its expert's conclusions created, at a minimum, a factual question of whether Benteler's PCBs reached the river, the *Kalamazoo* court found differently. The court found that KRSG's expert testimony was based on conjecture, speculation, and possibility. Thus, his testimony was scientifically unreliable due to its inadequate factual basis. In granting Benteler's motion for summary judgment, the court recognized that although the drawing of inferences from facts is generally the role of the jury, the court is required to determine whether an expert's conclusions rest on a reliable foundation. In this case, the court found that the expert's testimony left "a gap" that was "simply too wide to allow a jury to speculate on the issue of causation."[72]

[68]Id., at 1068.
[69]Id.
[70]Id.
[71]Id., at 1070.
[72]Id., at 1073.

In *F.P. Woll & Co. v. Fifth and Mitchell Street Corp.*,[73] however, a district court judge denied the plaintiff's motion for summary judgment based on plaintiff's speculation that the defendant or its tenants probably emitted contaminants during the defendant's ownership of a parcel of property. The court also denied the defendant's motion for summary judgment based on the defendant's failure to introduce any evidence that it did not contribute to the release of hazardous substances and the inference from the plaintiff's expert evidence that the defendant may have been culpable.[74]

Suit was brought against the defendant in *Woll* under CERCLA for cleanup costs for contamination to soil on property leased by the defendant to tenants during the late 1960s through the early 1980s. The defendant and its tenants engaged in manufacturing and industrial processes on the property that used and produced some of the chemicals now found in the soil. In support of its case, the plaintiff prepared and submitted reports that stated in part that because the defendant's tenants "conducted precisely the sort of operations that produce the kind of hazardous substances in a manner causing the kind of contamination [that] was commonplace in the tenants' industry at the time, the tenants more likely than not contributed to the release of hazardous substances during the time [the defendant] owned the property."[75]

The *Woll* court found no showing was made that the plaintiff's expert relied on data of a kind not ordinarily relied on by experts in the same field or that his methodology was not scientific. The court found that the expert set out good grounds for his findings and that his conclusions could be reached on the basis of the underlying data and studies that he considered.[76]

At the same time, the *Woll* court found that although the plaintiff introduced sufficient evidence for a jury to find the defendant responsible, the evidence was not conclusive. The court held "the weight ultimately to be accorded the expert's opinions is best left for trial where he will be subject for the first time to cross-examination, where the force of the underlying data can be better assessed and where the degree of professional certainty with which he expresses his opinions will be more apparent."[77]

A different result was reached in *Koch v. Shell Oil Company.*[78] In that

[73]1999 U.S. Dist. LEXIS 894 (E.D. Pa. February 4, 1999).
[74]Id., at *20.
[75]Id., at *19.
[76]Id., at *20–*21.
[77]Id.
[78]49 F. Supp. 2d 1262 (D. Kan. 1999).

case, the plaintiff alleged that he, his family, as well as his cattle, suffered injuries due to exposure to Rabon Oral Larvicide Premix manufactured by the defendant. The trial judge found that the expert testimony introduced by plaintiff was so speculative that it could not meet the reliability requirement of *Daubert*. Because the expert testimony was determined as neither relevant nor reliable, the judge granted the defendant's motion for summary judgment in light of plaintiff's failure to meet its burden of proof by showing causation.[79]

The plaintiff offered the expert testimony of four different medical doctors. In the case of each expert, the court, relying heavily on the *Daubert* factors, found the expert testimony inadmissible. Some of the facts on which the decision to exclude testimony were based included:

- Testimony of an expert that he could not reconstruct the calculations that he performed to reach his results and that there were probably mistakes in his calculations.
- Testimony of an expert that he had never submitted his methodology for peer review to determine its validity, reliability, and reproducibility.
- An expert's failure to demonstrate that his commonsense procedure/methodology had ever been submitted for peer review and failure to show that the procedure/methodology had ever been used again.
- The failure of every expert proffered by plaintiff to demonstrate a causative link between any of the plaintiff's illnesses and exposure to Rabon.[80]

MOTIONS TO EXCLUDE EXPERT TESTIMONY PURSUANT TO *DAUBERT*

Federal Court

Parties should consider bringing a *Daubert* motion as early in the process as possible. The disclosure requirements of Federal Rule of Civil Procedure 26 may give an early warning as to whether an opposing party's expert is vulnerable to a challenge. Under Rule 26(a)(2), all parties must disclose the identity of each expert witness who may testify at trial under FRE 702, 703, and 705. Each expert is required to prepare a report containing:

[79]Id., at 1272.
[80]Id., at 1267–1271.

- A complete statement of all opinions to be expressed and the basis and reasons therefor;
- The data or other information considered by the witness in forming the opinions;
- Any exhibits to be used as a summary of or support for the opinions;
- The qualifications of the witness, including a list of all publications authored by the witness within the preceding ten years;
- The compensation to be paid for the study and testimony;
- And a listing of any other cases in which the witness has as an expert at trial or by deposition within the preceding four years.[81]

These disclosures are to be made at the times and in the sequence directed by the court. In the absence of court direction or stipulations between the parties, the disclosures must be made at least 90 days before trial.

Accordingly, a party will have ample time before trial to evaluate whether expert testimony should be challenged. If there is a basis for challenging expert testimony, a party should file a motion in limine under FRE 104(a) before or at the outset of trial seeking exclusion of the testimony. FRE 104(a) allows the court to conduct preliminary inquiries into a witness's qualifications and the admissibility of evidence, among other things. The rule requires the court to make a "preliminary assessment of whether the reasoning or methodology underlying the testimony is scientifically valid and of whether that reasoning or methodology properly can be applied to the facts in issue. . . ."[82] The court will generally hold a formal *Daubert* hearing to evaluate the admissibility of the testimony. The proponent of the expert opinion testimony will be required to establish by a preponderance of the evidence that the testimony is admissible.[83]

A formal hearing is not required, however, so long as the court has made a proper determination on the record of the admissibility of the proposed expert testimony.[84] Some courts will act *sua sponte* in the absence of any motion to determine whether expert testimony is admissible. For example, in *Hoult v. Hoult*, the First Circuit Court of Appeals pronounced, "[w]e think *Daubert* does instruct district courts to conduct a preliminary assessment of the reliability of expert testimony, even in the absence of an objection."[85]

The Seventh Circuit Court of Appeals' holding in *Kirstein v. Parks Corp.*[86]

[81] F.R. Civ. P. 26(a)(2).
[82] *Daubert*, 509 U.S. at 592–593.
[83] Id., at 593.
[84] *Hopkins v. Dow Corning Corp.*, 33 F.3d 1116, 1123 (9th Cir. 1994).
[85] 57 F.3d 1, 4 (1st Cir. 1995).
[86] 159 F.3d 1065 (7th Cir. 1998).

is another example of a court's power to determine the admissibility of expert testimony on its own in the absence of a party's motion. In *Kirstein*, the Seventh Circuit held that *Daubert* does not require a court to make a specific form of inquiry, nor does it require the court to hold a hearing before ruling on the admissibility of an expert's testimony. The *Kirstein* court found that plaintiff's expert was not a chemist and therefore could not offer testimony that a chemical reaction between the products started a fire. The court further noted that, in the past, it has upheld a judge's *sua sponte* consideration of the admissibility of expert testimony.[87]

State Court

In many state courts, parties will still be required to bring a *Frye* motion instead of a *Daubert* motion when challenging expert testimony. Many states have decided not to follow the Federal Rules of Evidence. While some states have not yet addressed the question of whether *Daubert* supersedes *Frye*, other state courts have plainly decided to continue to follow the *Frye* test.[88] For example, in *People v. Leahy*,[89] the California Supreme Court affirmed that it would continue to follow *Frye* in determining whether new scientific techniques will be admissible, thus requiring that such techniques have gained general acceptance by members of the relevant scientific community before they can be considered by a jury.[90]

In California, the *Frye* test is only applicable to expert testimony based on new scientific techniques. Expert testimony based on scientific techniques that are not new, or testimony based on nonscientific methodologies, will not be subject to the *Frye* test, and must merely pass muster under the California Evidence Code. The California court in *Leahy* determined that *Daubert* did not apply to state court proceedings, because the holding in *Daubert* was based on the Federal Rules of Evidence and not on a particular problem with the *Frye* standard itself. In addition, the *Leahy* court found that, despite the similarities between the Federal Rules of Evidence and the California Evidence Code, *Daubert* remains inapplicable in California courts due to the relative timing of the court decisions and differences in the legislative directives of the two jurisdictions.

[87] Id., at 1067, citing to *O'Conner v. Commonwealth Edison Co.*, 13 F.3d 1090 (7th Cir. 1994).
[88] Cynthia A. Cwik, "Guarding the Gate: Expert Evidence Admissibility," *Litigation*, 25(4): 10.
[89] 8 Cal.4th 587 (1994).
[90] The California test is known as the "Kelly/Frye" test, named "Kelly" after a 1976 California state court decision following *Frye*.

Opening and Closing the *Daubert* Gate: In a Nutshell

The cases discussed previously provide some examples of how a court may react to attempts to admit expert testimony into evidence. A more inclusive list of factors that environmental practitioners should consider is as follows:

- *Kumho* and *Daubert* give an enormous amount of discretion to the trial court, and predicting what any court will require will be inexact. Because the trial court's determinations are subject only to an abuse of discretion standard on review, the environmental practitioner should assume that there will be only one chance to offer the expert testimony. Thus, every effort must be made to ensure the admissibility of the expert's testimony in the first instance. A well-prepared, *Daubert*-qualified expert will also send a message to the opponent that the case is credible, thus enhancing settlement value.
- Both the counsel and the expert must have a strong understanding of the methodologies behind the question at issue. With such knowledge, the proponent of the testimony will be that much better equipped to make certain that the expert has followed established and accepted methodologies in the relevant field, unlike in the *Freeport* case discussed previously. Once a qualified expert is selected, make sure that the expert is following through with the proper methodology and keeping accurate records to present to the court.
- Both the counsel and the expert must be able to clarify, explain, and validate the expert's methodology, the facts, calculations or data used by the expert, and the application of the expert's opinion to the facts of the case.
- Similarly, it is crucial that the expert (and counsel) understand the applicable legal standard of proof for the matter at hand. For instance, in *Paoli Railroad PCB Litigation,* discussed previously, one of the plaintiff's experts testified that PCBs contributed to one of the plaintiff's pregnancy losses and caused her to have a hysterectomy, but did not exclude other possibilities. The court noted that even if the proposed expert's testimony was admissible, such testimony that PCBs were a *possible* cause of the plaintiff's injuries did not have sufficient scientific certainty to survive summary judgment.[91]

[91]*Paoli*, 35 F.3d at 766.

- Part and parcel to the importance of understanding the science, technique, or specific knowledge at issue is focusing on the discreet question to be addressed by the expert. The opinion must be relevant; that is, it must fit the facts at issue and be helpful to the judge or jury.
- The importance of learning the science, technique, or specialized knowledge is also important when taking on the opponent's expert. While judges may exclude expert testimony *sua sponte* as in the *Kirstein* case, in most cases it will be left to counsel to move to exclude expert testimony under *Daubert*. Although the *Daubert* factors are not mandatory, an environmental practitioner should be prepared to answer any of them. In the *Curtis* and *Paoli* cases, for example, the court evaluated every factor and applied them to the experts' testimony.
- Consider how and why your expert is different from your opponent's expert, and be prepared to address those differences with both of them.
- Do not confuse *Daubert*'s tests of methodology with a license to attack the correctness of the conclusion. For instance, in *Westberry*, the court noted that the remedy for a shaky opinion or conclusion based on an accepted methodology is vigorous cross-examination, presentation of contrary evidence, and careful instruction on the burden of proof.[92]
- Just because an expert uses an approved methodology does not end the inquiry. The expert must still formulate his or her opinions with "the same intellectual rigor that characterizes the practice of an expert in the particular field."[93] Thus, the expert must consider all of the information that experts in that expert's field consider in arriving at conclusions—not just a single methodology in all cases.
- Experts testifying based on experience, but not education or formal training, are still subject to *Daubert* factors, including whether the methodology can be or has been tested or has a verifiable error rate.
- Modeling will be a fertile area for *Daubert* challenges. Be ready to address the source of the model; the extent to which it has been tested, peer reviewed, and accepted by the profession; how it has been modified and whether others in the profession have modified

[92]See also, *Daubert*, 509 U.S. at 596.
[93]*Kumho*, 526 U.S. 137, 119; S. Ct. at 1175.

it in a similar way; and its assumptions, inputs, error rates, and flaws.

- Be prepared to bring or resist a *Daubert* motion early in the process. For instance, if the opponent's expert appears vulnerable, consider making the motion together with a motion for summary judgment if the opponent's case hinges on the expert's testimony.
- In the appropriate circumstance, consider asking the judge to appoint his or her own expert to address reliability issues, and be ready if the judge suggests such an appointment *sua sponte*. Justice Breyer suggested this possibility in his concurring opinion in *General Electric Company v. Joiner*: "[A] judge could better fulfill this gatekeeper function if he or she had help from scientists. Judges should be strongly encouraged to make greater use of their inherent authority . . . to appoint experts. . . ."[94] Apart from assisting the judge in the *Daubert* process, the court-appointed expert can testify as well. Experience tells us that the court-appointed expert's opinion has an enormous impact on the outcome. For instance, as noted by Justice Breyer in that same opinion, a study of 58 cases with court-appointed experts revealed that only two of the cases resulted in decisions inconsistent with the positions of the experts, and those two involved bench trials decided by legal issues independent of the technical issues addressed by the expert.[95]
- Consider using a "peer review" expert to assist in making sure that the testifying experts on both sides pass muster under *Daubert*.
- Beware the expert whose methodology was prepared for litigation only.
- Investigate the organization behind an expert's diplomas, certifications, awards, and peer-reviewed articles. With *Daubert* may come litigation-driven diploma mills and organizations formed for the purpose of providing testifying experts the imprimatur of peer recognition, acceptance, and review.[96]
- Prior testimony does not equal peer review.
- *Daubert* factors may or may not be accepted or used in an arbitration, but a *Daubert*-ready expert will almost certainly be more credible to arbitrators.

[94] *Joiner*, 522 U.S. at 150.

[95] Id.

[96] See, J. Piller, "Beware the '*Expert*': Diploma Mills Help Marginally Qualified Experts Fudge Credentials," *Litigation News*, 24(5) (July 1999).

- Similarly, even in state courts that do not employ the *Daubert* factors, a *Daubert*-ready expert will be superior to one who is not, because all of the *Daubert* factors will certainly be relevant to cross-examination.

Additional Challenges to the Admissibility of Expert Testimony and Evidence

Although the *Daubert* relevance and reliability standard is grounded in FRE 702,[97] an expert's testimony may be challenged on many other evidentiary grounds as well. The *Daubert* Court itself recognized that a court assessing a proffer of expert testimony should also be mindful of other applicable Federal Rules of Evidence, including FRE 403 and 703. Because the *Daubert* standard is flexible and the courts have been given an enormous amount of discretion in determining the admissibility of expert evidence, these additional factors may also be relied on by courts in exercising their gatekeeping functions. At the same time, these Federal Rules of Evidence provide alternative bases for a party to challenge expert testimony and evidence and should not be overlooked.

ADDITIONAL CHALLENGES TO EXPERT TESTIMONY

Challenges Under FRE 403: Probative Value Versus Danger of Prejudice

Under FRE 403, "although relevant, evidence may be excluded if its probative value is substantially outweighed by the danger of unfair prejudice, confusion of the issues, or misleading the jury, or by considerations of undue delay, waste of time, or needless presentation of cumulative evidence."

In that regard, the *Daubert* Court noted that "[e]xpert evidence can be both powerful and quite misleading because of the difficulty in evaluating it. Because of the risk, the judge in weighing possible prejudice against pro-

[97]FRE 701, 702, and 703 are expected to be amended to make them consistent with *Daubert* effective as early as December 2000. Rule 702 would be amended to add language limiting expert testimony unless "(1) the testimony is based upon sufficient facts or data, (2) the testimony is the product of reliable principles and methods, and (3) the witness has applied the principles and methods reliably to the facts of the case." *Litigation News*, 24(6) (September 1999).

bative force under FRE 403 of the present rules exercises more control over experts than over lay witnesses."[98] An example of the court's FRE 403 power can be found in *Dombrowski v. Gould Electronics*, discussed previously. The *Dombrowski* court exercised its power to invoke FRE 403, finding that the expert's technique would most likely create "more confusion, would be misleading, and would possibly lead to an inappropriate decision."[99]

Challenges Under FRE 703: Bases of Expert Testimony

Although the *Daubert* decision was made pursuant to FRE 702, its pronouncement that trial courts must act as "gatekeepers" overlaps the requirement of FRE 703, which allows an expert's testimony to be challenged on grounds that the expert's opinions are based on improper facts or data. FRE 703 describes the sources of information on which experts may base their testimony:

> The facts or data in the particular case upon which an expert bases an opinion or inference may be those perceived by or made known to the expert at or before the hearing. If of a type reasonably relied upon by experts in the particular field in forming opinions or inferences upon the subject, the facts or data need not be admissible into evidence.

The first clause of FRE 703 permits an expert to base his opinion on facts within the expert's personal knowledge, as well as facts or data made known to the expert at or before a legal proceeding. The first clause is a codification of pre–Federal Rules of Evidence practice. The second clause allows an expert to base his opinion on facts or data of the type reasonably relied on by other experts in the expert's particular field in reaching their conclusions. The materials relied on do not need to be admissible into evidence. The second clause of FRE 703 has proven to be more controversial in the courts.[100]

[98]*Daubert* citing to Weinstein, *Rule 702 of the Federal Rules of Evidence Is Sound: It Should Not Be Amended*, 138 F.R.D. 631, at 632 (1991).

[99]*Dombrowski*, 31 F. Supp. 2d at 443.

[100]As discussed in footnote 97, FRE 703 is expected to be amended effective as early as December 2000. The presently proposed amendment to FRE 703 would limit the admissibility of evidence relied on by the expert by adding the following language: "Facts and data that are otherwise inadmissible shall not be disclosed to the jury by the proponent of the opinion or inference unless the court determines that their probative value in assisting the jury to evaluate the expert's opinion substantially outweighs their prejudicial value." *Litigation News*, 24(6) (September 1999).

Because FRE 703 is concerned with establishing the trustworthiness of expert opinion, the proponent of expert testimony must demonstrate that other experts in the same field as the testifying expert would reasonably rely on the same kind of information when conducting their own research. Some courts have held that the district judge must make a factual finding as to what data similarly situated experts find reliable, and that if an expert contends that he or she relied on materials that are of a type customarily relied on by other experts in the field, the expert's testimony will generally be allowed.[101]

In *In re Japanese Electronics*,[102] the trial court found that the plaintiffs' expert's findings were not grounded on the types of documents that would be relied on by other experts in the same field. The experts testified that other experts in their field customarily relied on similar information. Nevertheless, the court ruled that the only way the expert evidence could be admitted would be if the underlying documents themselves were admissible. On appeal, the Court of Appeals for the Third Circuit reversed, stating that the proper inquiry is not what the court finds as reliable, but what other experts in the same field customarily find as reliable. The court held that "once the court finds that the data relied on is such as experts in the field reasonably rely upon, the rigorous examination should be conducted in the cross-examination. . . ."[103] The U.S. Supreme Court granted *certiorari* on another issue in the case and reversed, but did not overrule the court of appeals' decision on the admissibility of the expert's testimony.

Still other courts have taken a more restrictive view of FRE 703, for example, the trial court in *In re Japanese Electronics*. These courts have conducted their own review of the reliability of the facts and data relied on by an expert, consequently disallowing expert opinion.[104] In this way, the line has been blurred between FRE 702 and FRE 703. FRE 703's requirement that an expert's opinions must not be based on improper facts or data is similar to language in *Joiner*: "[n]othing in either *Daubert* or the Federal Rules of Evidence requires a district court to admit opinion evidence which is connected to existing data only by the ipse dixit of the expert."[105]

[101] *In re Japanese Electronics*, 723 F.2d 238, 277.

[102] Id.

[103] Id., at 276–277.

[104] See, e.g., *Christophersen v. Allied Sign-Corp.*, *cert. denied*, 112 S. Ct. 1280, 939 F.2d 1106, 1113–1114 (5th Cir. 1991). See also *Ealy v. Richardson-Merrell, Inc.*, 897 F.2d 1159, 1161–1162, *cert. denied*, 498 U.S. 950 (D.C. Cir. 1990).

[105] *Joiner*, *supra*, 522 U.S. at 146.

Challenges to Documentary and Physical Evidence

Some of the most effective evidence in environmental disputes is documentary and physical evidence. In many environmental disputes, for example, expert testimony is based on documentary or physical evidence that demonstrates that a party has participated in past wrongdoing, such as contributing to the contamination of a given parcel of property. Documentary and physical evidence can be equally important when used as impeachment material against the opposing party's witnesses. Indeed, when considering conflicting evidence, judges and juries traditionally give the most weight to documents or physical evidence created at the time or directly linked to an alleged incident of wrongdoing. Thus, documentary and physical evidence may be the most effective and persuasive way of linking a party's past conduct to present accountability.

Finding a way to introduce this information into evidence can be a challenge. Often in environmental cases, for example, much of the evidence consists of historical documents, dating back several decades. Evidence may also consist of physical samples of soil, water, or the like allegedly containing hazardous pollutants on which an expert's findings may be based. Environmental practitioners must be prepared to overcome numerous potential objections to these kinds of evidence. At the outset, environmental practitioners should be prepared to face the same relevancy and FRE 403 objections as they face with the introduction of expert testimony. They also need to be prepared to lay the foundation for and authenticate any documentary or physical evidence that they plan on introducing to the court. Finally, with regard to documentary evidence, environmental practitioners should be prepared to overcome any hearsay objections in order to ensure its consideration by the judge or jury.

Relevance Requirement and FRE 403 Also Apply to Documentary and Physical Evidence

As with all other evidence, documentary and physical evidence must be relevant to the disputed issue. The judge determines relevance. Determinations of relevance generally hinge on the question of whether the evidence tends in any way to prove or disprove any question at issue in a dispute.

Similarly, as with expert testimony, a court may choose to exclude relevant documentary and physical evidence under FRE 403, if "its probative value is substantially outweighed by the danger of unfair prejudice, confusion of the issues, or misleading the jury, or by considerations of undue delay, waste of time, or needless presentation of cumulative evidence." In environmental disputes, the kinds of evidence at issue may be perceived as confusing, misleading, cumulative, or wasteful of time. A party should be prepared for an objection on FRE 403 grounds either from the opposing party or from the judge.

Foundation and Authentication

Before a court will consider documentary or physical evidence, its proponent must lay the foundation for its admission. In order to lay the foundation, the proponent of the evidence must show that the witness through whom the evidence is admitted is competent, the evidence is relevant, and the evidence is authentic or genuine.

FRE 901 requires the authentication of all evidence.[106] FRE 901(b) lists some of the techniques that are commonly used to authenticate documents, such as through the testimony of a witness with knowledge of the document or specimen, the testimony of a nonexpert witness as to the genuineness of the handwriting contained in a document, or the comparison of a document or specimen by an expert witness with other documents or specimens that have already been authenticated.

Chain of Custody

A commonly asserted foundational challenge to both documentary and physical evidence is the allegation that a document or specimen's chain of custody has been broken. In such a case, a party alleges that a document or specimen's integrity has been compromised due to its mishandling by individuals who cannot guarantee that the document or specimen has been maintained in its original form. If such a challenge is made, the proponent of the evidence will be required to show, with a reasonable amount of certainty, that the evidence has not been altered in any way. The proponent will be required to make an accounting of each link in an item's chain of custody to prove that the document or specimen's integrity remains intact. While a break in the chain of custody of evidence in an egre-

[106]FRE 901(a) states, "The requirement of authentication or identification as a condition precedent to admissibility is satisfied by evidence sufficient to support a finding that the matter in question is what its proponent claims."

gious case forms the basis for the exclusion of evidence, it most frequently affects the weight and not the admissibility of the evidence.[107]

United States v. Davis,[108] was an action under CERCLA in which the U.S. government sought to recover environmental cleanup costs. The defendant brought a motion in limine to exclude some of the government's evidence. One of the bases of the defendant's challenge was its allegation that there was an inadequate showing of a proper chain of custody for soil samples allegedly taken at the site and that the government's reports contained conflicting results for identical samples. In its defense, the government responded that the number of samples without a well-documented chain of custody was small in comparison with the total number of samples. The government also argued that the remaining samples were handled using routine government procedures following set government standards. Likewise, the government asserted that the variation in some of sample results was not significant in light of the total number of trustworthy samples taken.[109]

The *Davis* court ruled that, given the magnitude of the government's report, it was inevitable that some mistakes would be made. Minor mistakes in methodology did not taint the entire report. The defendant's remedy was not the exclusion of the evidence, but rather the ability to attack the mistakes in the report at trial and to question the weight to be given to the value of the government's report.

Best Evidence Rule

Another commonly asserted foundational challenge to documentary evidence is that the proponent of the evidence has not provided the "best evidence." Under FRE 1002, "[t]o prove the content of a writing, recording, or photograph, the original writing, recording, or photograph is required, except as otherwise provided in these rules or by Act of Congress."

The best evidence rule is meant to ensure that the judge and jury are presented with the most reliable evidence possible where there is a risk that evidence could be fraudulent or compromised. The rule only applies where a proponent is seeking to introduce the content of a document, pho-

[107]See, *United States v. Casto*, 889 F.2d 562, 568–569 (5th Cir. 1989), *cert. denied*, 493 U.S. 1092 (1990).

[108]826 F. Supp. 617 (D. R.I. 1993).

[109]Id., at 623.

tograph, or recording, so the rule is inapplicable, for example, where a document is introduced to refresh a witness's recollection. There are several exceptions to the best evidence rule, however. FRE 1003, for example, provides that exact duplicates of documents can be admitted to the same extent as originals. FRE 1004 provides that the original is not required where the original is lost or destroyed, not obtainable, in possession of an opponent, or not closely related to a controlling issue in a case. FRE 1005 provides that, under certain circumstances, the contents of official public records may be proven by the presentation of copies testified or certified to be correct by competent witnesses. Rule 1006 provides that the contents of voluminous writings, recordings, or photographs may be presented to the court in the form of charts, summaries, or calculations.

Environmental practitioners should be prepared to meet any challenges made under the best evidence rule when seeking to introduce documentary evidence. In most cases, the challenged evidence fits into one of the exceptions to the rule.

Hearsay

Documentary evidence may also be challenged as hearsay. FRE 802 provides that any out-of-court statement offered to prove the truth of the matter asserted by the statement is hearsay, and is inadmissible as evidence. Most documents offered as evidence, including the types of documents often relied on by environmental practitioners, constitute hearsay in that their contents are usually offered for their truth. Thus, the environmental practitioner must find a basis to admit these documents, as well as find a witness who can lay the foundation for their introduction into evidence.

While hearsay is generally inadmissible at trial, courts have developed and legislatures have codified several exceptions to the hearsay rule. First, the courts have interpreted FRE 703 to create a partial loophole for documents considered by expert witnesses. In addition, FRE 803 and 804 outline 24 exceptions to the hearsay rules, 16 of which specifically relate to documents and documentary evidence. Some of these exceptions are discussed next.

FRE 703 "Exception"

Although not an exception to hearsay under FRE 803 and 804, FRE 703 provides that experts may discuss and explain, on the stand, documents, facts, and data on which they relied in reaching their conclusions. The documents, facts, and data do not need to be admissible into evidence. The expert may discuss and explain the documents so long as they are of a type reasonably relied on by similarly situated experts in the same field as the testifying expert.[110]

Under FRE 703, a court will generally allow an expert to testify about the information underlying his or her testimony, but will not allow the information to be admitted for its truth. A court will generally only admit the expert's testimony regarding the underlying information for the purpose of laying a foundation, but not as substantive evidence.[111] However, during a typical examination of an expert witness, an expert will be required to discuss the data or facts that influenced his or her opinions. An expert's testimony concerning the basis of his or her conclusions may be just as effective as introducing the underlying materials themselves.

Even if a court allows an expert to testify about underlying documents that are otherwise inadmissible, and even if an expert effectively describes the contents of the inadmissible documents, the weight of the expert's testimony will undoubtedly suffer on cross-examination based on the untrustworthiness of the underlying research. Indeed, the impact of expert testimony can be greatly affected by its factual underpinnings. Therefore, experts and practitioners should make every effort to gain the admission of underlying documents through one of the exceptions to the hearsay rule described in FRE 803 and 804.

FRE 803 and 804: Exceptions to Hearsay

Most documentary evidence constitutes hearsay. Due to the inherent trustworthy nature of certain categories of documents, however, courts and legislatures have carved out whole categories of documents as exceptions to the hearsay rule that are therefore admissible into evidence. These exceptions have been codified in FRE 803 and 804.

There are several categories of documents that are often relied on by ex-

[110]See, footnote 100, regarding expected changes to FRE 703.

[111]While inadmissible material relied on by an expert may not be shown to the jury or formally received into evidence as an exhibit, a proposed amendment to FRE 703 is currently pending, which would make the introduction of the inadmissible material relied on by the expert possible. However, the amendment still would not allow the trier of fact to rely on the evidence for its substance. The proposed amendment to FRE 703 reads as follows: "If the trial judge finds the probative value of the information in assessing the expert's opinion substantially outweighs its prejudicial effect, the information may be disclosed to the jury, and limiting instruction must be given upon request, informing the jury that the underlying information must not be used for substantive purposes." Proposed Advisory Committee Note to Proposed Amendment to FRE 703, 156 F.3d p. Ct. R. 136 (1998).

perts in environmental disputes. These include historical documents, corporate documents, aerial photographs, and newspaper articles. There are several exceptions to the hearsay rule under which these types of documents may gain admittance. Some notable exceptions are the business records, public records, and ancient documents exceptions.

Business Records Exception Where corporate documents are at issue, experts and practitioners may seek admission of evidence under FRE 803(6), the business records exception to the hearsay rule. FRE 803(6) provides:

> [A] memorandum, report, record, or data compilation, in any form, of acts, events, conditions, opinions, or diagnosis, made at or near the time by, or from information transmitted by, a person with knowledge, if kept in the course of a regularly conducted business activity, and if it was the regular practice of that business activity to make the memorandum, report, record, or data compilation, all as shown by the testimony of the custodian or other qualified witness, unless the source of information or the method or circumstances of preparation indicate a lack of trustworthiness is not excluded by the hearsay rule even though the declarant is available as a witness.

The basic theory of the business records exception is that records kept in a routine manner normally possess a circumstantial probability of trustworthiness, and therefore ought to be received in evidence as an exception to the hearsay rule.

As stated in the language of the exception, the document being offered must have been created at or near the same time that the information it contains was gathered. Likewise, the document must have been kept in the course of a regularly conducted business activity. The requirements of FRE 803(6) must be shown by the testimony of the custodian or other qualified witness. It is typically necessary to have the testimony of a witness capable of explaining the recordkeeping procedures. While a witness who was responsible for the creation of the record has most readily been held to be a qualified witness, the witness need not have actually created the record in question or even have personal knowledge of its creation or of the matter created. Additionally, the witness need not have been in the employ of the business to which it relates at the time of its making, so long as the witness understands the system that was used in its creation.[112] Essentially, these requirements may be satisfied by the testimony of anyone

[112] 4 Weinstein's Evidence Section 16.07[2][d], pp. 16-32 (1999).

who is familiar with the practices of the business in question and the manner in which the record was prepared.

Public Records Exception Where public records are at issue, experts and practitioners may seek admission of evidence under FRE 803(8), the public records exception to the hearsay rule. The public records exception is similar to the business records exception, and provides that the following documents are admissible hearsay:

> Records, reports, statements, or data compilations, in any form, of public offices or agencies, setting forth (a) the activities of the office or agency, or (b) matters observed pursuant to duty imposed by law as to which matters there was a duty to report, . . . or (c) in civil actions or proceedings . . . factual findings resulting from an investigation made pursuant to authority granted by law, unless the sources of information or other circumstances indicate lack of trustworthiness.

Similar to the business records exception, the basic theory of the public records exception is that records kept in a routine manner normally possess a circumstantial probability of trustworthiness, and therefore ought to be received in evidence as an exception to the hearsay rule. The proponent of a public record must show that the record was authorized by law and kept in a public office where items of the same nature are kept. The document itself must have been created under a duty imposed by law or as required by the nature of the public office. The facts contained within the documents are then admissible into evidence. For example, in *United States v. Davis*, discussed previously, the defendant also sought to exclude the government's report because it contained hearsay. The government successfully gained the admission of the report under FRE 803(8).[113]

Ancient Documents Exception Where documents in existence for 20 years or more are at issue, experts and practitioners may seek admission of evidence under FRE 803(16), the ancient documents exception to the hearsay rule. FRE 803(16) provides that the hearsay rule does not exclude the admission of statements in a document in existence for 20 years or more and for which the authenticity has been established.

In order to establish an ancient document's authenticity, FRE 901(b)(8) provides that:

[113] *Davis*, 826 F. Supp. at 624.

> Evidence that a document or data compilation, in any form, (a) is in such a condition as to create no suspicion concerning its authenticity, (b) was in a place where it, if authentic, would likely be, and (c) has been in existence 20 years or more at the time it is offered.

In order for an ancient document to be authenticated, the document must be "in such a condition as to create no suspicion concerning its authenticity." Essentially, any evidence of erasures, discontinuity of handwriting, pagination or the flow of its contents, or any other change in authorship that is not explained as being consistent with the document's authenticity may give rise to suspicion. Attestation, recordation, or notarization provides substantial support for the foundation of a proffered document, if such is seemingly authentic.

The document must be located "in a place where it, if authentic, would likely be." This requirement may be satisfied by evidence regarding the actual location in which the document was found, the circumstances surrounding the discovery of the document, the individual finding the document or having custody of the document, or almost any evidence indicating that the document is what it purports to be. Documents found to be in the files of a party with adverse interests or in the files of the party opposed to the introduction of the evidence are routinely admitted. Basically, all that is required is that the document, when it is found or produced, came from the type of location where it might reasonably have been expected to be or from an individual who might reasonably have been expected to have it.

Additionally, the document must have been in existence for at least 20 years at the time of the offering. The age requirement can be met in a number of ways. It can be shown by a witness with knowledge of the age of the document, expert testimony, physical appearance of the evidence, or by the contents of the document itself together with evidence of the surrounding circumstances. A recordkeeper can testify that the document has been in his or her possession or in the possession of another for at least 20 years. An expert can testify to the age of the paper, ink, typewriting, or other identifying marks using either visual identification or scientific testing to make the determination. Courts are permitted to consider dates, seals, recordations, attestations, postmarks, and other identifying markings. In some instances, dates on the document alone have been held insufficient evidence of age. Indeed, the contents of the document itself may assist in the determination of its age. References to individuals deceased more than 20 years or signatures by notaries or employees no longer in the employ of the company or located within the building where the document was created for more than 20 years tend to support a finding of the requisite age.

These are only some of the exceptions to the hearsay rule that are commonly relied on in the environmental field to gain the admission as substantive evidence of documents relied on by experts in forming their opinions. Practitioners should consider all of the hearsay exceptions when faced with challenges of admissibility.

Chapter 12

Alternative Dispute Resolution Techniques

Michael Kavanaugh and Daniel Weinstein

Alternative dispute resolution (ADR) is now well established and has long been used in the resolution of legal disputes. Its earliest origins can be traced back to the biblical prophets (see Genesis). Modern methods on the ADR menu include dispute review boards, facilitation, formal fact finding, mediation, mini-trials, arbitration, and judicial reference.

The application of ADR methods to the resolution of environmental disputes has a shorter history compared to applications within other legal arenas. Nevertheless, ADR has proven to be an effective means of accelerating dispute closure and reducing overall transaction costs in many environmental cases. Opinions on the suitability and effectiveness of ADR techniques when applied to environmental disputes vary. Some proponents argue that the various ADR procedures have wide applicability [1–3], while a small minority argue that the process often has liabilities that restrict its application to a narrow range of issues [4]. Clearly, different ADR techniques have advantages and disadvantages, depending on the characteristics of the environmental dispute, and one selection is not necessarily the optimum method for all such disputes. Nonetheless, ADR techniques enjoy increasing popularity, and their use is likely to expand in the future as a prominent legal tool for resolving environmental disputes.

In fact, ADR techniques may be best suited to resolve certain types of environmental disputes where the inherent level of technical uncertainty in the facts of the case makes the outcome of a jury or judge trial highly uncertain and largely dependent on perceptions rather than the factual merits of a case. That is, a jury of our peers would not have a clue of what we are talking about, with the result being a "crap shoot."

This chapter provides an overview of ADR techniques with an emphasis on their application to the resolution of environmental disputes. Issues we will address include (1) ADR techniques, (2) selecting the appropriate ADR technique, (3) the role of neutral technical experts and forensic analyses in ADR processes, and (4) managing the process.

DESCRIPTION OF ADR METHODS

The relative success of different ADR methods depends on the type, age, and other characteristics of the dispute, as well as the preferences, resources, and personalities of the parties. Each method has distinct advantages and disadvantages, in part depending on the context of its application. Table 12.1 provides a brief summary of the seven most commonly used ADR methods in resolving environmental disputes.

Dispute Resolution Boards

A dispute resolution board consists of three or more experts retained by the parties to provide ongoing advice on the conduct of a particular project. This process is most widely used in the construction industry [5–7] but has had some limited application in environmental remediation projects, which have a significant construction component [8]. The intent of such a board is to avoid the adversarial environment that is typically encountered in construction projects. The principal parties, namely, the owner and the contractor, typically mutually agree on members of the panel who meet on a regular basis and who are asked to provide independent advice on major project issues[1] in an effort to avoid disputes before they arise. Decisions and recommendations of the board are nonbinding. While this method of ADR avoids litigation, there is the potential for delays in construction while the board, first, finds the time to meet (together) and, second, deliberates on the issues at hand. These delays can be insignificant when compared to the never-ending remedial investigations and public comment marathons. As such, environmental consultants, used as a resolution board, has viability. The downside, of course, is the subsequent inability of more than one consultant agreeing to anything.

[1]All too often however, scheduling three experts can be problematic, at best.

Facilitation (A Second Opinion)

In the early stages of a dispute, parties may seek advice from third-party neutrals to determine if the dispute can be resolved without litigation. Facilitation is not defined as a formal process; thus, diverse approaches have been used to achieve the goal of the facilitation process. Generally, the neutral would play a minimal role in the process, except to serve as a catalyst for informal exchanges between the parties. The third-party neutral would meet with the parties to the dispute and determine agreed-on ground rules for the process. Typically, no formal documents are exchanged, but facts may be presented to the neutral verbally by the parties, to determine the potential for early resolution of the dispute. Usually, this ADR technique would not involve an independent technical expert or experts assisting the facilitator, but instead relies on the ability and experience of the facilitator to assist the parties in assessing the potential for dispute resolution.

Fact Finding

This technique is a more formal approach to facilitation and early assessment of the potential for dispute resolution. A neutral fact finder is selected by the parties, with specific expertise in the type of environmental dispute or in the technical issues underlying the case, and rules of the process are determined by the parties. Alternatively, the neutral fact finder may retain qualified technical experts who can provide an independent assessment of the merits of the dispute from a technical perspective. Such experts may also generate new data for consideration by the parties, if there are issues at stake for which limited information is available. The fact-finding process must be agreed to by the parties and usually involves informal procedures, such as meetings without document exchanges or reports by independent technical experts to the parties. Parties typically request a nonbinding opinion from the neutral regarding the merits of each position in the dispute. The advantages of this method are early identification of the potential for use of more formal ADR techniques (e.g., mediation, mini-trials) to resolve the dispute without litigation. The major limitation of this approach is the fear of early disclosure of facts that may be detrimental to one of the parties. If the law is against you, argue the facts . . . if the facts are against you, consider another ADR option.

Table 12.1 Comparison of Alternative Dispute Resolution Methods in Resolving Environmental Issues

Technique	*Description*	*Applicability*	*Advantages*	*Disadvantages*
Dispute Review Boards	Panel of experts established to review and provide nonbinding recommendations on disputes during project implementation	Used in construction industry environmental projects; could be used in remediation projects	Dispute prevention; ensures best practices; nonadversarial	Cost; project delays; potential for antagonism between parties
Facilitation	Third-party neutral assists participants in nonsubstantive issues related to dispute	In early stages of dispute, parties may benefit from initial fact finding by neutral	Informal and flexible; low cost; useful screening for suitability	Premature revelations; inadequate documentation and therefore ineffective
Fact finding	Neutral assists parties in developing reliable information	Provides neutral compilation of relevant case information; independent analysis of facts of case	Early warning system for parties to assess suitability of other ADR steps; unbiased review of technical facts	May be considered as stealth form of discovery; uncertain costs

Mediation	Mediator or co mediator facilitates the mutual negotiation of dispute	Applicable to wide range of environmental disputes; required mutual interest in reaching settlement	Cost reductions compared to litigation; rapid resolution; can maintain business relation between disputants	Risk of failure; nonbinding; private transactions may not be suitable for public policy environmental disputes
Mini-trials	Parties present case before third-party neutral; follows rules of regular trial, but scope limited by agreement	Used near final stage of dispute prior to litigation when potential for settlement appears promising	Less expensive than regular trial; allows for cross-examination	Potential for high costs; decisions nonbinding
Arbitration	Arbitrator or arbitration panel decision provides binding on issues in dispute	Contractually mandated in some cases; best suited to resolution of narrow issues	Cost savings; can be completed quickly; formal process	No justification required of decision; no appeals except in rare cases
Judicial reference	Private decision maker adjudicates proceedings between parties	Parties must adhere to rules of evidence and civil procedure; last-resort option	Retains advantages of private process; parties select adjudicator; costs lower than regular trial; avoids jury	Costs higher than mediation, high costs if fails

Mediation

Mediation is the most popular ADR technique. It involves the use of a third-party neutral, with or without the assistance of comediators, who assist parties to resolve their disputes using a variety of facilitation techniques. The neutral acting as mediator holds informal meetings with the parties, both jointly and privately, to understand the positions of each party regarding the matter. The private meetings, known as caucuses, place maximum demands on the skills of the mediator to assist the parties in finding agreeable terms for settlement. Hearings are not considered evidentiary in nature. All proceedings are considered confidential, and information exchanged between the parties cannot be used in future trials (although the parties should commit that understanding to writing). Numerous articles and guidance documents describe the mediation process in depth [9–11]. The neutral can be a passive or active facilitator, either acting primarily as a conduit for information between the parties or playing a more active role in advising each party on the merits of case and using "rough justice" to articulate potential settlement conditions. The neutral must rely on noncoercive techniques[2] to promote fruitful dialogue between the parties, and the success of the mediation often depends on the interpersonal skills of the mediator. In some cases, neutrals rely on assistants or comediators who bring certain technical knowledge to the matter, to advise the lead mediator on the validity of the technical positions taken by the opposing parties or to suggest financial strategies that may promote settlement. With this method, the parties have the benefit of hearing how an independent third party (such as might be the case later with a judge or jury) views their position.

Mediation is suitable for all forms of environmental disputes. The decision to use this method hinges solely on the willingness of the parties to seek settlement. Mediation, which is usually a private process, may be more difficult for environmental disputes involving public land use decisions, because of the requirements for greater interactions with the public. However, with appropriate ground rules for the mediation process and attention to other details such as ex parte contacts, mediation of public environmental disputes is also feasible and suitable. Most independent studies of the use of mediation to resolve certain environmental disputes, for example, Superfund and other kinds of toxic waste litigation cases, have

[2]Sounds good, but the occasional reminder by the mediator of the alternative, that is, litigation and the costs associated therewith, can often "coerce" a settlement.

shown significant time and cost savings compared to resolution via litigation [2], although process failures have occurred [12]. The primary disadvantage to mediation is the risk of a failed process. When this occurs, costs for the mediation process, which can be substantial if significant document preparation is required, would thus be added to the costs of litigation. However, even in cases that do not settle, there are often secondary benefits gained. For example, the issues in dispute have been narrowed, and, of course, preparation has begun and weaknesses have been identified.

Mini-Trial

A mini-trial is a more formal and structured process, compared to mediation, with characteristics similar to an actual trial, but conducted by a neutral party in a private setting rather than in a court of law. Factual evidence and offers of proof are formally presented to representatives of each party who are capable of decision making, according to procedures agreed to by the parties. The neutral acts as mediator and facilitator during and following the mini-trial and also serves as the surrogate judge during the proceedings. Any settlement must still be mutually agreed to by the parties. Mini-trials are generally more costly than mediation, but because the process is more formal, the parties may find that their interests are better represented by this venue. Most agree that, unless the decision is binding, this approach is not worth the expense.

Arbitration

In stark contrast to mediation and mini-trials, arbitration is an adjudicative process in which the decisions made by the neutral are binding. By necessity, this technique has well-defined procedures (see JAMS–Endispute, Rules of Procedure, or other public sources on arbitration procedures) for the conduct of the arbitration process. The parties have considerable flexibility in setting the scope of the arbitration and defining the specific issues in dispute. Generically, the arbitrator or arbitration panel reviews or listens to evidence from each party and reaches a judgment. The process is private, and no written justification for the decision is required, unless otherwise agreed to.

Arbitration is more adversarial than mediation, and selection of this ADR technique often reflects failure of the parties to reach a consensus agreement. One or more issues may then require arbitration, where the parties prefer such a process to lengthy and costly litigation. Since one party will emerge the winner in an arbitration, the parties need to consider the cost/risk associated with the decision to use this option for settlement.

Judicial Reference

A final ADR technique is judicial reference. This method most closely resembles an actual trial, and the parties adhere to rules of evidence and civil procedure and have the right of appeal. The advantages of this technique compared to a trial in a public court of law are that the parties can (1) select the adjudicator, usually a retired judge, who may have special relevant expertise; (2) control the proceedings to avoid lengthy delays; and (3) conduct the proceedings in private to maintain confidentiality. Based on literature references, this technique does not appear to have been widely used to settle environmental disputes.

HISTORICAL ROLE FOR ADR IN ENVIRONMENTAL DISPUTE RESOLUTION

ADR techniques have been used to settle a wide range of environmental disputes over the past 25 years, since the apparent inception of ADR techniques applied to the environmental arena in 1974. It has been used in connection with the following environmental issues [1]:

Land use
Natural resource management and use of public lands
Water resources
Energy
Air quality
Toxics

Other areas of past applications of ADR include environmental policymaking, standard setting, determining development choices, and enforcing environmental standards [13]. Recent reviews of the use of ADR in environmental disputes have summarized other specific examples of ADR use, including the siting of controversial facilities (incinerators, landfills), the enforcement of contested cleanup plans at contaminated sites such as federal or state Superfund sites, and the drafting of detailed technical federal regulations [14]. It is indeed a dirty business.

Since that time, federal and state agencies have encouraged the use of ADR to settle environmental disputes [15, 3]. For example, in 1990, Congress passed the Administrative Dispute Resolution Act and the Negotiated Rulemaking Act. Both of these acts direct disputing parties to use ADR techniques for settlement of disputes with federal agencies. In February 1996, President Clinton issued Executive Order 12,988, 61 Fed. Reg. 4729, which directs government agencies to make the use of ADR a priority throughout the federal government [16]. In addition, many federal environmental statutes and regulations either explicitly or implicitly encourage the use of ADR for dispute resolution. Many state and local regulatory agencies have also encouraged the use of ADR. The proportion of environmental disputes in the United States that have been recommended for settlement via ADR since the early 1970s is unknown. Some have estimated that only 10% of environmental disputes are likely candidates for use of ADR resolution methods. However, many experts feel that the fraction is considerably higher and that, in fact, most environmental disputes could potentially be settled by these techniques [3, 14].

Legal and Statutory Framework

Federal statutes, state laws, and associated regulations establish the legal framework within which many environmental disputes arise. Additionally, common law doctrines of nuisance, trespass, negligence, and strict liability may set the stage for disputes caused by alleged damages or harm due to environmental releases of hazardous or toxic substances. Since the early 1970s, federal statutes and subsequent regulations have codified the government's positions regarding civil or criminal liability associated with actions causing harm to the environment. Each statute attempts to limit environmental impacts in a specific medium (e.g., Clean Air Act, Clean Water Act), impacts due to a specific category of wastes (e.g., Resource Conservation and Recovery Act [RCRA]), and impacts due to a specific category of chemicals in commercial use (e.g., Toxic Substance Control Act [TSCA], Federal Insecticide, Fungicide and Rodenticide Act^{IFRA}), or address past disposal practices that have damaged the environment (e.g., Comprehensive Environmental Response, Compensation, and Liability Act [CERCLA]).

Many other federal statutes and state laws have been promulgated to provide a comprehensive and complex legal and regulatory structure within which environmental disputes may arise. The following sections provide a brief overview of the history of the use of ADR to address disputes that have arisen due to alleged damages caused by the release of hazardous and toxic substances to the environment [4, 13–15, 17].

Superfund Disputes

The term "Superfund" was coined following the passage of CERCLA in 1980 to describe the magnitude of the probable costs associated with cleaning up the most contaminated sites in the United States. This multibillion-dollar program, supported by taxes imposed on the petroleum and chemical industries and cost recovery from potentially responsible parties (PRPs), has set the stage for environmental disputes with many millions of dollars at stake. Disputes can arise among all parties involved, including regulatory agencies, PRPs, property owners who are not PRPs, impacted citizens, nongovernmental organizations, and insurance companies. Within CERCLA and its amendments (Superfund Amendments and Reauthorization Act [SARA] of 1986), subsequent regulations and a plethora of EPA guidance documents, internal memoranda, and a growing body of case law lie the seeds of many unresolved environmental disputes, many of which have mutated to enormous proportions. It is generally acknowledged that ADR techniques are particularly well suited to resolving Superfund-related disputes [2, 8, 18]. Such disputes include (1) liability and costs of remedies, (2) cost allocations between PRPs, (3) cost reimbursement to the EPA, (4) acceptability of cost documentation, (5) consistency of remedy selection with the National Contingency Plan (NCP), (6) remedy selection, and (7) compliance with consent orders. Future issues amenable to ADR resolution will include negotiated conditions for site closure and long-term institutional controls at sites where cleanups are considered to be "technically impracticable" and some contamination may remain indefinitely. This latter issue is highly technically complex, and the use of ADR techniques to resolve disputes over this issue have distinct advantages compared to relying on decisions of judges or juries unfamiliar with the technical complexities of subsurface remediation. Of course, if the scientific evidence is not on "your side," you may prefer a judge or jury.

Selection of the appropriate ADR method for Superfund-related disputes depends on a variety of factors, including (1) EPA willingness to litigate, (2) identification of issues appropriate for meditation, (3) timing of the case, and (4) assessment of the types of parties involved. Other factors include the number of parties involved, the willingness of parties to participate, the amount in dispute, and the ability of parties to share the costs of mediation, also known as the "number of available deep pockets" [2, 18].

As can quickly be seen, determining the suitability of ADR for Superfund cases requires experience with the process and familiarity with the legal and technical issues in dispute. The complexity of the issues, from institutional, regulatory, and technical perspectives, demands an integrative approach to decision making and the appropriate constellation of PRPs, regulators, insurance companies, and neutral parties. These are the "stars" of the case. As the legal issues surrounding Superfund disputes become better defined through extensive litigation, the advantages of using ADR become more pronounced. It is likely that the role of ADR in settling Superfund-related environmental disputes will continue to expand.

Environmental Insurance Coverage

A second arena where ADR can play a more significant role is disputes arising between insurance policyholders, the "insured," and policy providers, the "insurers," regarding insurance coverage of damages resulting from land contamination. Policyholders look to insurance carriers to cover transactional and remediation costs incurred to address cleanup of sites under CERCLA, RCRA, or other applicable statutes and regulations. Such disputes are highly complex from an institutional, legal, and technical point of view. A typical case will involve a Fortune 500 company suing all primary and excess insurers providing coverage over different time frames for all costs incurred for events that occurred over decades, both with respect to the occurrence of events that caused the contamination as well as all costs associated with site cleanup, such as site characterization, analysis of alternatives, remedial design, and remedial actions. The case is further complicated by calculations of the time value of money and the projected future costs for ongoing cleanup activities. The legal and technical literature is replete with examples of such large, contentious, and complex insurance coverage disputes [19].

Legal complexities also characterize environmental insurance coverage cases. Typical issues that must often be resolved during litigation include the following:

- *What is the "trigger" of coverage*? Common theories of what constitutes a "trigger", that is, the point in time when an insurance policy should respond to a covered loss, include (a) the exposure theory, (b) the manifestation theory, (c) the injury-in-fact theory, and (d) the multiple- or continuous-trigger theory (when did what occur and how long did it last?).
- *Is environmental damage property damage*? Disputes arise over whether response or transactional costs are considered property damage claims, routine maintenance, or requirements under a permit.
- *Was the environmental damage caused by normal business practices and arguably expected and intended by the insured or was it fortuitous, that is, "unexpected" or "accidental"*? A common dispute arises over whether contamination occurred as a result of standard operating practices, for example, the operation of a wastewater treatment lagoon, or whether releases were accidental, that is, unintended.
- *Does pollution exclusion apply*? Pollution exclusion clauses came into practice in the mid-1970s, and either are generally "absolute," that is, all damages caused by environmental releases are excluded, or include provisions exempting "sudden and accidental" occurrences from the exclusion.

Resolution of these legal issues hinges on various legal theories, case law, and the skill of the legal team. It also depends on the effective use of experts knowledgeable in historical best management practices for the handling of hazardous and toxic wastes, fate and transport of contaminants in the subsurface, subsurface remediation decision making, and effectiveness and costs of selected remedial measures. For example, both the plaintiffs and the defendants may retain experts to predict when damage occurred due to releases to the environment, thus defining the date for initiation of policy coverage (i.e., the trigger date). Agreement on this date is essential to allocate damage between primary carriers and, thereafter, excess carriers. Needless to say, agreement between such experts is the exception rather than the rule.

Because of these legal complexities, environmental coverage disputes initiated in the 1980s have typically followed the traditional litigation cycle, leading to lengthy, costly cases. As the legal issues have been slowly clarified by decisions at the state and federal levels, the applicability of

ADR techniques has increased. The suitability of different ADR techniques to facilitate settlement in environmental insurance disputes depends on the maturity of the dispute, the number of parties in the dispute, and the willingness of the parties to resolve the dispute without continuing costly transactions. Mediation is an option for parties wishing to control both the process and the outcome of the dispute. Because of the technical complexities associated with the underlying facts of the matter, mini-trials followed by mediation may be more suitable. Finally, arbitration may be the most cost-effective approach if the extent or scope of the dispute can be limited by agreement of the parties [10].

Enforcement Cases

In 1987, the EPA issued a document providing guidance on the use of ADR in EPA enforcement cases [11]. This document describes ADR methods, addresses characteristics of enforcement cases that are suitable for ADR, establishes internal procedures for approval of the use of ADR, provides guidance on the selection of third-party neutrals, and includes guidance on process management, including draft procedures, protocols, and agreement for various types of ADR. The EPA's rationale for encouraging the use of ADR was to reduce the duration of litigation, thereby reducing transaction costs and case backlog within the Department of Justice (DOJ). The success of these efforts in achieving this goal is unknown, but ADR continues to be supported by DOJ, and subsequent policy statements have reinforced the DOJ's desire to increase the use of ADR techniques in the resolution of enforcement cases.

ADR in its most popular form, mediation, may not be suitable for all enforcement cases, but it is likely appropriate for the vast majority of these cases. Some of the factors that should be considered in evaluating whether a case should be nominated for ADR include (1) the ability of a neutral to conduct frank, private discussions that may improve the outcome; (2) the desire for more complete information exchange outside of an adjudicated setting; (3) numerous parties, such that a structured settlement process would be helpful; (4) the unwillingness of each party to take the lead role in an ADR process; and (5) a broad range of issues, such that creative solutions may be generated that will lead to win–win situations. The EPA's 1987 guidance suggests that cases pending for years without significant movement toward resolution are candidates for the use of ADR. Other suitable cases may involve those with significant impasses or those requiring significant resources that do not appear commensurate with the magnitude of the case. The biggest impediment to ADR is the right of public comment and/or the actions of special-interest groups in the process. While their intentions are, presumably, admirable, the result is often delay upon delay without resolution [16].

Regulatory and Siting Disputes

ADR has been applied to a wide range of regulatory and land use disputes of an environmental nature. Examples include (1) resource allocation, (2) rate setting, (3) siting of hazardous waste facilities, (4) negotiated rule making (e.g., endangered species, drinking-water standards), (5) siting of a flood control dam, (6) siting of a hydroelectric power facility, (7) water rights, (8) land management plans, and (9) natural resource damage disputes [4, 14, 17].

The suitability of ADR techniques to regulatory and siting disputes is controversial, in contrast to the general support for use of these techniques to resolve Superfund or environmental coverage disputes. Critics of ADR use in these applications identify a number of limitations, including [14]:

- Public regulatory agencies become less accountable.
- Public agencies and other vital interests may be left out or deliberately omitted.
- Regulatory standards may be overridden to secure purely local, site-specific deals.
- Nonaccountable actors (e.g., the mediator) gain undue influence.
- Powerful interests can impose their will on weaker interests.

Fundamentally, the success of ADR in resolving these disputes rests on the willingness of the parties to reach consensus and settlement and the skill of the third-party neutral or third-party neutral team. Nonetheless, these concerns regarding the suitability of ADR methods to resolve regulatory and siting disputes may limit the proportion of these disputes that are best resolved using ADR techniques.

TECHNICAL NEUTRALS IN ADR

In each of the ADR methods mentioned earlier in this chapter, the typical process begins with the parties selecting a third-party neutral to fill the appropriate role, depending on the chosen ADR method. This person is often a retired judge or lawyer who is skilled in the art of facilitation and mediation, but understandably often lacks technical expertise in the specific matters in dispute. In certain cases, the third-party neutral may choose to retain additional neutral experts acceptable to the parties to expand the ADR team. Examples include a neutral technical expert selected to advise the mediator on the technical and scientific merits of a particular dispute, or a financial or business expert who may assist in facilitation of the process and may provide particular expertise in financial structuring of potential settlement agreements.

The role of these technical neutrals, who may act as comediators or serve as participants in an arbitration panel, depends on the nature of the dispute and the willingness of the parties to accept the cost and the participation of additional third-party neutrals.

The primary advantage of this strategy is an independent assessment of the scientific and technical merits of the opposing arguments. A singular characteristic of environmental disputes is the high degree of uncertainty in the accuracy of the technical bases of the conflict, or simply, differences of opinion regarding same. Whether the dispute is a Superfund cost allocation matter, an insurance coverage dispute regarding pollution liability, or a toxic tort case, the level of technical uncertainty is high. Usually, the quantities and timing of release of hazardous and toxic chemicals are unknown and can only be inferred by site characterization data. The accuracy of this inference process depends on many assumptions regarding models of the underlying processes influencing the fate of these chemicals. Conflicts between experts over the values of parameters used to model the fate and transport of chemicals in the subsurface are the rule. In addition, much of the underlying science is still in development. Most models track a single chemical in the environment; yet, in almost all cases, multiple contaminants are present in multiple forms. The impact of these chemical mixtures on the environment and on human health is still uncertain. This, however, does not prevent the parties from presenting conflicting opinions, with certainty.

Perhaps the greatest element of uncertainty, however, is the inherent complexity of subsurface environments and the limited amount of sampling that can be achieved to characterize this complexity [20]. Environmental disputes arising from contaminant releases to the soil and groundwater usually rest on highly uncertain facts. Cost constraints limit the number of soil samples that can be collected and the number and depth of groundwater monitoring wells that can be installed. Typically, samples of soil and groundwater from subsurface zones suspected of being contaminated represent a very small volume of the contaminated zone, ranging from less than 1 part in 10 million to 1 part in 10,000. Certain important physical features of the contaminated zones, such as the hydraulic conductivity, density of preferential migration pathways, mineral content, or organic carbon content of the aquifer materials, may vary by two or more orders of magnitude over scales of centimeters to a few meters. Site characterization also occurs over a very short time frame relative to the elapsed time since the initial release of chemicals. Thus, data may be collected to show groundwater elevations and groundwater directions for a few years, but whether these snapshots in time reflect these characteristics (groundwater elevation and direction) of the aquifer several decades in the past remains a mystery. Accurate modeling of the fate and transport of chemicals in the subsurface is thus fraught with significant uncertainty and differences of opinion.

Finally, various chemical and biological processes that may affect the distribution and ultimate fate of chemicals in the environment are poorly understood quantitatively, although the literature continues to grow. Certain organic chemicals can undergo biologically mediated reactions to form more toxic compounds that are also more mobile in the subsurface than the parent compound (e.g., transformation of perchloroethylene [PCE] to vinyl chloride). The fate of certain pesticides is strongly dependent on the biogeochemical environment, and in certain conditions, some of these compounds can degrade to harmless byproducts. Alternately, their persistence can be lengthy.

Given these uncertainties, models, which must be used to infer, among other things, the time of chemical release, the ultimate fate of the chemicals, the effectiveness of proposed remedial systems, and the reconstruction of exposure dose, are inherently inaccurate and susceptible to manipulation by paid experts representing the interests of specific parties. Judgment is required to distinguish fact from fantasy and to provide an independent assessment of the validity of positions taken based on inherently incomplete information. Therein lies the potential value of a technical neutral in assisting the mediator or arbitrator to resolve the dispute [2, 21]. Another advantage of this approach is the potential to provide the parties with the option of a technically sound basis for the final settlement, if the parties all agree that the option is a more rational solution than leaving it to the vagaries of a judge or jury.

This approach is not without risks, however. A technical neutral may be unable to avoid controversies with experts retained by the parties to the dispute, thus exacerbating the adversarial environment and reducing the potential for settlement. A technical neutral might also not have the mediation and communication skills needed to interact effectively and credibly with each party or may undertake unwelcome alternative analyses of the facts of the case. Some mediators may also feel that adding technical expertise to his or her team may result in a more time consuming and expensive process, and thus prefer to rely primarily on negotiating skills to resolve the dispute regardless of the relative technical merit of the competing positions. These problems are best addressed by careful selection of the technical neutral with the appropriate stature, facilitation skills, and reputation for fairness to increase the likelihood of success. It should be made clear at the start of the process what value added can be expected from the addition of independent technical neutral(s) to the process. As always, if the law is in your favor, argue the law, if not, argue the facts.

CASE STUDIES

Given the large number of cases that have been settled using ADR techniques, it is surprising that only a few case studies have been described in detail in the literature. On the other hand, because mediation is a confidential and private process, it is understandable that public information is limited. In addition, each dispute using ADR has unique characteristics reflecting the type of dispute, the number of parties involved, the size of the gap between the respective positions of the parties, and the technical and scientific facts underlying the disputes. Thus, it is difficult to find more than cursory descriptions of the use of ADR techniques to settle environmental disputes.

To illustrate some of the advantages and problems related to the use of ADR in resolving environmental disputes, following is a brief description of two large environmental disputes successfully settled by using ADR techniques. The description of these cases are necessarily generic because mediation is a confidential process. However, sufficient details are provided to illustrate how ADR techniques can successfully overcome barriers to settlement.

Property Damage Suits

Property damage suits generally consist of certain common elements. One party or parties allege that their property has been damaged by accidental or intentional release(s) of toxic substances that have migrated onto or under their property via various pathways, including air, surface discharge, or migration beneath the property via the groundwater. The owner or owners of these properties demand compensation for diminution of property value and, in some cases, potential or actual exposure due to the presence of toxic substances. Resolution of these disputes can be accelerated using all of the ADR methods described previously. Selection of the appropriate method will hinge on a number of factors, including the magnitude or size of the dispute and the extent to which there is disagreement between the parties on the underlying scientific basis for the damage claim.

In a large property damage suit, mediation was undertaken to attempt to reach settlement and avoid a lengthy and costly trial set only a few months hence. The parties had been in dispute for a number of years, and significant resources had been invested in analyzing the characteristics of the sites in the area, with particular focus on the nature and extent of the contamination and the basis for the damage claims. In this case, a municipality had filed a suit against the manufacturer of organic chemicals used to control nematodes (worms) that attacked various agricultural crops. The chemicals had migrated through the soil to the groundwater, which had subsequently migrated into the municipal water supply well fields. The concentration of the insecticides at some of the water supply wells exceeded state drinking-water standards, and thus treatment was required, if the wells were to continue in use. Continued use was considered essential, because of the unacceptable cost of building transmission pipelines to transport water from other parts of the municipality to the affected areas. In addition, future impacts were expected due to the continued migration of these chemicals. The parties agreed to initiate mediation and established protocols for the ADR process.

At the initiation of the mediation, several hundred million dollars separated the demand from the municipality for property damage compensation and an initial offer by the PRPs. The parties agreed to a mini-trial format for the mediation and a scope of the mediation based on responses to eight questions addressing key issues in the dispute. Two independent neutral experts were retained by the mediator, with each party to the dispute recommending one. The parties then presented documents and verbal presentations in support of their respective positions on each of the eight questions developed by the parties (this almost sounds like the typical presidential candidate debate approach). The mini-trial required about ten days of joint presentations and two to three additional days of individual caucuses with the parties to facilitate settlement discussions.

Although there were numerous scientific and technical issues in the dispute, one issue, the number of water supply wells expected to be impacted in the future and the future cost of that impact, hinged on the fate and transport of the chemicals of concern. The experts for the plaintiff and defendant parties disagreed on the number of wells expected to be impacted in the future that would require installation of an expensive treatment system. The two neutral experts developed a scientifically sound, yet simple conceptual model that predicted the expected time and travel distance of the chemicals of concern. This model used an estimate of the half-life (defined as the time required for 50% of the chemical to be converted into nontoxic byproducts) of the main chemical of concern to estimate the number of wells likely to be impacted by continued migration. In addition, the neutral experts assessed alternative cost models provided by the parties' experts and estimated expected future cost impacts of continued contaminant migration. This information was then used to propose a present-worth cost of damages related to the release of the chemicals. Subsequent to the presentation of evidence on the eight questions developed by the parties, various caucuses were held to facilitate resolution of the dispute. While no agreement was achieved at the conclusion of the mediation, the parties ultimately reached a settlement shortly after the start of the trial. The primary basis for that settlement was the conceptual framework developed during the mediation process.

This case study illustrates the main advantage of mediation; avoidance of the time and expense of a trial (or at least a portion of it). The neutral technical experts were able to determine a simple, yet scientifically sound conceptual model that was agreed to by the parties and ultimately used to settle a number of similar cases. This avoided the significant cost and uncertainties of presenting this information to less technically informed audiences. This case also illustrates a situation in which the ADR approach was only successful after the parties had expended significant resources preparing for trial and thus is illustrative of a failed ADR process as well. However, it is probable that, in the case at hand, had mediation been undertaken at an earlier date, the case could have been resolved sooner at substantial savings in transaction costs.

Superfund Cost Allocation

As noted, Superfund cost allocation disputes are good candidates for ADR. Disputes are highly complex, both institutionally and technically. They involve numerous parties who are likely to disagree on numerous substantive issues, including the basis for cost allocation, the data on waste quantities being used to allocate cost, the timing and duration of waste disposal, and the impact of the different wastes on the final cost of Superfund cleanup. Other issues often in dispute include the relative contributions of individual parties in the PRP group and ongoing disputes between these parties outside of the cost allocation issues. In addition, the PRP group may be in dispute with the EPA or state lead regulatory agencies regarding justifications for the reimbursement requests from the regulatory agencies. Finally, significant dispute arises over estimate of future costs for Superfund cleanups. In other words, each PRP wants to pay as little as possible—as would we all.

At a major Superfund site, mediation was undertaken to resolve an ongoing dispute with numerous parties representing generators, transporters, and current or former site owners. This particular site consisted of several industrial facilities located around a former wetland in close proximity to a major river. Principal industries at the site included a battery recycling facility, a chemical plant producing highly toxic pesticides and insecticides, and other manufacturing facilities discharging treated wastewater to the wetland/lagoon. The primary focus of site remediation, however, was the battery recycling facility where remediation involved, among other things, management of disposed battery casings and residues from the secondary lead smelter use to recycle lead from the batteries. One unique aspect of this dispute was that an innovative technology for recycling the disposed casings had been proposed and this alternative was then compared to an on-site land disposal option for this waste. The PRP group was directed by the EPA to use this recycling remedy, in spite of vigorous arguments by the PRP group who supported the alternative land disposal option. The recycling process was installed, but failed, after an expenditure of more than $15 million. This failure was ultimately the subject of a lawsuit between the PRPs and the technology vendor. However, lingering behind that suit was an ongoing disagreement between most of the PRPs and one member of the PRP group regarding possible culpability for the EPA having selected the recycling remedy.

The mediator selected by the PRPs retained a neutral technical expert and a neutral business advisor to facilitate the mediation. The parties agreed to prepare a single brief summarizing the stipulated facts for the dispute. In addition, each of the parties provided a short written summary of its position on allocation of costs.

This Superfund cost allocation case exhibited the common characteristics of such disputes. Data were limited regarding the quantity of waste that had been deposited, although a summary tabulation of the number of batteries disposed of at the site had been prepared. Nonetheless, disputes arose over the accuracy of the waste-in estimate, and the PRPs disputed the allocation formula.

Mediation was successful in facilitating settlement agreements on cost allocation among all of the parties. However, the parties were unable to agree on the failed remedy issue. A major dispute arose over the actual costs of the failed remedy and whether the EPA could have been convinced to use a more reliable, and less costly remedy had additional information been provided by one of the parties regarding the nature and extent of soil and groundwater contamination at the site. It can be speculated that this issue was incapable of resolution because it represented more of a legal issue than one of fact, an all-or-nothing issue.

The parties agreed that resolution of this last remaining issue could best be addressed through an arbitration process. An arbitration panel was established, consisting of the three individuals who participated in the mediation, and the parties presented evidence in support of their position. The primary technical issue was whether the EPA would have changed its remedy recommendation for the battery casings if certain information had been revealed in a more timely manner. Specifically, information on the discovery of highly toxic 2,3,7,8-tetrachlorodibenzodioxin (dioxin) and related chemicals in the soil and groundwater in the vicinity of the buried battery casings could have influenced the EPA's decision on the remedy for the buried battery casings. This last issue was decided by the arbitration panel as an adjudicatory matter.

This case also illustrates the advantages of combining mediation and arbitration, with arbitration deciding a limited, but financially significant portion of a case. The technical neutral provided useful advice to the arbitration panel on the relative technical merits of the opposing positions. The technical issues were complex and touched on the fate and transport of dioxins via all relevant pathways, other potential sources of these compounds, the relative toxicity of dioxin compounds, and the potential impact of this information on the remedy selection process. Presenting this information to a judge or jury would have been costly, with the risk that a lay audience would have been overwhelmed with the technical complexity and would have reached conclusions based on nontechnical issues, such as the demeanor of the litigants and experts (as with every other case that goes to court). In this case, the matter was settled expeditiously, at a cost significantly less than the cost of litigation.

MANAGING THE PROCESS

Successful use of ADR techniques to resolve environmental disputes requires attention to the details of the process. Detailed guidance on implementation of each of the ADR techniques discussed in this chapter can be obtained from professional organizations such as JAMS–Endispute, the American Arbitration Association, or other independent organizations. The EPA's guidance on the use of ADR in enforcement cases [11] is an example of a government agency specifying the details of the process, with the stated goal of enhancing the likelihood of success. This guidance addresses the characteristics of enforcement cases suitable for ADR, internal procedures for EPA personnel to initiate ADR, procedures for selection of third-party neutrals, and procedures for management and implementation of ADR cases, among other issues. This section provides a brief summary of key issues that must be managed to realize the potential benefits in the use of ADR techniques.

Although the structural details of each of the ADR techniques vary, depending on the nature of the environmental dispute, there are common issues that arise. The ADR process can be initiated by any of the parties, but all must agree or be contractually bound. The first threshold issue that must be resolved is selection of the parties to be involved in the process and the characteristics of the individuals representing the parties. Once past this hurdle, the parties must then select a mutually acceptable third-party neutral and reach agreement on a strategic approach and process tactics. This means, at a minimum, agreeing on the scope of the process, that is, the issues that will be addressed, the duration of the process, the location of the proceedings, and the schedule and agenda of the meetings. If arbitration is the ADR method selected, more formal structuring is required, because arbitration is an adjudicative process.

A critical factor in the success of ADR is selection of the third-party neutral. The characteristics of viable candidates for this position include (1) demonstrated experience, (2) independence, (3) subject matter expertise, and (4) no conflict of interests. There are a few talented, qualified neutrals to handle these complex matters effectively and different organizations specializing in ADR services and various resources can provide listings of qualified individuals for environmental dispute resolution. The outcome of the selection process, as with retention of any professional service, depends primarily on the experiences and preferences of the parties seeking ADR services and the willingness of the parties to reach a compromise. The capabilities of the third-party neutral are particularly critical in non-binding forms of ADR, including fact finding, mini-trials, and mediation.

SUMMARY

The use of ADR techniques to resolve environmental disputes has expanded considerably since the mid-1970s. This reflects, in part, the expansion of the legal and regulatory framework addressing environmental problems and the growing body of experience in applying different ADR techniques to such disputes. It also indicates that there were few environmental disputes prior to the mid-1970s. Nevertheless, ADR has been shown to be an effective means of resolving private civil disputes in cost allocation cases arising from CERCLA, environmental coverage cases, and property damage disputes, among others. Environmental disputes involving both public and private parties, such as siting of controversial facilities or major land use decisions, may be less amenable to resolution through ADR because of public disclosure issues. Nonetheless, properly managed and organized ADR processes have been effective at achieving resolution of some major public environmental disputes.

In comparison to traditional litigation practices, involving lengthy and costly discovery and trial phases, with likely acrimonious results, ADR techniques offer the potential to reduce costs, accelerate settlement, and maintain existing business relationships beyond the dispute. ADR may not be appropriate for all cases, and in limited cases, the cost savings or acceleration of settlement may not be sufficient to balance some of the inherent disadvantages of early information exchanges between the parties. As the legal bases for disputes arising in the context of certain federal environmental statutes becomes more mature, it is likely that ADR techniques will be more widely used.

In certain complex technical environmental disputes, the use of a third-party neutral technical expert may add significant value to the process. Such an independent could advise the third-party neutral; could conduct the ADR process alone, depending on the nature of the dispute; or could act as comediator in a mediation process or as a member of an arbitration panel. The value added by this approach includes an independent assessment of the technical merit of opposing sides and the crafting of a settlement that should be based on sound science. If the technical neutral also possesses excellent negotiation and leadership skills, the overall mediation process may be enhanced. Such a technical neutral can also even the playing field if opposing parties have disproportionate resources invested in experts. The stature and experience of the technical neutral must match the seriousness of the matter, and careful selection of this person is essential. The potential value of adding such neutrals to the ADR team must be balanced against costs.

In the future, ADR use to resolve environmental disputes will continue to grow. Societal and business pressures demand less costly and more rapid resolution of these disputes. Finally, environmental disputes often transcend national boundaries. Thus, ADR may soon find an expanding role in resolving cross-border disputes, such as litigation under the North American Free Trade Agreement (NAFTA) over varying environmental standards in the NAFTA countries. What is also likely to change is the makeup of ADR teams resolving environmental disputes. In high-stakes disputes involving complex technical issues, a single third party neutral may find the use of additional subject matter experts with good negotiating and communication skills an essential component of a successful process.

REFERENCES

1. G. Bingham and L. V. Haygood, "Environmental Dispute Resolution: The First Ten Years," *Arbitration Journal*, 10–11 (1986).

2. L. Peterson, "The Promise of Mediated Settlements of Environmental Disputes: The Experience of EPA Region V," *Columbia Journal of Environmental Law*, 17(2): 327–380 (1992).

3. C. Stukenberg, "The Proper Role of Alternative Dispute Resolution (ADR) in Environmental Conflicts," *University of Dayton Law Review,* 19(3): 1305–1339 (1994).

4. M. Ryan, "Alternative Dispute Resolution in Environmental Cases: Friend or Foe?" *Tulane Environmental Law Journal*, 10(2): 387–414 (Summer 1997).

5. S.D. Ellison & D. W. Miller, "Beyond ADR: Working Toward Synergistic Strategic Partnership," *Journal of Management in Engineering*, 11(6): 44–54 (November–December 1995).

6. F. Muller, "Alternative Dispute Resolution (ADR) for Construction," *Construction Congress, Proceedings of the 1995 Conference. Sponsored by the Construction Division of the American Society of Civil Engineers*, 22–26: 132–140 (October 1995).

7. R. A. Rubin, "ADR, 25 Years of Progress," *Civil Engineers Influencing Public Policy, Proceedings of the Sessions Sponsored by the Construction Division of the American Society of Civil Engineers*, 21–29 (November 1996).

8. V. O. Manuele, "Alternative Contracting Strategies for Superfund Projects," *Proceedings of the U.S. Environmental Protection Agency Superfund XIV Conference and Exhibits*, 1: 177–186 (November–December 1993).

9. "A Guide to Mediation and Arbitration for Business People," American Arbitration Association (1996).

10. D. Castles and C. R. Orton, "ADR and Environmental Coverage Disputes: A Primer on Methods and Selection," *Federation of Insurance and Corporate Counsel Quarterly*, 46(3): 279–288 (Spring 1996).

11. "Guidance on Alternative Dispute Resolution in Enforcement Actions," United States Environmental Protection Agency, Office of Solid Waste and Emergency Response (August 1987).

12. L. F. Charla and G. J. Parry, "Mediation Services: Successes and Failures of Site-Specific Alternative Dispute Resolution," *Villanova Environmental Law Journal*, II, I: 89–97 (1991).

13. F. P. Grad, "Alternative Dispute Resolution in Environmental Law," *Columbia Journal of Environmental Law*, 14: 157–185 (1989).

14. J. Harrison, "Environmental Mediation: The Ethical and Constitutional Dimension," *Journal of Environmental Law*, 9(1): 79–102 (1997).

15. C. A. Sinderbrand, "Alternative Dispute Resolution in the Environmental Arena," *Wisconsin Lawyer*, 64(12): 25–27, 59–60 (December 1991).

16. L. J. Schiffer and R. L. Juni, "Alternative Dispute Resolution in the Department of Justice," *National Resources and Environment ABA Section of Natural Resources, Energy and Environmental Law*, 11(1): 27–30 (Summer 1996).

17. J. Brock, "Mandated Mediation: A Contradiction in Terms," *Villanova Environmental Law Journal*, II, I: 57–87 (1991).

18. J. B. Bates and R. L. Hines, "Using ADR to Resolve Superfund and Related Environmental Disputes," *Using ADR in Superfund Disputes*, 33–48 (November 1994).

19. S. Crable, "ADR: A Solution for Environmental Disputes," *American Arbitration Association, Dispute Resolution Journal*, 48(1): 24 (March 1993).

20. N. M. Ram et al., "Environmental Sleuth at Work," *Environmental Science and Technology*, 33(21): 464A (November 1999).

21. R. E. Tompkins, "Mediation, the Mediator and the Environment," *National Resources and Environment, ABA Section of Natural Resources, Energy and Environmental Law*, II, I: 27–30 (Summer 1996).

Chapter 13

Forensic Case Management

In its most elementary terms, forensic case management is the process whereby the expert ensures that his or her client has the best available information from which to make decisions pertaining to technical case issues. In the broadest sense, this involves assisting the client in understanding the technical issues of a case, contributing to tactical and strategic planning, identifying/selecting other experts, managing multiple experts, providing information/guidance in the document discovery process, managing technical documents, supporting the deposition process, and providing either expert testimony or work product guidance.

Managing the technical issues of a forensic case also requires that the expert understands the client's approach to managing the litigation. The client's approach, whether an industry or an insurance client, is usually based on a team concept and a defined case management plan. An example of this approach is illustrated in Sidebar 13.1. The most pivotal aspect of the client's approach to case management is when to choose, retain, and utilize experts.

TIMING IS EVERYTHING

The most critical time period in a forensic investigation is just prior to the cutoff of the discovery process. Unfortunately, in some instances, the expert is retained at the last minute (sometimes on the same day) when experts must be declared and discovery has already been closed. When this happens, it becomes impossible for experts to obtain additional documents and/or collect field data that they may feel are critical to the case and to

Sidebar 13.1 Client's Approach to Managing Litigation.[1]

The client, be it an industry or an insurance company, is the litigation manager and is responsible for assembling a litigation team, facilitating the development of a sound case management plan, and implementing the case management plan to bring the litigation to a successful conclusion. In general, the management philosophy incorporates a team concept and the development and implementation of a case management plan.

Team Concept

In litigation, the team concept implies that no one entity is always right (i.e., effective communication and compromise are required in order to develop case objectives). Depending on the type of litigation (insurance or noninsurance related), the litigation team may consist of the following entities:

- The client's litigation manager
- Coverage counsel (regional and/or local counsel, if insurance related)
- Local counsel (a firm located in the state where the lawsuit will be heard)
- Technical consultant (nontestifying) and/or expert witness

The function of the litigation manager is to outline the objectives in the case. Counsel's role is to develop potential legal arguments to advance the client's case. It is counsel's responsibility to work with the technical consultant/expert witness to evaluate and determine (1) the technical issues that must be addressed in the litigation and (2) what type of expertise and opinions may be required.

Case Management Plan

The case management plan is really an organic process by which new information is identified and incorporated into the case. At its core, this process consists of the following basic steps:

[1]Paul Burbage, P.E., Firemans's Fund, Novato, California, was a contributing author to this sidebar.

- Identify litigation team members.
- Develop litigation objectives with participation from the litigation team.
- Evaluate existing documentation and identify information and/or data gaps.
- Begin the process of identifying areas where expert testimony may be necessary.
- Prepare a list of sources where pertinent information is located and a schedule to systematically review each source (both technical and nontechnical).
- Evaluate newly obtained information and incorporate this information into the case management plan (i.e., litigation presents itsself).

their opinions. The experts must then rely on the technical competence of counsel to have (1) collected all of the appropriate documents and (2) asked all of the correct technical questions.

On the other hand, when an expert is retained at the beginning of an environmental case, the expert's case management responsibilities may entail any and/or all of the case functions described previously. The level of involvement is highly dependent on whether the expert will be providing testimony in a case or only technical guidance to counsel. In the latter case, all work product and observations do not have to be produced to the opposition because the expert's work will be considered privileged as attorney work product.

When experts are providing technical support as attorney work product, their involvement can be much more extensive. For example, an expert who is familiar with all of the technical issues may have considerable involvement with the strategic development of a case, assist in the response to interrogatories, provide coordination of multiple experts (including their preparation for deposition), attend the depositions of opposition witnesses and experts, attend settlement conferences, and coordinate the production of graphics for settlement or trial. In either instance (i.e., whether providing testimony or work product), the case management elements remain the same.

ELEMENTS OF FORENSIC CASE MANAGEMENT

In virtually all environmental cases, the process begins with documents. The universe of potential document sources is considerable. In an environmental case, however, the list of common sources and types of documents has been fairly well defined.[2]

Documents

During the course of a litigation, the forensic expert will accumulate a vast array of documents, including site documents, site investigation data, correspondence, deposition testimony, opposing expert opinions and reports, pleadings, findings of fact, communications from counsel, maps, and aerial photographs. In addition to these materials, an expert may generate written communications to counsel, take notes, develop models, model input and output files, and prepare calculations, reports, and, finally, opinions. As a testifying expert, these generated materials (as well as documents furnished by counsel) may eventually be produced to opposing counsel.

As a consequence, the expert must always be aware that all written material (hard copy) as well as computer files might be requested and reviewed by opposing counsel. It is therefore necessary to ensure that, during the course of the litigation, whatever information is committed in writing *does not* contradict an expert's final published opinions. In order to avoid this problem, some simple guidelines follow:

- Do not take notes. If notes are taken, only record facts.
- All written materials should be on a computer.
- Do not print or distribute draft reports or model runs.
- Do not write on case documents (e.g., a site investigation map).
- Talk to counsel on the phone. Do not write letters or send faxes or e-mails unless the correspondence only contains facts.
- Do not talk to opposing experts.

Having suggested these guidelines, it should be obvious that each attorney usually has a policy relative to handling expert-generated materials.

[2]However, with almost every case, some new or previously unidentified document source is discovered.

Thus, once retained, the first order of business should be to discuss counsel's guidelines related to expert-generated documents.

Documents Usually Furnished by Counsel Counsel should provide the following types of documents that are usually produced during the discovery process or, at the very least, counsel should be made aware of what documents may be required or anticipated.

- Company, department, division, and/or employee files and technical documents related to the site contamination (i.e., relevant documents that are produced to the opposition), including site investigations (see Sidebar 13.2)
- Relevant technical documents produced by the opposition (including information on their libraries and journal subscriptions)
- Files from all consulting engineering/laboratory firms that may have provided technical site information
- Files from engineering/construction firms related to facility construction (especially borings at building footings)
- Files from regulatory agencies that may have (or had) oversight authority (including city, county, state, and federal agencies)
- Files from local agencies, including city planning/building departments, sanitary districts, and fire departments (if relevant)
- Documents from underlying cases (if any)
- Documents from counsel's own independent searches for related documents (usually literature)
- Relevant documents collected by other experts[3] (i.e., on the same side of the case)

The forensic expert should ensure that counsel is aware of the following list of common sources and types of documents that are usually collected by the expert.

Documents Usually Collected by Experts It should be remembered that litigation almost always begins long after the site(s) has either been abandoned, sold, modified, razed, and, often, remediated (or at least in the process of remediation). Therefore, site visits, although important, may have

[3]Never forget to get copies of all documents published and produced by other experts. A "second reading" often yields surprisingly relevant information.

Sidebar 13.2 Site Investigation Report

The value of these reports lies in their description of the nature, location, and extent of the site contamination. Also, the investigative reports often include a history of site activities as well as the location of facilities, both current and historic. These reports also provide the basic information needed to perform either soil and/or groundwater modeling of contaminant fate and transport.

When using either historic or site data from a series of site investigation reports, it should be remembered that initial reports (including any errors contained therein) tend to be taken as gospel in later reports. Geotechnical firms do make mistakes. Therefore, historic information should be compared to other available sources if at all possible. In addition, reported analytical data, as plotted and/or modeled, can also contain mistakes. One should not always assume that site maps and graphics prepared by site consultants are correct. Always compare the basic data to their graphical representation. In other words, do not have blind faith in the output presented in site investigations. Another pitfall with consultant reports occurs when maps prepared by subsequently retained site consultants have different orientations, scales, symbols, or site features. This may require a careful review of each subsequent map to make sure that everything is in its proper place.

little or no significance in developing a pattern of forensic evidence. Thus, it is critical that relevant historical site information be collected. These documents in combination with documents provided by counsel will represent the pool of knowledge used to identify and characterize the environmental issues. The types of documents that are most commonly collected are:

- Aerial photographs (numerous, diverse sources)
- Government archives
- Sanborn maps
- Technical literature (government agencies, libraries, and universities)

- Technical association memberships and meeting minutes
- Land use history (historical society archives, libraries, city and county property records, newspapers)
- Newspapers
- County, state, and federal regulations

With regard to the collection of documents, the forensic expert has one primary document management problem: the efficient retrieval of documents from both counsel's and his or her own files. Thus, it is important that the forensic expert understand how law firms manage documents.

Law Firm Document Management

In general, law firms handle their case documents in three ways. The two most common methods are to store the case documents in a central repository to which all parties have access (i.e., either physical access or online), or to copy and distribute the documents to all parties. The least common method employed is to convert the paper documents into a full-text computer database.

Regardless of the management system selected by counsel, in most cases the expert will initially receive only copies of those technical documents that counsel has identified as "hot documents." As the expert reviews the initial documents and becomes familiar with the technical facts and case issues, it is the expert's responsibility to (1) request counsel to provide specific documents (or document types) and/or (2) review and tag documents in the central repository for copies in order to develop expert opinions.

Because the process of technical document review, the subsequent identification of relevant documents, and their retrieval is an iterative process, most experts are at the mercy of a law firm's document management system. Thus, when time is of the essence, almost always the circumstance in litigation, it is the authors' experience that full-text computer-based document management systems are far superior to handling coded databases or paper copies.

Thus, any expert who is fortunate enough to be involved with early document management decisions associated with a law firm's case management should stress the use of full-text computer-based systems. For a description of computer based systems and their potential costs, see Sidebars 13.3 and 13.4.

Sidebar 13.3. Digital Document Repositories.[4]

In order to produce a digital document repository, a vendor that can provide the necessary hardware and software to operate a digital repository must be selected. The documents must then be processed (staples and binding removed), scanned (converted to an image), and converted into text.

Two basic options exist for building a digital repository: (1) imaging the entire document collection and then conducting a searchable document review (i.e., at a computer workstation or through an Internet connection) or (2) conducting the document review manually and imaging only those parts of the document collection that are deemed responsive or relevant. The advantages and disadvantages of each option are also discussed.

The primary benefits of the first option are as follows:

- If the litigation objectives change, all of the documents can be searched relative to the new issues.
- Internet-based review makes the database accessible to a large number of users.
- An imaged collection (i.e., scanned) provides for easy storage.
- Production and distribution costs are much lower than paper copies.

The primary drawback is that imaging the entire collection can be cost prohibitive.

The primary benefit of the second option is that it saves the cost of imaging the entire collection. However, if other documents are required for distribution, the imaging process will have to be performed a second time. In this case, it can delay pertinent analyses by both counsel and experts.

In deciding which option to use, cost tends to be a key consideration. As a result, the following cost guidelines can be used to assist in the decision-making process:

[4]Paul Burbage, P.E., Firemans's Fund, Novato, California, was a contributing author to this sidebar.

- Estimate the number of pages in the repository by assuming that there are 2000 pages in a banker's box.
- Paper copies, on average, can be produced at $0.03/page.
- Scanning ([imaging] only) costs range from $0.15 to $0.20/page.
- Text conversion (optical character recognition [OCR]) costs range from $0.07 to $0.12/page. OCR costs are variable based on the quality of the original documents. Poor-quality originals will generate additional editing costs.
- CD-ROM production costs average $50/CD. This is typically how repository data are stored and distributed.
- Vendor project management costs average $75/hour. For cost, assume that 0.5 hour will be required per banker's box.
- Document coding costs range from $1.30 to $1.90/document.
- A simple cost analysis allows one to estimate the break-even point or the crossover between copying/distribution and scanning/OCR.

If counsel (and/or the client) has not previously employed computer database management methods, it should be appreciated that these methods are not limited to large productions but can be used on the smaller universe of technical documents that have been identified and indexed. Because some cases may require multiple experts (in different geographic locations), not only will reproduction costs be reduced, but the digital database can then be electronically transferred to both attorneys and experts when needed.

Cost Summary

Summarizing all of the costs associated with the development of a computerized (scan/OCR) database yields an approximate page cost of $0.30. Using the figure of $0.03 per page for copying, the crossover is at 10 copies per page. As an example, if a case involves five or more law firms and five or more experts, the cost is at the break-even point.

However, even if there are fewer than 10 copies per page, there can be significant savings resulting from the rapid ability to search, find, and print key documents during various stages of the litigation. This is particularly true if case objectives and issues change during litigation.

(*continues*)

(Sidebar 13.3, Continued)

Words of Wisdom on Vendors

Consider the following when choosing a vendor:

- Select an established vendor (e.g., more than five years of experience).
- When selecting software, consideration should be given to the type of software used to generate/operate the repository (i.e., proprietary software written by a vendor versus "off the shelf" software). The concern here is that not all repository vendors will be in business over the next few years. If proprietary software is used and the vendor has gone out of business, the repository owner may find itself in a position where the repository cannot be revised or updated.

A Note on Document Organization

There is a long history of using Bates numbers to code and subsequently retrieve documents. From an environmental case management perspective, however, date sequencing of documents is of much greater value due to the historic perspective of environmental forensics. Historic investigations are much more efficient when the relevant documents are sequenced by date.

After the original production is assigned a Bates number, a second copy of the relevant documents can be sorted by date. This set can then be scanned and converted to text (i.e., by OCR). Computer databases in this format can be reviewed very efficiently. From this database, the expert can select the key documents that will be used to support the technical case objectives. If the need arises, the computer will also retrieve Bates-numbered documents.

Sidebar 13.4 Historical Cost of Searchable Computer Databases

The cost of managing a digital database has decreased significantly in the last 10 years. If this continues at even half this rate, digital document management systems should be economically feasible for even relatively small document production within the near future. This trend is illustrated by the following case.

In 1989, a document production generated approximately one million documents with an average of two pages per document. Counsel for this case retained one of the first document management firms in the United States to create a searchable computer database. This was accomplished by typing the full text of each document into the database (including a quality control check).[5] At the end of one year, and after only some 200,000 documents had been entered into the system, a detailed review of the document management fees showed that the initial estimate of $1.00 per page had spiraled to approximately $7.00 per page.

Given these extreme costs, counsel was able to find a vendor that could scan and convert the documents by OCR and provide the necessary software and hardware to conduct full-text document searches. However, in order to operate an efficient document management system required proper document preparation prior to scanning. Subsequently, the Bates-numbered documents were coded to identify the source of the production and basic document type (i.e., invoice, correspondence, technical report, etc.). Prior to scanning, all handwritten documents (including those with marginalia) were date-coded and maintained as separate files. All documents were then sorted by date. During the sorting process, documents that were of poor quality (i.e., faded, portions not copied, etc.) and could not be converted accurately by OCR were also separated into files.[6] The completed database was then processed.

All of the scanned images were stored as a TIFF file on disks that held approximately 50,000 pages (each page was approximately 1 megabyte).

(*continues*)

[5]Each document, letterhead and all (including each and every attorney's name), was typed off-shore by two typists (for comparative accuracy) before entering the database.

[6]Note that handwritten documents, as well as documents of poor quality, were completely and accurately "scanned"; the text was just not converted by OCR.

(Sidebar 13.4, Continued)

The text generated by the OCR process was approximately 95% accurate. The handwritten and poor-quality documents (approximately 5000 pages) were not converted into text.

Because the database was sorted by date, when a word search identified a specific document and that document was displayed, it was very easy to view the earlier/later-dated documents around the selected document without changing disks. This provided a major time-saving advantage (with today's gigabyte hard drives this is not as crucial). This effort not only saved considerable time but was implemented for approximately $1.00 per page.

Based on this 1989 per page cost for scanning and text conversion, there has been a 70% per page cost reduction in the last 10 years (i.e., $0.30 per page in 1999; see Sidebar 13.3). If this trend continues, cost-effective digital management of litigation documents may be realized within the next three to five years.

In addition to full-text databases, some law firms choose to code documents by key words so that a computer-based document index can be searched for defined information. This requires that each document be read to find the key words (e.g., a site name, an employee name, an event, a date, etc.) so that a database can be developed, which can then be computer searched. The development of the database is time consuming. If new key words are subsequently identified as being important, the database must be modified or it will become less useful. At a minimum, many litigation databases simply rely on Bates numbers, which bear no relationship to either the chronological or the subject-specific grouping of documents. As a result, most coded databases are of little use to technical experts other than identifying documents with specific date ranges.

Expert Document Management

When developing specific opinions on case issues (e.g., soil contamination, air emissions, plant maintenance, surface water quality, etc.), it is recommended that the key documents for each issue be date sequenced into a file or binder (hard copy or computer based). As the discovery process progresses, new documents can be added accordingly, to expand the "history."

It is also recommended that a computer master list of each document be kept[7] so that, if necessary, documents can be easily searched and selected to form subcategories of documents for special purposes. An example of this might be the need to develop a separate timeline for articles published, meetings attended, or patents filed. The list is as endless as one wants to make it.

By managing documents in this fashion, the ability to factually support opinions and respond to a client's questions is not only more efficient (both in time and in money), it ultimately makes the deposition process much easier for the expert.

Document Review and Selection

In large environmental litigation cases (usually involving more that 50 boxes of documents), counsel will usually have nontechnical staff begin the review (usually on a carton-by-carton basis) and prepare a box index of the contents. Because document productions related to environmental pollution from industrial companies are extremely diverse, it is almost impossible to generate a useful "technical" box index. In the event a forensic expert is retained prior to the indexing process, it is recommended that the expert work with counsel to establish guidelines to help nontechnical staff identify relevant technical documents. The optimal approach, however, is to employ technically qualified individuals (and this means not only technically qualified but also legally qualified personnel) to review and index technical productions.

Document Discovery

Recognizing that key documents are often obtained only through discovery and that discovery is carried out through specific information requests, it would seem obvious that where technical issues are paramount, the technical consultant(s) should be, indeed must be, involved in aiding the development of the information request. All too often, the consultant is retained after the close of discovery or at a point so close to the cutoff date that valuable information may be lost.

When first retained, the expert should determine what the cutoff date is for the discovery process. If there is still time remaining, the expert should provide counsel with a list of documents that should be obtained.

[7]An abstract can also be included for each document in this list.

Even if the document requests have already been prepared, and perhaps already filed, the expert should conduct a review to ensure that the requests are written with the proper technical terms and are specific enough to ensure a comprehensive and focused request. The issue here is not the universe of documents so much as the completeness of all categories and sources of all documents.

Expert Opinions

Needless to say, we have ours and they have theirs. So, let us begin with theirs (also see Sidebar 13.5).

Opposing Expert Opinions

In reviewing expert reports, the first step is to compare opinions (i.e., if there is more than one expert). We are always amazed as to how often expert reports contain contradictory technical information.

Having reviewed the opinions, the next step involves reviewing the source information used to support the opinions. Some experts provide detailed supporting information that is well referenced[8] within the opinions, while others refer only to a list of documents that supports their opinions.

When reviewing the source information, it is most important to review not only the text but also the references used to support the source information. Often, a document supporting an opinion has within its contents other statements or references that contradict the opinion. Also, a key step (see Sidebar 13.5) is to carefully review the publications of the expert to determine if, at some former time, a different opinion was stated. This, of course, also includes prior deposition and trial testimony as well as former reports, declarations, and affidavits.

As a note of caution relative to an expert's list of publications, in some cases courts require that only the most recent publications be listed on a résumé (e.g., five years). Such a list is not adequate to fully assess past opinions. As a result, it may be necessary to conduct a library search to access additional prior publications.

[8]We have seen document lists so extensive that it begs the question, "Did you personally review all of these documents?" The follow-up questions focus on either who did the review or just how many hours it took to conduct the review. More than once, it has surfaced that the expert managed to review documents at the rate of 50 to 100 pages per minute—a task difficult even for a graduate of Evelyn Wood Reading Dynamics!

Sidebar 13.5 Some Notes on Opposing Experts

The key to undermining opposing experts often resides in their prior testimony experience and their publications. A thorough investigation of *all prior work* is mandatory in order to properly assess the "strength" of opposing experts. For those with lots of experience, it is common practice to list only the past five years of testimony and only the past five years of publications. Needless to say, previous testimony and/or publications (that contain contradictory information) can come back and "bite" an expert. Therefore, thoroughness of review is mandatory. Not to be overlooked, past declarations and affidavits as well as education (particularly the background of their former professors) often shows a pattern or uncovers areas that can greatly assist in neutralizing experts. Preparing appropriate questions (and of equal importance, answers to the questions) becomes a collaborative effort between counsel and expert. On occasion, it is possible for counsel's expert (or consultant) to attend the deposition of the opposing expert. Real-time input (the ability to recognize the appropriate follow-up question[s]) can be critical in achieving the desired deposition outcome. In addition, the presence of a consultant can also have a positive effect on the experts being deposed, such that they become less evasive to technical questions they could obfuscate with technical jargon.

Client Expert Reports

There are two basic types of expert reports: the abbreviated expert report (i.e., do not give away too much) and the comprehensive expert report (i.e., when they read this, they will want to settle the case).[9] The "expert" has little to do with the decision making, only the production of the product. Regardless of the extent of the expert report, declaration, or affidavit, there are two areas of concern: (1) the generation of an opinion without all available supporting documentation and/or information and (2) the coordination, or the lack of coordination, with other experts retained on the same case (i.e., having different areas of expertise and opinions).

[9]Some expert reports are produced in response to a technical issue not previously addressed by counsel. In this case, an expert's role is to rebut specific testimony. Rather than presenting an opinion, the consultant simply punches holes in the opposing expert's testimony.

Because some attorneys wait until the last minute to retain an expert, the expert may be required to generate an opinion within hours or days of being hired. An expert retained under these circumstances runs the risk of not completing a thorough review of all relevant information, so that the subsequent report, declaration, or affidavit may be factually incorrect. In these cases, it becomes difficult to defend prior work. Be prepared to present a strong defensible document, or suffer possible embarrassment later.[10]

Another problem area often arises when multiple experts are retained on the same case. Some attorneys feel that it is best to compartmentalize their experts, thereby protecting the integrity of each report or opinion. Although this may be considered a valid point of view, we have never participated in a case where cross-communication between experts became a problem. Indeed, it has been our experience that interaction often brings to light unknown information and facts that can then be used by all experts in preparing their opinions and reports. The net result is to eliminate the duplication of effort (and cost) as well as the potential of experts on the same case using contradictory facts and putting forth contradictory opinions. It is highly recommended when a consultant is retained in a case where counsel wants to compartmentalize the experts that counsel be reminded to seriously consider the alternative. Once again, we have never encountered a case where this apparent collaboration between experts hurt the case objectives. Indeed, the opposite has been the case, that is, a strengthening of opinions and a better understanding of the supporting information.

Draft Reports

If an expert report is required by the client, it is strongly recommended that draft reports not be published. Technical opinions are based on factual information that in many instances is supplied by the client. As a result of the litigation process, the expert usually receives this information (i.e., documents and deposition testimony) over a period of weeks or months. Therefore, opinions might be altered over time. If draft reports are published, the expert may be placed in the position of having to defend contradictory opinions.

[10]An expert could say "I'm not ready" to produce an opinion. In the real world, filing deadlines dictate the schedule. If the consultant chooses to accept such an assignment, the consultant must live with the consequences.

The Deposition Process

Fundamental to the deposition process, whether giving a deposition or taking one, is that the process is very similar to a doctoral oral exam. It is virtually impossible to get a perfect score. It must be understood that not every answer has to be perfect, just as not every question will solicit the desired or hoped-for answer.

Opposition Depositions

As a forensic expert, the deposition process involves not only your own deposition but also the depositions of opposition experts. In every case, the expert should offer counsel assistance in the preparation of technical deposition objectives, deposition questions, and likely follow-up questions for opposing experts. The expert should also provide counsel with an assessment of opposing expert's strengths and weaknesses, as well as prior publications related to deposition objectives.[11]

In most cases, attorneys will request that their experts prepare questions and/or suggest documents that should be used in the deposition. When the technical issues are really complex, counsel may take their expert (or a work product expert) with them to attend the deposition.

Regardless of how well prepared an attorney may be for a deposition, good experts can truthfully answer counsel's questions without really providing the answer counsel is seeking. If this occurs, it can only be countered by having your own expert present to rephrase follow-up questions (unless the attorney is well trained in the technical area of the questioning). All too often, we have seen missed opportunities because of the failure to take advantage of an opening with follow-up questions. Obviously, cost may be the issue in cases where technical issues are a minor part of the litigation. However, in technically heavy cases, the use of experts to attend an opposition expert's deposition may yield a significant savings.

The primary objectives of deposing an opposing expert include determining the opinion and the factual basis for the opinion. When the expert uses documents for support, it is very important to determine the speci-

[11]An expert should also keep a book on opposition experts (i.e., their publications, their deposition testimony, their opinions, and their methods for supporting these opinions). This type of information can be useful if the opposing experts are confronted in subsequent litigations.

ficity of the document(s) as opposed to the circumstantial strength of the document(s). In addition to these objectives, it is also important to ascertain the true expertise of the expert, as measured by prior testimony, experience, and prior publications (authored and coauthored), as well as published reports.

An Expert's Deposition

Experts often exhibit certain "failings" during a deposition, which counsel generally feel contribute to a poor performance. One major failing is the desire of the expert to say too much simply because he or she is smart and wants everyone to know it. Overstating leads to openings that would not necessarily be obvious to opposing counsel. An expert should stick to the subject and not provide any more information than is necessary to provide a truthful answer. Another bad practice is to go too far afield in an answer, thereby swimming (as it were) into areas outside of his or her expertise (e.g., into shark-infested waters).

The expert must learn to be alert for compound questions that appear to require a single "yes" or "no" but, in fact, might require a "yes-and-no" answer. Finally, establish a signal with counsel to warn of these problems (e.g., a swift kick in the shins usually works), to wrap up an answer, or to signal the need for a break (i.e., to rest the mind). Some simple rules for giving a deposition include:

- Limit opinions to only those that are strongly supported. Avoid providing opinions that are tangentially supported. They provide the best opportunity for opposing counsel to discredit testimony (including the good opinions).
- Never speculate.[12]
- Show extreme caution with hypothetical questions and state for the record that the answer is to a hypothetical and not related to the facts in the case. When this occurs, make opposing counsel give you all the hypothetical facts necessary to answer a hypothetical question. If counsel cannot provide all the hypothetical facts, do not answer the question.

[12]In point of fact, the following is a direct quote from trial testimony, "I don't speculate in the presence of very intelligent defense lawyers." This is good advice, even when being questioned by not very intelligent lawyers.

- Have copies of all the documents that support your opinions at the deposition. Mark or tag the documents so that key sections can be easily found.
- Be completely accurate regarding publications, prior deposition testimony, and so on. It is better to "not remember" or "not recall" than to test one's memory and give an incorrect answer.

The Development of Graphics and Exhibits

Prior to any mediation, settlement meeting, or trial, the single most important function of the expert is the development of graphics and exhibits for presentation of technical subjects to lay individuals. This process is both difficult and time consuming. As a consequence, the process should be started well in advance of scheduled presentations.

Most judges and juries have a gut understanding of technical issues. However, simple exhibits relating the issues to some "common experience" can go a long way to clarifying and simplifying complex subjects. Also, the side with the best exhibits often establishes the strongest foundation for case issues. Most assuredly, people (juries) react positively to a clear and well-demonstrated presentation that is both interesting and focused. The use of exhibit boards (large enough to see and understand), computer graphics (again with either large TV screens or high lumen projectors), large aerial photos, and the occasional true exhibit such as sand and gravel in a clear glass column (to demonstrate permeability) or an oil/water mixture (to demonstrate either floaters (LNAPLs) or sinkers (DNAPLs)) add a persuasive dimension to the testimony.

Of perhaps even more importance, especially in environmental cases, is the simple story telling of the site history. History can be interesting if the presentation is focused on key issues. Those who present the most entertaining picture often establish the basic credibility for the case.

Mediation/Settlement Meetings and Strategy

All experts should anticipate that they may be expected to participate in mediation or settlement meetings. The key to successful mediation (revolving around technical issues) is in the presentation. The stronger the technical presentation (i.e., good graphics supported by clear and easy-to-understand facts and issues), the more directly will the mediator "buy in" to the logic. Some guidelines for mediation presentations include the following:

- Conduct a comprehensive technical investigation (you do not want any surprises).
- Hire a well-known and respected technical consultant/expert to present the issues.
- Present *simple* graphics.
- Do homework on the opposing side's argument.
- Prepare for negotiation, and possibly the reduction of expectations.
- Have a realistic appreciation of the costs of going forward to litigation.
- With regulatory agency involvement, do your homework (regulators just might become allies and support technical and cost arguments).
- Prepare a book on the mediator: prior experience, prior mediation methods, biases.
- Suggest a technical ombudsman satisfactory to both sides, if this might help the mediator follow the technical issues.

In the final analysis, the mediation usually boilsdown to negotiating dollars. The technical issues usually represent posturing and window dressing (unless there really was only one side to the technical issues—in which case one of the parties is really fooling itself or has been sold a bill of goods by its consultant).

Trial Preparation and Expert Testimony

Just as in sport, war, debate, and education, successes are achieved with practice. A well-prepared group of attorneys and experts is mandatory to the success of the case. When preparing for testimony, the experts should not only be familiar with their own presentation, they should also be familiar with general case facts and, perhaps most important, have an understanding of case weaknesses.

The importance of technical issues may shift as a trial proceeds. With this shift comes the need to select previously thought to be unimportant documents on a time critical basis. Given an expert's familiarity with the technical issues, experts may be called on to determine which technical documents are needed. This is only half of the problem. The remaining problem is locating the required document in "quick time." As so often happens, documents have to be found during a brief recess and/or over lunch breaks. Considering that the typical environmental case might have a platform of several thousand "key" documents, their instant retrieval takes on a life of its own.

Based on our experience, having documents converted by OCR allows the most rapid search and retrieval. The second line of organization (and usually the most common) is to have key documents timelined, as well as broken down by specific category. Having these documents in binders allows for rapid searching and retrieval. Interestingly (and disturbingly), the approach of coding documents and using the computer to search and find seems simple but usually breaks down due to the fact that having identified the document in the computer does not necessarily correlate with one's ability to find the document in the storage boxes. At this point, it should be obvious that if the documents are scanned, the image can be printed out as required.

However, one must not forget the participants—counsel and experts. Trial preparation includes trial practice. Although one never wants to appear rehearsed, it is better to sound rehearsed than to lose one's way during trial. An expert is usually grilled by counsel to make sure that trial testimony is consistent with declarations, reports, and depositions. The expert should be prepared with supporting documents, exhibits, and other visuals, and have a clear understanding of their content and presentation sequence.

One of the saddest sights is to see an expert (i.e., unless, of course, it's an opposing expert) confused over the content of an exhibit or confused regarding some of the more simple facts regarding a site, site operations, site operating history, or site remediation.

The golden rule, as an expert, is to answer "yes" or "no" and then go on to explain your answer. To begin with a "may" or "possibly" invites controversy and can damage the expert's credibility.

Maintenance of Files

Once a case has settled or trial is over, the case documents must be retained in storage because of appeals (unless counsel specifically directs the expert to dispose of the documents). As a result, files and documents may have to be stored for a number of years. We have had experiences with cases reappearing (in a somewhat altered form) as much as 10 years later.

SUMMARY

In order to provide the best and most cost effective service to a client, there are a few simple guidelines:

- Ensure that all available information has been collected and reviewed.
- Maintain a well-organized document database (for opinions).
- Know the facts.
- Understand the opposition.
- Appreciate your weaknesses.
- Honesty is the best policy.
- Prepare, prepare, prepare.

Chapter 14

Environmental Extortion[1]

Patrick Sullivan, Tad Patzek, and Doug Zunkel

The initial site investigation was coincident with the removal of underground storage tanks (USTs) containing petroleum hydrocarbons. It was found that both the soil and the groundwater were acidic and contained highly elevated concentrations of trace metals that included arsenic, barium, cadmium, copper, lead, and zinc. The UST removal occurred as a result of a Phase I site assessment initiated by the property being evaluated for sale. The consultant carrying out the remedial investigation identified a paint manufacturer that once owned and operated at the site and, therefore, concluded that the party responsible for the soil and groundwater contamination was the paint manufacturer. As a result of this conclusion, the current property owner threatened to file a suit against the paint manufacturer unless the party paid for the site investigative costs to date, agreed to pay for any subsequent cleanup, or agreed to purchase the property—as is. Needless to say, based on limited information and faced with an apparent *fait accompli*, the paint manufacturer responded in the negative.

A PROPERTY SALE GONE WRONG

The site presented in this case study is located within an industrial section of the city of Bayside. Since the early 1970s, this site has been used as an automobile dealership (hereafter referred to as Bay Auto). In 1990,

[1]A case that involves the minimizing cost to the client when there are multiple contributors to a site contamination.

the bayfront properties (although primarily industrial) were appreciating in value due to nearby sporting and recreational development. In this improving economic environment, Bay Auto entered into a contract to sell its property. While in escrow, the prospective buyer commissioned a local engineering firm to complete a Phase I site assessment.[2] The Phase I assessment revealed that the nationally known ACE Paint Company (i.e., a "deep pocket"[3]) had once operated a factory on the Bay Auto property. As a result of this discovery, Bay Auto launched a Phase II site assessment to evaluate the extent of any soil and groundwater contamination with a different geotechnical firm.

The results of the Phase II study showed that the soil and groundwater were indeed contaminated with high concentrations of trace metals. As a result, the prospective buyer declined to close the deal. Bay Auto then contracted with another geotechnical firm[4] (hereafter referred to as GeoExperts) to complete a remedial investigation (RI). The GeoExperts RI was completed in late 1994. This report contained the following conclusions:

1. Based on our review of an environmental site assessment report and aerial photographs, the site was occupied by chemical manufacturing facilities over a period of at least 50 years, with operations ceasing in the early 1960s. Beginning in approximately 1920, the site was part of a larger property owned by several chemical companies and used for the production of lithophone (a white paint pigment consisting of a mixture of barium sulfate and zinc sulfide).
2. Barium sulfide was prepared from baryte ore and petroleum coke. Zinc sulfate was produced from crude zinc sources and sulfuric acid. During the process, metal impurities such as copper, nickel, and cadmium were removed from the zinc sulfate solution; lead peroxide was one of the reactants used in the purification process.
3. Elevated concentrations and types of metals and low pH [acid] conditions in soil and groundwater are consistent with materials and processes used to manufacture lithophone and related products. . . . Sulfuric acid was stored in a tank next to the zinc sulfate building. Large quantities of metal-rich solutions and sulfuric acid were handled daily for decades. . . . Wastes generated during several phases of manufac-

[2]An initial site investigation to identify past land use and conditions/activities that may have resulted in potential environmental hazard.

[3]A company that is still in business (or its successor) with a lot of money and/or insurance.

[4]As strange as it may seem, it is a rare occurrence to find that one engineering/geotechnical firm has completed all of the site investigations. Multiple geotechnical firms usually means that none of the site maps and dates are coordinated. As a result, more time and money must be spent in order to make sense of disorganized site information.

turing were disposed of on the property and have likely contributed to the elevated concentrations of metal in soil and groundwater.

The site investigation revealed that the soil above the bay mud (i.e., the old tidal flats) was extensively contaminated with arsenic, cadmium, copper, lead, zinc, and traces of silver. A number of the soil borings showed the presence of glassy and slaglike materials, reddish brown sandlike materials, carbonaceous materials, highly colored materials, and metallic agglomerates. In addition to the soil contamination, over one-third of the site had a zone of slag cobbles from 2 to 8 feet thick. The fill ranged in thickness from 10 to 14 feet and was deposited on top of a black silty clay (i.e., the bay mud) that was several feet thick.

An example of the representative chemical characteristics of the fill is shown in Table 14.1. Although massive slag cobbles were identified by the geologist supervising the site investigation, the fill was not characterized as being related to anything other than lithopone production, an early and, as subsequently seen, unfortunate and costly oversight.

In order to illustrate the contamination pattern at the site, GeoExperts generated a historic diagram of the ACE Paint Company facility from a 1950 Sanborn map (see Figure 14.1). This map could then be compared to the sitewide soil and groundwater contamination data collected from the RI. The soil and groundwater contamination patterns for arsenic, cadmium, copper, lead, and zinc are illustrated in Figures 14.2 to 14.6. Given that the direction of groundwater flow was approximately from the northeast to the southwest corner of the property and the location of the zinc

Table 14.1 Representative Fill Analyses

Sample Depth(ft)	*Element (mg/kg)*						
	Ag	*As*	*Cd*	*Cu*	*Ni*	*Pb*	*Zn*
2.5	7	1,600	53	470	58	2,300	7,700
2.5	7	170	27	460	22	3,100	1,800
4.0	39	1,200	150	1,700	87	14,000	19,000
5.0	9	660	64	650	99	5,000	17,000
5.0	65	350	110	460	20	14,000	4,500
7.5	14	260	880	280	4	240	3,000
8.0	4	210	12	260	23	1,300	3,600
11.5	22	170	17	1,700	28	14,000	8,400
11.5	19	310	17	3,300	33	4,600	16,000

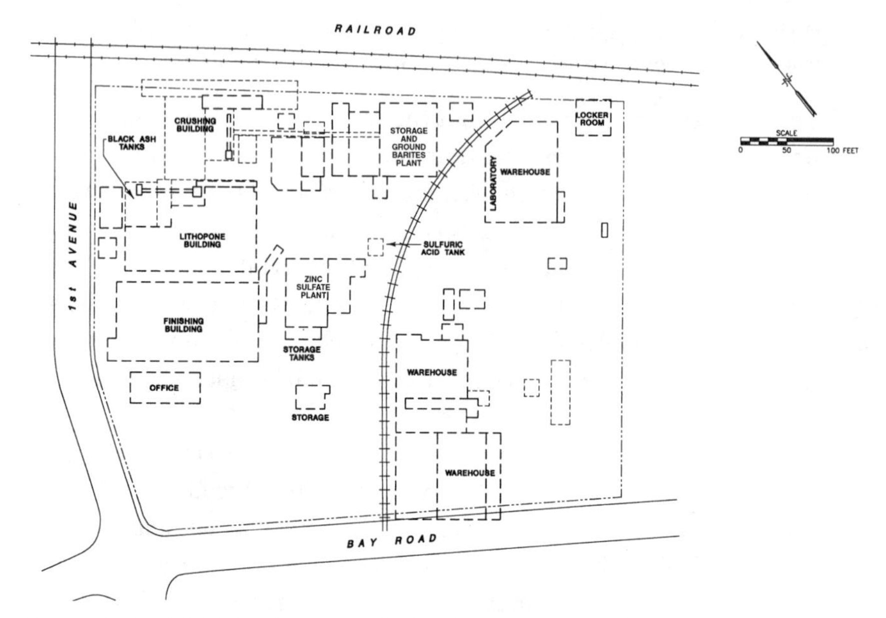

Figure 14.1 ACE Paint Company facility from a 1950 Sanborn map.

contamination in the vicinity of the zinc sulfate plant, the initial data suggested that ACE Paint Company was responsible for the contamination.

A Search for Owners and Operators

Given the site contamination, Bay Auto notified the local and state environmental agencies of its findings. Burdened with site investigation costs and potential remedial cleanup costs, Bay Auto began a search for past property owners.

Investigators for Bay Auto determined that the ACE Paint factory operated on the property from 1926 through 1954 when it was purchased by Industrial Chemicals. Industrial Chemicals continued to manufacture paint at this facility until the property was sold to a land developer in 1963. The developer razed the site and eventually sold the property to the current property owner (Bay Auto) in 1973.

Having failed to coerce either ACE or Industrial Chemicals into a purchase or settlement agreement, a suit was filed in federal court by Bay Auto against both ACE Paint Company and Industrial Chemicals, making the following argument:

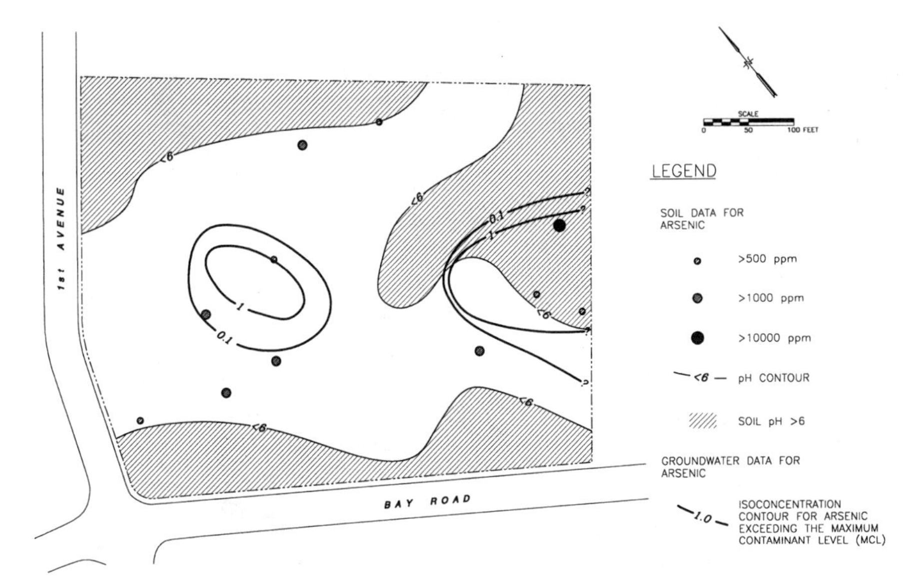

Figure 14.2 Soil and groundwater contamination pattern for arsenic.

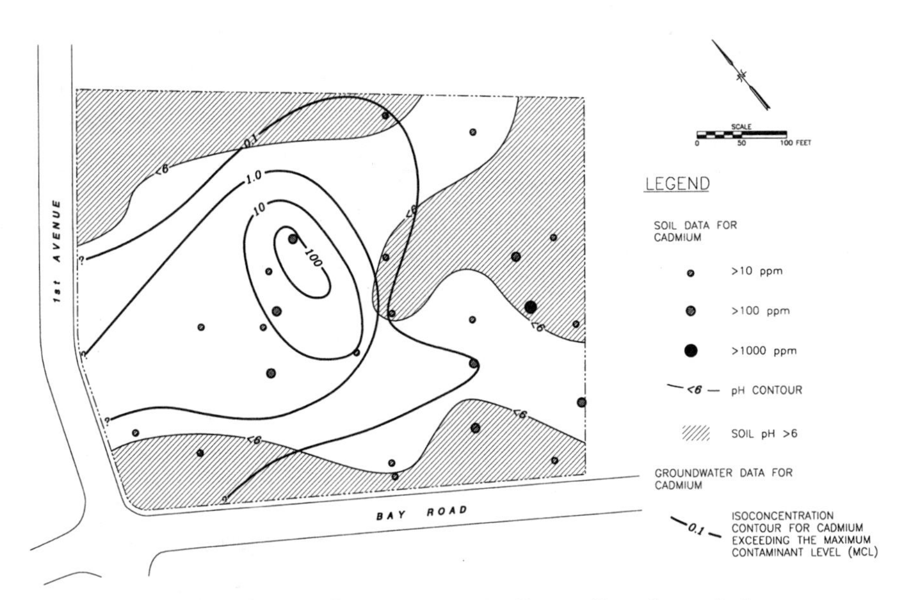

Figure 14.3 Soil and groundwater contamination pattern for cadmium.

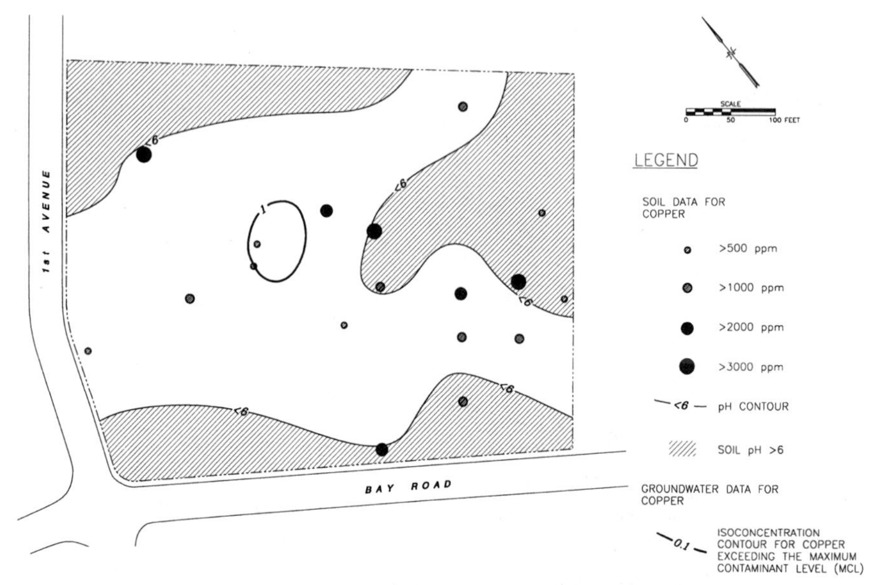

Figure 14.4 Soil and groundwater contamination pattern for copper.

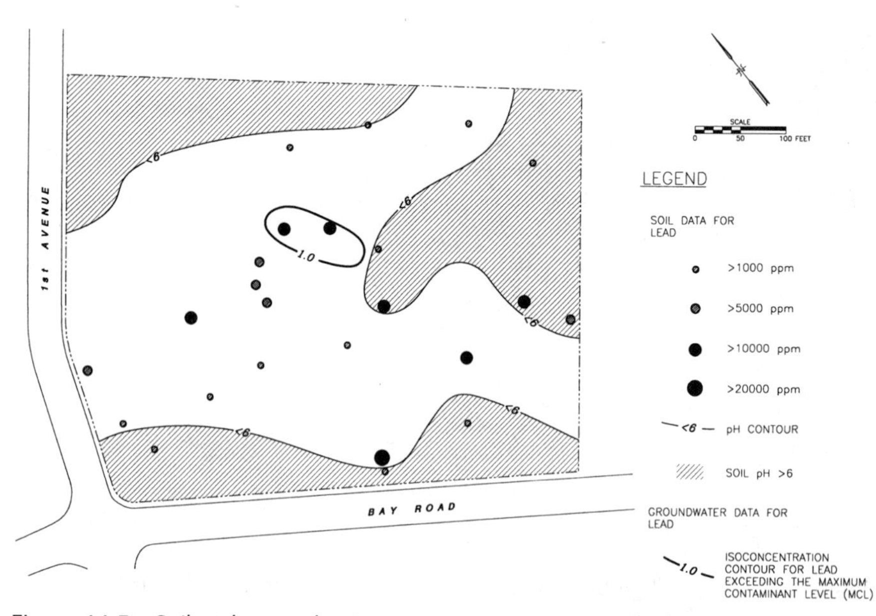

Figure 14.5 Soil and groundwater contamination pattern for lead.

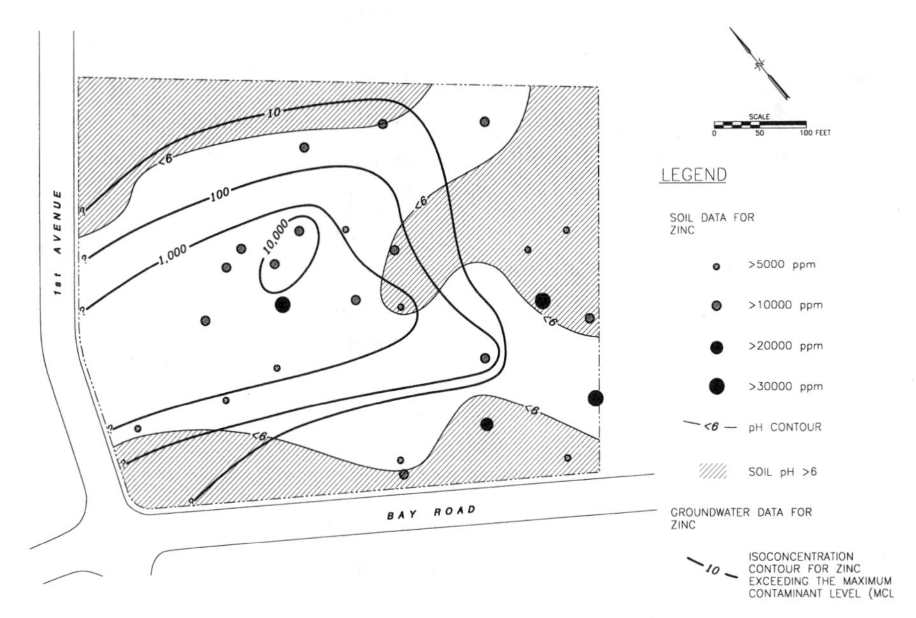

Figure 14.6 Soil and groundwater contamination pattern for zinc.

> From 1926 through 1963, two companies owned a chemical factory that manufactured zinc sulfate and other chemicals. During this time, zinc-sulfate solution leaked from tanks, pipes, pumps, and other equipment, contaminating the soil and groundwater with zinc and other toxic metals. . . . ACE Paint Company and Industrial Chemicals are liable under CERCLA as owners and operators of the facility during the time of disposal of a hazardous substance . . . are liable under RCRA as parties that contributed to the disposal of a "solid waste" that may present an "imminent and substantial endangerment" to the environment . . . are liable under state laws of trespass, nuisance, and equitable indemnity because their contamination of the property has interfered with the use of it by plaintiff Bay Auto. . . . Defendants' liability is joint and several under common law tort principles, and under RCRA and CERCLA case law.

Providing an Initial Defense

ACE Paint Company, through its counsel, retained a forensic consultant to evaluate the information provided by Bay Auto and to determine the extent to which ACE Paint Company was responsible for the soil and groundwater contamination. In this regard, the forensic consultant was hired to provide technical work product to counsel (i.e., the consultant would provide technical strategies and advice but no expert testimony). Furthermore, as the case information developed, the forensic consultant would identify the necessary technical opinions and recommend experts.

The initial forensic investigation of the Phase I report revealed that, prior to ACE Paint Company's occupation of the site, a chemical company at the same location manufactured sulfuric acid. Although Bay Auto (and its counsel) had this information, it was conveniently excluded from GeoExperts' RI and Bay Auto's complaint. Subsequently, an initial review of local property records, city and county documents, as well as old newspaper files showed that the industrial use of the site dated as far back as 1872. This more comprehensive industrial history of the property is outlined in Table 14.2.

The smelter and refinery operations (which were not mentioned by GeoExperts) would have treated ore containing lead, zinc, silver, gold, and various trace metals (i.e., cadmium, copper, lead, etc.). The waste generated by this process would explain the widespread distribution of metals in the soil at the site. Although GeoExperts' soil boring logs show a thick deposit of slag[5] in the center of the site, the RI suggested that the slag was from the roadbed of a past railroad spur.

In addition to the smelting operations, the manufacture of sulfuric acid from pyrite (which was also not mentioned by GeoExperts) would generate waste materials containing trace metals, acids, and acid-generating metal sulfides (see the 1912 Sanborn map of the operations in Figure 14.7). When comparing the ACE Paint Company operations in Figure 14.1 with the sulfuric acid operations in Figure 14.7, it is obvious that the sulfuric acid operations[6] directly overlapped with the ACE Paint Company zinc sulfate operations.

In addition, the distribution of sulfides (see Figure 14.8) was sitewide with the highest concentrations located in acid soils. This result would be expected given the acid potential of iron sulfide oxidation.

Based on the occurrence of the smelting and acid manufacturing operations at the site, it is highly likely that the widespread metal and acid contamination in both the soil and the groundwater occurred because of these operations. An aerial photograph, from a U.S. Geological Survey database, of ACE Paint Company operations (see Figure 14.9) showed that ACE Paint Company had disposed of its wastes into the indicated "waste

[5]A possible smelter waste material.

[6]According to the Sanborn map, the sulfuric acid plant had an earth floor.

Table 14.2 Industrial Ownership Timeline

Years of Ownership	*Owner*	*Type of Operation*
1872–1903	Precious Metals Smelting & Reduction Works	Production of gold, silver, and lead bullion from precious metals and zinc/lead ores
1906–1926	Pyrite Chemical Company	Manufacture of sulfuric, hydrochloric, and nitric acids
1926–1954	ACE Paint Company	Manufacture of lithopone and drilling mud
1954–1963	Industrial Chemicals	Same as ACE Paint Company

disposal area." This area was the property of Bayside Electric Company. This off-site disposal strongly suggests that the soil contamination on the Bay Auto property was not primarily due to the operations of ACE Paint Company.

Although the soil contamination did not appear to be from operations of ACE Paint Company, the groundwater contamination (i.e., trace metals and sulfuric acid) was definitely a possibility. However, it was also possible that the groundwater contamination could have occurred from leaks or spills from Pyrite Chemical Company. In addition to these liquid sources, there was an additional complicating factor. The solid wastes produced by smelting and acid manufacture could also have produced groundwater contamination as the result of acid rock drainage.[7] Thus, it turned out that there were multiple industrial processes that may have contributed to both soil and groundwater contamination.

In addition to the initial historic site data review, the extent of the soil and groundwater contamination was evaluated to determine potential remedial solutions that could be imposed by either local or state regulatory agencies. This review suggested the following:

[7]Acid rock drainage is created when iron sulfide minerals (e.g., pyrite) contact water and oxygen (e.g., in the zone of groundwater fluctuation) to form sulfuric acid. This process also releases trace metals along with the acid.

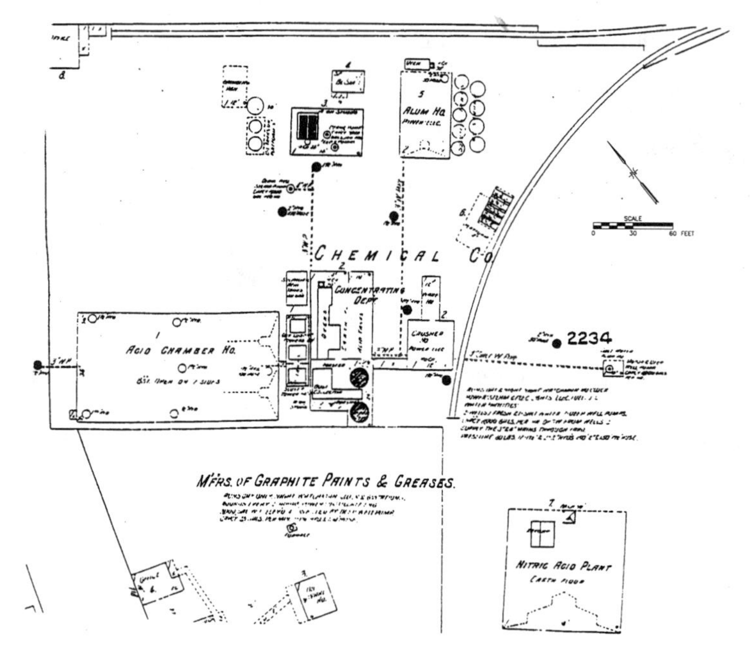

Figure 14.7 1912 Sanborn map of the sulfuric acid operations.

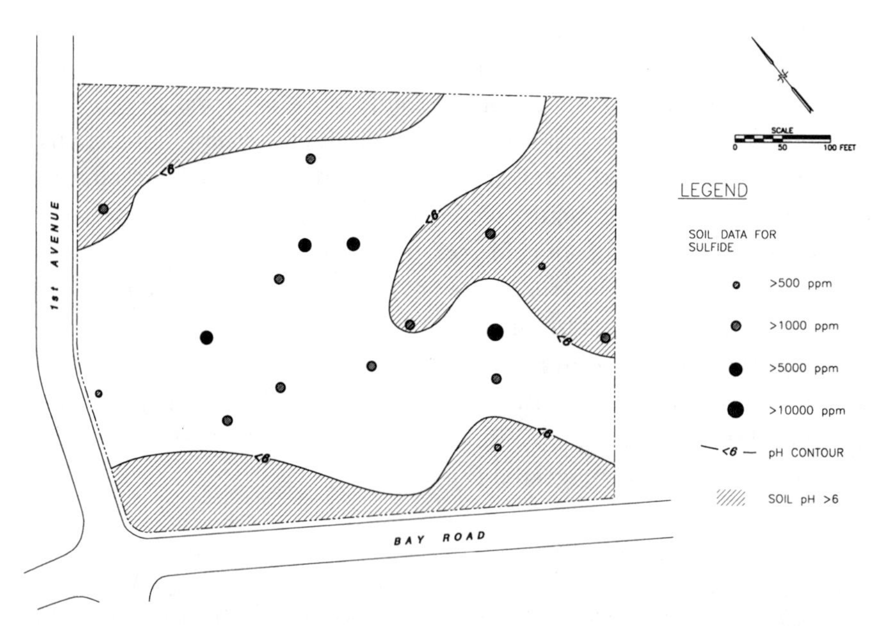

Figure 14.8 Soil sulfide data.

Figure 14.9 Site map of waste disposal areas and surface water drainage.

- Although the soil was contaminated with a wide range of toxic metals, as long as the site remained capped (i.e., covered with asphalt or concrete—which, by the way, was already in place), there was no danger to people at the site. With a deed restriction on the land use to limit excavations (i.e., underground parking lots, swimming pools, etc.), there would be no health hazard or need for any remedial action.
- The contaminated groundwater was essentially static and not moving toward the bay and there were no off-site groundwater wells (i.e., down-gradient drinking water, industrial or agricultural wells). However, some of the contaminated groundwater did appear to intersect a concrete stormwater drainage structure that did discharge to the bay (refer to Figure 14.9). Under a worst case scenario, it was possible that an impermeable barrier might have to be placed on the up-gradient side of the storm drain (i.e., that side of the storm drain in contact with the contaminated groundwater) to limit any potential trace metal discharge to the bay. Again, the

need for groundwater cleanup was not apparent at this site. Based on the newly "discovered" site history and the minimal remedial action at the site, counsel for ACE Paint Company arranged a meeting with counsel and representatives of Bay Auto.

A Difference of Opinion: Round 1

The representatives for Bay Auto included a hydrologist from GeoExperts, the president of Bay Auto, and counsel for Bay Auto. As the technical representative for ACE Paint Company, the forensic consultant presented the following information/opinions:

- The pre-1926 industrial ownership history (see Table 14.2).
- Soil borings in the GeoExperts RI indicated the occurrence of slag and fill (i.e., containing arsenic, cadmium, copper, lead, and zinc) distributed above the natural bay sediments (see soil boring lithology example in Figure 14.10), giving a clear indication that wastes from the pre-1926 activities were used as fill material on the site. The bay mud boundary with the fill is at 8.5 feet.
- Groundwater contamination was consistent with acid rock drainage and/or spills of sulfuric acid and possible spills of zinc sulfate.
- Soil and groundwater contamination was no threat to employees/customers of Bay Auto nor was there any demonstrated discharge of trace metals to the bay, suggesting that there would be no groundwater remediation.

At the conclusion of the presentation, the president of Bay Auto angrily pounded the table and accused the forensic consultant of being a lying whore for ACE Paint Company (in fact, he said that all consultants were liars—except his own, of course).

The offended GeoExperts hydrologist countered that GeoExperts' historical research did not uncover the pre-1926 activities. The hydrologist suggested that the slag could have been from a railroad spur that was buried on site after the site was razed and that there was no evidence of any pyrite in the soil, so how could there be any acid rock drainage? In addition, GeoExperts argued strongly that the concentration of trace metals in the groundwater could not have occurred from acid rock drainage and, further, that the groundwater contamination was an obvious threat to the bay. GeoExperts continued to maintain that local and state regulators

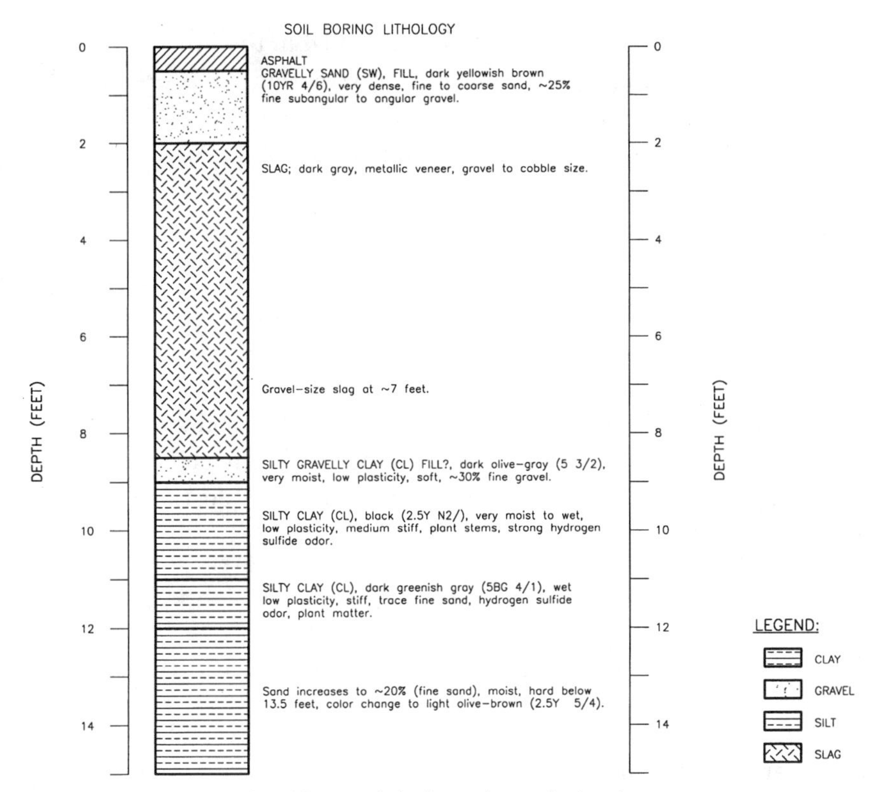

Figure 14.10 Typical soil boring lithology through the slag area.

would require extensive remedial action (estimated by GeoExperts to be approximately 6 million dollars). However, one of GeoExperts' younger team members mumbled that perhaps the forensic consultant for ACE Paint Company was right about minimal or no remediation.

It was clear from this exchange of opposing opinions that the site information collected by GeoExperts (i.e., trace-metal analysis of soil and groundwater) was gathered to address potential cleanup issues but was presented in such a way as to compel ACE Paint Company and Industrial Chemicals to pay for any damage. The data, however, were not collected in such a manner that would allow the development of any allocation theories. Thus, the following issues demanded resolution:

- What were the contamination characteristics and sources associated with the ACE Paint Company operations?
- What was the origin of the buried slag?

- Could smelter waste and pyrite be identified in the fill?
- Could the existing groundwater contamination have originated from acid rock drainage?
- Was contaminated groundwater contributing trace metals to the bay?
- What was the most likely remedial solution?

In other words, there were significant data gaps which, for the moment, allowed a difference of opinion.

In order to potentially resolve some of these issues, Bay Auto's counsel agreed to let ACE Paint Company's consultant collect soil and slag samples from soil boring cores stored by GeoExperts. At this point, it was the forensic expert's job to provide some answers to these questions.

FILLING IN THE DATA GAPS

Characterizing the Manufacture of Paint

The ability to differentiate between the potential sources of both soil and groundwater contamination from the manufacture of paint and lead/zinc ore processing can only be accomplished by characterizing both activities. As a result, the historical records from ACE Paint Company and Industrial Chemicals were reviewed as well as historical literature (e.g., libraries, government archives, industrial/chemical association documents).

The Search for Pyrite

In order to quantify the disposal of both smelter waste and pyrite waste from the pre-1926 industries, experts were needed who could provide (1) guidance (and, if necessary, opinions) on the historical aspects of both ore smelting/refining and sulfuric acid manufacturing and (2) a positive identification of the mineral wastes at the site. Therefore, a consulting mining/metallurgical engineer was retained to provide an evaluation of the historic mineral operations, and a process mineralogist was hired to identify the presence of any mineral waste on the site.

The objective of subsampling of the soil cores collected by GeoExperts was to select samples that would be representative of the fill. Based on

GeoExperts' soil boring logs, subsamples would be taken from a range of slag and purple/red fill.

Characterizing the Soil and Groundwater Contamination

Figure 14.11 shows the subsurface location of the slag[8] identified in the GeoExperts soil borings (up to 8 feet thick). As discussed previously, it was possible that the acid and trace metal contamination in the vicinity of the ACE Paint Company zinc sulfate building was the result of the manufacture of zinc sulfate. However, it was hypothesized that leaks from the manufacture of sulfuric acid (through earthen floors) could have contacted the buried slag and created the same condition. In either case, the distribution of sulfate in the groundwater, as shown in Figure 14.12, is centered in the vicinity of these historic operations. To test this hypothesis, laboratory leaching studies using slag samples and sulfur acid were conducted.

Figures 14.6 and 14.9 illustrate the relationship between the distribution of the groundwater contamination (as represented by zinc) and the groundwater flow direction and location of a subsurface storm drain that discharges to the bay. In order to assess the potential movement of trace metals in the groundwater and their potential discharge to the subsurface storm drain, it became necessary to model the groundwater flow as well as any effects that tidal oscillations may have had on groundwater movement. Since a three-dimensional model of the site's subsurface was to be generated, it would also be possible to estimate the mass and distribution of trace metals in the fill.

Defining the Threat to the Bay

Although Bay Auto provided no evidence of surface water contamination from the defined groundwater contamination, it was a possibility. If there were a continuous source of groundwater trace metal contamination reaching the surface water, it would need to be characterized in order to assess potential remedial solutions. As a result, ACE Paint Company's forensic consultant undertook studies to (1) evaluate surface water quality upstream and downstream of the site during both high tide and low tide, (2) evaluate the integrity of the concrete storm drain as a barrier to groundwater

[8]This figure also shows the relative location of the smelter slag to the 1900 fill boundary.

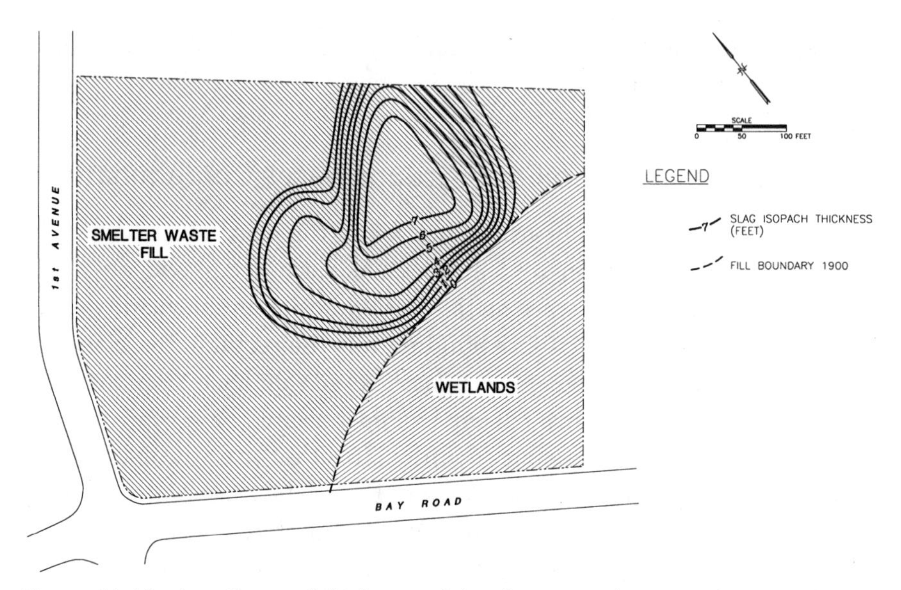

Figure 14.11 Location and thickness of the slag area relative to the 1900 fill boundary.

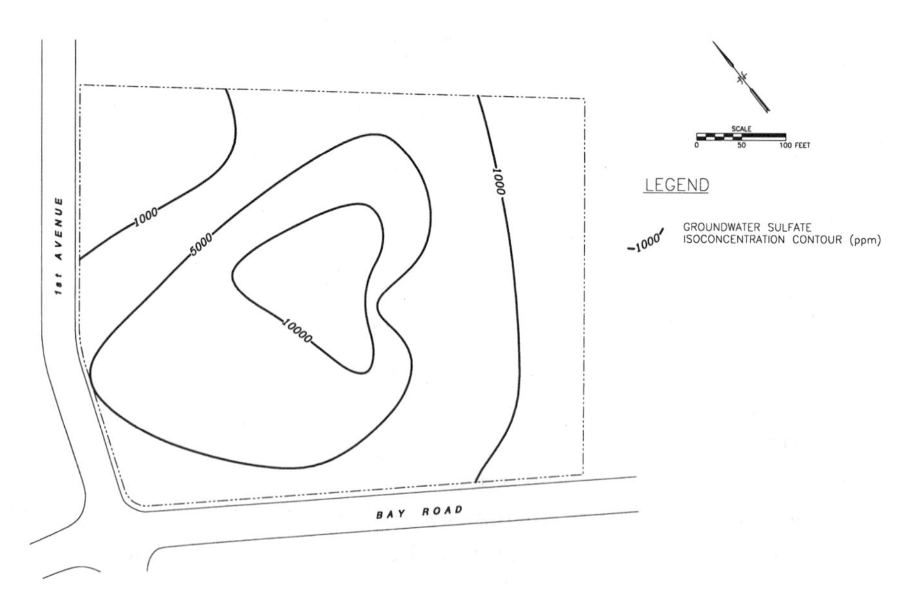

Figure 14.12 Sulfate concentration distribution in groundwater.

flow, and (3) discuss potential remedial options with the appropriate regulatory agencies (i.e., define the potential monetary damages).

With the preceding investigations under way, it was now possible to begin the process of defining the scientific basis of each party's position.

EXPLAINING THE PIECES OF THE PUZZLE

ACE Paint Company and Industrial Chemicals Manufacturing

The paint manufactured by ACE Paint Company and Industrial Chemicals was a white opaque compound called lithopone. Lithopone was manufactured by combining solutions of barium sulfide (BaS) and zinc sulfate ($ZnSO_4$) to form a precipitate composed of approximately 29% zinc sulfide (ZnS) and 70% barium sulfate ($BaSO_4$). The process was carried out in a building obviously called the lithopone building (see Figure 14.1).

The $ZnSO_4$ component for the lithopone was produced from the following process:

- Zinc residuals from smelters located in Idaho were the primary raw zinc sources for the process.
- The raw zinc was then placed into wooden tanks and filled with water. Concentrated sulfuric acid (H_2SO_4) was then added to oxidize the raw zinc and form $ZnSO_4$. The resulting $ZnSO_4$ solution was passed through a filter press to remove the impurities (i.e., primarily silica/silicate minerals and lead—precipitated by sulfate). This residue (called "mud") was neutralized, filtered, and discarded onto the adjacent property (see Figure 14.9).
- The $ZnSO_4$ liquid was then oxidized (i.e., using sodium carbonate and potassium permanganate) to remove iron and manganese. These oxide residues were also disposed of with the "mud." If required, the liquid was oxidized a second time with lead peroxide (these solutions would be alkaline and lead would precipitate). The $ZnSO_4$ liquid was then treated with zinc dust to convert soluble copper, nickel, and cadmium into solid metals (i.e., the trace metals were deposited onto the Zn dust). This trace-metal-enriched zinc was filtered from the solution and sold to recover the trace metals (usually the cadmium).

- The pH of zinc solutions was always greater than 5.1 (by measurement and equilibrium conditions).

The BaS for the lithopone was produced by the following process:

- Barite ore from Battle Mountain, Nevada, contained 90% barium sulfate ($BaSO_4$), 0.75% barium carbonate ($BaCO_3$), 0.25% iron, and silica. This ore was heated with a carbon source and oxygen to produce a black ash (BaS, 76.0%; $BaCO_3$, 12.0%; $BaSO_4$, 1.5%; and silica, 10.0%).
- The black ash was then leached with hot water to extract the BaS, while the remaining residue (insoluble silica, sulfate, carbonate, and sulfide phases) was discarded to the tailings pond (see the northeast portion of Figure 14.9).

Based on these process descriptions, the zinc "mud" and barium black ash residues that were generated during the years of paint manufacturing were contaminating the adjacent properties not owned by Bay Auto. Bay Auto, being a good neighbor, informed the adjacent landowners of ACE Paint Company's and Industrial Chemicals' historic disposal practices. These landowners immediately filed suit against ACE Paint Company and Industrial Chemicals. Although the further investigations that this action triggered are not discussed in this chapter, the outcome is provided.

The positive result of identifying these past disposal practices was that the evidence now suggested that the zinc and barium wastes were indeed not disposed of on the Bay Auto property. In other words, there had to be another source responsible for the trace metals found on the Bay Auto property—the initial "straw man" had been blown away through Bay Auto's "good-neighbor policy."

Ore Processing Operations and Identification

During the time period that smelters operated at this site, smelting activities were actually located adjacent to the bay. Figures 14.13 to 14.15 show a comparison of the bay shoreline relative to the site in 1885, 1900, and 1931, respectively. Thus, as ore was processed at the site, the waste products were disposed of onto the tidal flats of the bay until the area was completely filled.

Prior to discussing the site history, however, it is necessary to provide a brief explanation of (1) smelting and reduction works operations and (2) the manufacture of sulfuric acid.

An Introduction to Smelting and Ore Refining

Lead is relatively abundant in nature and occurs primarily as lead sulfide (PbS), lead carbonate ($PbCO_3$), and lead sulfate ($PbSO_4$). These minerals often contain significant amounts of silver and gold. Lead can also be associated with arsenic and antimony as complex silver- and gold-containing sulfides or sulfates.

If the primary lead-bearing mineral in the ore is lead sulfide, a two-step process is required to obtain lead metal: high-temperature oxidation by roasting or sintering to convert the lead sulfide to lead oxide, followed by high-temperature reduction of the lead oxide to lead metal in a shaft furnace.

Some metallic iron is generally added to the furnace charge to reduce residual lead sulfide in either the roasted sulfide or the carbonate/sulfate ore. This metallic iron forms iron sulfide, which remains free or becomes matte (i.e., that waste consisting chiefly of a combination of sulfur and iron). During the reduction process, copper, arsenic, and antimony oxides in the ore are reduced and these metals tend to associate with the lead metal. Zinc and iron oxides are not reduced and remain in the liquid slag with the oxides of silica, calcium, magnesium, and aluminum. Unoxidized sulfides remaining in the furnace charge either remain free or combine to form matte. Unoxidized arsenides and antimonides tend to occur in speiss (i.e., a combination of arsenic or antimony and metallic iron). If silver and gold are present in the ore, they also stay associated with the liquid lead metal. The liquid lead metal, termed bullion, produced in the reduction step requires refining to remove impurities such as copper, antimony, and arsenic before it can be sold for commercial application. If present in sufficient quantities, gold and silver are recovered and refined for sale.

In the 1860s, two predominant processes were in use for smelting lead—the reverberatory furnace and the ore hearth. The blast furnace was in its early stage of development at that time. Both the reverberatory furnace and the ore hearth technologies were quite crude and environmentally abusive by today's standards but were capable of treating both sulfidic and carbonate/sulfate ores. Both oxidation to remove

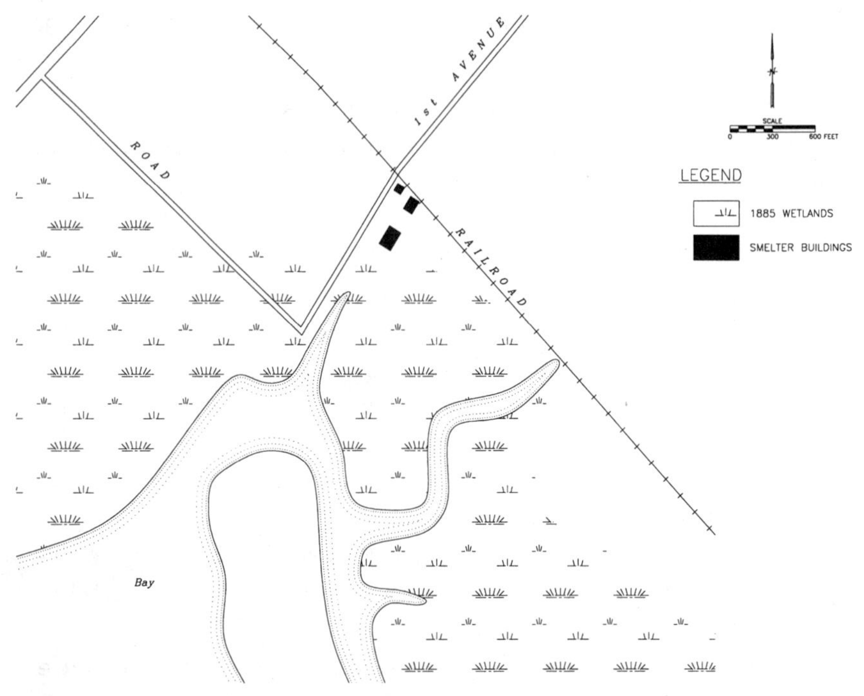

Figure 14.13 Relationship of site fill relative to 1885 wetlands boundary.

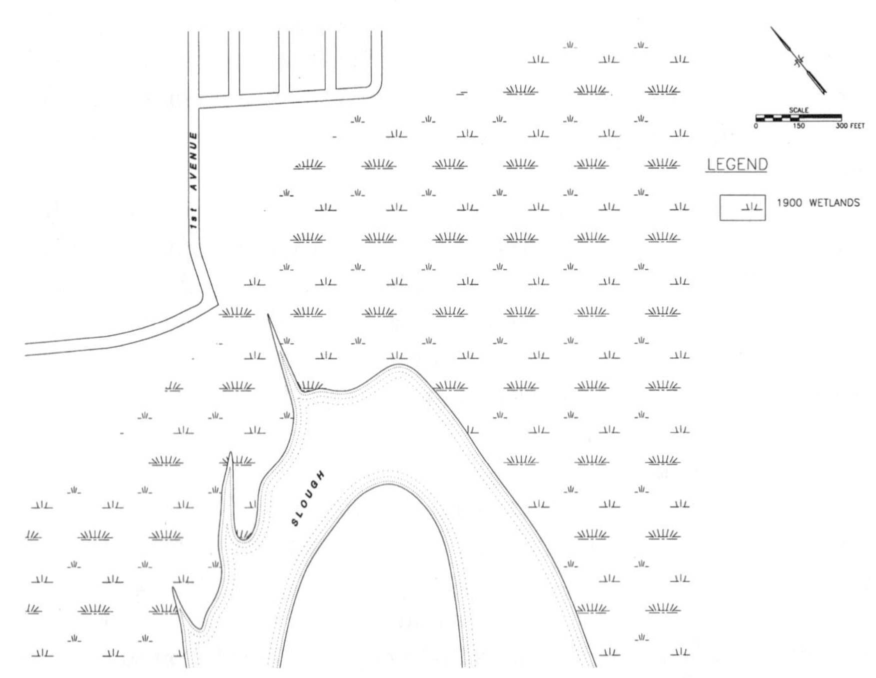

Figure 14.14 Relationship of site fill relative to 1900 wetlands boundary.

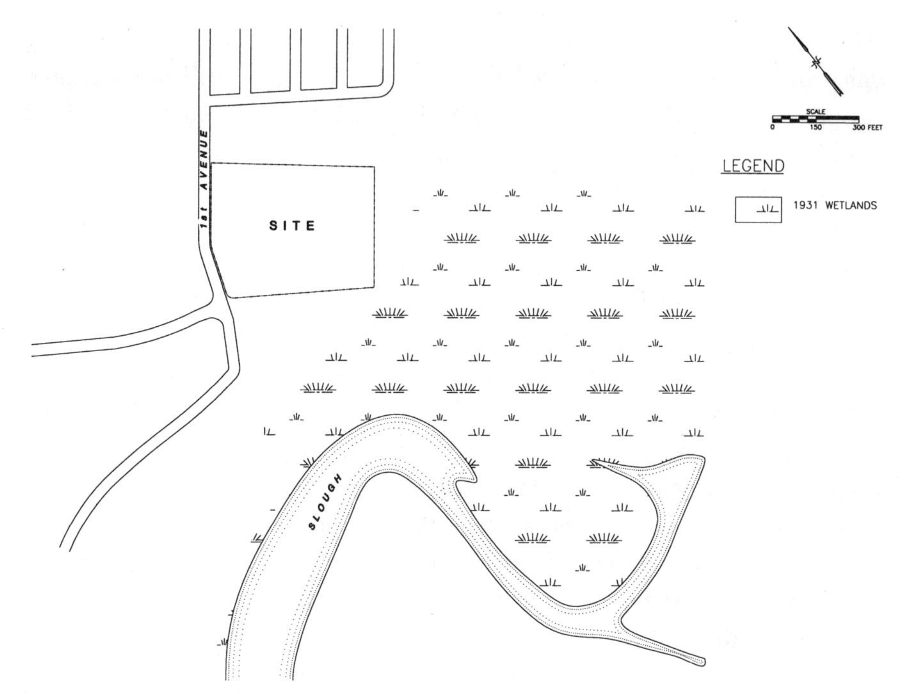

Figure 14.15 Relationship of site fill relative to 1931 wetlands boundary.

sulfur and reduction with either coal or charcoal supplemented by metallic iron were the key features of these single-furnace, small-scale, batch processes. Molten metallic lead and slag (i.e., a mixture of waste minerals) were produced by the furnaces. Lead was tapped into a lead well. Slag was periodically tapped into ladles and resmelted or discarded, depending on the lead content, which could be as high as 20 to 30%.

Both processes were relatively inefficient, leaving a significant amount of lead metal, unreacted lead sulfide and lead oxide, and unreacted coal or charcoal in the slag. Other sulfides in the feed were also inefficiently reduced and reported in the slag and other furnace products. Unreacted copper sulfides formed matte with iron sulfides; unreacted arsenic sulfides formed speiss. These materials were entrapped in the slag during slag–metal separation prior to slag disposal.

The blast furnace began to emerge as a commercially viable process in the early 1870s, first as a cupola-type furnace and later as a conventional water-jacketed furnace. The feed to the blast furnace had to be primarily oxidic since carbon does not reduce sulfides. If the ore or concentrate was primarily lead carbonate or sulfate, it could be smelted directly. If the ore was primarily sulfidic, it had to be desulfurized by heap, stall, or kiln roasting prior to the 1890s or by sintering in the 1890s and later. The direct smelting ore or calcine/sinter, following roasting or sintering, was fed to the top of the blast furnace with coke and metallic iron. The smelting reactions produced molten metallic lead and slag, which were tapped either continuously or intermittently. Air for combustion of the coke (the blast) was provided by air introduced through a series of pipes near the bottom of the furnace. The blast furnace was significantly more efficient in terms of lead recovery than either the reverberatory furnace or the ore hearth because far fewer sulfides were present in the feed and higher smelting temperatures were used. Carbonate and sulfate ores generally contained small amounts of sulfides and sulfide ores were desulfurized prior to reduction. Thus, fewer sulfides were present in blast furnace slag. Some matte and speiss were produced in the furnace and entrapped with metallic lead and unreduced sulfides in the slag due to incomplete slag–metal separation prior to slag disposal. Coke was completely consumed in the blast furnace so unreacted coke was generally not present in the slag. Coal was unsuitable for use in the blast furnace due to its low porosity and reactivity.

The liquid slag from any of the furnaces, generally containing up to 10% lead and 15% zinc, was either water granulated and transported to the slag dump for disposal or transported as molten and dumped on the slag pile where it cooled more slowly. The water-granulated slag was glassy and less reactive than the more crystalline slowly cooled slag.

It is obvious from the smelting process that the type of waste generated, and its constituent compounds, is dependent on the specific smelting processes employed and the extent of any secondary treatment to extract metals. However, the chemical composition of any waste is ultimately a function of the parent ore that is processed (even ores can be "fingerprinted").

For example, the ore from one of the major lead–zinc–silver-producing areas in the world, the Coeur d'Alene region of Idaho, is primarily composed of galena (PbS), sphalerite (ZnS), tetrahedrite [$(Cu, Fe, Zn, Ag)_{12}Sb_4S_{13}$], chalcopyrite ($CuFeS_2$), pyrrhotite (FeS), pyrite (FeS_2), arsenopyrite (FeAsS), and magnetite (Fe_3O_4) [1]. In addition to the metals in these minerals, sphalerite is a host mineral for both Cu and Cd, galena is a host mineral for Ag, and pyrite is a host mineral for As [2]. Thus, it should be no surprise to find these same elements in the smelter waste that utilized these ores. Smelter waste from the Bunker Hill Region of Idaho was reported to contain the compounds shown in Table 14.3.

Table 14.3 Bunker Hill Slag and Dust Analyses

	Concentration Range (mg/kg)	
Element	*Slag*	*Dust*
Arsenic (As)	172–571	5–1,130
Cadmium (Cd)	1–100	6–483
Lead (Pb)	130–7,880	163–15,800
Zinc (Zn)	14,000–43,900	362–35,400

Data from the Bunker Hill site file, Idaho Department of Environmental Quality.

Given the wide range of chemical composition and solid phase properties of smelter wastes (i.e., glassy or massive, production of pyrrhotite/pyrite during smelting, particle size distribution), a leachate containing elevated trace metal concentrations could be generated. However, the generation of a leachate containing elevated concentrations of trace metals is highly dependent on the characteristics of the smelter waste and the environment into which it is deposited.

Producing Sulfuric Acid from Pyrite

A description of the manufacture of sulfuric acid by the chamber process is summarized next [3]: The chamber process for the manufacture of sulfuric acid takes its name from the lead chambers that constitute the chief essential part of the apparatus.

The normal product of chamber plants is sulfuric acid of 50° to 60° Baumé. The chamber process cannot, however, make high-strength acid. The essential parts of a modern chamber plant are:

- Burners of some kind for the production of SO_2
- Dust settling apparatus, except in those cases where brimstone is burned to make SO_2
- Glover tower
- Chambers
- Gay-Lussac towers
- Acid circulating apparatus
- Fans and flues
- Apparatus for introducing the oxides of nitrogen

The gas produced in the burners is derived from the oxidation of elemental sulfur, iron sulfide, iron-copper sulfides, zinc sulfide, or mixed sulfides. When necessary, this gas is drawn through some form of dust settling apparatus to remove the greater part of the dust that would otherwise contaminate the acid.

The gas mixture next enters the Glover tower at a temperature of 800 to 1000°F. It is in this tower that the gas is brought into intimate contact with a mixture of 60° Baumé sulfuric acid, which is carrying N_2O_3 in solution. The hot gas with its SO_2 reacts with the acid and removes the N_2O_3, forming some H_2SO_4 and converting practically all of the N_2O_3 into NO, a gas, which continues with the main gas stream. The gas mixture from the Glover tower is conducted into the chambers, usually from 3 to 10 in number, in series. Any given portion of the gas occupies from one to two hours in passing through the set of chambers. Steam or atomized water is introduced at various points. Sulfuric acid is formed by the reactions among SO_2, oxygen, the nitrogen oxides, and water. The sulfuric acid collects in the bottoms or pans of the chambers.

When these reactions have gone on for the proper length of time and the gas finally reaches the end of the last chamber, the SO_2 percentage has been reduced to less than 1/10 of 1%, and the nitrogen oxides are practically all in the form of N_2O_3. It is extremely important that the SO_2 percentage be reduced below 1/10 of 1% or else the recovery of the nitrogen compounds will be incomplete. It is almost as essential that the SO_2 percentage not be less than 2/100 of 1% for the same reason, as well as because of increased corrosion of the lead.

By properly regulating the amount of water or steam introduced, the acid made in the chambers is kept at approximately 50° Baumé. It is not permissible to allow the continued formation of acid of much greater concentration because of the tendency of such acid to take into solution some of the nitrogen oxides, in which case they are no longer available for reaction with SO_2.

From the chambers, the gases pass into the Gay-Lussac towers. Their function is to recover the N_2O_3. The acid issuing from the Gay-Lussac towers, carrying usually from 1 to 2% N_2O_3, is elevated to the top of the Glover tower and the N_2O_3 is therefore reintroduced into the system.

The process of roasting or burning pyrite is accomplished using a Herreshoff furnace. The burning of pyrite results in the following reaction:

$$4FeS_2 + 110_2 = 2Fe_2O_3 + 8SO_2$$

The burned pyrite solid waste of Fe_2O_3 commonly contains copper and is called "pyrite cinder." In addition to the cinder, "flue dust" from the burners and "mud" from the Glover tower are also waste products of the chamber process.

The Bureau of Mines reported in 1920 that the other constituents of pyrite ore had a deleterious effect on the manufacturing processes of sulfuric acid. Volatile metallic compounds such as arsenic and selenium could contaminate the acid and reduce its value.

Pyrite Chemical Company also manufactured nitric acid. The production of nitric acid from sulfuric acid and saltpeter is described by the following reactions:

$$2NaNO_3 + H_2SO_4 = Na_2SO_4 + 2HNO_3$$
$$NaNO_3 + H_2SO_4 = NaHSO_4 + 2HNO_3$$

The resulting residue is a mixture of sodium hydrogen sulfate and sodium sulfate that was known as "niter cake."

The Identification of Ore Processing Waste

None of the studies completed by GeoExperts identified any ore processing waste other than the buried slag deposit. Using the same soil samples collected by GeoExperts, samples that had the greatest probability of containing ore processing waste (i.e., highly colored samples) were sent for analysis by a process mineralogist.

A trained process mineralogist experienced in the analysis of mineral phases that have been modified by industrial processes (i.e., smelting and roasting) can determine, from the analysis of soil samples, the presence of coke/coal, smelter slag, and pyrite cinders in the fill at the site. Indeed, the process mineralogy summary showed the following soil and fill characteristics:

- Coke/Coal: From the coke particles, about 40% show proof of smelter residue association due to intimate intergrowth with or inclusions of slag, sulfides, iron oxides or metal alloys. The unreacted residual coal present in all soil samples contains noticeable amounts (4–6%) of ultrafine iron sulfides (pyrite and marcasite). . . . Iron oxide particles associated with the coke residuals from the smelter operation contain very high amounts of lead (9–30%) and zinc (3.3–5%) and lesser amounts of copper.

- Smelter Slag: The glassy smelter slag is a major source of the heavy metals, copper, lead and zinc. The concentration of the metals in the vitreous silicate matrix of the slag particles ranges from less than 0.01% to 14% Pb, less than 0.01% to 6% for zinc, and less than 0.01% to 0.4% copper, based on semiquantitative SEM–EDX analysis. The residual iron sulfides, pyrite and pyrrhotite (which are physically encapsulated in the slag), also contain substantial concentrations of copper, lead and zinc.
- Pyrite Cinders: The pyrite cinders consist almost exclusively of the typical roasting products, i.e., porous hematite and magnetite with minor amounts of iron sulfates. Only traces (<0.5%) of residual (i.e., unoxidized) pyrite were found in the cinder material.

Using these technical data, as well as historical documents collected from library archives and historical associations, a site history was developed.

Site History Relative to Ore Processing Lead smelting and refining at the site began in late 1872 or early 1873. The smelter/refinery initially treated approximately 20 tons per day of ore containing primarily lead carbonate, silver, and gold. The ores assayed approximately 30 to 50% lead with significant (20–40 ounces per ton) silver and minor gold values. The smelting process used initially was most likely several small ore hearths using coal and iron as reductants. Some of the lead bullion was cupeled to recover gold and silver, which were further refined on site. Following cupeling, the lead oxide was resmelted in an ore hearth or small cupola to produce silver- and gold-free lead bullion, which was cast for sale and/or converted into white lead. The balance of the furnace bullion plus custom and purchased bullion was refined using the conventional kettle process. In later years, however, several small blast furnaces replaced the ore hearths.

The slag from the smelting furnaces was discarded adjacent to the smelting facilities in the bay tidal flats, with the likely intent of filling in the area for future commercial use. Other smelter and refinery residues, such as dross, dust, matte, and speiss, were likely disposed of in the same area as the slag. As much as 125,000 to 150,000 tons of slag were disposed of on site.

Sulfur dioxide–containing gases, dust, and fumes from the ore hearths and blast furnaces, refining kettles, cupels, and coal- or oil-fired boilers driving the blowers were vented to the atmosphere through relatively short flues and stacks with minimal treatment or capture. These gases contained up to 1 to 2% SO_2; the dusts and fumes contained up to 20 to 30% lead and small amounts of arsenic and antimony as both sulfides and oxides.

A significant portion of the lead bullion produced was converted into white lead for paint pigments. Details of the specific process employed are not known with certainty. However, the most likely technology used was corrosion of small cast-lead trellises piled between layers of glazed earthen pots containing acetic acid. These layers were housed in large rectangular brick chambers into which air was injected. A layer of manure was placed between each layer of pots and lead to generate carbonic acid and heat. In 2 to 4 months, the stacks were disassembled and the white lead was collected, ground, washed, dried, and packed for sale. The uncorroded lead was recycled to the refinery for reprocessing. The residual lead carbonate, spent acids, and manure were likely disposed of on the slag dump.

Operations of this general type of lead smelter/refinery producing lead for sale and/or conversion to white lead probably continued on the site with little modification other than installation of blast furnaces. Vigorous operations were reported in 1874 and 1888, and somewhat curtailed activity between 1891 and 1892. Operations were purported to have continued at the site until at least 1902.

Refractory gold and silver ores were processed in an area adjacent to the smelting operations. A chloride-based hydrometallurgical–pyrometallurgical process was most likely used. The final salable products were metallic gold and silver. However, the ores treated contained refractory pyritic, arsenical, and antimonial minerals and resulted in the generation of arsenic/antimony-containing flue dusts. When sulfur dioxide was vented to the atmosphere, the flue dust that was captured was likely disposed of. The flue dust that was not captured was vented to the atmosphere via a smokestack. The leach residues were likely discarded in a tailings disposal area in a remote area of the site, probably in the bay tidal flats. That the captured flue dust was commingled with the leach residue for disposal is quite probable. Mercury-containing gases and residues were likely produced in these operations as a result of the use of amalgamation to recover native gold and silver from some of the ores and to assist in gold and silver refining operations.

With ore smelting and refining ending at the site in approximately 1903, the site was apparently idle for three years. In 1906, sulfuric acid manufacturing was begun by Pyrite Chemical Company, which operated at the site until 1926. Pyrite for the sulfuric acid process was extracted locally and was composed primarily of pyrite (iron sulfide) and chalcopyrite (copper and iron sulfide). In addition to the copper in the pyrite ore, a major trace element that occurs in pyrite and chalcopyrite is arsenic. Based on the equipment and buildings identified on the 1912 Sanborn map, it is likely that Pyrite Chemical produced sulfuric acid using the following process:

- Pyrite and chalcopyrite were crushed and roasted (the concentrating department had two Herreshoff roasting furnaces[9]). The products from the roasting process would have been a sulfur dioxide gas stream and an iron oxide cinder.
- The sulfur dioxide gas stream was filtered to remove particulate matter (i.e., sulfates of zinc, copper, lead, calcium, and magnesium; oxides of arsenic, antimony, tellurium, selenium, and silica). Arsenic could be further removed by passing the gas from the roaster over heated aluminum sulfate (i.e., alum).
- Sulfur dioxide was collected and concentrated.
- Liquid sulfur acid could have been further treated with hydrogen sulfide gas and then filtered to remove arsenic (precipitated iron and arsenic sulfides).

There are no records that would suggest that the pyrite cinders were ever removed from the site. However, pyrite cinders were identified on site from samples taken from soil borings taken from the southern portion of the site (i.e., consistent with the fill history of the site; compare Figures 14.14 and 14.15).

Pyrite Chemical Company also produced nitric acid. However, it is not known if any of the wastes from these products were disposed of on the Bay Auto property. Given the other major sources of contamination, it was not possible to specifically identify the presence of nitric acid or aluminum sulfate wastes (see alum and nitric acid locations in Figure 14.7).

Redefining the Soil and Groundwater Contamination

Using the data collected by GeoExperts, both soil and groundwater data were replotted, recalculated, or remapped, providing a more accurate picture of the site contamination.

[9]These furnaces were reported to have been in a warehouse during the years that ACE Paint Company operated at the site.

Soil Contamination

Because the relatively coarse-textured fill is on top of the naturally fine-textured black silts and clays of the bay, there is a fairly distinct boundary that marked the lower limit of the fill. As a consequence, the concentration of various trace metals could be plotted as a function of depth above and below the lower limit of the fill. Of all these plots, the metal of most interest was lead (see Figure 14.16).

Lead in both soil and groundwater that is dominated by sulfate is not mobile. Thus, any soluble lead will precipitate as lead sulfate. Therefore, lead that may have been released at the soil surface during the years of operation by ACE Paint Company should remain in the upper few feet of the surface fill (in addition, any liquids that would contain lead were in sulfate solutions). However, the plot of soil lead showed a fairly uniform distribution of lead (i.e., both high and low concentrations from the top of the fill into the bay mud).

This result demonstrated that the lead in the soil was the result of the *in situ* fill (i.e., it was in the fill when the fill was placed at the site). The lead in the fill could not have possibly been distributed throughout the soil from spills at the soil surface. In other words, the soil contamination throughout the fill did not result from the surface operations of ACE Paint

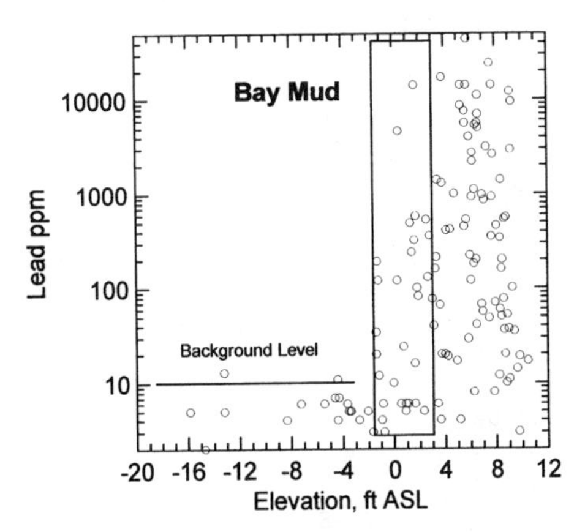

Figure 14.16 Soil lead distribution above and below the bay mud boundary.

Company. In addition, the operations that handled trace metals were isolated to the zinc sulfate building, whereas trace metal contamination occurs sitewide.

The concentrations of both zinc and cadmium plotted as a function of depth above and below the lower limit of the fill shows a much different distribution when compared to lead (see Figures 14.17 and 14.18). These data clearly show that both zinc and cadmium are mobile in acid leachate. Thus, groundwater contaminated with zinc and cadmium have migrated below the bay mud.

The fill samples collected by GeoExperts (as represented by the soil boring logs) show a continuous horizontal distribution of dark gray slag that ranges in thickness from approximately 1 to 8 feet. This deposit of slag occurs in the central to northeastern portion of the site, which is adjacent to the 1900 southern boundary to the wetlands (as shown in Figure 14.11). In addition to the slag deposit, slag fragments (cobbles to sand size) were described as occurring in the fill of almost all of the fill samples. In other words, smelter slag is distributed all over the site.

Conveniently, no chemical analyses were conducted by GeoExperts on any slag samples (i.e., what you don't know can't hurt you). However, the forensic consultant did have the slag samples analyzed (see Table 14.4).

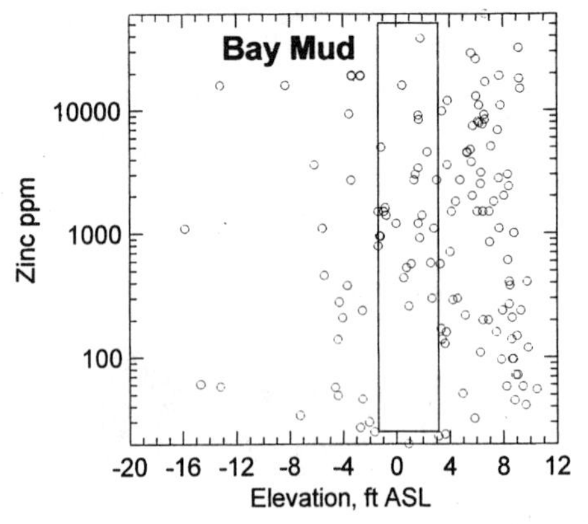

Figure 14.17 Soil zinc distribution above and below the bay mud boundary.

These data clearly illustrate the importance of slag as the source in the fill, whereas standard EPA methods for trace-metal analysis could not distinguish sources of the contamination.

The distributions and masses of trace metals in the fill have been estimated from order statistics of soil samples and geological logs, and are reported in Tables 14.5 to 14.9. All metals in soil samples from the site are lognormally distributed to a good approximation; that is, the logarithms of their concentrations in soil have normal distributions [4]. The distributions of zinc, lead, barium, and copper indicate more than one parent population (there are one or two significant breaks in their order statistics) as shown in Figures 14.19, 14.20, 14.21, and 14.22, respectively. In these cases, if feasible, a separate order statistic was generated for each part between the breaks and reported in Tables 14.5 and 14.6. The soil metal samples seem to be representative of their parent populations. Hence, the samples may be used to estimate the total metal masses underneath the site.

It is interesting to note that the upper bounds on the soil metal concentrations (the mean plus three standard deviations), predicted from the lognormal distributions, are very close to the highest concentrations seen in the samples. This suggests an oversampling of soil regions with very

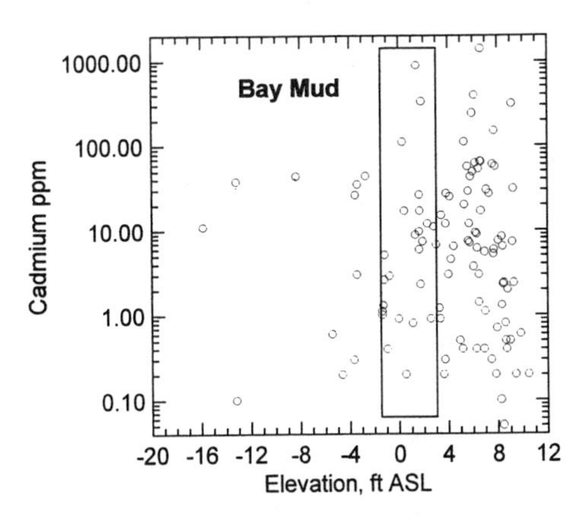

Figure 14.18 Soil cadmium distribution above and below the bay mud boundary.

Table 14.4 Slag Analyses

Slag Type	*Element (mg/kg)*						
	Ag	*As*	*Cd*	*Cu*	*Ni*	*Pb*	*Zn*
Massive 1	<1	16	43	2,900	3	43,500	21,000
Massive 2	<1	11	29	1,530	7	7,510	28,900
Glassy	<1	40	3	51	<1	522	804

Table 14.5 Summary Statistics of Lognormal Distribution of Metals in Soil

Metal	*Ln Mean*	*Ln STD*	*Number of Samples*
Barium, >412 ppm	7.7769	1.4781	59
Zinc, >55 ppm	7.2114	1.7706	119
Zinc, above BM	6.9426	2.0050	101
Zinc, all data	6.8029	2.0317	134
Zinc, below BM	6.3755	2.0840	33
Barium, all data	6.1342	1.8126	135
Copper, >35 ppm	5.7002	1.3927	68
Lead, >7 ppm	5.5462	2.3501	102
Barium, <412 ppm	4.8887	0.6581	78
Lead, all data	4.7056	2.6579	129
Copper, all data	4.6507	1.7736	108
Nickel	3.5753	0.7886	130
Zinc, <55 ppm	3.5627	0.3489	15
Arsenic	3.3755	2.1378	125
Copper, <35 ppm	2.8666	0.3373	40
Antimony	2.2446	1.6559	54
Cadmium	1.6386	2.1790	107
Mercury	−0.8662	1.5515	87

Table 14.6 Summary Statistics of Soil Metal Concentrations

Metal	*Mode (ppm)*	*Median*	*Mean, μ (ppm)*	*STD, σ (ppm)*	*Highest Probability*	*Highest Exp μ + 3σ (ppm)*	*Mass (metric tons)*
Barium, >412 ppm	268.27	2,385	7,111	19,973	3.37E−04	67,030	763
Zinc, >55 ppm	58.92	1,355	6,496	30,462	7.97E−04	97,883	697
Zinc, above BM	18.59	1,035	7,728	57,163	1.43E−03	179,218	830
Zinc, all data	14.52	900	7,092	55,411	1.72E−03	173,325	761
Zinc, below BM	7.63	587	5,151	44,889	2.86E−03	139,817	138
Barium, all	17.27	461	2,385	12,096	2.47E−03	38,672	256
Copper, >35 ppm	42.98	299	788	1,924	2.53E−03	6,560	85
Lead, >7 ppm	1.02	256	4,055	64,022	1.05E−02	196,119	435
Barium, <412 ppm	86.11	133	165	121	5.67E−03	529	18
Lead, all data	0.09	111	3,781	129,231	4.64E−02	391,473	406
Copper, all	4.50	105	505	2,379	1.04E−02	7,642	54
Nickel	19.17	36	49	45	1.93E−02	184	5
Zinc, <55 ppm	31.22	35	37	13	3.45E−02	78	4
Arsenic	0.30	29	287	2,809	6.27E−02	8,715	31
Copper, <35 ppm	15.69	18	19	6	7.12E−02	38	2
Antimony	0.61	9	37	142	1.01E−01	462	4
Cadium	0.04	5	55	591	3.82E−01	1,829	6
Mercury	0.04	0.42	1.40	4.45	2.04E+00	15	0.2

Table 14.7 Mass of Fill Between the Bay Mud and Current Surface

Parameter	*Value*	*Mass (metric tons)*
Volume (ft^3)	2.04354E+06	
Volume (m^3)	5.78666E+04	
Grain density (g/cc)	2.65	
Average porosity	0.3	
Bulk density (g/cc dry soil)	1.86107343	107,343
Low bulk density (g/cc)	1.6796608	96,608
High bulk density (g/cc)	2.04118077	118,077

Table 14.8 Mass of Metals in Fill Above Bay Mud Up to 1 ft Beneath Current Surface

Metal	*LND Mass (metric tons)*
Ba	781
Zn<BM	830
Zn<BM	138
Pb	435
Cu	87
As	31
Cd	6
Sb	4
Hg	0.2

LND, lognormal distribution, mean concentration based; BM, bay mud.

Table 14.9 Mass of Slag Up to 1 ft Below Surface

Parameter	*Value*	*Mass (metric tons)*	*Zn in Slag (metric tons)*	*Cd in Slag (metric tons)*	*As in Slag (metric tons)*	*Pb in Slag (metric tons)*
Volume(ft^3)	5.56575E+05					
Volume (m^3)	1.57604E+04					
Grain density (g/cc)	2.65					
Average Porosity	0.3					
Bulk density (g/cc dry soil)	1.86	29,236	729	1	0.4	746
Low bulk density (g/cc)	1.67	26,312	656	1	0.4	671
High bulk density (g/cc)	2.04	32,159	802	1	0.4	820

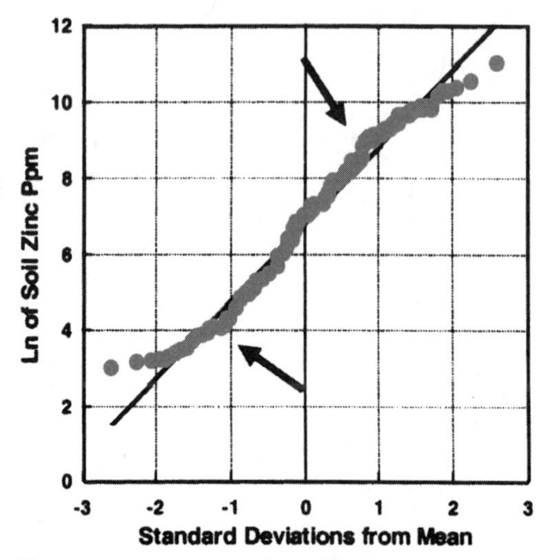

Figure 14.19 Normality of zinc concentrations in all soil data showing two breaks in the slope.

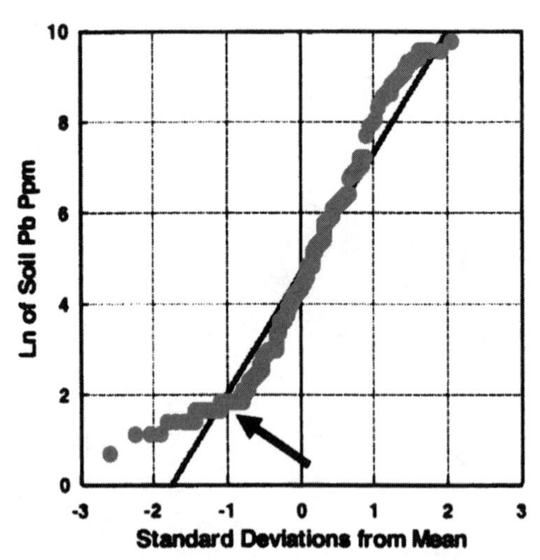

Figure 14.20 Normality of lead concentrations in all soil data showing a shape break at 7 ppm.

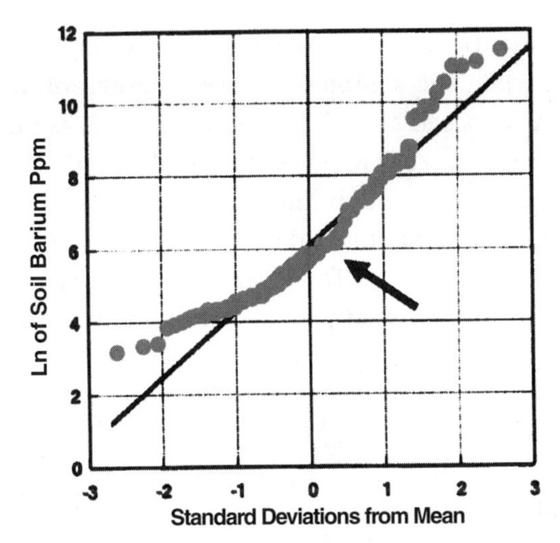

Figure 14.21 Normality of barium concentrations in all soil data showing a shape break at 412 ppm.

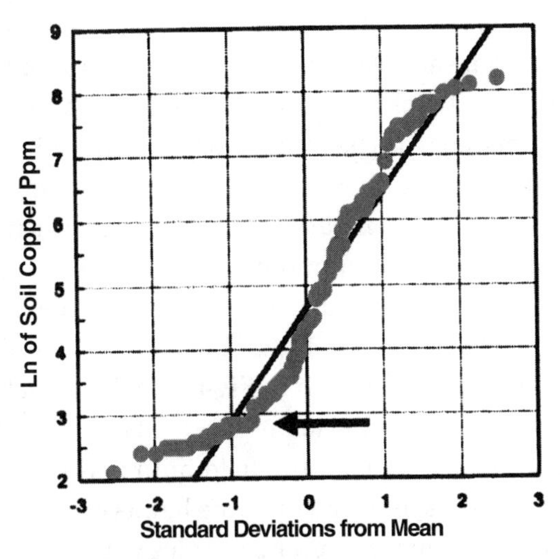

Figure 14.22 Normality of copper concentrations in all soil data showing a shape break at 35 ppm.

high metal concentrations, and may reflect a sampling bias. The sampling bias is expected when the samples are taken preferentially at the "suspect" locations, as opposed to the random ones. Lead is an exception; its highest predicted concentration is close to 200,000 ppm, reflecting perhaps more variability in its soil distribution than that of any other metal.

If the spatial distribution of metal concentrations at the site is ignored and the data are treated as a batch of numbers, then the mean of each lognormal distribution can be used to determine the total mass of the corresponding metal. With the exception of zinc, all the measured metals are trapped in shallow fill above the bay mud level. From the geologists' logs, it follows that there are roughly 107,000 tons of fill at the site (Table 14.7). The fill volume was calculated using Surfer. The zinc concentrations below the bay mud level have been found in a volume that contains about 25,000 tons of soil.

By multiplying the mean concentration of each metal by the appropriate soil volume, it was possible to obtain statistical estimates of the total metal masses underneath the site (see Table 14.6 and Table 14.8). Because solid slag is present across the central part of the site, and very few slag samples have been analyzed, it was felt that it was important to augment the soil masses of zinc, lead, and cadmium with their estimated masses in the slag region (Table 14.9). By comparing Tables 14.8 and 14.9, it is obvious that there may be as much zinc and lead trapped in the slag cobbles as is trapped in soil. An obvious weakness of the preceding statistical analysis is that it disregards the spatial distributions of metal concentrations. In other words, this analysis averages out the possible regions of significantly higher metal concentrations, such as the central solid-slag fill region, rich in zinc, lead, and cadmium.

The experimental variograms discussed in this section are based on the sampling points shown in Figure 14.23. Based on these sampling points, the areal zinc distribution as shown in Figure 14.24 has a clear east–west (EW) trend in the areal distribution of vertically averaged logarithms of zinc concentration in soil. In fact, the field data show two clear bands of high zinc concentrations separated by three bands of low ones. The largest high-concentration band in the center coincides with the solid-slag region. This means that the finer (sand-sized) slag particles were picked up by the soil analysis and the slag cobbles were not. Note the large central slag-rich region.

Similarly, Figure 14.25 shows that the vertical log mean of lead concentration has an areally nonuniform distribution with a large band of

Figure 14.23 Sampling points referenced map for kriged contour maps.

higher concentration running EW through the central slag-rich region. The areal distribution of lead is positively correlated with that of zinc, but it is not identical.

Figure 14.26 shows the vertical log mean of barium concentration at the site. Because barium was dispersed as a result of different industrial activity, its areal distribution is quite different from those of lead and zinc.

Table 14.8 lists the estimated masses of all of the most prevalent soil contaminants. Based on the available data, it appears that 800 metric tons each of zinc and barium are scattered in the fill above the bay mud. There are also about 140 tons of zinc below the bay mud. Plenty of lead, about 400 metric tons, is scattered throughout the site, followed by copper at 80 metric tons. Of course, slag, Table 14.9, may contain even more lead, an estimated 700 tons, and zinc, 700 tons.

These data also demonstrate that this mass of metals did not occur from surface spills of acid solutions containing low-ppm concentrations of trace metals.

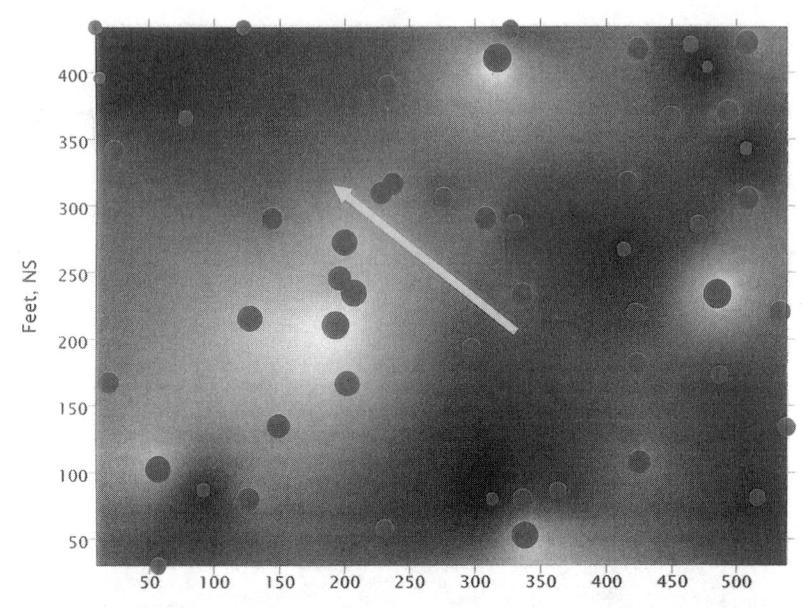

Figure 14.24 Kriged contours of the vertical log mean of zinc concentration in soil. The sampling wells are denoted by circles whose areas are proportional to the log-mean zinc concentrations. The arrow points north. It is obvious that the mean zinc concentration shows EW trending bands of low–high–low–high–low values.

Groundwater Contamination

A profile of each monitoring well revealed that two-thirds of the wells were completed in the bay mud with the remaining third completed in the fill. As a consequence, the hydrologic characteristics and trace-metal contamination distributions are different. The groundwater level contours (all obtained by linear kriging of the well data) for the fill above the bay mud are given in Figures 14.27 and 14.28. The trace-metal distributions and pH (all obtained by linear kriging of the well data) for the bay mud are given in Figures 14.29 to 14.34.

These data show that the zinc and cadmium plumes are centered near MW-11; however, copper (and some nickel) is centered near MW-16. This is a significant difference since zinc, cadmium, copper, and nickel have similar geochemical properties. This strongly suggests separate sources for these metals. This is not consistent with groundwater contamination from

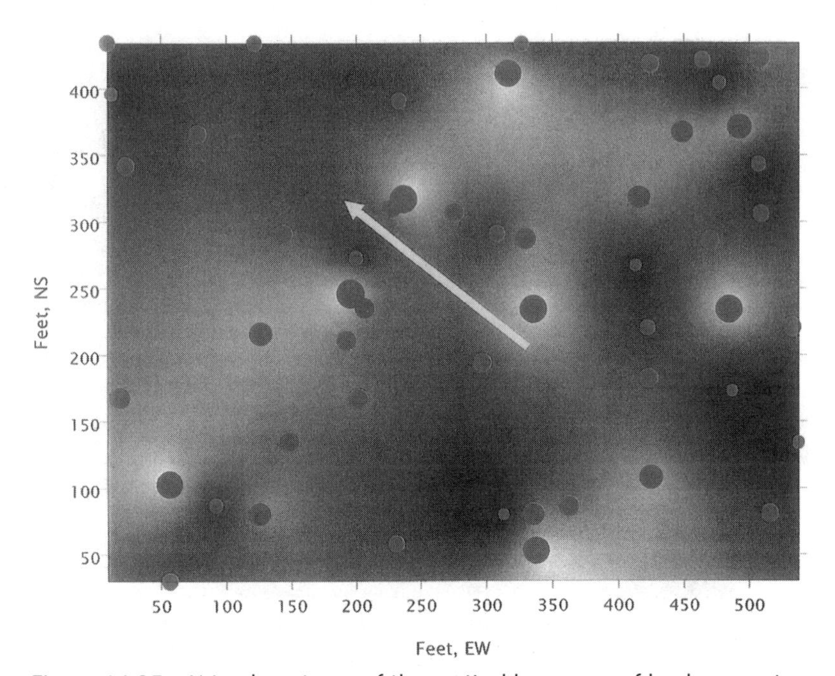

Figure 14.25 Kriged contours of the vertical log mean of lead concentration in soil. The sampling wells are denoted by circles whose areas are proportional to the log-mean lead concentrations. The arrow points north. The mean lead concentration shows spatial nonuniformity.

"zinc sulfate" spills. In addition, the arsenic plume is up gradient (with no correlated pH pattern; i.e., an acid pH can result in lower arsenic concentrations) and located in a region of pyrite cinders (i.e., high in arsenic).

Even given these plumes, the real issue centered on knowing if the plumes were moving off site and whether they were contaminating, or potentially could contaminate, the bay. Because there was the potential for contaminated groundwater to intersect the concrete stormwater drainage structure that discharges to the bay, it became necessary to determine if any of the groundwater wells were influenced by tidal changes. As a consequence, a tidal study was conducted. Results from the tidal study were used to calculate the hydraulic conductivities for the site (0.32 foot per day for the fill and 0.219 for the bay mud). These data, when combined with the groundwater gradient and flow direction, suggest that an insignificant amount of groundwater is being discharged from the property (i.e., less than 300 gallons per day).

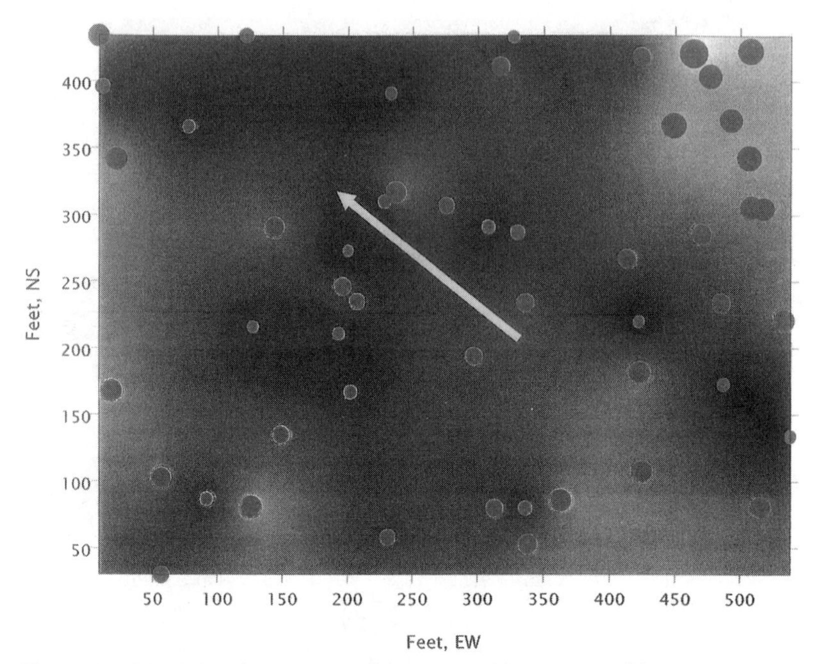

Figure 14.26 Kriged contours of the vertical log mean of barium concentration in soil. The sampling wells are denoted by circles whose areas are proportional to the log-mean barium concentrations. The arrow points north. The mean barium concentration shows spatial nonuniformity, which is inversely correlated with that of zinc in the central slag region (see Figure 14.24).

In addition, the tidal study also indicated that only one off-site well a djacent to the stormwater channel had a significant response (i.e., communication of groundwater with the bay). The combination of all these factors indicates a low probability that any contamination that might be transported off site would actually discharge to the bay.

Sources of Groundwater Contamination As discussed previously, if spilled "zinc sulfate" solutions do not adequately explain the groundwater contamination, how could the other source(s) of contamination? There are two other possible explanations (either separately or in combination).

The first possible source is from the oxidation of iron sulfides from the smelter waste, pyrite concentrate, and/or pyrite cinders in the fill. In the

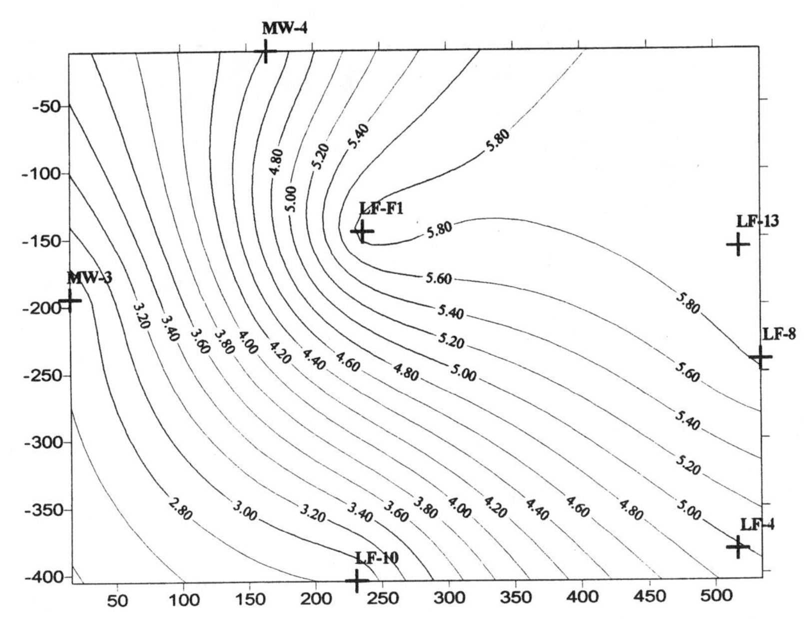

Figure 14.27 Water level contours above the bay mud horizon obtained by linear kriging.

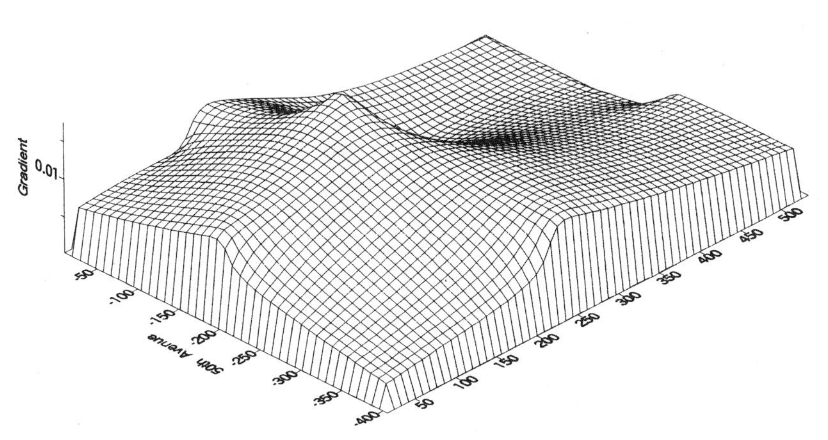

Figure 14.28 Gradient of water level surface above the bay mud horizon.

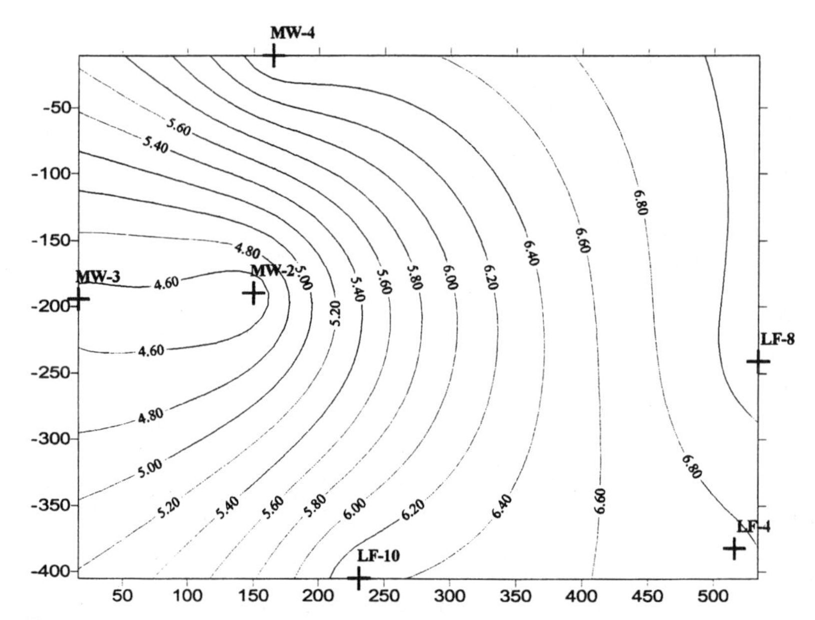

Figure 14.29 pH distribution in groundwater above the bay mud horizon obtained by linear kriging.

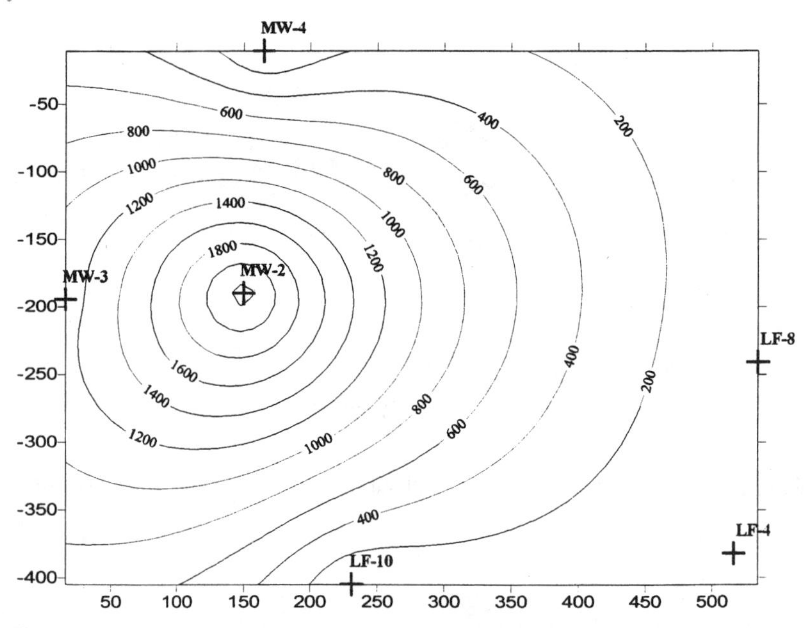

Figure 14.30 Zinc concentrations in groundwater above the bay mud horizon obtained by linear kriging.

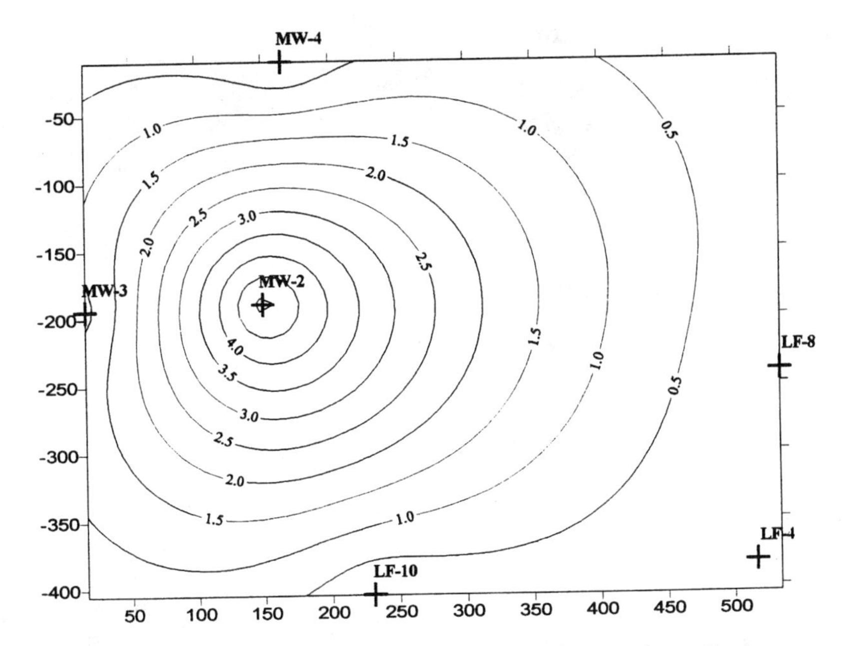

Figure 14.31 Cadmium concentrations in groundwater above the bay mud horizon obtained by linear kriging.

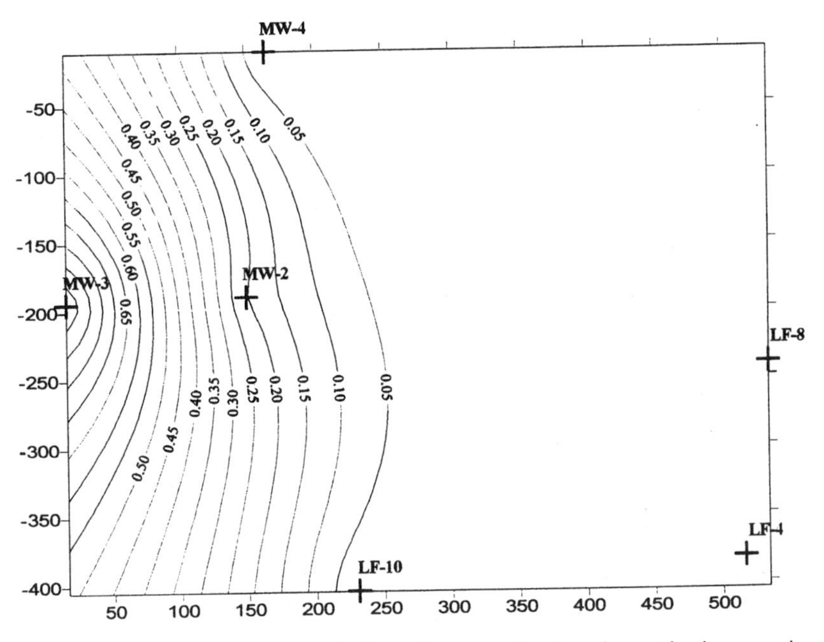

Figure 14.32 Copper concentrations in groundwater above the bay mud horizon obtained by linear kriging.

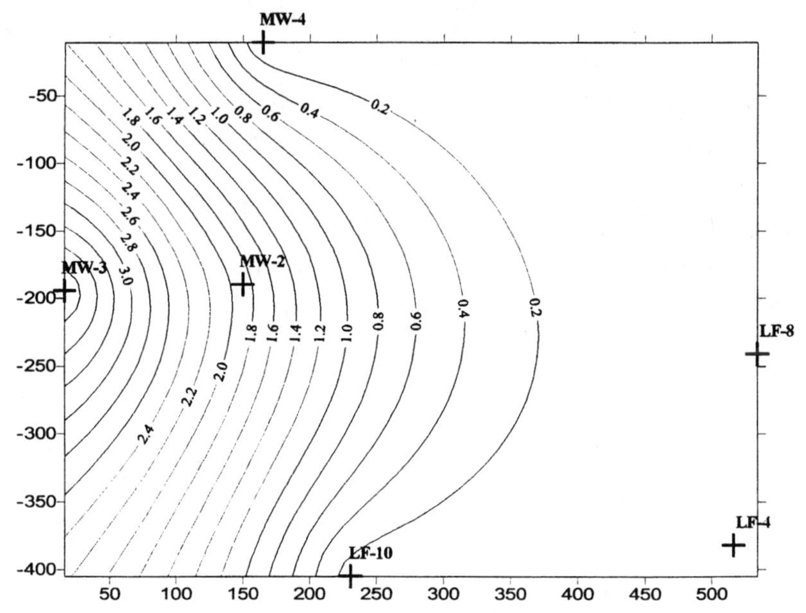

Figure 14.33 Nickel concentrations in groundwater above the bay mud horizon obtained by linear kriging.

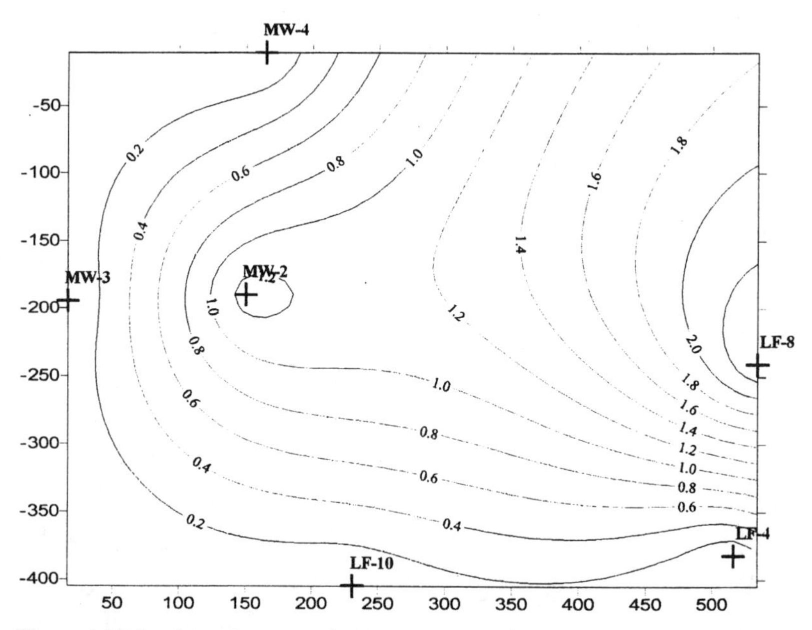

Figure 14.34 Arsenic concentrations in groundwater above the bay mud horizon obtained by linear kriging.

presence of water (H_2O), the oxidation of pyrite occurs at exposed surfaces. When the sulfur in pyrite (FeS_2) is oxidized, sulfate (SO_4^{2-}), ferrous iron (Fe^{2+}), and acid (H^+) are all released to solution. In an environment with abundant oxygen, the ferrous iron can be oxidized to ferric iron (Fe^{3+}). The rate of this reaction is slow but can be significantly increased by the presence of iron and sulfur oxidizing bacteria. These organisms are most active between a pH of 2 and 5. Once generated, the ferric iron can precipitate (e.g., form solid phases with iron and sulfate and/or with hydroxide species; the hydroxide species exhibit yellow to red colors), remain in solution, or further oxidize pyrite. As a result of ferric iron oxidizing pyrite, the pH decreases more rapidly. In general, however, once conditions allow the rapid conversion of ferrous to ferric iron, the pH can decrease to less than 3.

Other sulfide minerals[10] that are associated with pyrite can also be oxidized, which results in the release of metals (e.g., As, Cd, Cu, Pb, and Zn) into solution. The resulting combination of acid and soluble metals from the oxidation of sulfides is usually termed "acid mine drainage."

The second possible source of contamination may have occurred as a result of the release(s) of sulfuric acid in the area of the Pyrite Chemical Company sulfuric acid plant. Since the plant was documented as having had an earthen floor, the potential for spilled sulfuric acid to leach trace metals out of the fill is significant. In addition, the slag deposit identified by Geo-Experts is directly beneath the former sulfuric acid plant.

In order to evaluate this potential source of groundwater contamination, simple leaching experiments using slag samples from the site were leached using a modified EPA Toxicity Characteristic Leaching Procedure. These tests were performed using the following solutions:

A = 2 M H_2SO_4
B = 2 M H_2SO_4 + 0.5 M H_2O_2
C = 1500 ppm Cl + 550 ppm SO_4 at pH 5
D = 1500 ppm Cl + 550 ppm SO_4 + 0.5 M H_2O_2 at pH 5

Two massive slag samples and one fill sample (containing granular slag) were leached for up to 120 hours. The results of this study are summarized in Tables 14.10 to 14.12. These data demonstrate that the groundwater

[10]The common sulfides are sphalerite (zinc sulfide, ZnS), galena (lead sulfide, PbS), and chalcopyrite (copper iron sulfide, $CuFeS_2$).

Table 14.10 Fill Sample[a]

	Leaching Solution Element Concentrations (mg/L)					
	A	*B*		*C*	*D*	
Element	*48 hr*	*120 hr*		*48 hr*	*120 hr*	
As	43.5	73.5	16.9	0.016	0.035	0.146
Ba	0.773	0.030	0.048	0.135	0.144	0.321
Cd	1.24	2.14	1.01	0.058	0.030	0.172
Cu	68.4	499	50.6	0.019	0.015	1.02
Ni	4.80	2.39	1.44	0.087	0.017	0.20
Pb	5.75	5.56	2.98	2.02	0.60	10.8
Zn	1,050	1,780	631	5.79	1.18	23.4

[a] Total metals data: As, 35 ppm; Cd, 7 ppm; Cu, 2,400 ppm; Pb, 14,000 ppm; Zn, 7,500 ppm.

contamination could have resulted from the sulfuric acid spills leaching the slag in the fill.

Although it is important to identify the sources of contamination, the real costs associated with potential site remediation alternatives will depend on the extent to which any contamination is discharged to the bay.

Identifying Discharges to the Bay It was the contention of GeoExperts that trace metals were being discharged to the bay from the storm drain system. However, GeoExperts conducted no studies to determine if, in fact, trace metals were being discharged to the bay from the storm drain system or whether these concentrations exceeded the applicable standards.

According to the Environmental Protection Agency National Water Quality Ambient Criteria for marine environments, the trace-metal concentrations (mg/L) listed in Table 14.13 should not be exceeded. Given these standards, ACE Paint Company's forensic consultant conducted surface water sampling within the storm drain and also up gradient of the Bay Auto site. Of the four sampling events, the only metal that exceeded the standards was zinc. The highest concentration detected was 0.68 mg/L. However, there were also multiple sources of zinc detected up gradient of the Bay Auto site, which suggests that the Bay Auto site may not be a contributor.

As a result of these data, two hydropunch groundwater samples were retrieved approximately 50 feet in the down-gradient direction of the ground-

Table 14.11 Massive Slag Sample 1

	Leaching Solution Element Concentrations (mg/L)					
Element	*A 48 hr*	*B 120 hr*		*C 48 hr*	*D 120 hr*	
As	**	*	*	0.007	0.010	0.005
Ba	**	0.071	0.079	0.064	0.079	0.080
Cd	**	1.37	1.66	<0.005	<0.005	0.038
Cu	**	1.48	98.6	0.041	0.043	2.32
Ni	**	0.309	0.393	0.280	<0.015	0.108
Pb	**	3.49	<0.01	0.32	0.48	2.68
Zn	**	1,960	2,230	3.98	0.88	86.5

* No data reported.

** Unable to extract liquid from gel formed during leaching.

water flow from the Bay Auto site. The zinc concentrations were reported to be 0.038 mg/L and a nondetect. Based on all of the surface and off-site groundwater data, a meeting was held with the regulatory agency to discuss remedial actions at the site.

At this meeting, the forensic consultant was told that due to the site location (i.e., an industrial area with multiple sources of zinc contamination) and the fact that the groundwater was brackish (i.e., non–drinking water source), it was highly probable that no groundwater cleanup or source re-

Table 14.12 Massive Slag Sample 2

	Leaching Solution Element Concentrations (mg/L)					
Element	*A 48 hr*	*B 120 hr*		*C 48 hr*	*D 120 hr*	
As	31.6	295	43.5	0.262	0.173	0.048
Ba	0.083	0.192	0.018	0.177	0.203	0.094
Cd	2.79	6.66	1.25	0.145	0.155	0.344
Cu	106	237	59.5	0.044	0.022	0.046
Ni	0.315	0.827	0.279	0.048	0.095	0.116
Pb	9.38	12.4	2.97	2.90	3.92	8.77
Zn	1,670	3,110	732	5.68	4.25	21.2

Table 14.13 EPA National Water Quality Ambient Criteria for Marine Environments

Element	*4-day Average*	*1-hour Average*
Arsenic	0.036	0.069
Cadmium	0.009	0.043
Copper		0.003
Nickel	0.005	0.140
Lead	0.008	0.075
Zinc	0.086	0.095

moval would be required at the site. However, based on the surface water samples, it had to be demonstrated that there was no pathway for contaminated groundwater to reach the bay (or, alternatively, provide an engineering solution to ensure that there would be no pathway).

A DIFFERENCE OF OPINION: ROUND 2

With the completion of the previously discussed data that were collected in order to clarify site issues, a second meeting was held to discuss technical issues and possible settlement. A presentation of all of these new facts fell on the deaf ears of Bay Auto's president and counsel. It remained their opinion that ACE Paint Company and Industrial Chemicals were responsible for the entire soil and groundwater contamination of the site.

Counsel for Bay Auto again suggested that instead of continuing the "war," ACE Paint Company could simply purchase the property (at this point, the suggested purchase price had "escalated"). Although this was a potential option, there was not enough information concerning property values to make an economic decision. The only issue that was resolved was that both parties would now proceed to obtain expert opinions and that, at the completion of depositions, there would be a further attempt to mediate the dispute.

DEFINING THE EXPERTS

The experts retained by both sides were uniquely different. Counsel for ACE Paint Company and Industrial Chemicals retained experts with the following expertise:

- Environmental history
- Industrial chemistry
- Ore processing and chemistry
- Process mineralogy
- Acid rock drainage
- Hydrology and modeling groundwater and contaminant transport
- Remediation/regulatory issues
- Aquatic toxicity and risk assessments
- Land appraisal

Counsel for Bay Auto retained experts having the following expertise:

- Hydrology
- Geochemistry
- Remediation/regulatory/aquatic toxicity
- Land appraisal

From these lists, it became clear that Bay Auto would not address site related industrial processes prior to the acquisition by ACE Paint Company.

Ultimately, opinions were developed by all of the experts. These opinions were followed by rebuttal opinions, which were followed by depositions. After the completion of this process, there was no basic difference between each party's original position. These positions are:

Bay Auto Position

- The soil and groundwater contamination was caused by ACE Paint Company and Industrial Chemicals.
- The bay is being contaminated by the site.
- Both the soil and the groundwater must be remediated.

ACE Paint Company Position

- The soil contamination was the result of ore processing waste.
- The groundwater contamination is a combination of acid rock drainage and the release of sulfuric acid.
- No groundwater or soil remediation will be required.

Basically, the experts for Bay Auto ignored all of the site history prior to the operations of ACE Paint Company. These experts actually gave testimony that the ore processing operations did not contribute to the site contamination. Given this major disparity, both sides prepared for mediation.

THE MEDIATION

Initially, Bay Auto was still demanding damages of 6 million dollars. The mediation began early in the morning with Bay Auto's rather biased presentation, offering no factual basis to establish damages. Both ACE Paint Company and Industrial Chemicals followed not only with a complete history, but also with relevant surface water trace metal data clearly showing no demonstrated discharge to the bay. In addition, it was the position of ACE Paint Company that the site could be closed with no demonstrable cleanup and at a minimum cost of $350,000 to comply with regulatory guidelines.

The mediator, as was his role, attempted to convince both sides to modify their positions on damages. After several sessions, there appeared to be no movement by either side. It was at this point that ACE Paint Company offered to buy the Bay Auto property at its fair market value and deal with any potential remediation. Bay Auto accepted the offer.[11]

SOLD AT LAST

Bay Auto finally sold its property (approximately two years later). However, the value gained from the sale was offset by the considerable legal and consultant costs incurred as a result of filing suit.

ACE Paint Company subsequently found a new tenant for the Bay Auto property and eventually closed the site (as well as those portions of the adjacent properties that contained ACE Paint Company waste) under the regulatory guidelines without the need for any remedial action other than monitoring and deed restrictions. Currently, the property has appreciated greatly in value and, in effect, represents an overall profit (even after legal and consulting costs) to ACE.

LESSONS LEARNED

1. When filing a lawsuit with a profit objective, make sure the relevant facts are known (it helps to hire qualified consultants and to listen to them rather than telling them what they should do).

[11] ACE Paint Company also purchased the adjacent properties.

2. Know the most probable remedial solution (i.e., damages) prior to making a claim for damages (it could save everyone a lot of money).

REFERENCES

1. J. D. Ridge, *Ore Deposits of the United States, 1933–1967, Volume II*, The American Institute of Mining, Metallurgical, and Petroleum Engineers, Inc., New York: (1968).

2. V. M. Goldschmidt, *Geochemistry*, The Clarendon Press, Oxford: (1954).

3. P. DeWolf and E. L. Larson, *American Sulphuric Acid Practice*, McGraw-Hill, New York: (1921).

4. T. W. Patzek and P. J. Sullivan, "Statistical Characterization of Metal-Contaminated Fills," paper presented at the Society of Petroleum engineers, Western Regional Meeting, Los Beach, California, 25–27, June 1997.

Chapter 15

A Lesson in Communications

The ability to effectively communicate technical information in a clear and simple manner can be, and often is, the determining factor in winning an environmental forensic argument. This case history illustrates how an inexperienced mediator allowed a strong advocate to "control" the mediation process, thereby creating a mediation environment that skewed communication between defendants and, unknowingly, obstructed input from the regulatory authority. Because the plaintiff's expert was able to exert such an inordinate degree of influence on the mediator's understanding of the key technical issues and damage estimates, the mediator moved away from the technical issues and focused on "resolution of the dollars." It was in the twelfth hour of the mediation that a forensic consultant was retained to solve this "dysfunctional" condition and rescue one of the defendants from inevitable defeat.

AN OPPORTUNITY TO IMPROVE PROPERTY VALUES

A group of run-down industrial–commercial lots, owned by Lake Front Redevelopment Corporation (LFRC) and located along a lakefront, were gradually being surrounded by "higher" use (i.e., restaurants, townhouses, boating facilities, etc.) activities. These LFRC properties had a long and revolving history of industrial occupancy with little hope of changes in the land use pattern (i.e., without sufficient remedial funds to affect a conversion). A site map with parcel locations is found in Figure 15.1.

This all changed, however, when a large diesel oil spill occurred on the property leased by ACME Boat Yard (shown as Area B on Figure 15.1). The

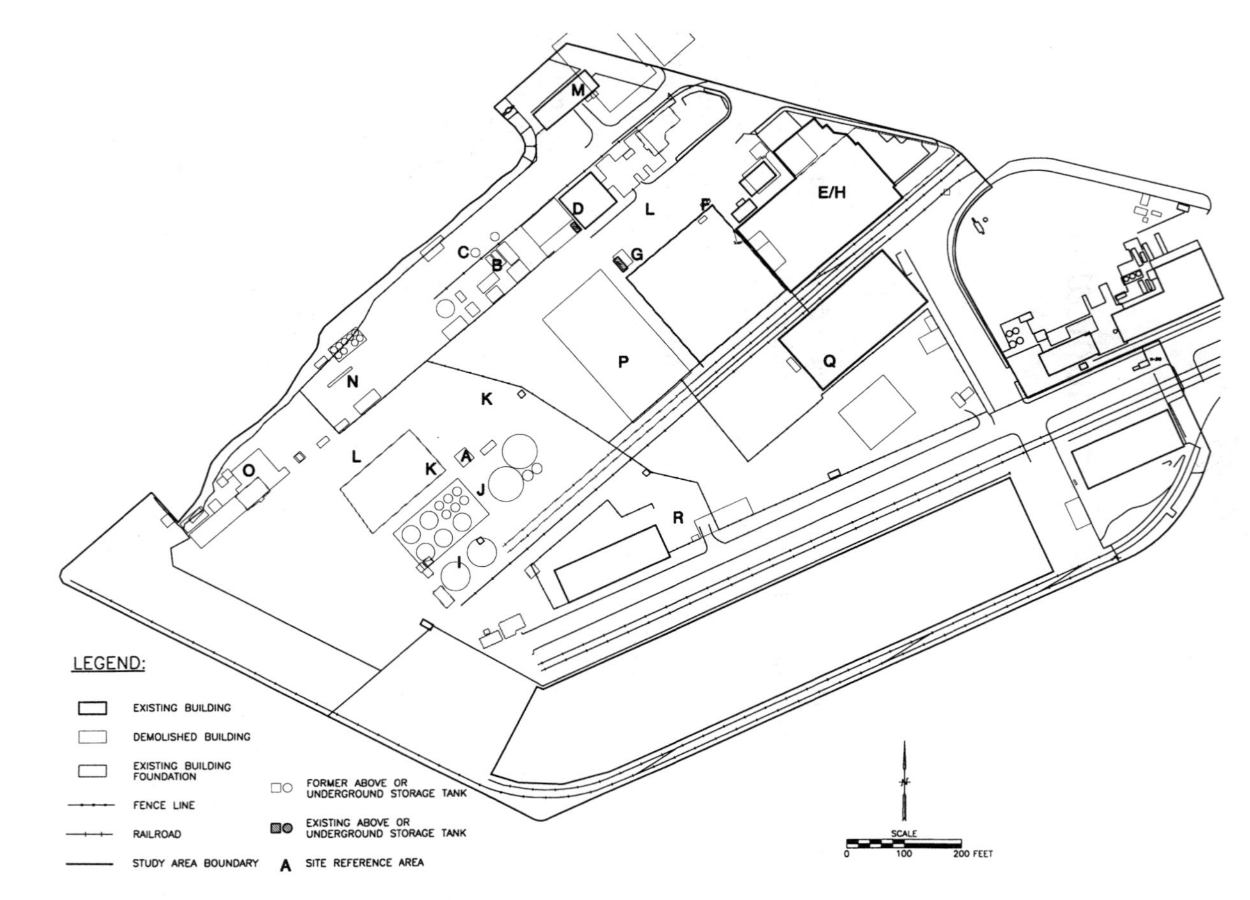

Figure 15.1 Overall property site.

County Health Department required an assessment and subsequent remediation to prevent further discharges to the lake. LFRC retained a local consulting firm, GeoExperts, to investigate the extent of the soil contamination and groundwater contamination (if any) and to clean up the diesel oil discharge to the lake. With the source of the spill controlled, the discharge to the lake ceased. This was then followed by remediation of the site whereby the contaminated soil was excavated and removed. Groundwater monitoring wells were located beyond the ACME property boundaries to characterize any groundwater contamination. The data collected from the monitoring wells showed that contamination from the diesel oil spill had migrated throughout the area, primarily by means of the gravel bedding of the area sewer system. The utility lines are shown in Figure 15.2.

As a result of this discovery, LFRC had GeoExperts expand its study to include a general soil and groundwater investigation of all of LFRC's industrial–commercial properties along the lakefront, presumably to define the extent and sources of the groundwater contamination. These site investigations revealed that all of the LFRC properties had some degree of soil and groundwater contamination (i.e., either from a source on the property or from up-gradient sources). Because of this areawide contamination, coupled with a group of small, unsophisticated tenants and former tenants, LFRC saw an opportunity to redevelop the lakefront tract at the expense of its tenants.

Prior to this incident, the county redevelopment plans focused on the attractiveness of the lakefront properties with the desire to upgrade the land use into a multiuse development (i.e., recreation, residential, and retail/commercial). However, the potential cost of terminating all of the existing leases and remediating any contaminated properties for residential development posed a significant obstacle to the plan. With the potential areawide remediation, LFRC had the opportunity to shift major redevelopment costs to its tenants and the tenants' insurance companies using the following approach:

- Using the results of the site investigations, the LFRC consultant (GeoExperts) developed remedial plans for each property and estimated the remedial cost for each property.[1]

[1]A site plan with tenant locations is shown in Figure 15.1, and the associated monitoring data are found in Tables 15.1 to 15.6.

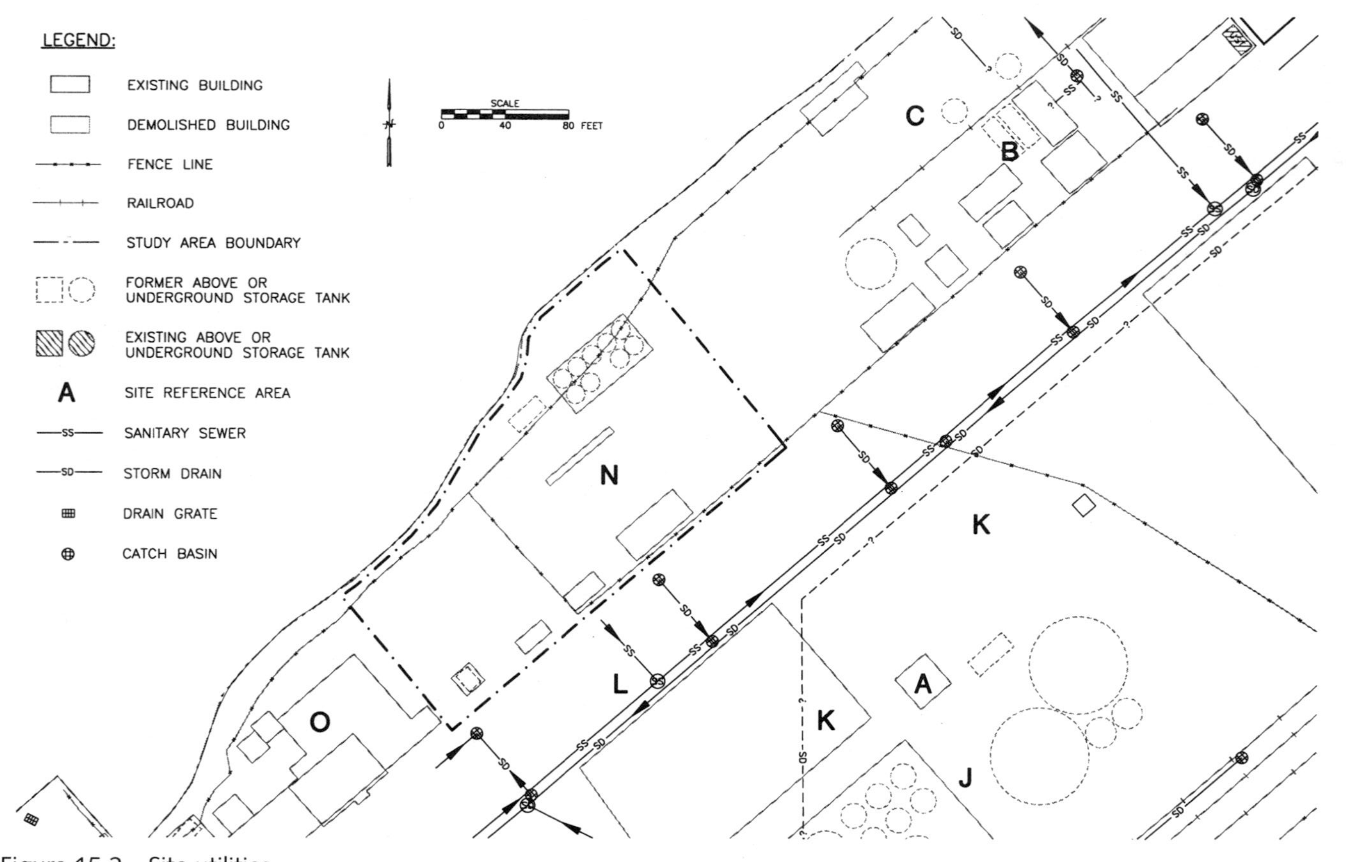

Figure 15.2 Site utilities.

- Based on the tenant history of each property and the chemical use history of each tenant, GeoExperts allocated a percentage of the estimated cleanup cost to each tenant of each property.
- Because the estimated cleanup cost calculated by GeoExperts vastly exceeded (i.e., a gold-plated estimate) the likely regulatory cleanup requirements, tenants that agreed to pay the LFRC cleanup cost allocation (prior to any regulatory ruling) would be contributing to a cash fund that would far exceed any mandated cleanup cost.[2]
- LFRC could then use these excess funds to further develop the infrastructure of its property to the planned "higher purpose."

This assumption, however, was based on the tenants' agreeing to pay the LFRC proposed allocation (or at least a major portion of it). As expected, the majority of the tenants refused to pay LFRC their allocated percentage. Threatened with a lawsuit by LFRC to "recover" its site investigation costs and remedial costs, the tenants agreed to a mediation of the LFRC allocated costs.

A "DYSFUNCTIONAL" MEDIATION

Reflecting on the fact that the tenant group, including former tenants, was, in the main, small companies, and recognizing that LFRC was the "giant" as well as the owner, the tenants (and their counsel) were somehow convinced by counsel for the LFRC that each allocation claim could best be resolved by conducting a separate mediation between LFRC and each tenant or former tenant of each property. The mediator being presented with this approach also agreed with the LFRC proposed mediation structure. As a result, each tenant was precluded from sharing information, sharing investigative costs, sharing strategy, and sharing settlement results with other tenants (for all of the properties).[3]

One of the former tenants, Mid-States Oil (referred to as Area N in Figures 15.1 and 15.2), wanted an independent evaluation of GeoExperts'

[2]One might ask just how these tenants could be so naive as to agree to settling at a higher cost. Obviously, they were not, and ultimately the dispute ended up in mediation. However, during the process of attempting to resolve the cost/allocation issue with the remaining tenants, a settlement was reached with ACME regarding the oil spill on its site, giving LFRC a "cash infusion" and thereby helping to fund the cost of LFRC's attack on the other tenants.

[3]This can most appropriately be considered a "divide and conquer" strategy by LFRC.

assigned allocation cost of $3.5 million (i.e., its allotted share—according to LFRC). Thus, Mid-States Oil retained a local consulting firm to review GeoExperts' site investigation as well as selected remedial plans in order to estimate the most likely remedial cost for the property. The evaluation included a review of the field investigation data collected by GeoExperts. The site-specific soil contaminants are shown in Figure 15.3. When this evaluation was completed, the consultant informed Mid-States Oil that the most likely remedial cost related to "hot spots" on its site (chemicals and concentrations are shown in Figure 15.4) was not $3.5 million but $75,000 (a far cry from the $3.5 million demand).

Possessed with this analysis, Mid-States Oil with its counsel, insurers (several additional deep pockets), and consultant had its first mediation with LFRC, its counsel, and GeoExperts. During the meetings, both LFRC and Mid-States Oil presented their detailed analysis and cost allocation (i.e., based on GeoExperts' site investigation data).

The mediator was singularly unimpressed with the weight of Mid-States Oil's technical arguments. The presentation of the facts by Mid-States Oil was so poor that the mediator was convinced that GeoExperts' proposed remedial actions and cleanup cost allocation were correct. As a result, the mediator attempted to pressure Mid-States Oil to agree to make a settlement offer close to the LFRC demand. After two separate mediation attempts, the result was the same. The Mid-States Oil mediation team (now including five insurance companies) was completely frustrated with the mediator's inability to understand the "obvious." At this critical point of the mediation, Mid-States Oil and its insurance companies decided to fire the consultant and retain another forensic (technical) expert who could, hopefully, present its argument in a more persuasive and convincing manner while making an independent determination of the estimated cleanup cost.

A NEW BEGINNING AND A SHORT FUSE

A new forensic expert was retained to lead the technical battle against LFRC; however, the "final" mediation was scheduled to occur in a scant 15 calendar days (with no possibility of an extension). Thus, the new expert had to immediately evaluate the GeoExperts report, conduct a site visit, and talk to the County Health Department.

Based on the site investigation, GeoExperts alleged that there were at least three underground storage tanks (USTs) that were still located on

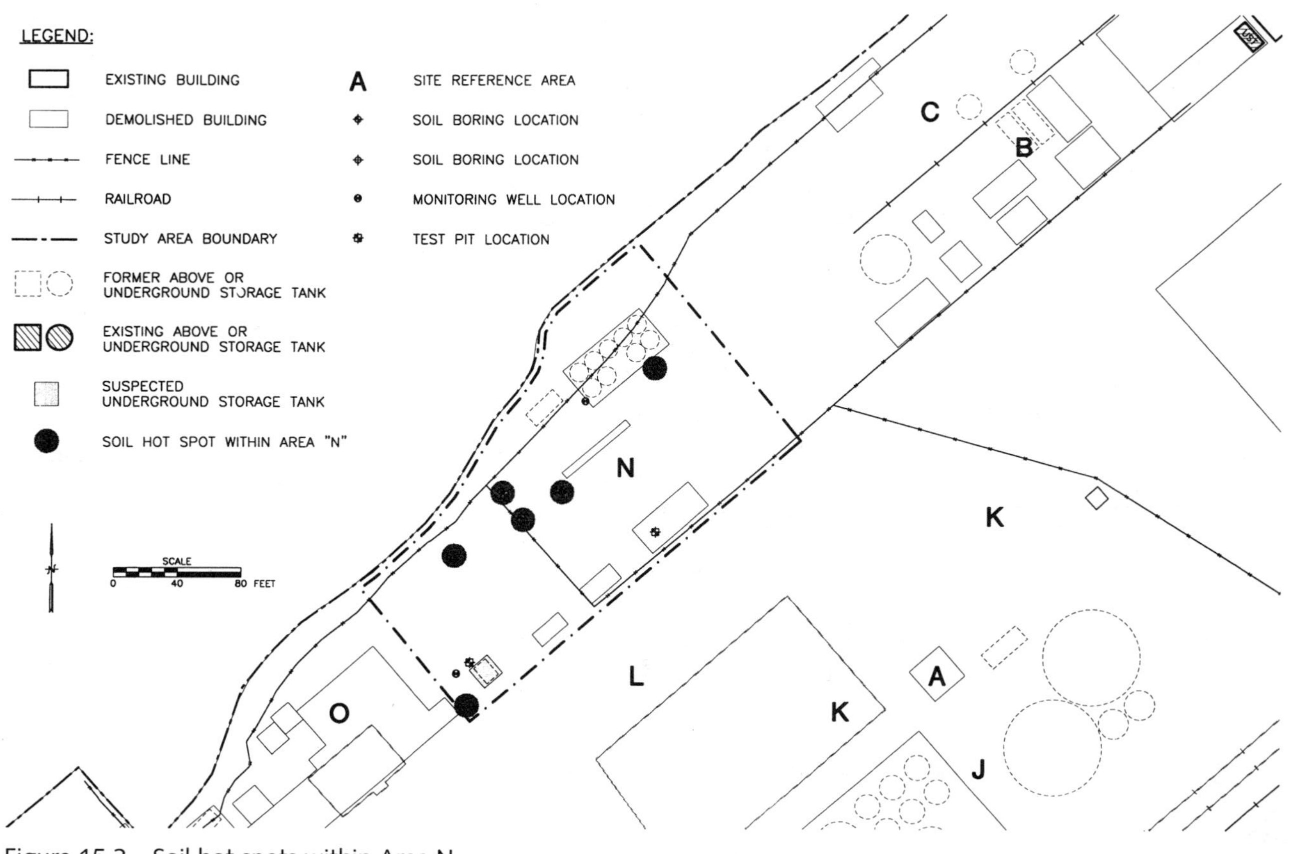

Figure 15.3 Soil hot spots within Area N.

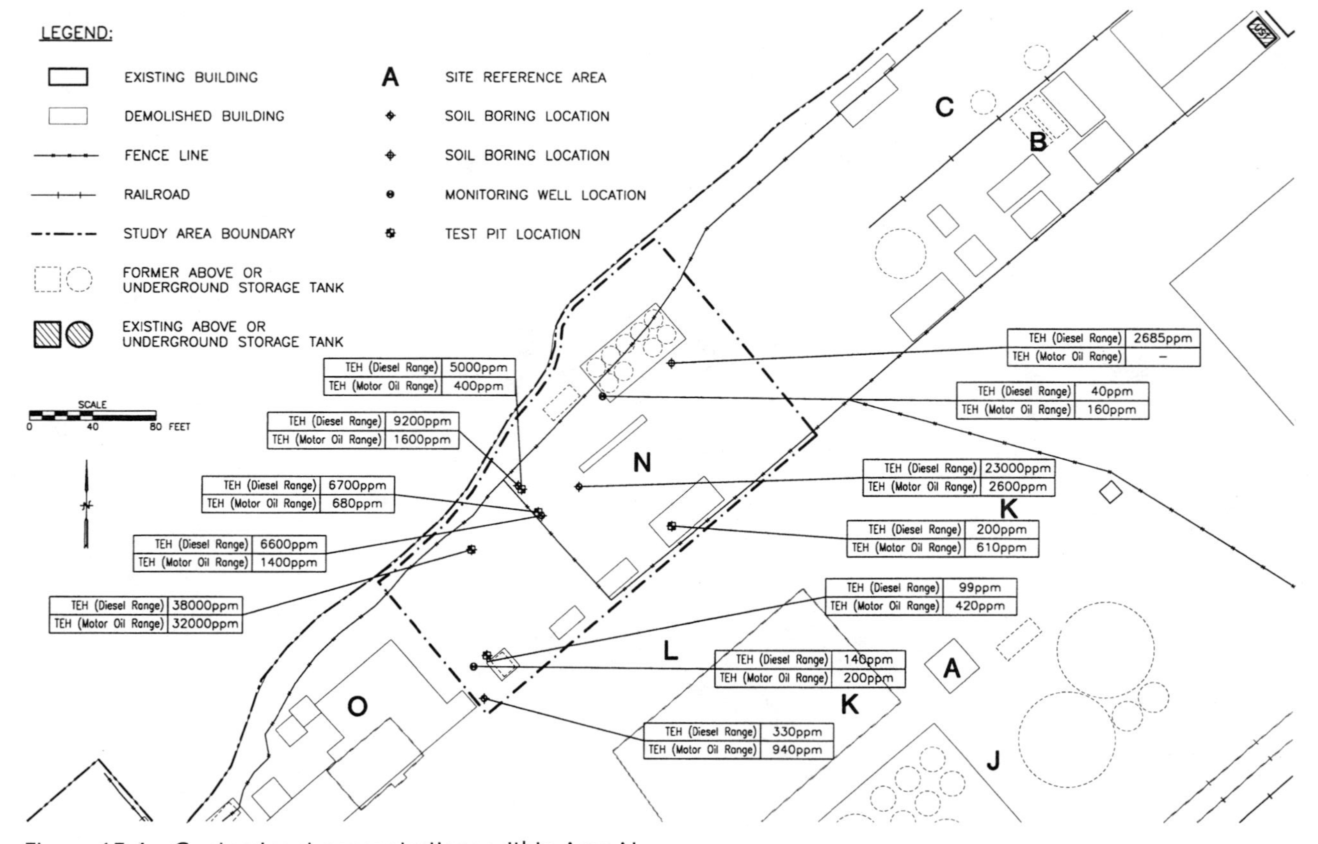

Figure 15.4 Contaminant concentrations within Area N.

the former Mid-States Oil property.[4] Using a limited number of soil samples from the "alleged" UST location, GeoExperts estimated the horizontal and areal distribution of the contaminated soil (estimating that the entire site had to be excavated). GeoExperts then assumed that this "contaminated soil" would have to be disposed of at a Class I hazardous waste landfill (an expensive assumption). Thus, the estimated cost of the soil cleanup was based on (1) inadequate site data and (2) an unjustified assumption of Class I disposal.

GeoExperts also assumed that some residual oil from the original spill (i.e., from ACME Boat Yard [Area B]) would be in the shallow groundwater aquifer beneath the Mid-States Oil site. Based on this assumption, GeoExperts proposed an extensive groundwater monitoring program with the anticipation that a groundwater cleanup would be required (including the removal of free product).[5] GeoExperts had also proposed a very expensive bulkhead wall surrounding the lakefront on the entire LFRC property to presumably make sure that no residual diesel oil or other contaminants would enter the lake. Each tenant was assessed a prorated share of the bulkhead cost.

In summary, in addition to the site soil excavation, GeoExperts also allocated the cost of groundwater monitoring, groundwater cleanup, and the prorated share of the installation of a bulkhead[6] to Mid-States Oil without any evidence that it had contributed any contamination to the groundwater. Because the groundwater contamination originated as a result of the ACME Boat Yard spill, one has to wonder how the projected cleanup costs were allocated to the other tenants (including ACME Boat Yard). If the tenants had communicated during the mediation, this issue might have been quickly resolved.

A review of the site conditions, as reported by GeoExperts, showed that the few identified hot spots[7] on the Mid-States site might require a limited excavation.[8] The forensic expert estimated the volume of soil to be ex-

[4]There were no site records or known operations that would support this conclusion.

[5]A separate diesel oil phase floating on the groundwater.

[6]Granted the bulkhead wall would help control any groundwater contamination reaching the lake; however, it would stabilize the lake shore property boundaries and would be a significant improvement to the property value.

[7]Elevated concentrations of petroleum hydrocarbons.

[8]GeoExperts proposed the disposal of contaminated soil to an expensive Class I landfill. Since the GeoExperts investigation of the Mid-States Oil site, a magnetometer study was completed. This study found no buried USTs on the site.

cavated and determined the disposal cost (based on an estimate of the contaminant levels in the soil) for shipment to a Class III landfill (not Class I as indicated by GeoExperts). The forensic expert's projected cleanup cost was remarkably close to the Mid-States Oil first consultant's estimate. Thus, with the basic site review completed, it was time to meet with the County Health Department site program manager to discuss the proposed remediation.

AN ABSENCE OF REGULATORY INPUT

As the meeting started, the Health Department project manager expressed his pleasure at finally talking to one of the tenant's consultants from the LFRC property. It seems that no one, including the mediator,[9] had bothered talking to the project manager. As a result, the mediator had no first-hand knowledge of the County Health Department's remedial objectives; he only had GeoExperts' account of what the County Health Department would require.

The project manager summarized the site conditions and remedial actions that the County Health Department would most likely require:

- The diesel oil spill at ACME Boat Yard had been remediated to state cleanup criteria (soil, surface water, and groundwater).
- ACME Boat Yard was required to continue monitoring groundwater at the property boundaries.
- There was no corrective action plan for this LFRC site property. No site development plan had been submitted to the County Health Department for environmental review, nor had an ecological risk assessment been completed.
- The additional site investigations (i.e., all of the properties owned by LFRC) conducted by GeoExperts were not required by the county, and GeoExperts (in his opinion) had conducted "the most expensive" site assessment he had ever reviewed. In fact, he commented that everything that the consultant did was "gold plated."
- It was also the County Health Department's opinion that the bulk-

[9]This is not too surprising in that GeoExperts assured the mediator that it was in regular contact with the county, and since none of the tenants seemingly thought to visit the county, thereby raising no objection, the mediator had no reason to independently contact the county.

head wall was not required as long as the monitoring data continued to demonstrate that no oil was leaving the site and entering the lake.[10]

- The Mid-States Oil site presently contained sufficient monitoring wells to meet the agency's monitoring requirements and no new wells would be required. The regulator further commented that only limited sampling of the monitoring wells would be necessary.
- Should any USTs be found on the LFRC properties, they could be dealt with through closure, on the site, without having to involve the entire LFRC-owned tract.[11]

The meeting with the County Health Department's project manager clearly brought to light the intentions of LFRC: Extort as much money as possible from the property tenants. A review of the County Health Department files revealed that GeoExperts had estimated the total site cleanup costs to be a staggering $42,000,000, with approximately $3,500,000 allocated to the Mid-States Oil site.[12]

DEVELOPING A MEDIATION STRATEGY

The reason the previous mediations were so "one sided" and unsuccessful was because the mediator did not understand how environmental regulatory agencies operated, nor did he educate himself as to the County Health Department's proposed requirements. It had been the forensic expert's experience that many mediators are ignorant of environmental technical issues and are reluctant to learn[13] and therefore fall prey to detailed consultant reports supporting what appears, on the surface, to be an environmental issue of major consequence. Once having been taken in by the

[10]It was the project manager's opinion that the bulkhead only provided "stability" for the proposed "redevelopment" and served no environmental purpose.

[11]It was subsequently shown that the alleged USTs did not exist. They either were never there in the first place or had been removed at an earlier date.

[12]The cost included Mid-States Oil's share of the "gold-plated" remedial investigation.

[13]A good mediator will recognize his or her own faults and take appropriate steps to deal with the ignorance issue. One solution is to ask each party in the dispute to recommend one or more technical consultants who, after selection by the mediator and acting in a neutral role, can assist the mediator in walking through the technical issues. Once a mediator has been through the environmental jungle, the situation changes and the expertise of the mediator allows a much better understanding and therefore less reliance on the weight of document.

overwhelming weight of evidence,[14] mediators can be prone to focus on "resolution of the dollars" rather than on the technical merits of the mediation.

Thus, in order to solve this communications problem, it was important to involve the County Health Department in the upcoming mediation. However, most regulator professionals (and their agencies) prefer to remain neutral in disputes between parties trying to allocate remedial costs. Hoping this would not be an issue, the site project manager was asked if he would be willing to participate in a conference call during the mediation to address the site-related issues. He agreed.

THE FINAL MEDIATION

The critical step was to develop a strong, technically sound, simple presentation that, in no more than half an hour, would demonstrate to the mediator (as well as the LFRC attorney and GeoExperts project manager) the position of the Mid-States Oil team and present a realistic estimate of necessary site-specific cleanup. The series of view graphs (in addition to Figures 15.1 to 15.4) that were prepared for the mediation included Figures 15.5 and 15.6 and Tables 15.1 to 15.6.

The purpose of this presentation was to compare the Mid-States Oil site to other parcels to demonstrate relative contamination. Having demonstrated this through the presentation of both figures and tables, a final summary (Table 15.6), showing a detailed cleanup cost analysis for site N was prepared and presented.

At the end of the presentation, the mediator asked the GeoExperts project manager if she had a rebuttal to the presentation. She had little to say, except to remind the mediator that the regulatory agency had authorized and supported the work effort, as well as the projected cleanup and related costs. It was at this point that the forensic expert's meeting with the regulator became the pivotal point in resolving the cost disparity.

[14]Thick investigative reports with lots of field data and extensive cost information, representing the view of a public agency pointing the finger at money-grubbing industry," can sell a mediator on a high-monetary-settlement resolution. The weight of evidence is based on document thickness supported by "technobabble." When one also adds the element of substantial dollars, the argument appears compelling if not adequately countered. Indeed, attorneys occasionally rule out the presentation of alternative solutions that appear to be too inexpensive, feeling that a mediator, judge, or jury would question the cheaper alternative regardless of its technical merit.

Soil TEH/diesel

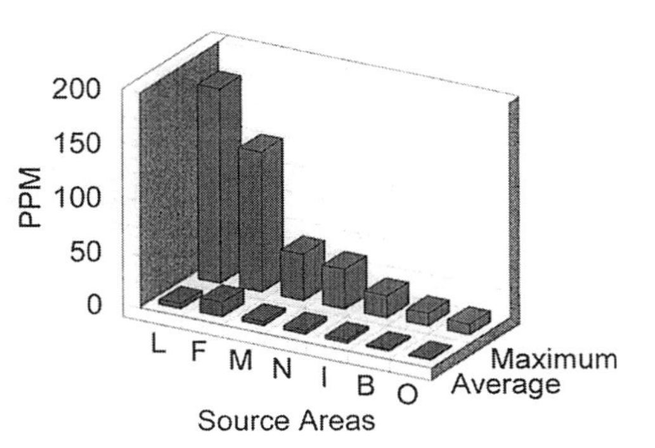

Soil TEH/oil

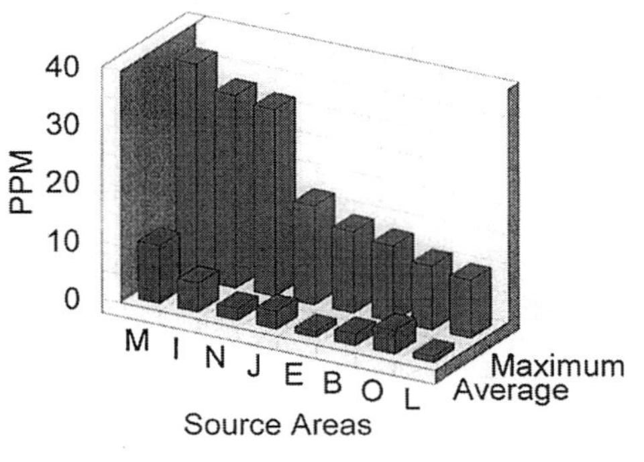

Figure 15.5 Comparative soil contamination.

Groundwater TEH/diesel

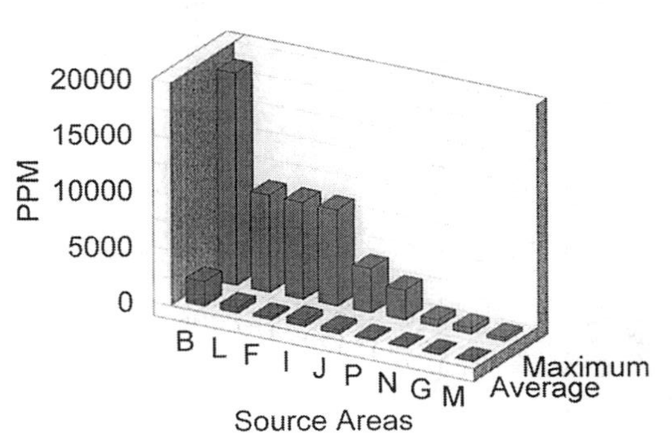

Groundwater TEH/oil

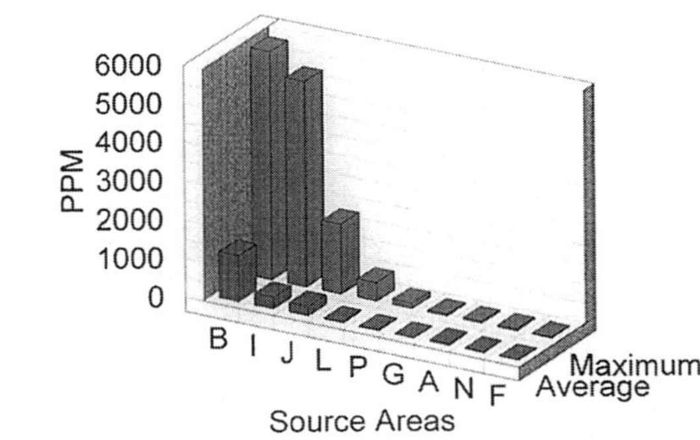

Figure 15.6 Comparative groundwater contamination.

Table 15.1 Soil Hot Spots Within Site N (Areas with TEH as Diesel >500 ppm)

Site Area	*Sample Location*	*Sample Depth (ft)*	*Concentration (ppm)*
Area of dispensing rack	SCI-43	4.5	9,200
	SCITP-4	5 (sidewall)	5,000
	SCITP-4	5	3,600
	SCITP-3	2.5–3	6,700
	SCITP-3	3.5–4	4,900
	SCI-44	2	1,300
	SCI-44	4.5	6,600
	SCI-45	5	23,000
Area SW of dispensing rack	SCTIP-19	2.5	38,000
Area near ASTs	RMA-25	5.5–6	2,685

Table 15.2 Soil Hot Spots Within Site N (Areas with TEH as Diesel >50 ppm)

Site Area	*Sample Location*	*Sample Depth (ft)*	*Concentration (ppm)*
Area of dispensing rack	SCI-43	4.5	9,200
	SCITP-4	5 (sidewall)	5,000
	SCITP-4	5	3,600
	SCITP-3	2.5–3	6,700
	SCITP-3	3.5–4	4,900
	SCI-44	2	1,300
	SCI-44	4.5	6,600
	SCI-45	5	23,000
	SCI-45	8.5	95
Area SW of dispensing rack	SCITP-19	2.5	38,000
Area near UST#(H-204)	SCITP-14	4	99
	SCIMW-24	6	140
	SCI-12	6.5	330
Area near ASTs	RMA-25	5.5–6	2,685
Area underneath H-224	SCITP-26	3	200

Table 15.3 Overall Soil Contamination Within Site N (All Concentrations in ppm)

Site Area	*Sample Location*	*TEH as Diesel Concentration*	*TEH as Motor Oil Concentration*
Area of dispensing rack	SCI-43	9,200	1,600
	SCITP-4	5,000	400
	SCITP-4	3,600	1,800
	SCITP-3	6,700	680
	SCITP-3	4,900	210
	SCI-44	1,300	3,200
	SCI-44	6,600	1,400
	SCI-45	23,000	2,600
	SCI-45	95	56
Area SW of dispensing rack	SCITP-19	38,000	32,000
	SCITP-19	33	18
Area near UST (H-204)	SCITP-14	99	420
	SCIMW-24	20	140
	SCIMW-24	140	200
	SCI-12	330	940
Area near ASTs	SCIMW-2	40	160
	RMA-25	2,685	549
Area underneath H-224	SCITP-26	200	610
	SCITP-26	2	17

The forensic expert proposed that a call be placed by the mediator to the regulator to confirm GeoExperts' reporting of the County Health Department's remedial recommendations. LFRC's counsel strongly opposed this step, pointing out that it would be a wasted effort in that GeoExperts had already expressed the regulator's opinion and concurrence with the proposed remedial plans. Of course, the LFRC team had no knowledge of the forensic expert's visit with the regulator. The mediator, when presented with the opportunity of speaking with the regulator, thought that the idea was excellent. Once again, LFRC's counsel and GeoExperts opposed the phone call on the grounds that they had already interfaced with the regulator on several occasions; that they had conveyed the regulator's position to the mediator on several occasions; that, therefore, nothing would be gained by an additional call. The mediator rejected the argument and placed the call.

Table 15.4 Probable Groundwater-Contaminated Source Areas (TEH as Diesel Levels >100,000 ppb)

Reference Area	*Sample Location*	*Sample Date*	*Concentration (ppb)*
F	RMA-5	11/18/96	8,668,000
F	9AV-B13-W1, 2	3/1/93	2,000,000
F	9AV-UST-2	2/12/93	1,000,000
F	MW-6	5/6/97	620,000
F	9AV-B16-W1, 2	3/2/93	310,000
F	SCI-23	5/31/96	248,000
F	MW-4	9/4/96	240,000
M	BH-2	3/29/95	300,000
G	SCITP-23A	4/26/97	8,700,000
G	SCITP-24A	4/26/97	520,000
G	SCI-35	8/30/96	230,000
B	SCITP-6	1/28/97	19,000,000
P	RMA-22	11/22/96	2,689,000
J	SCITP-11	2/4/97	4,000,000
J	RMA-14	11/20/96	440,100
J	SCI-6	5/22/96	240,000
I	SCI-2	5/22/96	5,300,000
I	Manhole	10/16/96	910,000
I	Manhole	5/13/96	720,000
N	SCITP-3	1/27/97	590,000
N	SCI-45	1/23/97	490,000
N	RMA-25	11/22/96	248,500
N	SCI-43	1/23/97	190,000

Based on the forensic expert's presentation, the mediator questioned the County Health Department project manager on each issue. The answers, not surprisingly, completely supported the forensic expert's presentation. LFRC's counsel and GeoExperts' project manager (sitting in on the call) were shocked by these revelations. Thus, the newly baptized mediator censured the LFRC representatives for the "inaccuracy" of their earlier presentations.

THE SETTLEMENT

It was profoundly gratifying to the Mid-States Oil team when the mediator addressed the LFRC team and asked why they had not proposed set-

Table 15.5 Maximum Groundwater TEH as Diesel Levels Found in Probable Source Areas

Reference Area	*Sample Location*	*Concentration (ppb)*	*Relative Percentage of Areawide Maximum*
F	RMA-5	8,668,000	18%
M	BH-2	300,000	1%
G	SCITP-23A	8,700,000	18%
B	SCITP-6	19,000,000	38%
P	RMA-22	2,689,000	5%
J	SCITP-11	4,000,000	8%
I	SCI-2	5,300,000	11%
N	SCITP-3	590,000	1%

Table 15.6 Summary of Remedial and Investigative Costs

Remedial Costs ($34,695)

1. Excavation/disposal of on-site soils (six locations >500 ppm TEH)
   $24,737
2. Removal of two 1,000-gallon USTs (assuming their presence)
   $9,958

Investigative Costs ($56,000)

1. Soil borings drilled and sampled on site
   $9,000 (six borings)
2. Pits excavated and sampled on site
   $6,000 (six pits)
3. Soil and groundwater analyses
   $21,000 (34 TEH/BETX samples, 25 VOC samples, and 16 metal samples)
4. Report preparation
   $20,000 (2 reports)

Total cost: $90,695

tling the action with a number close to the estimate prepared by the Mid-States forensic expert. In the final analysis, the settlement discussion had moved from a number somewhat less than $3.5 million to somewhat more than $75,000. This represented "one hell of a move" and an apparent victory for Mid-States Oil and its insurance carriers.

The parties settled within several weeks, at a cost to the Mid-States Oil team significantly below the LFRC demand, but significantly higher than the forensic expert's projection. The added settlement amount essentially consisted of the anticipated litigation costs had the matter gone to trial.

LESSONS LEARNED

1. It should be clear from this case history that all projected remedial costs are not necessarily appropriate just because an engineering consultant developed them.
2. Property owners with access to remedial funds (i.e., settlements from other responsible parties or insurance companies in excess of the estimated remedial costs) have been known to make property improvements with these funds (e.g., upgrade roads, improve utilities, raze buildings, and build sea walls) in the name of remediation.
3. Regulators, by the very nature of their responsibility (to address contamination issues), rarely suggest "less" and usually concur with "more." In other words, a geotechnical consultant may propose a "gold-plated" investigation or remediation. The agency's concern is that the plan address the contamination; it does not involve itself in the monetary aspects of the investigation or remediation. Because a regulator has not voiced an objection to a "gold-plated solution" does not mean an agency had "ordered" the proposed work. This is a very common and costly mistake.
4. Regardless of time constraints, homework prior to any mediation is the key to success. In this case, however, the key to success was establishing the appropriate lines of communication necessary for an effective mediation.

Chapter 16

Allocation, Allocation, Allocation

One of the most difficult environmental allocation problems that can arise is when there are multiple parties that have all contributed the same contaminant to a regional groundwater plume. In other words, the individual contaminant plumes become commingled. This situation becomes even more complex when the parties have similar types of operations and employ similar solvents. When this occurs, the issue becomes one of attempting to sort out each contributor and determine the appropriate "share allocation" for the cleanup. In this case, all of the contamination occurred from one industrial park. Given this circumstance, it would be logical to assume that each party's operational size and site longevity should bear a close relationship to the amount of contamination contributed to the groundwater. However, in spite of the obvious, the manufacturing behemoth of the park claimed that the other adjacent facilities contributed the vast majority of the contamination.[1]

A POLLUTION-FREE INDUSTRY AND GROUNDWATER CONTAMINATION

In its infancy, the electronics industry was hailed by many as the inauguration of the clean or pollution-free enterprise. Indeed, considering the care shown in the production of electronic components, and a total absence of the smokestack discharges normally found with smokestack industries, on the surface, this assumption appeared correct.

[1]In this case, the behemoth took the approach that the best defense is a good offense. Needless to say, each of the other parties on the site employed a similar tactic.

It is well known that electronic components have to be clean in order to function properly. Thus, electronic manufacturing facilities employed several highly efficient chlorinated hydrocarbon solvents in the cleaning processes. It was the use of these solvents that was responsible for the site contamination. Considering just how much attention was being paid to cleanliness, one must wonder why so little attention was paid to the handling, storage, and disposal of these solvents.

The site in question was an industrial park housing 12 past and present companies, located on nine parcels within the park. Each of the site tenants manufactured similar electronics, all used similar solvents, and all were subsequently shown to have a history of leaks. By the time the contamination was discovered, an adjacent property housing a school and health care center was already underlain by a "commingled" solvent plume that could only have had its origins from the up-gradient industrial park. Thus, there were two separate but totally entwined problems: (1) the allocation of cost for the investigation and cleanup of the down-gradient property and (2) the allocation of on-site cleanup costs, on a parcel-by-parcel basis and among present and former occupants. As a result, a gigantic make-work program was initiated for environmental firms, laboratories, attorneys, consultants, and regulatory agencies.

Recognizing the complex nature of the situation, the parties did take several steps, which ultimately assisted in resolving most of the conflicts. Initially, the potentially responsible parties (PRPs) formed a joint committee and selected one engineering firm to investigate and define the extent of the contamination. As part of that process, the PRPs settled on an arbitrary allocation for cost reimbursement, which was subject to final adjudication of the contaminant allocation. Finally, the PRPs agreed to a mediation process in order to resolve shared responsibility.

THE INDUSTRIAL PARK INVESTIGATIONS

The complexity of the industrial park environmental characterization was compounded by the number of individual site investigations as well as the number of consultants. There was one overall site investigation[2] with the purpose of (1) determining the extent of the contamination, (2) identifying

[2]The geotechnical firm retained for this investigation will be referred to herein as the "primary consultant."

the potential sources that have impacted the adjacent off-site properties, and (3) assessing the potential risk to the off-site properties. In addition to the overall investigation, each current site occupant[3] retained a consultant to investigate its site as well as other adjacent sites so that an "allocation position and defense" could be developed. An overview of the industrial park is shown in Figure 16.1. The potential contaminant source locations are shown in Figure 16.2, and the location of PCE sources is shown in Figure 16.3. Total VOC groundwater concentrations are shown in Figure 16.4. TCE and PCE groundwater concentrations are shown in Figures 16.5 and 16.6.

The primary consultant had the fundamental tasks of locating and placing the monitoring wells, determining the pattern(s) of contamination within the aquifer, interfacing with the regulatory agency, and defining the potential off-site remedial actions (including cost). In addition, the primary consultant prepared and distributed a monthly activity report to each PRP, each PRP's consultant, and each PRP's counsel (a truly massive amount of paper).

With the general contamination pattern having been identified, each PRP's consultant attempted to "explain" the source of the groundwater contamination beneath his or her client's property. For each PRP's property, the contamination could have had its origin from on-site sources or off-site sources from up-gradient or adjacent properties. Thus, each PRP's consultant, all accessing the same sitewide data, developed groundwater contamination "patterns" whereby, in the main, the contamination stemmed from its neighbors rather than from its own operations. As a result, the "technical finger pointing" in the informal meetings was intense. Each PRP took the position that it had never contaminated the groundwater beneath its property or that its contribution was insignificant compared to its neighbors or to any prior occupants. A typical comparison of PCE soil contamination at different sites is shown in Figure 16.7.

Due to the complex nature of the contamination, there were two separate cost allocations that had to be resolved. First, sitewide investigative

[3]During one of the investigation phases, and as costs continued to mount, the industry steering committee explored the possibility of replacing its primary consultant with another firm. Requests were made to each participating company to make recommendations. This proved to be a futile exercise in that virtually every firm, qualified to handle the assignment, was already involved, either directly or as a subconsultant, on some phase of the ongoing investigations. As a result, in spite of general dissatisfaction with the performance of the primary consultant, it was retained by default.

Figure 16.1 Industrial park overview.

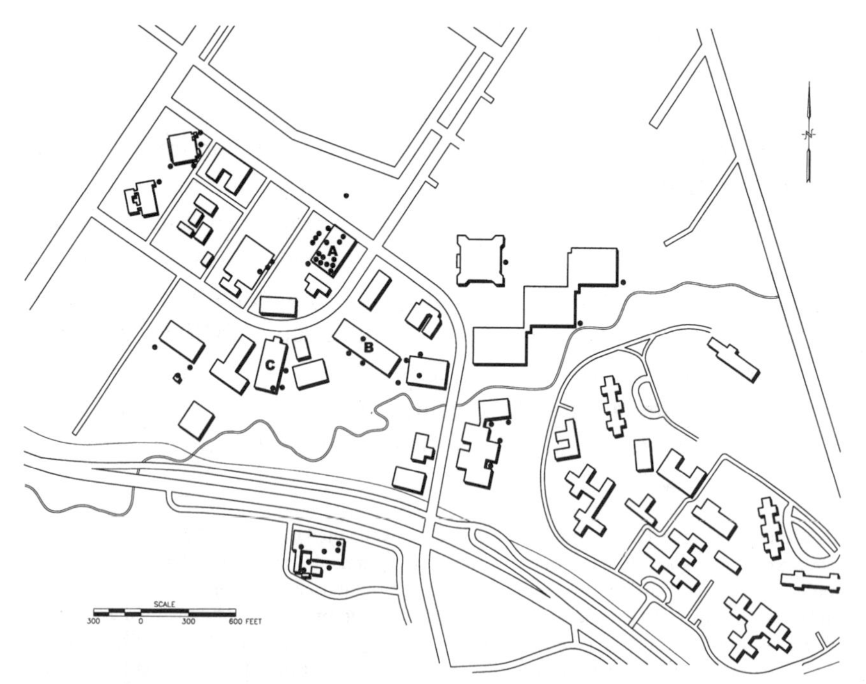

Figure 16.2 Initially identified contaminant locations.

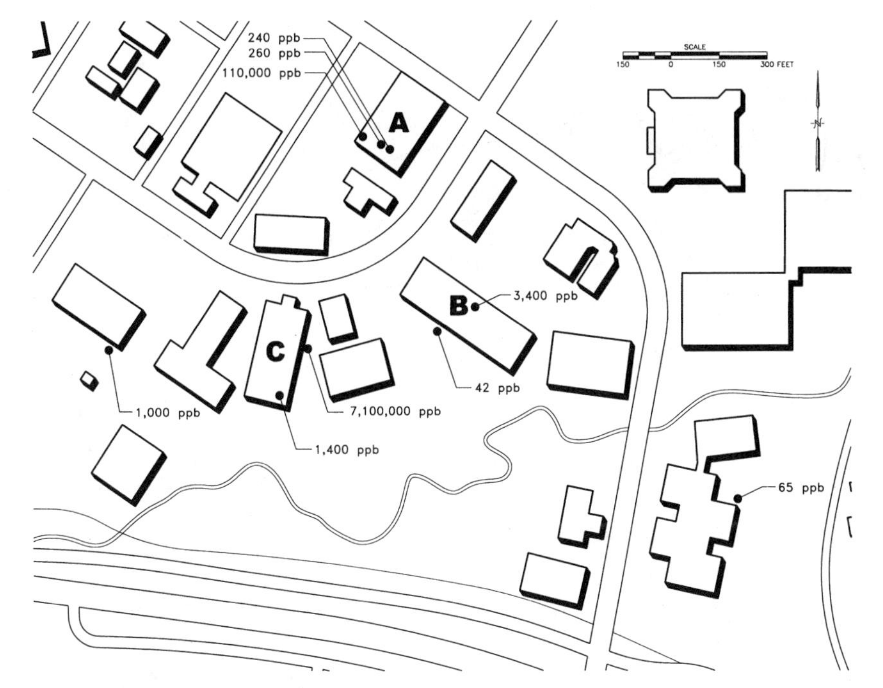

Figure 16.3 Locations of PCB contamination.

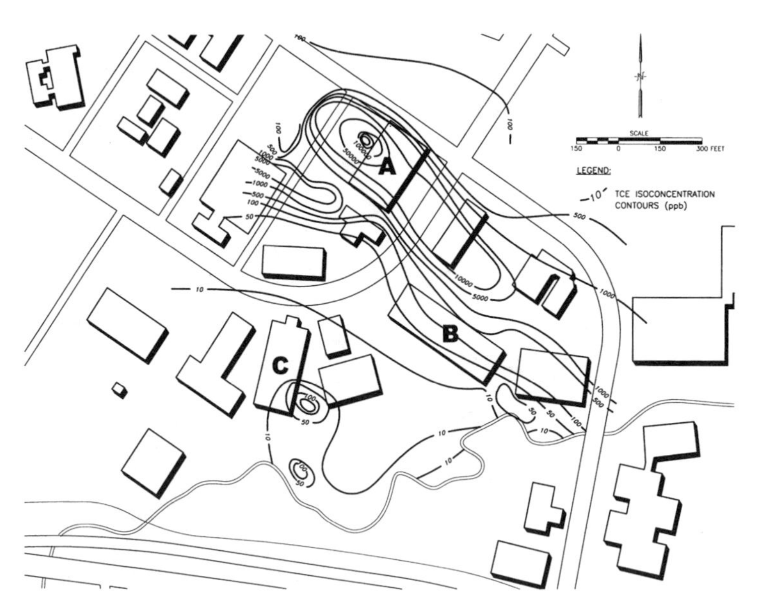

Figure 16.4 Initial VOC plume pattern.

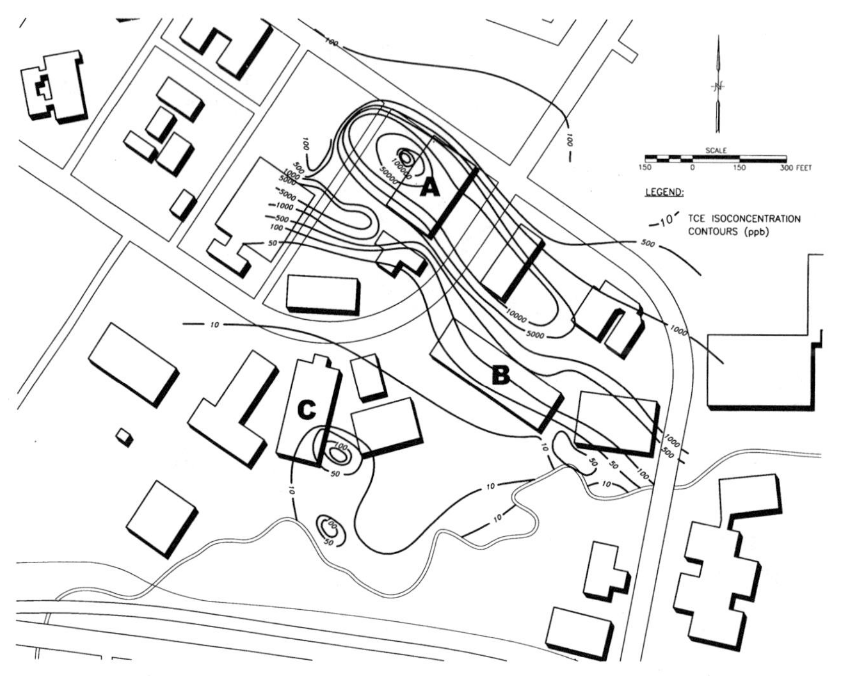

Figure 16.5 TCE groundwater isoconcentration contours.

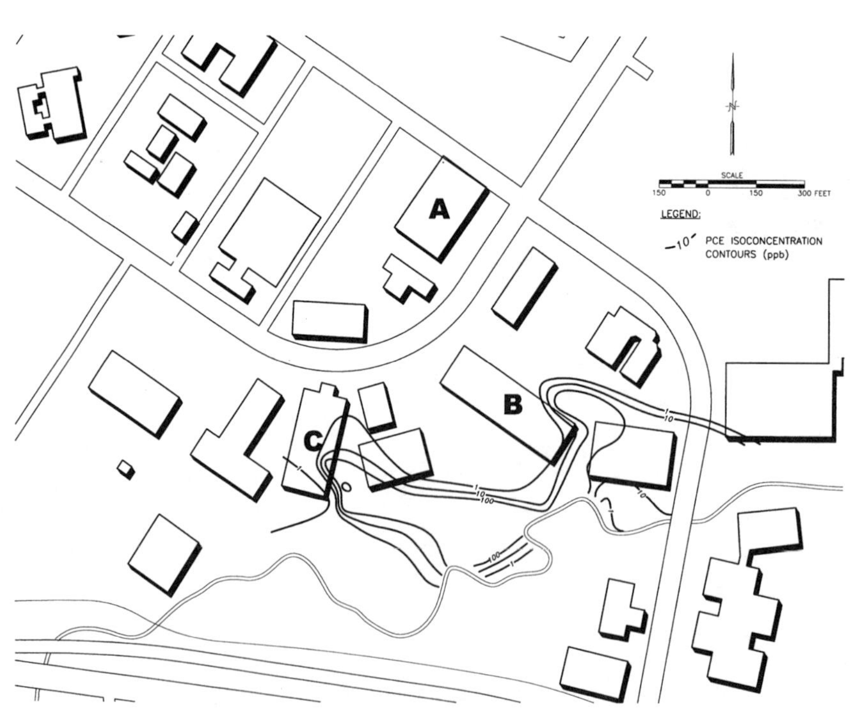

Figure 16.6 PCE groundwater isoconcentration contours.

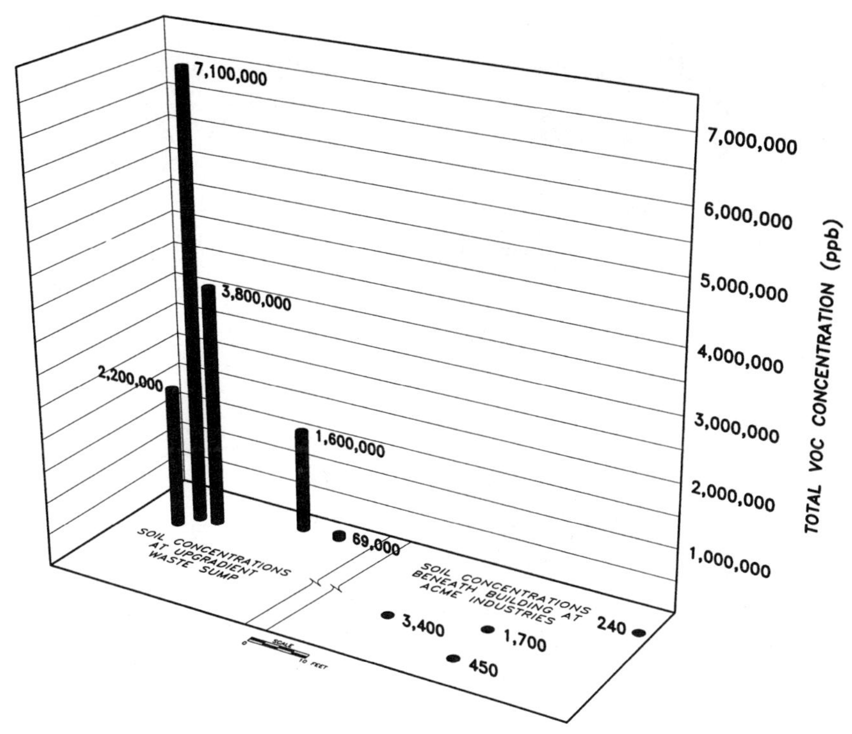

Figure 16.7 Illustration of up-gradient and down-gradient soil VOC concentrations.

and remedial costs had to be allocated among individual parcels. Having "resolved" the allocation against each parcel, the "parcel" cost had to be further allocated among the past and present tenants of each parcel.

ALLOCATION AMONG PARCELS

A forensic expert was retained by Medical Electronics, which was the former occupant of a 4-acre parcel identified in Figure 16.8. Medical Electronics occupied this parcel for approximately 10 years, beginning in 1960. The building housed office space, research and development, and the assembly of electronic medical equipment. However, after Medical Electronics stopped operations at the site, ACME Industries utilized the same building for another 30 years. The operations that might have contributed to site contamination are shown on Figure 16.9.

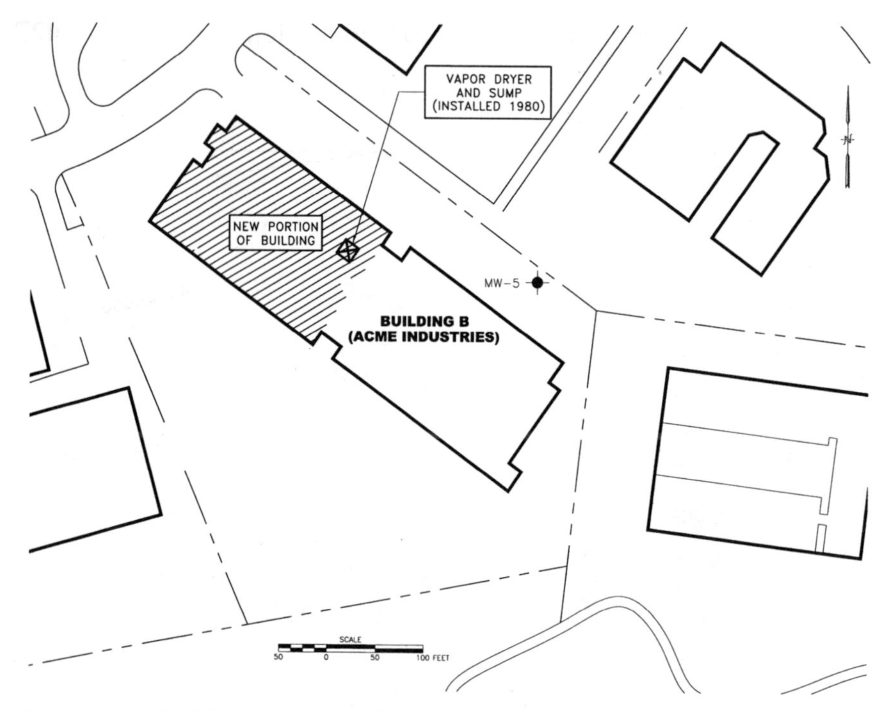

Figure 16.8 Building B ACME Industries.

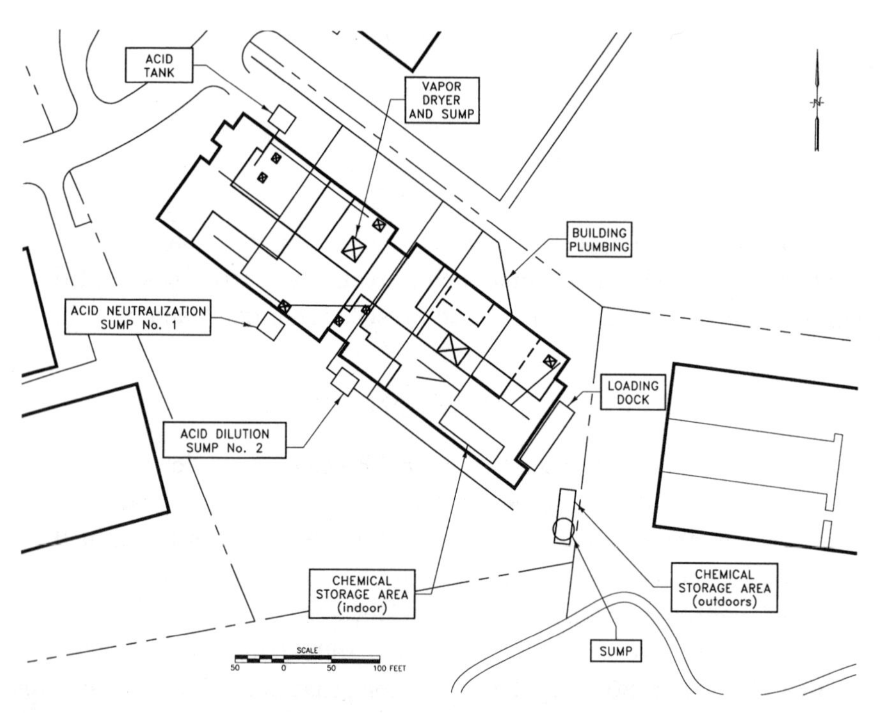

Figure 16.9 Potential contaminant sources associated with Building B.

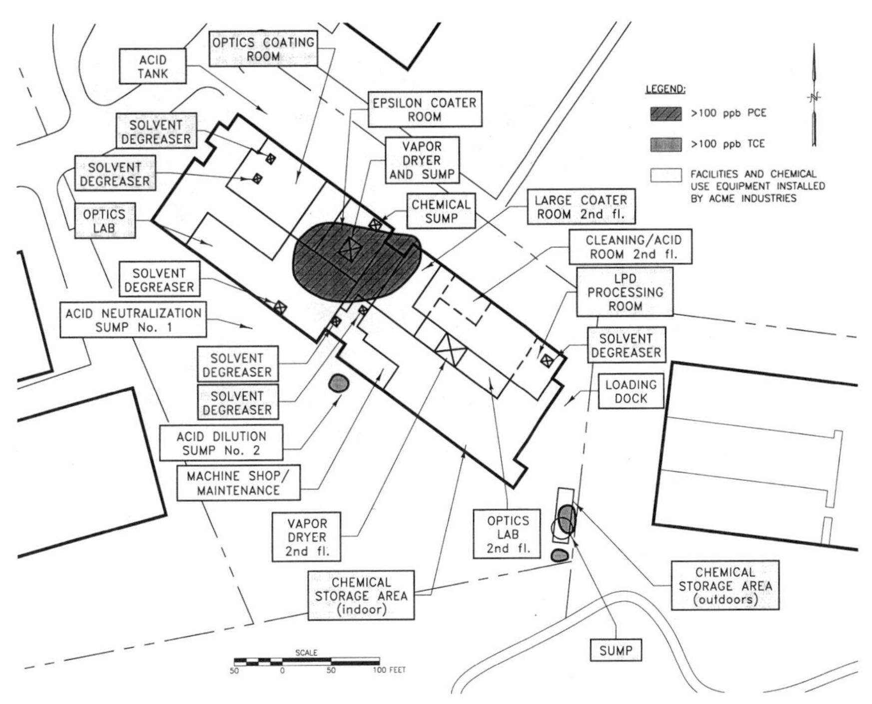

Figure 16.10 PCE and TCE identified under Building B.

Thus, when considering the parcel allocation issue, the forensic expert's initial task was to review all of the currently available site investigative documents and determine the nature of the contamination under the Medical Electronics/ACME parcel. This included additional soil borings (to determine the chemicals in the soil), as well as more extensive sampling of the site monitoring wells. The results of this investigation (for TCE and PCE) are shown in Figure 16.10. With this review completed, the next task consisted of obtaining all of the past chemical records for both Medical Electronics and AMCE Industries during their site occupancy. Relative to Medical Electronics, this included finding former employees and determining from them just what chemicals[4] had been used on the site and how they had been used. The employee interviews did reveal that chemicals such as trichloroethylene (TCE), 1,1,1-trichloroethane (TCA), xylenes,

[4]There were no past records related to chemical use; therefore, an attempt was made to obtain employee records, locate former employees, and conduct interviews in an effort to verify chemicals used and quantities employed.

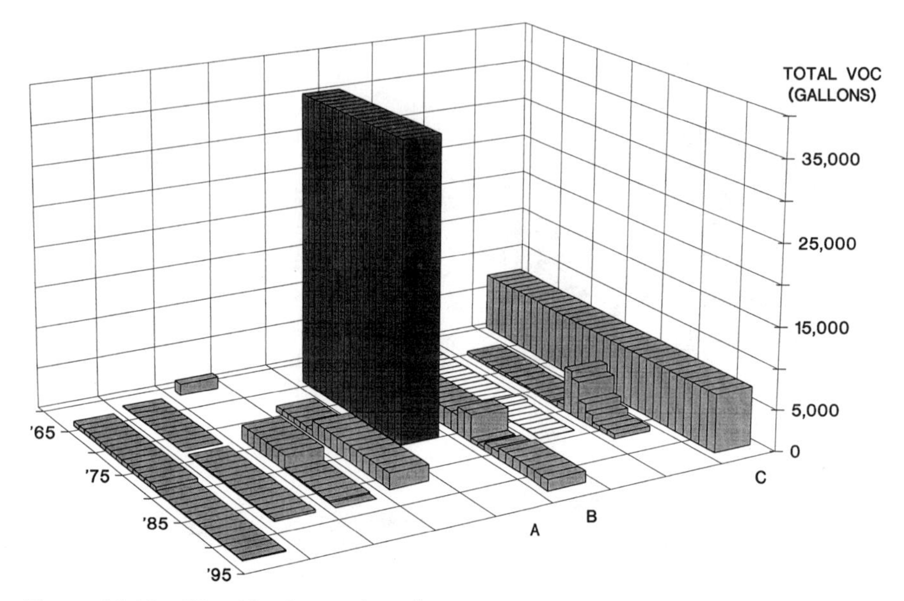

Figure 16.11 Sitewide chemical use history.

toluene, methylene chloride, and freon had all been used at the facility. These employees also provided their best estimates of the amount of each chemical that was used during the years of occupancy.

At the request of the state regulatory agency, each PRP was requested to prepare and submit a chemical use history to the agency. All of the chemical use histories were then made available to all parties. As a result, the sitewide chemical use history information became a major factor in developing the final groundwater remedial cost allocations.

Simple Illustrations

In order to illustrate the relative contribution of Medical Electronics to the site contamination, several simple illustrations were developed. The first illustration showed the chemical use history for each PRP on a common axis, as shown in Figure 16.11.[5] The second illustration was an estimate of the total mass of volatile organic compounds (VOCs)[6] in the groundwa-

[5]For purposes of simplicity, only the three principal sites have been designated—as Facility A, Facility B, and Facility C.

[6]Calculated from plume isoconcentration contours and estimates of the thickness of the saturated aquifer.

ter beneath each parcel. The total VOC values were then compared to the chemical use history and typical chemical loss estimates (mass) for the industrial processes utilized in electronics manufacturing. The chemical use, as shown in Figure 16.11, can be compared to the mass calculation found in Table 16.1.

It can be seen that the mass allocations showed a close resemblance to the relative chemical use histories of each PRP. These simple illustrations became the basis of Medical Electronics' argument for allocation (see Sidebar 16.1). Interestingly, each site party produced its own allocation and these are shown in Table 16.2.[7] The forensic expert's allocation is given in Table 16.3.

The Mediation

The large conference room in which the mediation was held was filled with each PRP's company representatives, its consultants, and its counsel (both in house and outside) for the stated purpose of arguing their positions in the hopes of securing a minimum allocation. Fortunately, the mediator had established a protocol whereby the initial discussions were limited only to technical consultants who debated technical issues. While this round of all-day discussions did not resolve any allocation issues, the mediator did begin to see a pattern of technical strengths and weaknesses. The technical consultants' meeting was then followed with each PRP making a formal presentation, supported by charts, graphs, video projections of plumes, and even models of the site. Questions (a polite term for heated cross-examination) were allowed after each presentation. With the completion of each PRP's formal presentation, PRPs were then allowed "rebuttal" presentations. The rebuttal meetings consumed several very long days and generated short tempers and both physical and mental fatigue.

In the final analysis, after all of the detailed and sometimes eloquent presentations and subsequent arguments, the simple chemical use histories shown in Figure 16.11 prevailed. Even though the technical solution had been accepted, the exact percentage for each parcel allocation still had to be agreed on by each PRP. This is another area where the mediator earned his money in that he had to go "one on one" with each PRP, argu-

[7]Interestingly, and not unexpectedly, the total contribution fell far short of 100%. Each party managed to allocate "to itself" a small contribution (in some cases zero contribution) and allocated to others the major contribution.

Table 16.1 Comparison of Total Site VOC Use to On-Site Contaminant Mass

Site	*VOC Use*	*Contaminant Mass*
A	63.2%	60.5%
B	4.2%	1.0%
C	24.6%	18.8%
All others	8.0	19.7%

Table 16.2 Allocations by Site Parties

Site	*Allocation*
A	30%
B	8%
C	6%
All others	15%
Total	59%

Table 16.3 Allocations Proposed by the Forensic Expert

Site	*Allocation*
A	70%
B	4%
C	19%
All others	7%

Table 16.4 Final Allocation (by Mediation)

Site	*Allocation*
A	67%
B	4%
C	15%
All others	14%

Sidebar 16.1 Role of Chemical Use History

Many sophisticated technical arguments were presented by the individual parcel consultants. However, the forensic expert's rather simple position prevailed: Based on the fact that all parties essentially shared the same VOCs, the contamination allocation could be simply resolved by looking at the chemical use histories. In simple terms, the allocation was based on the amount of chemicals used, assuming that each operation had the same pattern of chemical loss. The mediator both appreciated and endorsed this approach. This approach is further supported by the fact that many state regulatory agencies, following the lead of the EPA, have established water and air chemical constituent discharge standards using quantity (i.e., mass units).

Under the EPA Superfund program, Non-Binding Preliminary Allocations of Responsibility (NBAR), the EPA may consider such factors as volume, toxicity, and mobility of hazardous substances contributed to the site. In commingled waste cases, the first step in the allocation phase of an NBAR is to allocate 100% of the responsibility among generators based on the volume (i.e., mass) each contributed.

ing contribution, in a round-robin marathon. Finally, all PRPs agreed with the final allocation, as shown in Table 16.4.

The allocation for the major contributor turned out to be within 4% of the forensic expert's proposed allocation. Fortunately, the forensic expert's client walked away with a lower allocation than originally allocated and essentially the allocation proposed by its forensic expert. It was the other PRPs that had shifted allocations one way or another, among themselves. At this point in time, the regional groundwater cleanup cost was estimated to be $42,000,000.

Once all parties signed off on the allocation, the major contributor proposed to set up the groundwater treatment system on its site, allow all parties to pipe their extracted groundwater to the central system (at an agreed-on operating cost allocation), and take responsibility for system operation as well as regulatory reporting.

ALLOCATION AMONG PARCEL OCCUPANTS

Although the forensic expert was successful in gaining a low allocation percentage relative to the groundwater cleanup, there still remained the soil cleanup[8] of the parcel that was still occupied by ACME Industries. With the settlement of the overall sitewide groundwater allocation, allies suddenly became enemies,[9] and historic site operations that would have contaminated soil became the central issue of an upcoming mediation.

In preparing for this mediation, five alternative approaches to obtaining a technical allocation for the soil contamination, which would be both valid and supportable, were considered for presentation. These alternatives included allocation by (1) length of site occupancy, (2) chemical use history, (3) chemicals used, (4) contaminant age,[10] and (5) contaminant location(s). The allocation by each method is given next.

Medical Electronics occupied the site from 1960 to 1971, while ACME Industries' occupancy ran from 1972 through 1987 (the building, with the addition put on by ACME is shown in Figure 16.8). On this basis the allocation became:

Medical Electronics: 41%
ACME Industries: 59%

Medical Electronics' chemical use had to be based on employee interviews (as stated earlier) and this approach seemed to indicate a combined solvent use of approximately 300 gallons per year. ACME Industries, being the more recent occupant, had better records and its chemical use averaged 753 gallons per year. On this basis, the allocation became:

Medical Electronics: 28.5%
ACME Industries: 71.5%

According to Medical Electronics' former employees, Medical Electronics never used PCE. ACME Industries made the same claim. These claims

[8]Soil vapor extraction of perchloroethylene (PCE).

[9]During this process, both Medical Electronics and ACME Industries worked together closely in joint defense of their parcel allocation.

[10]Chlorinated hydrocarbons are known to degrade over time. Thus, as an example, a recent release of PCE will show up only as PCE. However, with the passage of time, breakdown products such as TCE, DCE, and vinyl chloride will be detected if microbial conditions are favorable. By looking at the amounts and ratios, one can "guesstimate" the relative age of the contamination.

contradicted the fact that the contamination beneath the site was largely PCE, as shown in Figure 16.10. With no documented PCE use by either company, no allocation could be made.

The soil boring data under the building showed only PCE, with no decomposition products (such as TCE, DCE, or vinyl chloride). If the discharge of PCE had occurred many years ago, one would expect to find some evidence of microbial degradation and therefore decomposition products. Similarly, if, as ACME initially argued, PCE in the soil was the result of upward mobility of contaminants from the underlying groundwater, one would expect to find evidence of other VOCs along with the PCE. Since Medical Electronics had left the site by 1972, the lack of decomposition products suggested that the source of the PCE release had to have been more recent than 1972. Thus, on that basis, the allocation would be:

Medical Electronics:	0%
ACME Industries:	100%

Because there was insufficient data to conclusively argue the contaminant age method of allocation, it was not included in the mediation. However, it was brought up so that the issue was not “lost” on the mediator.

The soil sampling data showed a general pattern of PCE contamination centered on several locations (see Figure 16.10). Because of this distribution, the possibility existed for associating company-specific operations to the PCE contamination. Just as with the contaminant age approach, there was insufficient soil data to propose an allocation based on contaminant location.

Preparation for the Mediation

These methods were combined and numerically weighted to reflect the sum of the individual method’s allocations, as shown in Table 16.5.

Based on this weighted method of allocation, it was Medical Electronics’ position that ACME should be responsible for approximately 79% of the soil cleanup costs.

The First Mediation

The methods of technical allocation were presented along with the weighted allocation percentage. Needless to say, ACME Industries was not sold on

Table 16.5 Weighted Allocations

Allocation Method	*Weight*	*Medical Electronics*	*ACME Industries*
Time of occupancy	30%	12.3	17.7
Amount of chemicals used	30%	8.5	21.4
Contaminant used	40%	0.0	40.0
Total		20.8%	79.1%

the approach. Its position was that ACME did not use PCE and that either (1) Medical Electronics did use PCE and it has remained in place all these years (without decomposition) or (2) PCE in the soil originated through volatilization of PCE from the groundwater that arrived on site from some up-gradient source. Because ACME Industries had no method of allocation (i.e., other than asserting that Medical Electronics was responsible), it was not possible to arrive at an allocation that would satisfy both parties.

A Site Investigation

Based on the mediation, it was clear that the only technical allocation method that would influence ACME Industries' position would be evidence of PCE contamination associated with past manufacturing operations. In other words, was there any PCE contamination beneath any building sumps, drains, vapor dryers, degreasers, or chemical storage areas associated with either Medical Electronics or ACME Industries operations? In order to answer this question, additional soil samples had to be taken in the identified source areas (hot spots).

Since the building had been modified several times (again, reference Figure 16.8), even locating specified sections of the building flooring and piping proved to be a major undertaking. When a recent set of as-built drawings was found, it showed a vapor dryer and sump that were located[11] in a portion of the building that was added subsequent to Medical Electronics' occupancy (see vapor dryer and sump locations in Figure 16.9).

[11]The location of the sump also coincided with one of the two areas of identified soil PCE contamination.

When these drawings were reviewed with ACME Industries, it continued to deny the existence of the sump and, thus, the use of a vapor dryer.[12] Considering that the as-built not only showed the sump but showed a vapor dryer as well as the manufacturer and model number of the vapor dryer, the position taken by ACME seemed somewhat ridiculous. However, ACME did allow an inspection of the building, with the condition that all costs were to be borne by Medical Electronics.[13]

The initial search of the building revealed no sump. However, there was a faint outline of a concrete patch approximating the size of a sump, and a former floor plan indicated the location of a vapor dryer in that same area. Although ACME Industries strenuously denied that a sump existed, it agreed to allow the forensic expert's geotechnical consultant to drill at the location.

Several soil borings through the floor identified the sump location (again, reference Figure 16.10). The borings showed that between the top of the concrete floor and the bottom of the concrete sump the void space had been filled with sand. Several samplings of the sand as well as samplings of concrete cores from the sides and bottom of the sump were taken for VOC analysis. The analyses identified PCE in all of the samples.

When faced with the existence of the sump, ACME Industries acknowledged that it "may have" used a vapor dryer, but it continued to insist that it did not use PCE in the vapor dryer. ACME still clung to the theory that the PCE contamination under and within the concrete sump originated from the groundwater and not the other way around.

With the sump source area identified, an effort to pinpoint additional VOC sources was undertaken. This became somewhat more difficult in that although the facility plans showed a number of potential sources (e.g., former chemical storage area, floor drains, solvent degreasers, sumps, etc.) in the building, there did not appear to be any soil contamination in these areas. However, there were subsurface drain pipes near the area showing PCE in the soil (i.e., any contamination from the pipes could have originated from either party). Rather than spending more money on random soil boring studies through the building flooring, both parties agreed to at-

[12]A call to the vapor degreaser manufacturer revealed that the unit required periodic cleaning and recharging. This consisted typically of draining the unit, with the sump being necessary to "catch" whatever solvent leaked out during the draining process. The manufacturer stated that it was common to lose 5 to 10 gallons during the draining and refilling process.

[13]This, of course, "changed" during the mediation process and during final settlement. The "embarrassed party" agreed to pay the costs of the investigation as part of its allocation.

tempt an allocation between source locations (note the TCE hot spots in Figure 16.10) by estimating the mass of VOCs within each of the soil contours.[14] Using this approach, the forensic expert allocated approximately 11% of the contamination to the TCE locations (including an estimate of how much of the PCE might have come from subsurface drain pipes) and 89% to the sump. However, there was still no agreement as to which party was responsible for the contamination.

The Second Mediation

With the sump identified and both parties agreeing that the major source of the PCE soil contamination (which occurred in a portion of the building never occupied by Medical Electronics) was indeed under the sump, one would have assumed that an easy settlement would be forthcoming. However, with a projected $2,000,000 of soil remedial costs added to previous expenditures of $1,500,000 for site investigative costs, neither side wanted to accommodate the other. At the end of a grueling day, and after repeated threats by the mediator, who regularly pointed out the projected costs of going to trial, the parties agreed on a settlement. It appeared as if both sides had simply tired of arguing and wanted to go home.

Based on the evidence, Medical Electronics settled at a greater amount than expected, but less than the projected investment in continued litigation. Medical Electronics agreed to accept responsibility for the piping source, the TCE sources, and 50% of the investigative costs. In the final allocation, the costs were distributed to each party using the following formula:

Medical Electronics (27.7%) = 0.5($1.5M)+0.11($2.0M) = $ 970,000
ACME Industries (72.3%) = 0.5($1.5M)+0.89($2.0M) = $2,530,000

When compared to the previous allocation methods, the final percentages were fairly close, indeed almost identical to the cost allocation based on chemical use history.

[14]Why use soil contours? Why not use groundwater plume data? Reflecting on the fact that the groundwater was contaminated by many sources and that many of the plumes had commingled, it was hardly possible to look under our parcel and attempt to sort out plume characteristics unique to each parcel. Since the soil was directly contaminated by the parcel activities, this seemed to be the best alternative for estimating and allocating.

DECREASING REMEDIAL COSTS

The forensic expert, on behalf of Medical Electronics, has continued to monitor the remedial activities, both sitewide and at the former Medical Electronics site. As of 1999, the projected sitewide groundwater cleanup cost is now estimated at $35,000,000, savings of $7,000,000 over the original projected cost. In similar fashion, the parcel soil cleanup cost is now projected at $1,200,000, which is $800,000 under the original estimate.

LESSONS LEARNED

1. If one naively accepts each consultant's estimate of his or her client's allocation, then one quickly discovers that nobody contaminated anything (i.e., the contamination was simply a freak of nature).
2. Although not explicitly discussed in the chapter, it was readily recognized by all site parties that not only were the regulatory oversite costs high, but they also seemed to be never ending.
3. It pays to explore every avenue for estimating allocation.
4. When everything is said and done, the technical issues represent, at best, a starting point in the final settlement.
5. A good mediator is worth every dollar paid, regardless of the hourly rate.

Chapter 17

Deflating Environmental Costs

Although the allocation of remedial costs between past and present tenants of a contaminated property is usually driven by technical issues, occasionally a situation arises that calls into question the business acumen of certain parties to a dispute. Even when the technical issues are well defined, once the attempt to settle enters the legal arena, the monetary stakes can be raised significantly. As each party prepares a position to force a favorable settlement, it usually becomes necessary for the forensic consultant to separate fact from fiction in order to place inflated and unrealistic settlement demands into perspective.

THE PARTIES AND THEIR PROBLEMS

A specialty chemical company, Unique Chemicals, had operated at its location for several decades. The area was a historic manufacturing neighborhood replete with small commercial and industrial operations. Unique Chemicals sold the business to Chemical Consultants, which continued to operate the facility for approximately 10 years. In 1989, Chemical Consultants was served a notice by the County Environmental Health Department (CEHD) for a permit violation stemming from the storage of some 90 drums of hazardous wastes[1] outside its warehouse facility. The office complex, warehouse facility, and drum storage areas (northwest side

[1]These wastes essentially consisted of inks and dyes, solvent cleaning compounds, volatile and semivolatile hydrocarbons, and color pigments that contained a wide range of compounds, including benzene, toluene, ethylbenzene, xylenes (BTEX), petroleum hydrocarbons (diesel fuel and gasoline range of compounds), oil, grease, polychlorinated biphenyls (PCBs), and chlorobenzene.

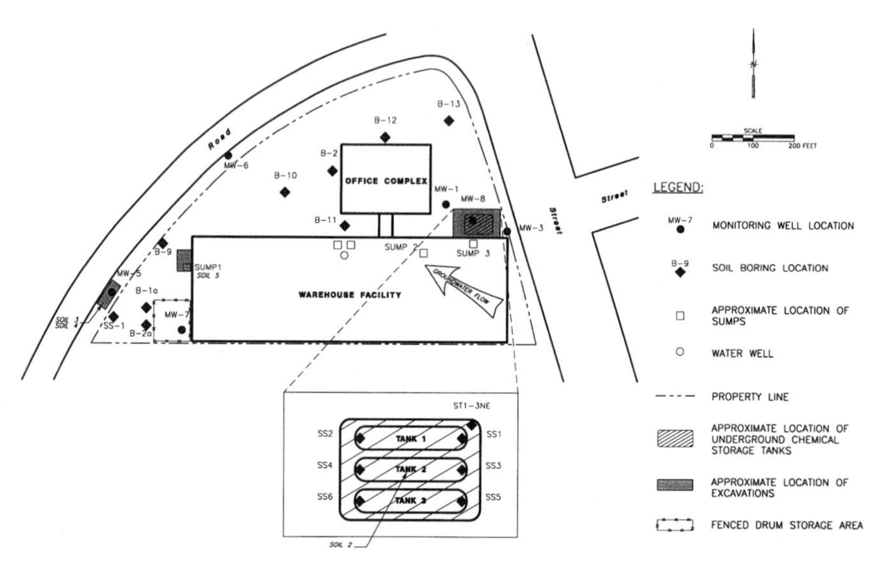

Figure 17.1 Site location and investigation.

of warehouse facility) are shown in Figure 17.1. Unfortunately for Chemical Consultants, it was also attempting to sell the business at the time that the notice of violation was received.

The CEHD required that the drums be shipped to a licensed recycling company or a Class I landfill, that a minimum of seven soil borings and one groundwater monitoring well be situated at the site, and that soil and groundwater samples be analyzed for a range of organic contaminants. Chemical Consultants retained a geotechnical firm to provide an estimate of the proposed work. The soil boring and monitoring wells are shown in Figure 17.1. The initial work plan estimated that the soil and groundwater sampling would cost approximately $44,000.

The analytical results of the soil boring samples showed the presence of petroleum hydrocarbons, including gasoline, diesel fuel, oil, and grease. The groundwater samples (with the exception of MW-1 and MW-8, which were not installed during the initial investigation) contained no detectable organic compounds. Unfortunately, and not uncommonly, by the time the initial work was completed, the cost of the investigation and subsequent disposal of contaminated soils had escalated to $110,000. It now also appeared that the sale of the site to the new owner would go through. However, as part of the due diligence process, the new owner's consultant

Table 17.1 Soil Analyses Surrounding the UST

Sample Location	*MIBK (mg/kg)*
East end—tank 1	600
East end—tank 2	0
East end—tank 3	180
West end—tank 1	20
West end—tank 2	3300
West end—tank 3	5000

"discovered" some pipes[2] protruding from the north side of the building. Chemical Consultants claimed to be unaware of these pipes and their intended purpose.

Upon further investigation, the pipes were found to be connected to three underground storage tanks (USTs).[3] The location of these USTs, as well as their description, is found in Figure 17.1. Chemical Consultants then retained a second geotechnical firm to sample the contents of the unknown USTs. The results of the analysis identified the chemical methylisobutylketone (MIBK), including samples taken from MW-1 and MW-8. Because Chemical Consultants alleged ignorance of the USTs, it declined knowledge of and responsibility for the chemicals found in the tanks. Chemical Consultants also declined responsibility for any soil and groundwater contamination associated with these tanks.

With the UST discovery, the sale of the site was delayed and another series of soil and groundwater investigations were proposed to determine if there was any contamination associated with the USTs. Chemical Consultants retained a third geotechnical firm to perform the UST investigation. The analyses confirmed that both the soil surrounding the USTs and the groundwater beneath the USTs were contaminated with MIBK. The results of the soil analyses are shown in Table 17.1 and the groundwater data are found in Table 17.2.

Thus, with the previous site investigation cost and the specter of addi-

[2]Which disappeared into the ground.

[3]Why were these tanks never discovered or discussed during the earlier sale by Unique Chemicals?

Table 17.2 Groundwater Data in the Vicinity of the USTs

Monitoring Well	*MIBK (μg/L)*
MW-8	8,300
MW-3	ND
MW-7	ND
MW-1	110,000

ND, Not Detected.

tional investigative and remedial costs associated with the USTs, Chemical Consultants filed a lawsuit against Unique Chemicals in order to recover its present and future costs.

THE FORENSIC INVESTIGATION

Almost a year after the UST investigation (the result of which was the removal and disposal of the three tanks, as well as localized excavation and disposal of the surrounding and underlying soils), counsel for Unique Chemicals retained a forensic expert to review the technical case issues and recommend a course of action. In order to provide guidance to Unique Chemicals, the following objectives had to be accomplished:

- Determine, if possible, the source of the MIBK found in the USTs.
- Establish the extent of any additional soil contamination, as well as groundwater contamination, and determine the likely remedial actions.
- Provide input to the CEHD in order to support the least cost remedial alternative for the site that was consistent with CEHD policy.
- Determine the extent to which each party contributed to the site contamination (i.e., allocate responsibility).

Therefore, the initial steps of this investigation were to (1) review all of the geotechnical reports prepared to date (see Sidebar 17.1 on the use of multiple consultants) and (2) visit the CEHD to evaluate potential remedial solutions.

Sidebar 17.1 Problems with Multiple Consultants

During the history of the site investigations, Chemical Consultants retained the services of three separate geotechnical firms. Its reasons for changing geotechnical firms are not known (e.g., cost, not qualified to perform expanded tasks, disagreed with the conclusions, etc.). However, when multiple firms are used, the forensic expert should be aware of two common problems.

The first problem pertains to data reporting and the concept embraced by many consultants that just because the data are published they are correct. For the uninitiated, it is important to realize that each consultant usually copies the boiler plate work done by a predecessor firm. Thus, by the time a third or forth consultant has copied the information, errors that were never caught have now become gospel. As with so many cases, there appear to be numerous engineering firms and consultants following in each other's footsteps. When a site has had multiple consultant reports, the forensic expert is well advised to verify the accuracy of those key facts that will be used to support an opinion.

Other problems can arise from the graphical representation of contamination data. For example, the individual sample results may not have been plotted properly. Consequently, the sample data should be checked to determine if the mapped representation is consistent with the analytical results. In addition to this major problem, maps from different reports are almost always at different scales and will depict different site features. Thus, the forensic expert may have to replot and/or compile composite map(s) from different consultant reports. It is not uncommon to find a building, tank, sump, and/or storage area inconsistently located when comparing two or three different consultant reports. In a recent site investigation, some 17 monitoring wells were plotted on a site map, but when the surveyor's log was reviewed it was found that all of the wells had been misplotted.

Geotechnical Evaluation

The initial review of the geotechnical documents showed that (1) the three USTs had been successfully removed (although no attempt had been made to determine whether there were leaks in the tanks or piping), (2) the tanks had contained MIBK, (3) the underlying soil (which was excavated along with the tank removal) was contaminated with both MIBK and benzene, and (4) the underlying groundwater was contaminated only with MIBK. These investigations also included the additional soil samples (see Figure 17.2), as well as the placement of additional monitoring wells (see Figure 17.3). The results of these additional tests are shown in Tables 17.3 and 17.4.

According to the CEHD, most of the contaminated soils were removed upon removal of the USTs, and this was confirmed by the new soil data (Table 17.3). Therefore, the only meaningful site-related contamination that remained was in the groundwater, and was potentially subject to remediation.

Although the groundwater was contaminated, a review of the available local and regional groundwater data suggested that the groundwater at this site might not have to be remediated based on nonattainment policies (see Sidebar 17.2 on the application of groundwater nonattainment eval-

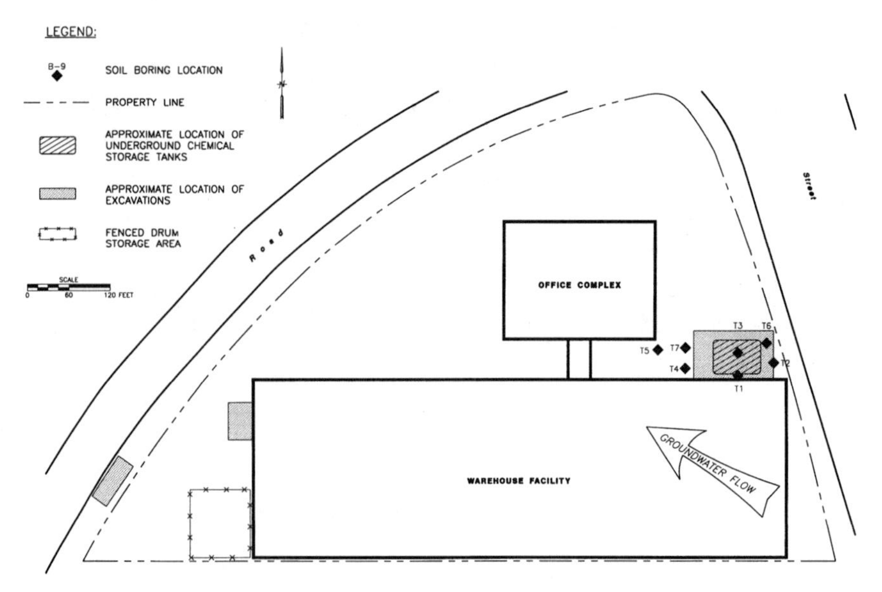

Figure 17.2 Additional soil boring locations.

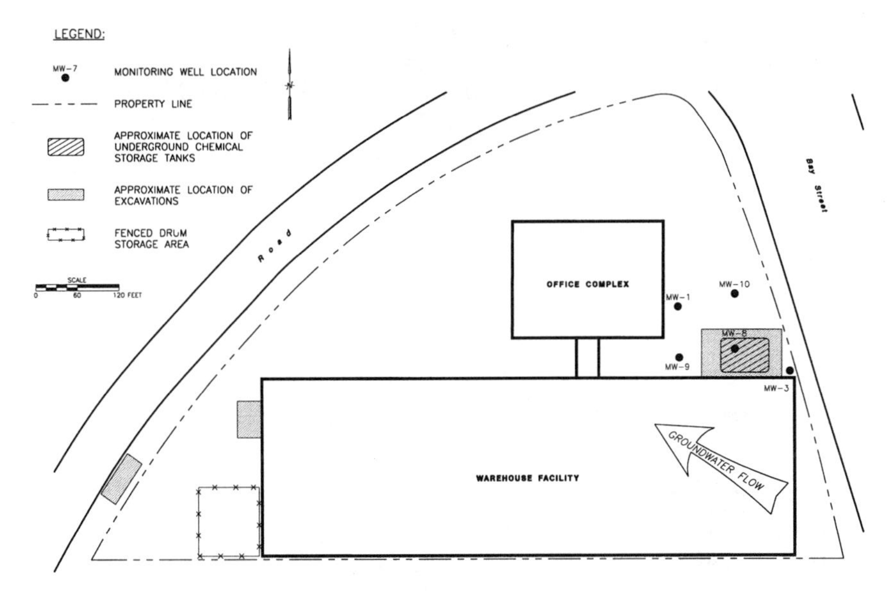

Figure 17.3 Additional groundwater monitoring wells.

uations). Thus, it was necessary to demonstrate to the CEHD that the groundwater contamination at the site was not a threat to human health or the environment. As part of the demonstration, groundwater data were collected and tabulated to determine both the fate and the transport of the MIBK. The collected data are found in Table 17.5.

In support of this position, the following site characteristics and facts were presented to the CEHD in order to obtain a nonattainment designation for the groundwater contamination:

Table 17.3 New Soil Data

Location	*Depth (ft)*	*MIBK (mg/kg)*
T1	8.0	ND
T2	6.0	ND
	8.5	ND
T3	8.0	ND
T4	9.0	10
T5	5.0	ND
	9.0	ND
T6	7.5	6
T7	7.5	ND

Table 17.4 New Groundwater Data

Location	*Depth (ft)*	*MIBK (ppb)*
MW-9	8.5	ND
MW-10	9.5	ND

- There was no explicit standard for the cleanup of MIBK.
- The MIBK-contaminated groundwater was at a sufficient depth to preclude any MIBK vapor phase accumulation in surface building structures (i.e., no risk associated with future land use development).
- The total dissolved-solids concentrations in the groundwater precluded the use of the regional groundwater for any beneficial use (i.e., domestic, agricultural, or industrial).
- Because of the depth of the groundwater, there was no actual or potential discharge to the ground surface.
- The estimated transport rate of the MIBK plume was on the order of 100 feet per year. At this rate of movement, it would take approximately 30 years to reach the nearest down-gradient irrigation well (assuming no MIBK decay). Based on the estimated half-life of MIBK[4] and the MIBK concentrations in the groundwater, it was predicted that the down-gradient irrigation well would not be impacted.
- The entire neighborhood was dominated by light industry and manufacturing, with no foreseeable land use changes predicted by the County Planning Agency and no areawide cleanup anticipated.

Based on this information, the CEHD, recognizing that the site characteristics fell within the county's nonattainment guidelines, was willing to accept a nonattainment proposal. As a result of this ruling, no groundwater cleanup was required. However, Chemical Consultants was ordered to conduct quarterly monitoring, and a deed restriction was placed on the property to ensure that there would be no residential land use.

[4]The published degradation rates for MIBK range from 1 day to 28 days, a very short half-life. Needless to say, the published information is, at best, an educated guess and should be used advisedly.

Table 17.5 Groundwater Measurements

Date	*Monitoring Well*	*MIBK (ppb)*
7/89	MW-1	ND
9/89	MW-1	90,000
	MW-3	ND
	MW-5	ND
	MW-6	ND
12/89	MW-1	110,000
12/90	MW-8	57,000
9/91	MW-8	150,000
6/93	MW-8	100,000
1/94	MW-8	840
4/94	MW-8	14,000
12/95	MW-8	8,000
	MW-10	11
	MW-1	74,000
6/96	MW-8	4,500
	MW-10	ND
	MW-8	1,500
12/96	MW-1	42,500
	MW-8	400

Chemical Use History

Although Unique Chemicals conceded that it had stored solvents in the USTs, it had no record of MIBK use. It also maintained that the USTs had all been drained and then filled with water prior to its leaving the site. Chemical Consultants contended that not only did it not know about the USTs, it never used MIBK in its operations. In the absence of chemical use records for either Unique Chemicals or Chemical Consultants regarding MIBK, only the analytical data available (the original drums of stored chemical wastes and the UST analyses) could be used, and therefore the weight of MIBK use appeared to lean toward Unique Chemicals simply because it had used the tanks. The lack of definitive chemical use records was a major disappointment. Based on the available information (or lack thereof), Unique Chemicals' forensic expert advised counsel that there was no factual information supporting an allocation of any of the MIBK contamination to Chemical Consultants. Therefore, it was recommended that Unique Chemicals attempt to settle the case.

Sidebar 17.2 Nonattainment Zones

The general concept of a nonattainment zone is based on the premise that, even though a groundwater resource is contaminated, it may not have to be remediated if the following conditions exist:

- The water resource is not a potable source of drinking water and there is no path of release that would pose a threat to the environment (i.e., groundwater discharge to the earth's surface into a body of water or as a seep).
- The water resource cannot be used for industrial or agricultural purposes (i.e., no beneficial use).
- The contaminant of concern will degrade naturally over time.
- Land use patterns in the contaminated area are not expected to significantly change in the long term.

Nonattainment policies provide both state and local regulatory agencies an acceptable remedial management option for polluted groundwater for those sites that are determined to have limited or no potential risk.

AN EXAGGERATED CLAIM

Putting aside the initial site investigation and related cleanup costs, by the completion of the UST removal, Chemical Consultants and/or its counsel determined that its damage claim was worth 5 million dollars. This was a ridiculous sum considering that (1) the land use was not expected to change (i.e., no real loss of property value), (2) there was only \$507,000 of claimed expenses, and (3) the remaining soil cleanup was minor and the groundwater would only be monitored.

Unfortunately, a great deal of posturing is a necessary part of the settlement process. As a result, the actual and projected costs tend to take on an exponential character. For example, based on alleged new soil data, Chemical Consultants' expert concluded that an additional 5000 cubic feet of soil had to be excavated. The projected cost of the excavation and disposal was between \$400,000 and \$500,000. This turn of events required

Unique Chemicals' forensic consultant to once again meet with the regulatory agency to discuss this newly proposed remediation. After a review of this so-called new data,[5] the county's project manager agreed that the existing soil data did not support a requirement for additional excavation and that capping the site would be sufficient to deal with the problem. He also agreed that if Chemical Consultants continued to insist that the excavation was necessary, he would have no objection and, indeed, may require Chemical Consultants to do so. Based on the county's response, it was pointed out to Chemical Consultants that if it did not prevail in its action, it would be responsible for the cost of the excavation and disposal since it proposed the excavation as being necessary. Within a week, Chemical Consultants informed the CEHD that it had made a "mistake," that the original soil excavation estimate was correct, and that no additional excavation would be required.[6]

Putting the proposed soil excavation and disposal issue aside, even by including projected groundwater monitoring costs,[7] there still remained a serious gap between Chemical Consultants' demand and Unique Chemicals' understanding of the expected remedial costs.

SETTLEMENT PREPARATION

With the site technical issues defined (i.e., the extent of soil remediation and projected monitoring requirements), the final issue to be resolved was the determination of the actual amount that Unique Chemicals should tender to Chemical Consultants. This evaluation required a two-step process. The first step consisted of matching invoices against payments to determine if supporting documents actually existed, while the second step was to determine if the expenses were appropriate.

Chemical Consultants claimed that its total expenses were $507,000, excluding legal fees, but could only produce invoices totaling $385,000. This left an undocumented claim of $122,000. With the invoice review completed, the next step was to determine if the claimed costs that could be documented were pertinent to the site remedial actions.

[5]In fact, there were no new analytical data. Chemical Consultants' expert simply assumed a need for additional excavation.

[6]Attempting to bluff your opponent into accepting a higher settlement amount is a common strategy. However, this settlement method should be used cautiously when regulatory agencies are involved in the process.

[7]Soil remediation and monitoring costs were estimated to be $30,000.

A further review of the claimed documented costs and the associated work activity showed that approximately $87,000 of the total $385,000 was related to the cleanup of Chemical Consultants' drum storage area. These costs had no connection to the costs associated with the soil cleanup and monitoring of the UST contamination (i.e., for which Unique Chemicals appeared to be responsible). Thus, going into the initial settlement meeting, it was the position of Unique Chemicals and its consultant that only $298,000 of the documented claim was valid.

THE FIRST SETTLEMENT MEETING

The demand presented by Chemical Consultants consisted of the fees paid to consultants, the costs associated with a senior member of Chemical Consultants overseeing the two-year work activity, the future soil and monitoring costs, and, lastly, but most interestingly, its legal fees. The numbers added up as follows:

Engineering costs	$510,000
Oversight (management)	125,000[8]
Future remediation	30,000
Legal fees	535,000[9]

As a result, Chemical Consultants' claim was now reduced from $5,000,000 to $1,200,000. Unique Chemicals took the position that, in spite of no absolute proof, Unique Chemicals would take responsibility for the MIBK-related contamination and, therefore, was willing to pay the documented expenses of $298,000, as well as the projected future remedial costs of $30,000, for a total of $328,000. This was considered unacceptable to Chemical Consultants. With the Unique Chemicals offer declined, a round of depositions was completed in which technical arguments and costs were defended by each party.

[8]The presentation by Chemical Consultants claimed that a vice president spent an average of 8 hours per week "overseeing" the work, 52 weeks per year for a period of 2 years. (Such commitment is admirable.)

[9]This is an astounding number. First, the legal fees exceeded the entire site investigation, remediation, and monitoring costs. Second, these fees also exceeded the combined legal and consulting fees incurred by Unique Chemicals.

Table 17.6 Comparison of Settlement Costs

Category	*Unique Chemicals*	*Chemical Consultants*
Litigation support	$ 65,000	$ 90,000
Legal fees	510,000	270,000
Total	$675,000	$360,000

A MORE REALISTIC SETTLEMENT

When the dust finally cleared, a settlement amount of $450,000 was agreed on.[10] Based on this amount, if indeed Chemical Consultants paid its engineering consultant his bill of $385,000, this would have left approximately $65,000 with which to settle its legal fees.

In summary, the original claim by Chemical Consultants began at $5,000,000, was decreased to $1,200,000, and was settled for $450,000. Unique Chemicals' forensic consultant fees over an 18-month work period amounted to $90,000. Unique Chemicals paid its counsel approximately $270,000. A comparison of expenditures for each party is illustrated in Table 17.6.

Chemical Consultants spent approximately $675,000 and Unique Chemicals spent approximately $810,000 ($450,000 in settlement and $360,000 in litigation) arguing over a $293,000 cleanup. In the aggregate, the excess cost exceeded the actual cleanup cost by $1,192,000! Although the forensic investigations resulted in a lower net cost to Unique Chemicals (i.e., approximately $400,000 less than the $1,200,000 claim), there is a need to apply more common sense and less advocacy when solving environmental allocation problems. Finally, although Chemical Consultants appeared to prevail, it only received a total of $450,000 against an expenditure of $968,000, a resultant short fall of approximately $518,000! So, who won?[11]

[10]The invoices clearly supported the engineering consultant's fee of $385,000. However, it was argued that the $125,000 "management oversight fee" was bogus, especially in light of the fact that the facility was out of production during this entire period and that the owner had retired. Finally, nowhere in the settlement meetings or agreements was there even a discussion of reimbursement of litigation costs.

[11]In this case where the remedial costs were fairly small, there is no clear winner. However, in multimillion-dollar cases, the consultant and legal fees are well justified.

LESSONS LEARNED

1. A company that does not know the environmental history of its property has little appreciation for its potential future liabilities.
2. Participation in the legal arena substantially increases nonremedial costs.
3. Regulatory agents do review work plans and do yield to well-framed and supported arguments.
4. Costs are not always what they appear to be.
5. Legal fees can, and sometimes do, stand in the way of a reasonable and expeditious settlement.

Chapter 18

Guilt by Association

In May 1993, Mountain Utility Company (MUC) began investigating soil and groundwater contamination from historic gas manufacturing practices at its facility. The results of this investigation showed an extensive amount of contaminated soil and groundwater. Based on the groundwater flow direction at the MUC site, it was determined that contamination was most likely migrating beneath the adjacent undeveloped property, which was owned by the county of Grass Hills. As a consequence, MUC installed five soil borings on the county property and collected one soil and five groundwater samples. The analytical data showed that both the soil and the groundwater were contaminated with polynuclear aromatic hydrocarbons (PAHs) and petroleum hydrocarbons.

Because the county's property had never had any industrial tenants that could have generated PAHs, the county blamed MUC for its contaminated property. However, MUC denied any responsibility for the contamination on the county's site. Furthermore, the utility confidently stated that "our name" cannot be associated with the contamination.

THE HISTORY OF MANUFACTURED GAS IN GRASS HILLS COUNTY

Operating History

Westfork Gas Company began gas manufacture from coal in approximately 1855. It produced about 10,000 cubic feet of gas a day. By 1886, there was a new 60,000-cubic-foot gas holder on the site. The gasworks consumed 80 tons of coal and 4 tons of lime a month, producing tar and

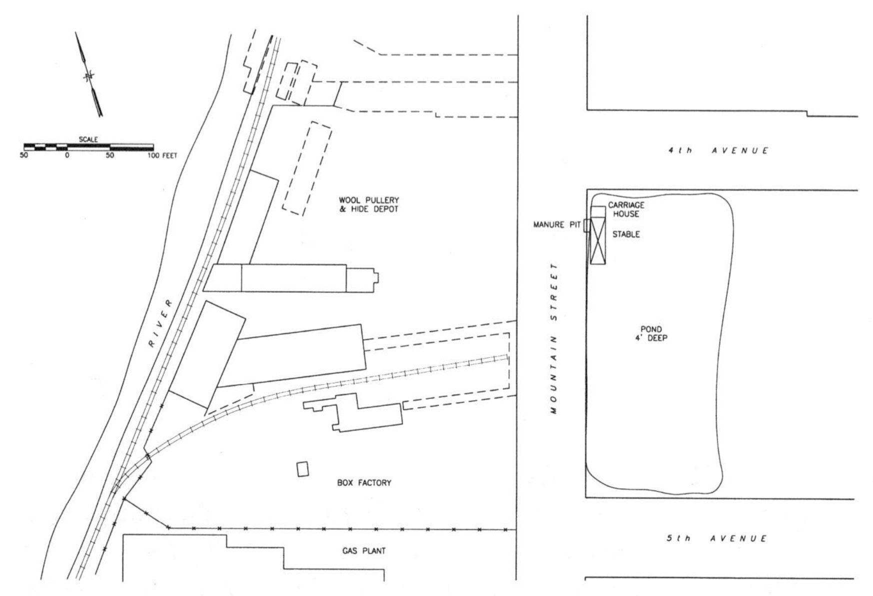

Figure 18.1 An 1895 map showing the relative locations of the gas plant to the pond.

coke as byproducts. Competition for Westfork Gas Company arrived in 1872 in the form of Coal Gas Light and Heating Company. In 1875, these companies merged to form Grass Hills Gas Light and Electric Company.

The Sanborn map of 1895 (see Figure 18.1) shows Grass Hills Gas Light and Electric Company (i.e., the gas plant) to the northwest of Mountain Street between 5th and 6th Avenues. To the northeast of Mountain Street between 4th and 5th Avenues is a pond.[1] By 1910, the pond had been filled in and the area was used as a lumberyard (see Figure 18.2).

The Grass Hills Gas Light and Electric Company plant remained exclusively a coal gas facility until it was rebuilt in 1901. At that time, the plant was converted to a Lowe-type oil gas generator. In 1920, the Grass Hills Gas Light and Electric Company facility was purchased by Mountain Utility Company. Gas production continued until 1930 when manufactured-gas production ended. The gas plant was converted to a natural-gas distribution center until it was closed in 1985. With the end of all gas-related operations, MUC continued to use the property as a maintenance and service facility.

[1]The county's property is approximately located in the same vicinity as the pond. The property boundary is shown in Figure 18.5.

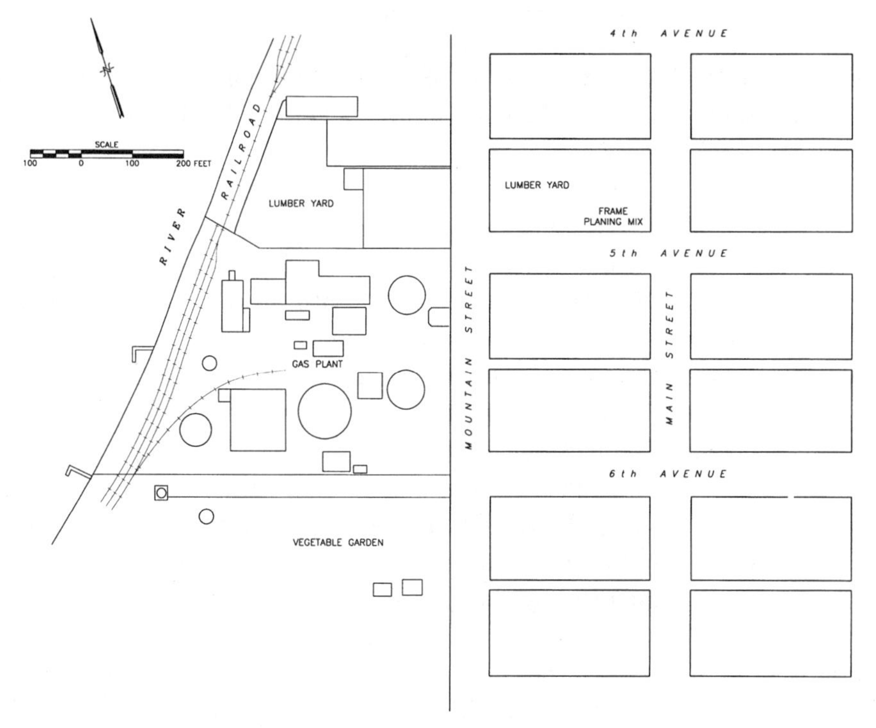

Figure 18.2 A 1910 map of the gas plant and lumberyard.

Waste from Coal and Oil Gas Manufacturing

The thermal conversion of both coal and oil fossil fuels into "manufactured gas" generates a vast array of chemical byproducts. The most easily identifiable byproducts are coal tar, coke, lampblack (powdered carbon), wood shavings from tar scrubbers, and bright blue masses of ferric ferrocyanides from gas purification. Of these materials, it is the black odorous coal tar that has been mainly responsible for soil and groundwater contamination.[2] Coal tars are a unique combination of polynuclear aromatic hydrocarbons (PAHs),[3] benzene, toluene, ethylbenzene, xylenes (BTEX), cresols, and phenols.

[2]Petroleum hydrocarbon (i.e., gasoline, diesel fuel, and heavy oil fractions) feedstocks and coal tar–contaminated lampblack are also common contaminants at manufactured-gas sites.

[3]These common compounds are acenaphthene, acenaphthylene, anthracene, benzo[*a*]anthracene, benzo[*a*]pyrene, benzo[*b*]fluoranthene, benzo[*g,h,i*]perylene, benzo[*k*]fluoranthene, chrysene, dibenzo[*a,h*]anthracene, fluoranthene, fluorene, indeno[1,2,3*c,d*]pyrene, naphthalene, phenanthrene, and pyrene.

THE COUNTY'S SITE INVESTIGATIONS

From June 1993 through October 1996, the county completed several site investigations. Figure 18.3 shows an overlay of the 1895 and 1910 Sanborn maps with the location of soil boring and groundwater monitoring wells. Figure 18.4 shows a geologic cross section along the northeast side of the site. This cross section shows a predominately sandy aquifer overlain by a silty clay. On average, the depth of the former pond was approximately 10 feet below the ground surface with an estimated maximum depth of 17 feet. Directly above the silty clay were black (contaminated with coal tar and petroleum) to brown wood chips and fill. The fill consisted of sandy to clay soil and debris consisting of broken glass, brick, and wood.

Based on the site investigation data, the State Environmental Protection Agency (SEPA) began to evaluate remedial options for the county's property. During this process, MUC made a technical presentation on the sitewide contamination to both SEPA and the county. The presentation made a detailed chemical-by-chemical comparison of the contamination[4] on both sites. MUC contended that its analysis of the data demonstrated that the contaminant on the county's property were not from the utility's manufactured-gas operations. MUC gave no explanation for the origin of the contamination on the county's property other than suggesting possible oil/fuel spills and the fact that the city of Grass Hills operated, during the last turn of the century, an incinerator for animal carcasses near the county's site. In addition, MUC contended that the wood chips were from the lumberyard operations (see Figure 18.2).

Based on the site data, SEPA ordered the county to remove all of the contaminated fill and soil (to groundwater) and to implement a soil vapor extraction and treatment system. As a result, the county filed a suit against MUC in order to recover its remedial costs. The county then retained a forensic expert to evaluate MUC's claim that the contamination on the county's site was not from gas manufacturing practices.

THE FORENSIC INVESTIGATION

A complete review of the analytical data collected on the county's site revealed the following characteristics:

[4]The analysis was based on both concentration and specific chemical associations.

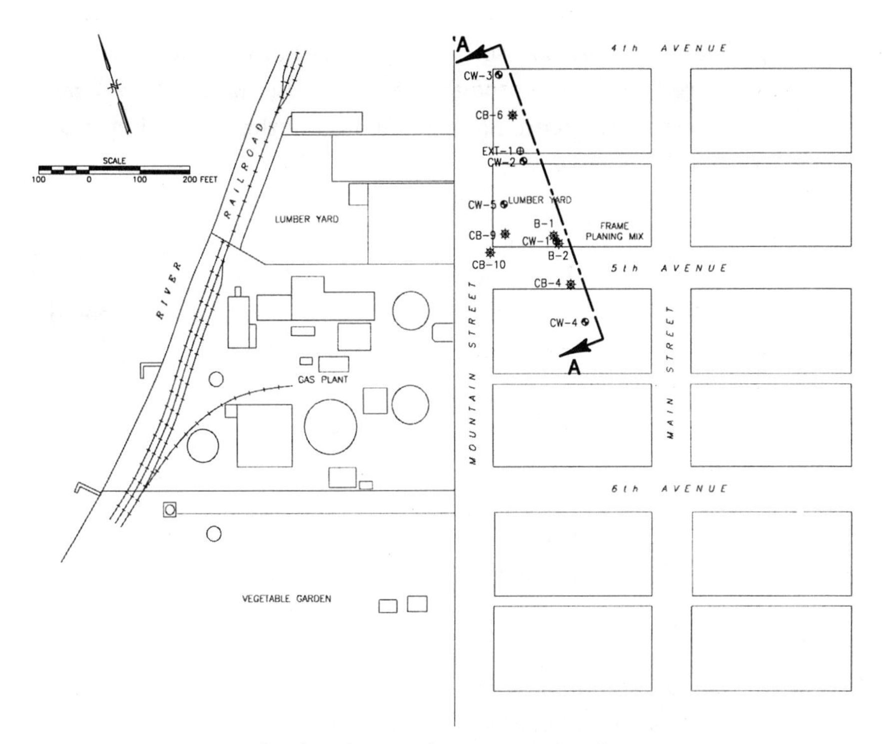

Figure 18.3 A map of soil and groundwater sampling locations.

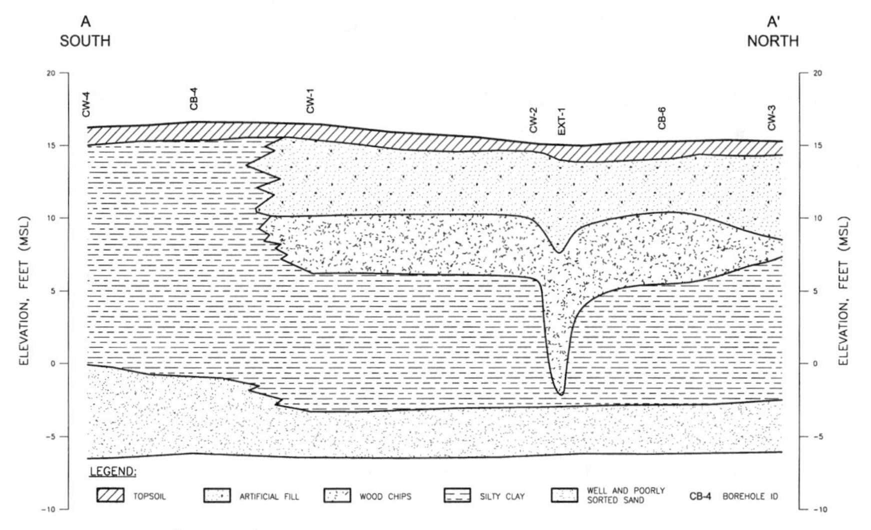

Figure 18.4 The pond cross section.

- The geologist logging the soil borings was not experienced with manufactured gas waste products. As a result, manufactured gas waste characteristics were not documented in the soil borings (except for the wood chips[5]).
- All of the PAH and petroleum hydrocarbon contamination was near the fill/pond bottom boundary.
- Soil samples taken within the surface fill (0–5 feet deep) were not contaminated. Thus, the mass of the subsurface contamination could not have originated from a surface release of PAHs or petroleum hydrocarbons.
- There were no analyses for cyanide, which is normally associated with purification waste (e.g., wood chips).

All of this information, when combined with the Sanborn maps, strongly suggested that prior to 1910 both coal tars and petroleum hydrocarbons were dumped into the pond and backfilled with wood chips and fill/debris[6] from the MUC site. This same information failed to indicate the presence of other identifiable manufactured gas waste (e.g., lampblack, coke, coal, ferrocyanides) being on the county's property.

Because remediation of the county's site was scheduled to begin within several months, the forensic consultant suggested that the site be sampled (prior to the excavation in order to preserve the stratification of the pond) to (1) collect evidence of manufactured-gas waste and (2) fill in some of the analytical data gaps. The field study was approved and implemented in the summer of 1998.

Nineteen soil borings were installed at the site. During their installation, numerous borings were found to contain extremely fine noncohesive black masses. Several fill samples also contained bright blue agglomerations. Selected fill and soil samples were characterized by a process mineralogist, as well as analyzed for relevant chemical compounds.

The Process Mineralogy

The microscopic analysis of the nine samples submitted for characterization contained a mix of soil minerals, coal, coke, lampblack, hydrocarbon combustion slag/ash with iron oxidation phases, lime-related residuals,

[5]Which MUC argued were lumberyard waste (see Sanborn maps).

[6]Several brick buildings were razed on the MUC site sometime between 1906 and 1908.

brick, and wood chips. The distribution of manufactured-gas waste and byproducts in each sample is given in Table 18.1.

Lampblack Characteristics

The material corresponding to lampblack carbon was reliably identified by high-power optical microscopy and chemical analyses. The lampblack carbon had the following two modes of occurrence in the samples:

- Ultrafine (<20–<1 μg) dust particulates occurring as coatings, pore space fillings, and microagglomerations with coal, coke, slag, ash, lime residuals, spent liner materials (containing mullite, an aluminum silicate inherent in brick constituents), and wood chips were disseminated throughout the silty soil minerals.
- Coarse-grained agglomerations up to 2 cm in diameter were also found in the samples.

Chemical analyses of the samples showed that all of the carbon present was organic carbon with a specific gravity of 1.23 g/cm^3 (extremely close to typical lampblack).

Wood Chips

The wood chip sample was soaked with coal tar/petroleum hydrocarbons with attachment of particulates (coal, coke, lampblack, and slag/ash). Many of the wood chips were either coated or completely permeated by an intensely blue ferrocyanide pigment.

Radiogenic Carbon Dating

Because MUC argued that the materials on the county's site could have come from the incineration of animal carcasses near the last turn of the century, two samples with a high organic carbon content were carbon-dated. The analysis showed that over 99% of the carbon in the two samples was of geological age.[7] The first sample was 39,590+5800/−3300 C-14 years B.P. (C-13 corrected), while the second was 41,580+9870/−4300 C-14 years B.P. (C-13 corrected). These data clearly establish that the waste materials were not of recent origin.

[7]The measured ages for the carbon are close to the maximum ages (i.e., about 50,000 years) that can be measured confidently with this method.

Table 18.1 Microscopic Modal Analysis of Manufactured-Gas Waste Constituents (Concentration in vol %)

Sample	*Coke*	*Coal*	*Lampblack*	*Wood Chips*	*Slag/Ash*	*Lime Residue*
1	14.6	3.4	5.5	1.5	51.0	24.0
2	3.6	2.0	10.1	34.0	12.0	5.0
3	3.0	3.0	2.0	24.0	30.6	1.0
4	30.2	4.3	0.4	2.0	42.9	1.1
5	2.0	1.2	0.2	18.1	6.0	2.0
6	4.9	2.2	1.4	4.6	26.3	0.2
7	5.0	2.0	0.5	30.3	13.2	—
8	5.0	—	46.0	—	22.0	27.0
9	0.5	—	0.1	91.4	4.2	—

The Chemical Analyses

With the completion of the soil sampling, the individual chemical constituents were compared with the chemical data from the MUC site. Based on the manufactured-gas process, there were two basic chemical groups that were selected for comparison between the sites.

Petroleum Hydrocarbons

In the early 1900s, MUC used petroleum crude oil as the raw feedstock for the manufactured gas. Thus, it would be expected to find these compounds on the MUCs site. In environmental studies, when petroleum crudes are analyzed today, the two common individual analytical protocols identify the following hydrocarbon compounds:

Number of Carbon Atoms	*Chemical Classification*
<10	Total petroleum hydrocarbons as gasoline
10–40	Total petroleum hydrocarbons as diesel fuel

Thus, the occurrence of petroleum hydrocarbons can be identified by determining the concentrations of total petroleum hydrocarbons as gasoline (TPHg) and total petroleum hydrocarbons as diesel fuel (TPHd).

All of the TPHg and TPHd data[8] were compiled for each site and their distributions determined. The frequency histograms for all of the data sets were not normal distributions. The log distributions, however, were normally distributed. Using the log distributions, the mean and maximum concentrations of the TPHg and TPHd are compared in Table 18.2. Using a T-test comparison of the means (95% confidence interval), no significant difference was found between the county and the MUC TPHg means and TPHd means.

These data demonstrated that the mean TPHg and TPHd concentrations are greater on the MUC property. This would be anticipated since leaks and spills associated with the manufactured-gas process would normally occur on site. Thus, it is not unexpected that the disposal of petroleum-polluted soils from the MUC site to the county's property would yield similar distributions (i.e., no significant difference).

[8]In parts per million (ppm).

Table 18.2 Total Petroleum Hydrocarbon Data Comparison (ppm)

Constituent	*Significant Difference In Means*	*County*		*MUC*	
		Mean	*Maximum*	*Mean*	*Maximum*
TPHg	No	178.4	18,000	284.9	8,070
TPHd	No	138.5	5,000	716.6	77,000

PAHs and BETX Compounds

As with the petroleum hydrocarbons, the frequency histograms for all of the PAH and BETX data sets were not normal distributions. The log distributions, however, were normally distributed. Using the log distributions, the mean and maximum concentrations of the PAH and BTEX compounds detected on both sites are compared in Table 18.3. Using a *T*-test comparison of the means (95% confidence interval), there was a significant difference between the county and the MUC PAH means and BTEX means.

These data show that the mean PAH and BTEX concentrations were greater on the county's property. Thus, the significant difference between the means and the higher concentrations on the county's property was expected since the manufactured-gas byproducts were intentionally dumped into the pond (i.e., high concentrations).

Cyanide Analysis

No cyanide analyses were completed on the MUC property. As a result, there was no basis to compare the data collected on the county's site. However, the bright blue agglomerations that were collected from the county's site contained in excess of 1000 ppm cyanide.

Summary

An illustration of all of the soil boring data (i.e., mineralogy and chemical analyses) is given in Figure 18.5. Both the process mineralogy and the chemical characteristics of the sites clearly demonstrate that the county's property was used by MUC for the disposal of its manufactured-gas byproducts. Because the forensic investigation was successful, no further sampling was required and the remediation of the county's property began.

Table 18.3 PAH and BTEX Data Comparison (ppm)

Constituent	*Significant Difference In Means*	*County*		*MUC*	
		Mean	*Maximum*	*Mean*	*Maximum*
Acenaphthene	Yes	29.6	1,680	1.05	1,600
Acenaphthylene	Yes	19.2	1,500	5.01	2,000
Anthracene	Yes	23.5	1,300	0.93	790
Benzo[*a*]anthracene	Yes	20.9	1,060	2.57	1,100
Benzo[*a*]pyrene	Yes	17.6	915	3.52	2,000
Benzo[*b*]fluoranthene	Yes	14.9	948	1.59	1,300
Benzo[*g,h,i*]perylene	Yes	10.6	337	1.22	6,600
Benzo[*k*]fluoranthene	Yes	12.2	453	1.26	560
Chrysene	Yes	18.2	960	1.84	1,400
Fluoranthene	Yes	48.0	1,740	3.51	6,000
Fluorene	Yes	38.0	1,510	2.92	1,320
Indeno[1,2,3-*c,d*]pyrene	Yes	38.4	246	1.55	1,600
Naphthalene	Yes	122.8	13,200	1.43	32,000
Phenanthrene	Yes	83.7	4,560	1.25	7,400
Pyrene	Yes	46.5	1,830	2.92	7,800
Benzene	Yes	2.47	350	0.31	203
Toluene	Yes	5.91	650	0.31	316
Ethylbenzene	Yes	3.74	750	0.17	501
Xylenes	Yes	9.72	1,200	0.47	1,140

AN ARCHEOLOGY INVESTIGATION

The excavation and removal of the contaminated PAH and BTEX wood chips and fill provided an opportunity to collect any artifacts found in the pond fill. When an artifact was found, it was placed in a plastic bucket and

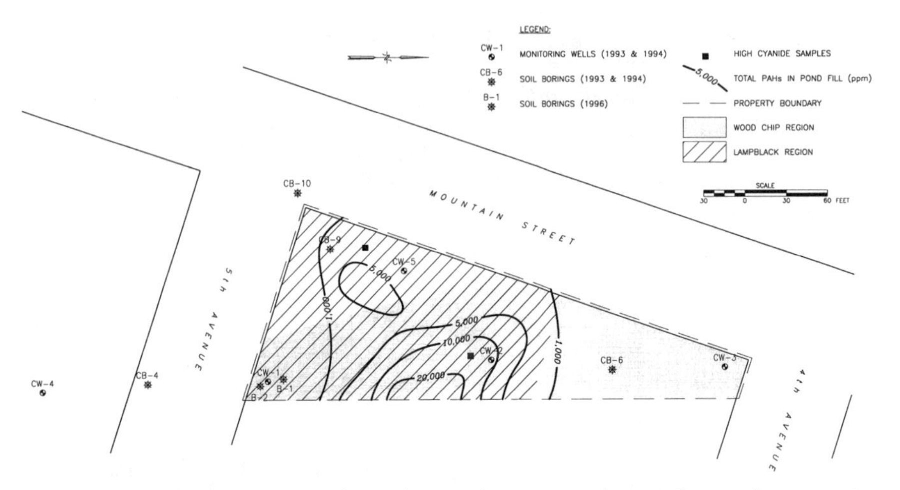

Figure 18.5 County property boundary with associated manufactured-gas waste and contamination distribution.

saved for future inspection. With the completion of the excavation, all of the recovered artifacts were displayed by the county.

On the day of the scheduled inspection, all of the interested parties (representatives of MUC and the county) assembled under a viewing tent. Each individually numbered item was placed on brown paper, with its corresponding number written on the paper, and photographed. In the morning session, waste samples and historic artifacts were examined. Some items that were presented included:

- Telephone in a wooden box
- Brass railroad lantern
- Ceramic and glass bottles (with cork/metal stoppers)
- Bricks
- Buggy parts and spring
- Metal gears
- Pitch fork with square tines

Photographs of some of these artifacts are shown in Figure 18.6. In the afternoon session, more brick, broken glass, and bottles were shown. Based on the nature of the artifacts, it was clear that the waste disposed of into the pond came from the late 1800s to early 1900s.

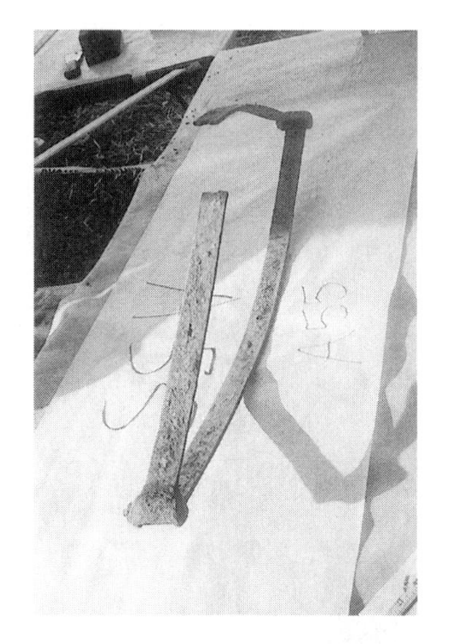

Figure 18.6 Artifact examples showing broken bricks, ceramic beer bottles, and a buggy spring.

Near the end of the afternoon inspection, three slightly crushed gas meters were "unveiled" (see Figure 18.7). Upon examination, one gas meter was identified as a "light meter" manufactured by American Meter Company of Philadelphia. This was conclusive proof that the waste pond contained both chemical waste and debris from the historic operations of the manufactured-gas plant. Best of all, a small brass plate riveted to the "light meter" proclaimed the owner to be Grass Hills Gas Light and Electric Company.

Figure 18.7 A gas meter and identification plates.

LESSONS LEARNED

1. Data collected for remedial investigations are not neccessarily specific enough to assist in the absolute identification of a responsible party.
2. Utility experts argue that coal tar is a byproduct and that it is "always" sold and never dumped as a waste. This site demonstrates that at least one manufactured gas producer did dump its "byproduct."

Chapter 19

An Issue of Disclosure

In the years following passage of the Resource Conservation and Recovery Act (RCRA) in 1976, the U.S. Environmental Protection Agency (EPA) developed regulations that were intended to implement the various provisions of this act. One of these provisions resulted in the development of a new type of insurance—environmental impairment liability (EIL) insurance. What made this insurance different from the traditional commercial general liability (CGL) insurance was that it was designed to fill the gap left open in CGL insurance, namely, coverage for damages resulting from gradual pollution events. However, it is important to recognize, from a coverage perspective, that many of the same exclusions (found in CGL policies) otherwise still applied and, as such, only one of many "gaps" was intended to be filled.

When EIL coverage was first offered by the insurance industry in the early 1980s, many industrial companies applied for this coverage. As part of the EIL application, a company was required to disclose (1) any known soil or groundwater pollution and (2) any environmental regulatory action that had or would result in remedial actions (in other words, a "claim" according to insurance terminology).[1]

When a company that had an EIL policy became aware of pollution on its property that occurred from what the company deemed to be a gradual release of the pollutants, it filed a claim with its EIL insurance company to cover the cost of any required cleanup of the resulting pollution. Depending on the nature of the pollution, the insurance company often initiated a

[1]This is equivalent to the requirement that when applying for medical insurance all prior medical conditions must be disclosed.

forensic investigation, relative to the insured's application, to determine if the applicant had any prior knowledge of the pollution. The case presented in this chapter describes a forensic investigation of an EIL claim that contrasts the issue of "disclosure" relative to the insured's "prior knowledge."

CLAIM BACKGROUND

An electronics manufacturer, Hi-Tech Industries, utilized numerous hazardous substances, including trichloroethylene (TCE), which was used to clean electronic components. With the implementation of RCRA, in 1980 Hi-Tech Industries filed a Part A application with the EPA for an interim hazardous waste management permit. This permit was necessary since Hi-Tech generated hazardous waste as defined by the RCRA statutes.

Hi-Tech Industries had applied for EIL insurance in 1982 and received its policy in 1983. As part of its application, a 1983 environmental survey of the facilities was prepared by an outside consultant.[2] This 1983 survey described waste handling and disposal practices at the facility and concluded that "this site complies with all U.S. EPA permit regulations." The role of these environmental survey companies is discussed in detail in Sidebar 19.1.

A Potential Claim

In the summer of 1982, the county health department identified TCE in several of the county's water wells, necessitating closure of the wells, and so notified the State Department of Environmental Quality (SDEQ). Following up on the reported contamination, the SDEQ sent a notice of investigation letter to several manufacturers in the area, including Hi-Tech. As further investigative work was carried out by both the county and the SDEQ, a regional TCE groundwater plume was delineated in the spring of 1984, with Hi-Tech being within the footprint. At this point, Hi-Tech initiated a "voluntary" site investigation to determine if it was a source of the TCE groundwater contamination. Because Hi-Tech considered the cost of the investigation (and potential remediation) as a potential claim, it informed its EIL carrier of the SDEQ's notice (failing to mention its voluntary action).

[2]Coincidentally, after the policy was issued, another facility report was prepared (in 1984), which concluded that there was a moderate probability that groundwater under the site had become contaminated from past site operations.

Side-Bar 19.1 EIL Environmental Engineering Surveys

In the early days of EIL insurance, a number of firms began businesses to provide facility evaluation reports conducted by qualified engineers. It was assumed that these reports would represent a comprehensive assessment of a facility's hazardous waste management and chemical use practices, and that this would be accomplished by conducting a site inspection, reviewing facility documents, and interviewing facility personnel. Based on this approach, it was the intent of the report to ascertain and evaluate potential environmental risks from chemical pollution.

In concept, this was a good idea. The implementation, however, was less than satisfactory. The key investigator responsible for implementing the report structure and risk evaluations was usually a biologist with little, if any, engineering or manufacturing training or experience. To the biologist, the term environmental risk was synonymous with exposure. Thus, if a site could be potentially polluted as a result of hazardous waste or chemical use practices, so long as there were no biological organisms (i.e., fish in a stream or a cow in the field) beyond the property boundaries that could be exposed, the facility was deemed a "low risk." In other words, soil contamination and groundwater contamination (and the subsequent cost of remediation) were not considered or included as part of the environmental risk. Soil contamination and groundwater contamination were considered only in the context of how they impacted an exposure assessment.

To further simplify the procedure, the site inspections, document reviews, and employee interviews were conducted on a minimal budget (and there were no interviews with regulatory agencies). Employee statements were taken at face value with no follow-up, and the final reports incorporated standard report boiler plate, so as to minimize costs. Even during site visits, when advised that certain portions of the facility were "off limits," this was deemed to be okay by the investigator. These practices, combined with exposure as a guiding light, resulted in virtually every EIL environmental assessment survey awarding the applicant's facility a rating of low risk. The old adage "you get what you pay for" was never more truly demonstrated.

The Initial Site Investigation

In 1984, Hi-Tech Industries retained the services of GeoExperts to begin the site investigation. The investigation was to focus on a former unlined pond and a stormwater lagoon. Soil samples from several borings in the pond area indicated TCE concentrations below 50 parts per billion (ppb). During its investigation, GeoExperts discovered that a number of dry wells[3] had been used at the site for the disposal of wastewater. As a result, a more extensive soil and groundwater investigation was conducted. With this additional phase completed, GeoExperts concluded that the primary source of TCE groundwater contamination beneath the site originated from the unlined pond.

Upon reflection, one has to wonder why it took approximately six years to complete (1990) these additional investigations, at a cost of some $2 million. Unfortunately, this is not an uncommon occurrence as one investigation begets another and another.

Remedial Requirements

Based on GeoExperts' investigations, the SDEQ required that Hi-Tech Industries install a groundwater pump-and-treat system. This system was estimated to cost $7.5 million, based on a projected operating period of 30 years.[4] During this period, Hi-Tech continued to inform the state that it was complying in a "voluntary" manner. However, Hi-Tech Industries submitted a claim to its EIL carrier for the estimated cost of the remediation. When the insurance company reviewed the claim file, it determined that Hi-Tech Industries may have known about the release of TCE from the unlined pond and the groundwater contamination before the effective date of its EIL insurance policy. Consequently, it refused to pay Hi-Tech's claim.

THE LITIGATION

Hi-Tech Industries sued its EIL insurance company for its anticipated remedial costs. The insurance company's counsel retained a forensic consultant to help determine when the TCE release occurred and to what extent Hi-Tech Industries had been aware of the release and potential contamination.

[3]A dry well is a well installed into permeable soil that is not saturated (i.e., is above groundwater).
[4]This estimate was based on both capital and operation/maintenance costs.

The Forensic Investigation

The first step of the investigation was to visit the SDEQ to review all documents relevant to the site. This initial visit took place in 1998. The site file documents showed that Hi-Tech Industries had purchased the site and manufacturing operations from Electrical Components Corporation (ECC) in 1969 and that ECC had been on the site since the mid-1950s. These SDEQ files also revealed that, in the summer of 1982, the city had notified the SDEQ that several of its wells were shown to contain TCE and that these wells were only three-quarters of a mile down gradient from the industrial park (occupied by Hi-Tech, as well as several other electronic component manufacturing facilities). The files also revealed that the SDEQ had notified Hi-Tech, in writing, of the well contamination.

The forensic consultant discussed this information with the insurance company counsel, and the result was that over a period of months the consultant received a series of document productions, each consisting of four or five cartons of site documents that had been obtained from Hi-Tech Industries. These document productions, not unexpectedly, were both disorganized and duplicative, so the first task was to sort through this morass of documents, organize the documents, isolate the key documents, and organize them in chronological order.

All of the site documents, no matter how mundane looking, were reviewed by technical professionals familiar with the handling and disposal of hazardous wastes.[5] The key documents identified by this review were then organized and combined with the SDEQ documents into a chronological annotated timeline. These key documents were reviewed to determine the historical site activities and establish the occurrence of any site contamination of which Hi-Tech Industries may have known.

The Environmental Site History

One of the more interesting documents concerned a study, commissioned by ECC in 1969, of the facility's liquid waste disposal practices. This study found that wastewaters from the plant were contaminated with numerous chemicals, heavy-metal ions, and poisons. It noted that these waste-

[5]Review of site documents by a technical expert is critical to any determination of contamination and site history. In all too many cases, lawyers have tried to cut corners by having nontechnical staff do this initial document review. Although this may initially save money, in many cases the same documents will have to be reviewed again by technical personnel, or key documents may be overlooked.

waters were discharged to the previously described unlined waste pond. According to the report's author, "this pond creates several problems . . . [because] toxicants such as hexavalent chrome will seep into the ground where they can contaminate water supplies . . . [and] this body of water will inhibit future construction on this site."

About this same time, ECC acknowledged, in several internal technical documents, that TCE was being used as a degreaser in several of its manufacturing processes. These documents also stated that contaminated TCE solvent should be drained and discarded from facility degreasers on a biweekly basis. However, there was no discussion as to the method for disposal.

By 1970, ECC had sold the plant to Hi-Tech Industries. Documents revealed that Hi-Tech began phasing out the use of TCE shortly after acquiring the facility. As a result, TCE use ended by early 1972. Hi-Tech documents also revealed a detailed and well managed program to control the use and disposal of solvents (i.e., suggesting that ECC, not Hi-Tech, was the source of the on-site TCE).

In 1980, some two years before Hi-Tech Industries applied for EIL insurance, Hi-Tech's environmental manager had his staff collect soil and sludge samples from the pond and from an adjacent sump. For some unexplained reason (but most likely cost), the pond samples were only analyzed for heavy metals, while the sump samples were analyzed for both metals and volatile organic compounds (VOCs). The results showed that several VOCs, including TCE, were present in the sump samples. Heavy metals were present in both the pond and the sump.

Although Hi-Tech did not conduct any groundwater tests at that time, it did discuss the "pros and cons" of conducting on-site groundwater monitoring. However, Hi-Tech concluded that such monitoring would be costly and was not currently required by the SDEQ. At the same time, however, its site consultant did report to Hi-Tech that drilling logs indicated that the area around the pond was *not* underlain by impermeable materials and, thus, contaminants reaching the first aquifer would ultimately reach and enter local drinking water wells. Faced with these concerns, Hi-Tech Industries proceeded to "dry up" the waste pond, remove the contaminated soil and sludge on the bottom of the pond, and ship these materials for off-site disposal. However, it made no attempt to actually determine if the pond had contaminated the underlying groundwater. Hi-Tech did notify both the SDEQ and the regional office of the EPA regarding its removal actions at the former waste lagoon. Both the SDEQ and the EPA concurred with Hi-Tech's action.

In 1981, a Hi-Tech internal memo, which described the history of solvent use by both ECC and Hi-Tech, stated that, even though TCE was highly volatile, if "disposed of by dumping on the ground, a potential does exist for groundwater contamination." Yet, again, Hi-Tech chose *not* to determine if such contamination had occurred due to its or ECC's past use of TCE.

Placing all of this information into a time perspective, Hi-Tech began site investigations in 1980, was warned of possible groundwater contamination in 1981, read (in the local press) that the county found trace amounts of TCE in several public water supply wells in the summer of 1982, and yet, in spite of this knowledge, just six months later in December 1982, Hi-Tech applied for EIL insurance. In its application, Hi-Tech assured the insurance company that it was not aware of any current contamination problems at the site. After all, contaminated soils and sludges associated with the waste pond had already been removed and disposed of off site.

A subsequent review of the title transfer documents between Hi-Tech and ECC showed that Hi-Tech had given a complete warranty to ECC, thereby "presumably" assuming all responsibility for subsequent contamination. Interestingly, once Hi-Tech received notice from the SDEQ in 1984, Hi-Tech did discuss the liability issue with ECC and, in fact, notified ECC of the situation.

Based on these case documents, the following site environmental history was prepared for the insurance company's counsel:

- Hi-Tech Industries knew that ECC had used TCE during its 15 years on the site and that Hi-Tech Industries ceased using TCE within 2 years of acquiring the facility.
- Hi-Tech Industries was aware that ECC had probably discharged its wastes (containing TCE) to an unlined lagoon.
- Hi-Tech's environmental consultant informed Hi-Tech, as early as 1980, that soil and sludge from both the lagoon and the sump were contaminated with TCE. It also warned Hi-Tech that there was danger of groundwater contamination from the past disposal practices.
- The county notified the SDEQ and the local newspapers that wells down gradient from the industrial park were contaminated with TCE. Hi-Tech documents showed that the company was aware of this information in 1982 (i.e., prior to its application for EIL insurance).

Again, in spite of this knowledge, Hi-Tech Industries applied for its EIL insurance with clean hands and no alleged awareness of any actual or potential contamination at its facility. Based on this information, the insurance carrier felt that it was time to begin settlement.

SETTLEMENT NEGOTIATIONS

Hi-Tech had initially demanded that the insurance company pay for all its investigative and remedial costs (which by this point was in excess of $9 million). Armed with the environmental site history, counsel for the EIL insurance company attempted to bring all the relevant facts to the attention of company decision makers by arranging an informal mediation. Senior corporate counsel and risk management personnel for Hi-Tech Industries attended along with its outside counsel. The EIL insurer attended along with its environmental consultant.

The timeline prepared by the forensic consultant was graphically displayed and discussed by counsel. Each document used to support the "prior knowledge" presentation was produced for Hi-Tech's representatives. Counsel, in turn, explained the significance of this information and its impact on insurance coverage, the misrepresentation in the application and, thus, the grounds for recession. In a matter of several hours, chief general counsel for Hi-Tech openly criticized its outside counsel for not advising him of the "most relevant information," and then proceeded to discuss settlement of the case.

While Hi-Tech Industries did not concede misrepresentation,[6] it did agree to a nuisance value settlement in the range of 10% of its actual cost of remediation.[7]

LESSON LEARNED

A careful review of historical site documents by a competent technical expert, in most cases, will reveal evidence of nondisclosure, if it exists.

[6]Far be it for anyone to admit "guilt."

[7]Again, it is important to point out that, even when one's position is correct, there can be a considerable "ongoing cost" in trying to achieve the perfect settlement. Therefore, a settlement offer, modest as it may be, is still necessary.

Chapter 20

Settling a Claim

There are contaminated sites where the responsible party has been clearly established and has a legitimate insurance claim for the reimbursement of remedial costs. When this condition exists, the consultant to the insurance company should provide its client with an unbiased assessment of the case facts so that the client can attempt to achieve a reasonable and early cost settlement. The worst case scenario occurs when an early settlement is not attained and as such, over time, both legal fees and investigative costs ultimately inflict a greater "financial" pain.

A SMALL FAMILY BUSINESS

A small family business, located on an 8-acre parcel approximately one-half mile from a small harbor, manufactured coil windings since the 1950s. As with many operations during this time period, perchloroethylene (PCE), trichloroethylene (TCE), and 1,1,1-trichloroethane (TCA) were employed as solvents. The solvent handling practices, both storage and disposal, were accomplished with a minimum cost to the business. The fenced drum storage area had an earthen floor with no liquid containment structure to protect soil and groundwater from chemical leaks or spills,[1] while solvent waste from the facility's degreasers was disposed of into a series of four shallow dry wells.[2] The site and the facility location are shown in Figure 20.1. The locations of the storage (disposal) area and dry wells are shown in Figure 20.2.

[1]This area was referred to as the "disposal area" in subsequent remedial investigations.

[2]This is essentially an underground injection well that is above groundwater.

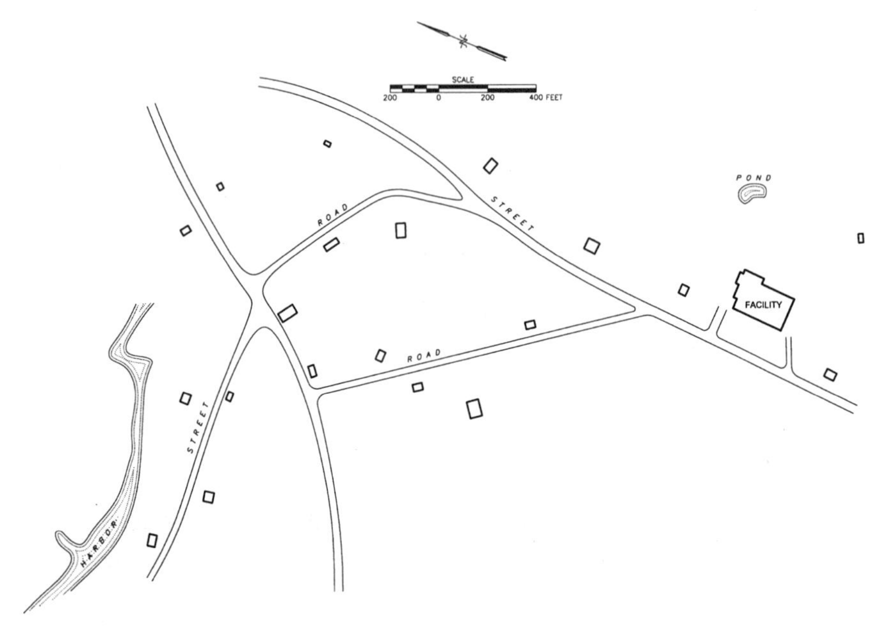

Figure 20.1 Site overview.

In the early 1970s, the small family business was sold to a large corporation, Electric Enterprises (EE), and continued to operate under the new management. Groundwater contaminated with PCE, TCE, and TCA was initially discovered in the vicinity of the EE facility in 1982, when a local well was tested by the County Health Department (CHD). A more comprehensive testing program, conducted by the CHD in 1984, confirmed a widespread pattern of chlorinated solvent contamination that had migrated into the nearby harbor. The monitoring well locations are shown in Figure 20.3.

The initial site investigation (undertaken by EE) began in 1988, and by 1992 it was determined that the plume had spread some 2800 feet down gradient and was now approximately 600 feet wide. Monitoring well locations, as well as the 1-ppb plume contour line, are shown in Figures 20.2 and 20.3. Along the plume axis were some 35 residences that depended on shallow wells (40–60 feet deep) for their water supply. There were no other businesses in the area that could have been the source of the chlorinated solvents in the groundwater.

Because of the threat to the groundwater users, the state required that the county provide all of the homes in the impacted residential neighbor-

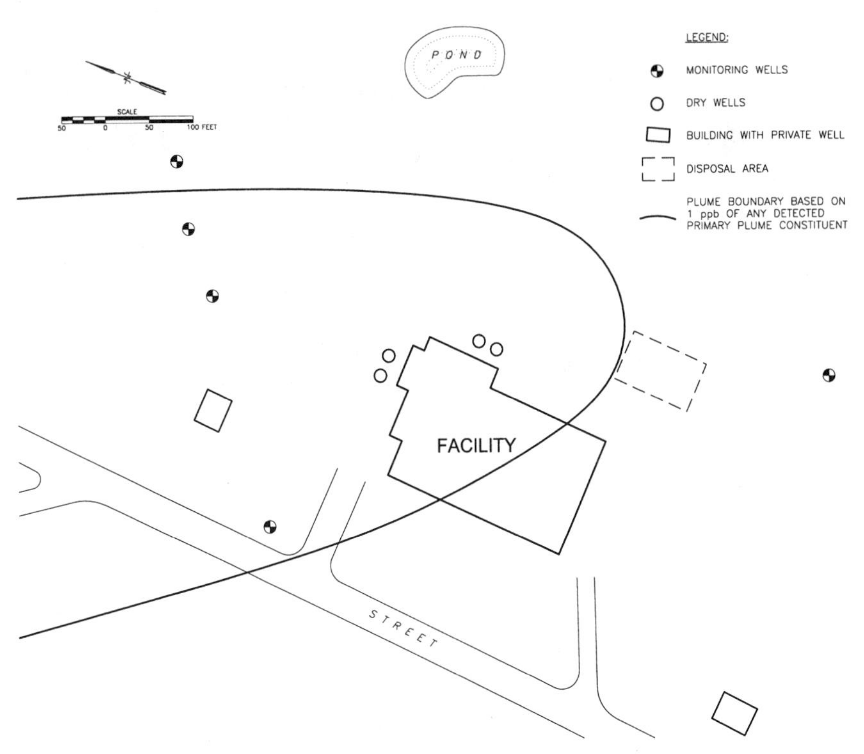

Figure 20.2 Dry wells and disposal area.

hood with municipal water service. Based on the existing data, the U.S. Environmental Protection Agency (EPA) ordered EE to carry out a full-scale remedial investigation. The first phase of the investigation was completed in early 1993; however, additional site studies were ordered by the EPA. By 1996, initial soil excavation and removal activities had taken place at the facility, with approximately 365 cubic yards of soil and sludge being removed from the area of the four dry wells. Finally, in 1997, the groundwater remediation plan was implemented. A total of ten groundwater extraction wells were developed, seven being on the site itself with three located down gradient on an adjacent property. It was anticipated that these wells would pump an average of 615 gallons per minute (gpm) from the underlying aquifer and would operate for approximately 20 years. Concurrently, the state established a requirement for a second soil excavation and on-site stabilization of 21,700 cubic yards of contaminated soil and sediment.

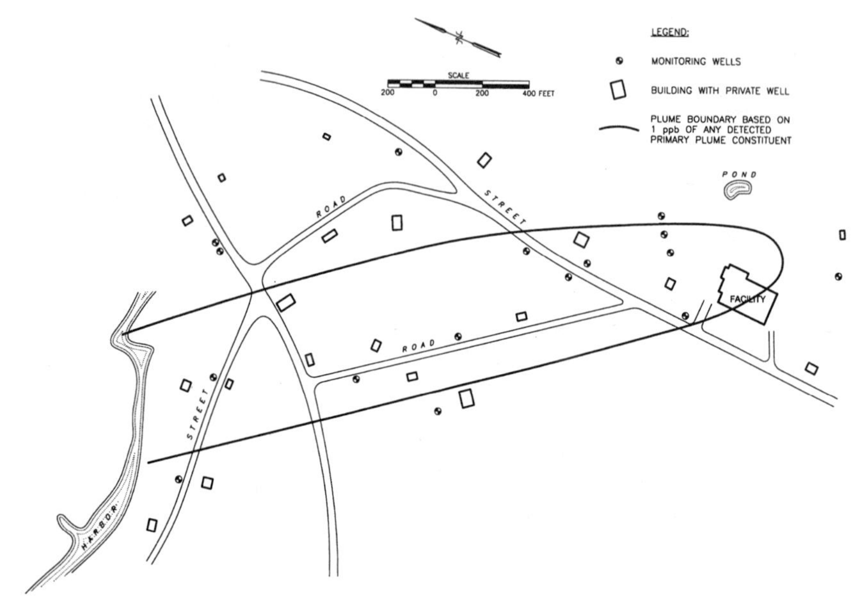

Figure 20.3 Monitoring wells and contaminant plume.

AN INSURANCE CLAIM

Electric Enterprises had obtained an EIL insurance policy for its facility, in 1981, in the amount of $14,000,000. As a result of the extensive and ongoing site investigative costs, EE filed a claim with its insurance company. Counsel for the insurance company subsequently retained a forensic expert to evaluate the claim.

Based on a review of all the remedial investigations to date, it was suspected by the forensic expert that the amount of solvent released to the groundwater environment, as well as the potential for a dense nonaqueous phase liquid (DNAPL) at the site, had not been evident to the site investigators. Clearly, the presence of a DNAPL would be critical in evaluating the effectiveness of any groundwater extraction and treatment system, as well as the time required to meet the anticipated remedial objectives. At this time, the site investigation was still ongoing and, thus, no Record of Decision (ROD) had been issued.

As a consequence, the forensic expert proposed to counsel that it would be advisable to model the plume in an attempt to verify the source characteristics. An expert hydrologist was retained to model the chlorinated hydrocarbon plume. The results of the modeling effort revealed that a DNAPL did exist and that there was upwards of 85,000 gallons of waste solvent in the source area around the dry wells. The model also demonstrated that the dry wells were indeed the source of the contamination. Given this volume of solvent in the aquifer, coupled with the relatively low solubilities of the VOCs, the initial 20-year cleanup estimate (assumed by the site environmental consultant and initially concurred with by the forensic expert) was now deemed (by the forensic expert) to be inappropriate. However, for purposes of settlement, the forensic expert suggested that cost analyses continue to be based on the 20-year projection for settlement purposes. The projected costs for both on-site and off-site remedial actions are shown in Table 20.1.

Table 20.1 The Settlement Analysis Projection

Cost Category	*On-Site Costs*	*Off-Site Costs*
Remedial design		
Predesign	$ 106,245	$ 25,089
Soil design	81,710	0
Groundwater design	221,504	110,752
Project administration	32,074	16,036
Subtotal	$ 441,533	$ 151,877
Remedial action		
Soil removal/treatment	$2,184,050	0
Groundwater (capital cost)	959,647	707,374
Project administration	277,997	74,669
Subtotal	$3,421,694	$ 782,043
Annual operation and maintenance (O&M)		
Groundwater	$ 243,610/year (Years 0–20)	$146,924/year (Years 0–13)
Subtotal	$4,872,200	$1,910,012
Additional groundwater actions		
Hot spot treatment	$ 150,000	0
Technical demonstration	50,000	0
Subtotal	$ 200,000	0
Grand total	$8,935,427	$2,843,932
Present worth	$6,725,532	$2,095,263[a]

[a] Present worth assumes that the annual groundwater O&M costs are subject to a 5% discount rate.

These projected groundwater remedial costs, when combined with the previous site investigations and soil remediation expenses ($3+ million), totaled approximately $12,000,000. In other words, if, indeed, the time period subsequently moved from 20 years to 30 years or longer, the potential remedial costs would most likely exceed the $14,000,000 EIL policy limit.

At the time this estimate was made, neither the state, the EPA, nor the remedial investigation consultant had raised the issue of a DNAPL. Based on the remedial investigation documents available to date, it appeared that the regulatory agencies assumed that the removal of the soil underlying the drum storage area, coupled with excavation of the dry-well areas, would eliminate the chlorinated hydrocarbon source and, therefore, the pump-and-treat system would deal with the residual groundwater contamination.

Given the fact that the remediation would probably never reach the anticipated groundwater cleanup goals, even in 30 years, and since the current remedial projections were for only 20 years, the forensic expert recommended to counsel that they settle the case as soon as possible.[3] As a result, the insurance company was prepared to make what it felt was a reasonable settlement offer ($4.0 million less the deductible, which would have covered all of the investigative costs to date) in the hope of heading off a protracted settlement process.

AN ATTEMPTED SETTLEMENT

By the time of the initial settlement meeting, EE's consultant, having completed his modeling effort, had also recognized the presence of a DNAPL. Thus, EE's position was that there would be an extended cleanup time at the facility, with the possibility that there was more than one "source of contamination."[4] As a result, EE refused the initial settlement offer (i.e., $4.0 million less the deductible).

A second meeting was called, and at that time a settlement based on the forensic expert's present value estimate of $8,820,795 was offered by the

[3]However, unknown to the forensic consultant and counsel for the EIL insurance company, the remedial consultant completed a groundwater model to verify the 20 year extraction and treatment program. The model indicated that the plume boundaries and concentrations showed little or no change (even after the soil and sludge sources were removed), demonstrating that a DNAPL continued to act as the source of the contamination.

[4]If the insured could prove that there was more than one source of groundwater contamination, the policy limits could be increased, based on the number of sources.

insurance company. Again, EE declined the offer, arguing that the potential policy coverage was in excess of its projected remedial cost[5] and that the true remedial costs were unknown and, thus, there was again no need to reach a settlement.

A third meeting proved to be the charm. The insurer's forensic expert explained that the data collected to date (by EE's consultant) did not support multiple sources of contamination at the site and, therefore, EE would not be able to prove its claim to multiple sources. As a result, there was only one policy limit at issue. Even then, the insurer pointed out that, without a settlement, and even if EE prevailed at trial, the subsequent payments would be spread out over the life of the remediation. Thus, it would be in the best interest of EE to accept, in settlement, a mutually agreed on present value of the projected cost with a subtraction of the deductible obligation, thereby saving perhaps millions in legal fees if they had to go to trial.

Electric Enterprises considered this proposal and was ready to accept. However, when EE conferred with its accountants and the Internal Revenue Service (IRS),[6] it was advised that if EE accepted the insurance proceeds in advance of actually paying for cleanup expenditures, the proceeds would be considered as taxable income at the time of receipt. This event again stalled the settlement.

Because of this impasse, both counsel, consultants, EE, and the insurance company began a joint process of resolving the settlement issues. Based on joint problem solving activities, the following resolution, and therefore settlement, was obtained:

- A procedure was developed for EE to demonstrate (to the IRS) that the significant investigative and remedial costs incurred to date (which had been spread out over many years) would now be addressed by the cash settlement, and thus a minimum tax consequence could be achieved.
- Based on site-specific data, the state and the EPA agreed to a reduced cleanup scope[7] of work.

[5]Electric Enterprises' consultant had not costed out the groundwater treatment for the full 30 years, so total costs were still present valued at a number under the maximum policy limit.

[6]The impact of a cash settlement, based on present valuing of future costs, on the company tax status had not been considered in previous settlements.

[7]At this point, both the state and the EPA recognized that the threat to the community was minimal, the impact on the harbor was de minimus, and the full cleanup of the DNAPL was "impossible."

- The agreed-on settlement amount would be paid in exchange for a release of the next policy year.

In the end, legal and financial analyses, coupled with consulting expertise, were essential in minimizing the actual costs incurred by the responsible party (EE) and, thus, by the insurer.

LESSONS LEARNED

1. All parties do not succumb to the "initial offer" regardless of how large (or fair) it appears to be.
2. Investigative data alone do not necessarily allow all parties to recognize the "extent" of a problem.
3. Regulatory agencies are not always "blind" to a reasonable technical argument.
4. The IRS can influence settlement negotiations.

Chapter 21

A Cost of Doing Business

After operating a sanitary landfill for approximately 25 years, the city of Quite Valley discovered extensive groundwater contamination under the landfill, as well as the migration of landfill gases onto adjacent properties. After extensive examination, the U.S. Environmental Protection Agency (EPA) required that the source of the groundwater contamination and gas migration be either eliminated or controlled. These remedial actions were projected to cost approximately $22,000,000. City officials determined that these costs were unanticipated business expenses that should be covered by their insurance and therefore filed a claim with their insurance carriers for the cost of the remedial work. The insurance companies denied the claim. This case describes the consequences of the lawsuit in which the city attempted to recover the unanticipated remedial cost of the landfill operation.

SOLID WASTE DISPOSAL

Open Dumps

The disposal of solid waste (garbage[1] and refuse[2]) from municipal and commercial sources has been an ever-increasing problem since the early 1900s. Solid wastes were commonly disposed of into open dumps or used to fill depressions in the earth (e.g., gravel pits, barrow pits, quarries, etc.). Waste disposed of in this fashion often led to fires, and also became a

[1]Organic waste.
[2]Paper, glass, metals, textiles, construction debris, etc.

habitat for disease-causing insects and vermin. Not to be overlooked was the odor, dust, and blowing paper associated with these open dumps.

In the early 1930s, several municipalities began disposing of waste using a more controlled and engineered approach referred to as "sanitary landfills." This involved covering the solid waste with at least 6 inches of soil at the end of each day's waste disposal activities. Using this approach, two methods of landfill were commonly employed and these are described in detail in the following section.

Sanitary Landfills

The first method involved excavating a trench below the ground surface and stockpiling the soil for covering the solid waste placed into the trench (called a "trench landfill"). Once all of the land allocated for trenching had been used, additional waste would be placed on top of the trenches (i.e., above the ground surface) and covered with compacted soil (called an "area landfill"). By employing both methods of landfilling, open burning was restricted, odor and blowing paper were controlled, and habitat for both insects and vermin were eliminated.

Within the landfill environment, solid waste (particularly organic materials) would decompose, forming methane gas. Precipitation and surface runoff could percolate through the landfill, creating leachate (which contains the decomposition products and other soluble fractions within the waste) that could, depending on site geology, either seep out of the sides of an area landfill and/or through the bottom and sides of the disposal trenches.

One result of passage of the Solid Waste Management Act of 1965[3] was that open dumps were eventually banned. As a result, the only solid waste land disposal facilities that could be operated were sanitary landfills. With the passage of this law, federal technical guidelines were promulgated to regulate solid waste disposal. As a result, many states adopted mandatory operating and closure procedures that would limit leachate generation (e.g., cover a landfill with a thick soil mantel and vegetation) and control landfill gas migration (e.g., vent gases through the top of the fill).

[3]Note that while the sanitary landfill method had its origins in the mid-1930s, and was recognized as a much superior method of dealing with municipal wastes, it took 30 years for the political/regulatory regime to ban open dumps.

Even with these controls, landfills were perceived by community residents as, at best, a necessary evil. It follows that, as the population of urban areas grew, the amount of refuse grew, creating an ever-increasing demand for waste disposal facilities. However, most communities were unwilling to have a landfill in near proximity.

Not in My Backyard

By the early 1970s, city and county planning agencies found it more and more difficult to obtain zoning approval for new sanitary landfills. As a consequence, existing landfills were allowed to expand (i.e., if there were adjacent properties that could be utilized) in close proximity to residential areas. If or when there was no room left for expansion, landfills were usually permitted to grow "vertically" into small mountains above the landscape.

Once the landfill had reached its maximum capacity, the facility had to be closed (i.e., provide final cover, surface water diversions, erosion and cover maintenance, and gas mitigation). If the city or county agency that owned and/or operated the facility had not collected the necessary fees for closure during the operating life of the landfill, the funds had to be obtained from other sources (i.e., taxes, budget surpluses,[4] insurance, etc.). The city in this case chose to obtain its funds from its insurers.

THE CITY'S HISTORY OF SOLID WASTE MANAGEMENT

The County Health Department issued a permit to operate a landfill in 1939. Aerial photographs of the Quite Valley City (QVC) landfill site showed that operations began in 1940. Based on the aerial photographs of the facility, the trenching portion of the landfill's history is illustrated in Figure 21.1. Prior to beginning any landfill operations, the land elevation was 265 feet above mean sea level (MSL).

The landfill trenches were initially dug to a depth of approximately 10 feet. With the completion of each trench, area landfilling began. This mode of operation continued until about 1968. At that time, it became clear to the city planners that the available area left for trenches would be ex-

[4]Realistically, when was the last time that a municipality or county actually had a "budget surplus"?

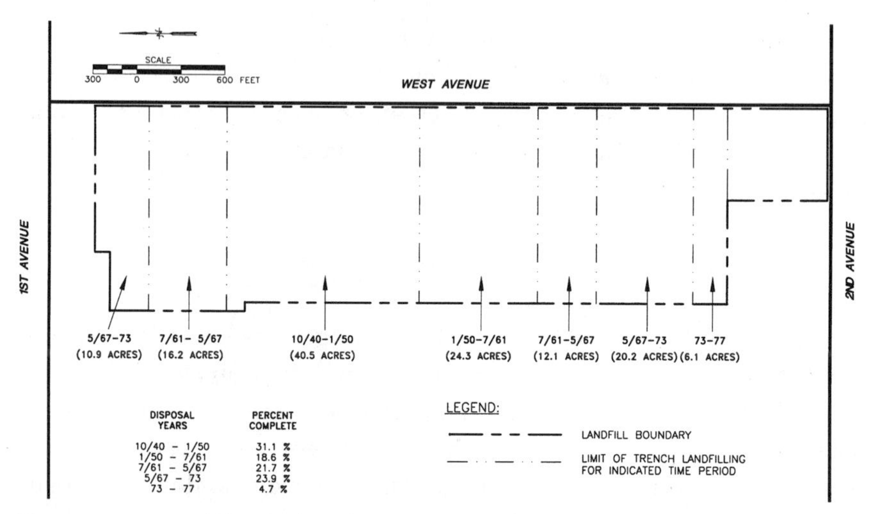

Figure 21.1 Landfill trenching history.

hausted in several years. Thus, they began the process of purchasing adjacent properties and pursued the county's zoning commission for approval of the expansion.

Until the process could be completed, the QVC landfill increased the trench depth to 35 feet. However, after nearly a year of studies and hearings, the county did not approve the expansion of the QVC landfill. As a result, once the remaining area for trenching was used up, landfilling could only be continued by increasing the height of the fill.

In the fall of 1972, the city counsel adopted the proposed state minimum standards for solid waste handling and disposal facilities. This required the operator to perform the following:

- Apply a 2-foot layer of final cover material.
- Monitor, collect, treat, and dispose of leachate.
- Monitor landfill gases and take necessary action to control such gases if they create a hazard.
- Provide surface drainage.
- Ensure that no solid waste is deposited in direct contact with groundwater.
- Grade the site to prevent ponding.

With these new regulations in place, the city of Quite Valley necessarily increased the per-ton cost of waste disposal at its facility. This incremental fee was earmarked to ultimately pay for these new standards.

Geologists from the County Health Department reviewed previous soil boring data from the QVC landfill in the summer of 1975 (see Figures 21.2 and 21.3).[5] As a result of this review, the city was instructed not to dig trenches deeper than 25 feet below the natural land surface at the site (i.e., 240 MSL). Trenches could only go below this elevation if it could be demonstrated that a natural barrier separating the landfill from the underlying groundwater body existed or that a synthetic liner had been installed. The city was not able to demonstrate that there was a natural barrier below the 240-foot elevation nor did it install any liners in the trenches. As a consequence, the trench depth remained at 25 feet.

By 1977, the area remaining for trenches had been exhausted.The elevation of the landfill now ranged from 280 to 283 feet above MSL. The county modified the QVC landfill permit to allow a maximum elevation of 296 feet (which translates into a 31-foot-high pile).

When the State Environmental Protection Agency sampled residential water wells in early 1982, it found between 0.1 and 1.0 part per billion (ppb) trichloroethylene and perchloroethylene in the well water. When confronted with this discovery, the city of Quite Valley purchased the adjacent properties where the contaminated wells were located and began an investigation to determine if there was any landfill gas migration off of the landfill property. Because of the contamination, the city was required to produce a landfill closure plan. The city's consultant, GeoExperts, produced a site closure evaluation for the QVC landfill. This report recommended a source for the required 2 feet of final soil cover and proposed methods by which to control gas migration from the site. However, for some (unexplained) reason, GeoExperts failed to estimate the associated costs.

By the middle of 1983, a preliminary landfill gas study had identified the presence of methane, as well as vinyl chloride, approximately 100 feet outside of the landfill boundary. As a consequence, permanent gas monitoring wells were installed within the QVC landfill, as well as along the property boundaries. Due to the potential hazards, the city was ordered by the State Environmental Protection Agency to eliminate all sources of water that could contribute to further biological decomposition (within the landfill)and therefore the continued formation of landfill gas. In addition, the city installed a bentonite slurry wall (i.e., a trench excavated to 25 feet

[5]As can be seen from Figure 21.3, the trench depth exceeded a less permeable clay layer in the dominantly sandy soils.

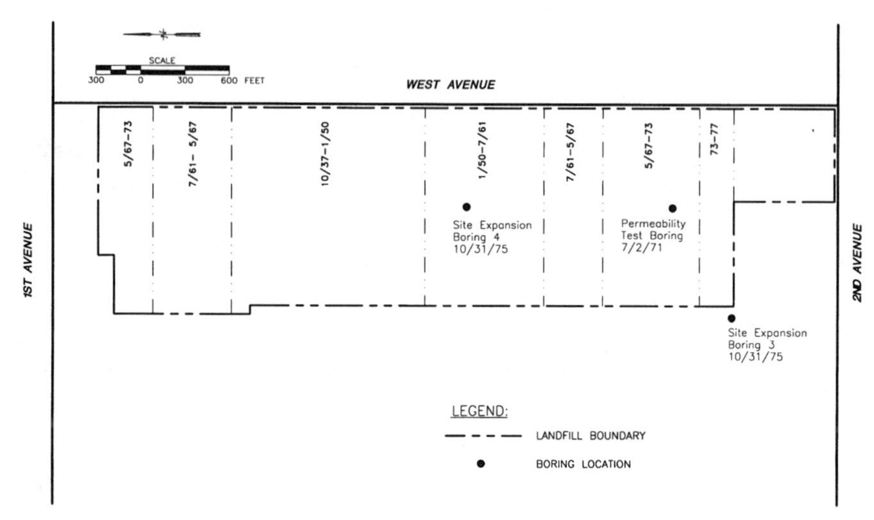

Figure 21.2 Landfill soil boring locations.

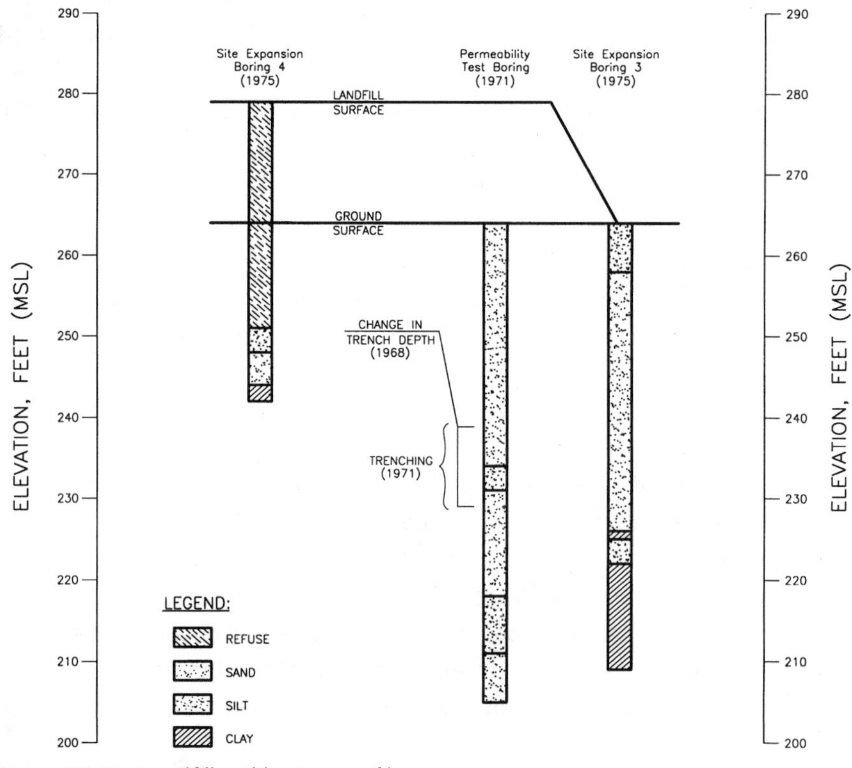

Figure 21.3 Landfill soil boring profile.

and filled with bentonite clay) around the QVC landfill as a further step in stopping gas migration.

Even in the face of groundwater contamination and the hazard associated with gas migration, the QVC landfill continued to operate (because there was no other cost-effective disposal facility in the area at the time). Thus, in 1986, the county permitted the QVC landfill to increase the elevation of the landfill to 320 feet above MSL. New groundwater monitoring wells were also installed at the boundary of the QVC landfill. Analysis of the groundwater showed that all of the boundary wells were contaminated with volatile organic compounds (see Figure 21.4). These data also suggested that the groundwater contamination may have originated from the deep trenching (see Figures 21.2 and 21.3).

By the end of 1987, the QVC landfill reached its permitted elevation, and landfill operations finally ceased. GeoExperts submitted a grading and drainage plan for the landfill, projecting site closure in 1988. This closure estimate, including final cover and drainage control, was $8,840,200, which did not include gas control or operation and maintenance (O&M) costs. This deficiency was corrected several months later when GeoExperts provided the following estimates:

Gas control	$ 575,000
O&M ($125,600/year @ 30 years)—PNW[6]	2,567,000
Subtotal	$3,142,000

Thus, the total closure cost, as estimated by GeoExperts, reached $11,982,200.

Because the city of Quite Valley failed to begin implementation of closure activities, in late 1989 the EPA added the QVC landfill to the National Priorities List (i.e., Superfund). The EPA then ordered that a vacuum system be added to the methane barriers and that a gas extraction system also be installed, since it was determined that the gas barrier installed by the city was not effective.

In 1990, the city of Quite Valley signed an EPA Consent Order to conduct a Remedial Investigation and Feasibility Study. As a result of these investigations, which were completed in 1993, a final cover, drainage con-

[6]PNW (present net worth) includes the addition of a 20% contingency fee and a PNW factor of 0.6.

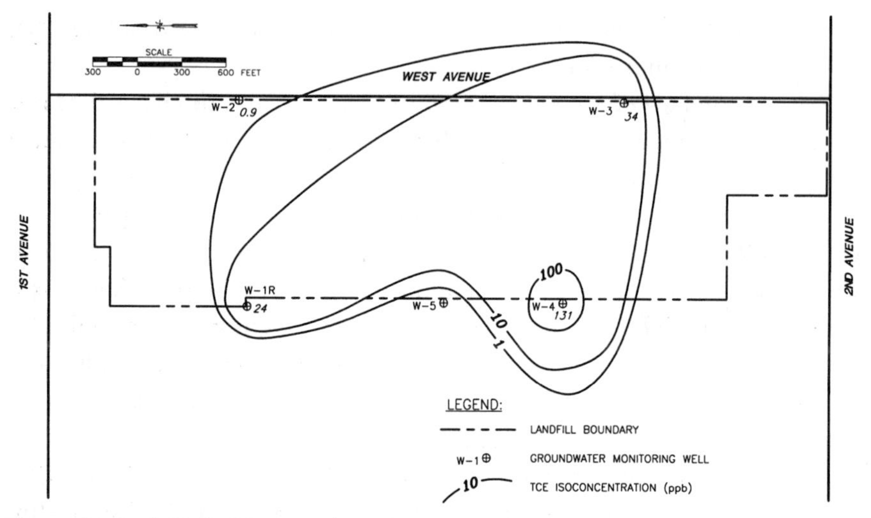

Figure 21.4 Initial TCE groundwater contamination contours.

trol, and gas migration control system were recommended. These remedial action costs were summarized as follows:

Final cover	$12,335,000[7]
Gas control	4,250,000
O&M—PNW 30 years	5,671,000
Total	$22,256,000

This final remedy was incorporated into an EPA Record of Decision in 1994. At this point, the closure costs had more than doubled since the estimates made in 1988.[8] The city of Quite Valley[9] filed claims with its past and present insurance companies for the "unexpected" costs "forced upon them" by the EPA.

The insurance companies denied coverage for the claims and so the city of Quite Valley sued to recover the remedial costs.

[7]This includes not only the 2 feet of soil cover and vegetation, but also an additional geotextile liner to control landfill gas recovery.

[8]Once again, "time is money."

[9]The city also sued those local businesses that disposed of waste into the QVC landfill to pay for these costs.

THE FORENSIC INVESTIGATION

After several attempts to reach a settlement with the city of Quite Valley, the insurance joint defense group retained a forensic consultant to review the landfill's site history and define the technical issues the joint defense group had to address relative to the "unexpected" damages claimed by the city.

After a complete review of the site history and remedial documents, the consultant conducted an all-day seminar on the QVC landfill situation for the joint defense group. The objective of the meeting was to educate the group on the technical terminology and technical issues relative to the environmental characteristics of sanitary landfills and to provide guidance on the specific technical issues pertaining to their coverage issues. The key conclusions presented to the joint defense group are summarized next:

General Knowledge of Landfill Problems

- The fundamental knowledge related to the biological decomposition of organic materials in the landfill environment, the formation of leachate, and the generation of landfill gases was well known prior to the Solid Wase Management Act of 1965.
- The fundamental knowledge of how to control the generation of leachate, prevent or substantially reduce groundwater contamination, and control landfill gas migration was also known prior to Solid Waste Management Act of 1965.

Site-Specific Issues

- In 1972, the city agreed to operate the QVC landfill under the state's minimum standards for solid waste handling and disposal facilities (these requirements have been summarized in Sidebar 21.1).
- The state's closure standards were essentially identical to the EPA Record of Decision remedial action requirements.
- The city implemented gas migration controls (unsuccessfully, because the barrier trenches were not deep enough) prior to any EPA orders.

Sidebar 21.1 Landfill Closure Regulations

Section 1: Precipitation and Drainage Controls

(a) Waste facilities shall be designed and constructed to limit, to the greatest extent possible, ponding, infiltration, inundation, erosion, slope failure, washout, and overtopping under the precipitation conditions specified in this article for each class of waste facility.
(b) Precipitation on landfills or waste piles which is not diverted by covers or drainage control systems shall be collected and managed through the leachate collection and removal system, which shall be designed and constructed to accommodate precipitation conditions specified in this article for each class of waste facility.
(c) Diversion and drainage facilities shall be designed and constructed to accommodate the anticipated volume of precipitation and peak flows from surface runoff under the precipitation conditions specified in this article for each class of waste facility.
(d) Surface and subsurface drainage from outside of a waste facility shall be diverted from the waste facility.
(e) Cover materials shall be graded to divert precipitation from the waste facility, to prevent ponding of surface water over wastes, and to resist erosion as a result of precipitation with the return frequency specified in this article for each class of waste facility.

Section 2: General Closure Requirements

(a) Partial or final closure of new and existing classified waste facilities shall be in compliance with the provisions of this regulation.
(b) Closure shall be under the direct supervision of a registered civil engineer or a certified engineering geologist.
(c) Landfills shall be closed pursuant to Section 3 of this regulation.
(d) Closed waste facilities shall be provided with at least two permanent monuments installed by a licensed land surveyor or a registered civil engineer.
(e) Vegetation for closed waste facilities shall be selected to require minimum irrigation and maintenance, and shall not impair the integrity of containment structures including the final cover.

Section 3: Landfill Closure Requirements

(a) Cover requirements:
  (1) Closed landfills shall be provided with not less than two feet of soil as a final cover.
  (2) The cover shall be designed and constructed to function with the minimum maintenance possible.

(b) Grading requirements:
  (1) Closed landfills shall be graded and maintained to prevent ponding and to provide slopes of at least three percent. Lesser slopes may be allowed if an effective system is provided for diverting surface drainage from covered wastes.
  (2) Areas with slopes greater than ten percent, surface drainage courses, and areas subject to erosion by water and wind shall be protected or designed and constructed to prevent such erosion.

(c) Throughout the postclosure maintenance period, facility owner shall:
  (1) maintain the structural integrity and effectiveness of all containment structures, and maintain the final cover as necessary to correct the effects of settlement or other adverse factors;
  (2) continue to operate the leachate collection and removal system as long as leachate is generated and detected;
  (3) prevent erosion and related damage of the final cover due to drainage; and
  (4) protect and maintain surveyed monuments.

Section 5: Gas Control

Where the Enforcement Agency has cause to believe a hazard or nuisance may be created by landfill decomposition gases, they shall so notify the owner. Thereafter, the site owner shall cause the site to be monitored for the presence and movement of gases, and shall take necessary action to control such gases.

- The city collected fees for the site closure but spent the funds on other projects (e.g., similar to the federal government using Social Security monies for the general budget).

Conclusions

In order to continue operating the QVC landfill, the city had to operate under the mandated state closure requirements. As a result, all of the closure costs were recognized (anticipated) to be business expenses (i.e., the cost of doing business). Only the magnitude of the closure costs was "unexpected."[10]

A SETTLEMENT MEETING

At a settlement meeting intended to resolve the claim, the joint defense group made a presentation based on the facts provided by their forensic consultant. Shortly after the completion of the presentation, the joint defense group offered the city $1,000,000 against the city's claim of $22,256,000. The city accepted the settlement offer.

LESSONS LEARNED

1. When evaluating the cost of environmental damages, the costs associated with normal business operations (i.e., the cost of doing business) should be clearly delineated and separated from the total remedial cost.
2. This case history clearly demonstrated the validity of documenting and separating the cost of business associated with county- or state-permitted waste management units.
3. Remedial costs are not always consistent with accidental damages covered by insurance.
4. A clear and factual presentation of information can lead to a successful settlement, thereby saving the heavy expense associated with protracted legal action (in simple terms, not everyone is foolish enough, even with $23,000,000 at stake, to enter a battle when defeat appears obvious).

[10]This was not the first time (nor will it be the last time) that the definition of "unexpected," as related to insurance coverage, is cost driven as opposed to contamination driven.

Chapter 22

A Sudden and Accidental Event[1]

When a company seeks to have its past and present insurance companies pay for the cost of a remedial action, the company may determine that the insurers will only cover contamination that resulted from a "sudden and accidental" event.[2] If the company can identify or point to an "accidental event," it may claim that this accident caused a portion or perhaps all of a site contamination, even if the claimed event does not necessarily fit the facts. The case presented in this chapter pertains to such a claim.

BACKGROUND TO A PROPERTY SALE

In the history of most companies, a time arrives when it becomes expedient to move to another location, close the doors, or sell the company. In the case cited here, Argon Pharmaceutical Company[3] decided to sell its Midwest laboratory to Xenon Corporation. By state law and as a condition of the sale, Argon had to conduct a site investigation to identify any soil/groundwater contamination. Argon retained the services of its longstanding engineering consultant, Universal Engineers.[4]

[1]Or maybe not.

[2]For example, an above-ground chemical storage tank ruptures when it is struck by a truck and the released chemical contaminates the soil.

[3]This facility originally began operations as General Laboratories in 1970 and was purchased by Argon Pharmaceutical in 1984.

[4]Interestingly, Universal Engineers, which had designed the Argon wastewater treatment facility, had never conducted an extensive environmental site investigation at any site prior to this assignment. However, business being business, it did agree to conduct the site investigagion.

Argon utilized a wide range of both inorganic and organic hazardous chemicals in its manufacturing process. The major organic chemicals were acetone, chloroform, freon, toluene, and trichloroethylene (TCE). Argon was also permitted under RCRA as a generator and storer of hazardous wastes. As a result, there were numerous locations around the facility that utilized both hazardous substances and stored hazardous waste.

The Initial Site Investigation

The initial site investigation conducted in 1987 focused on identifying potential areas of soil and groundwater contamination that could be associated with the chemical units and RCRA storage areas. The results of the groundwater sampling are shown relative to the RCRA and chemical storage units in Figure 22.1. This figure also shows the location of (1) the freeze dryers that utilized TCE and (2) the maintenance shop[5] that serviced the freeze dryers.

Having completed its investigation, Universal Engineers, with the concurrence of Argon, filed a facility report, which included the following statement, "There are no known spills or discharges of hazardous substances or wastes on site." However, since the investigation identified both soil and groundwater contamination, a proposed remedial action plan was developed by Universal Engineers.

The Remedial Action Plan

As with all remedial action plans, Universal had to identify the source[6] of the groundwater contamination so that the chemicals in the soil could be remediated prior to implementing a groundwater cleanup. During this process, Universal discovered (or perhaps "was informed") that an "alleged spill" of TCE had occurred in 1983.

According to several employees, some 20 gallons of TCE had been spilled on the soil near a drum storage pad (see Figure 22.1). Based on this incident, Universal conducted additional soil vapor (OVA) and soil sampling in the area of the "alleged spill." The data and sample locations are shown

[5]The maintenance shop was also an area of TCE use and storage.

[6]The term "source" as used here represents the location of the contamination, not its initial origin.

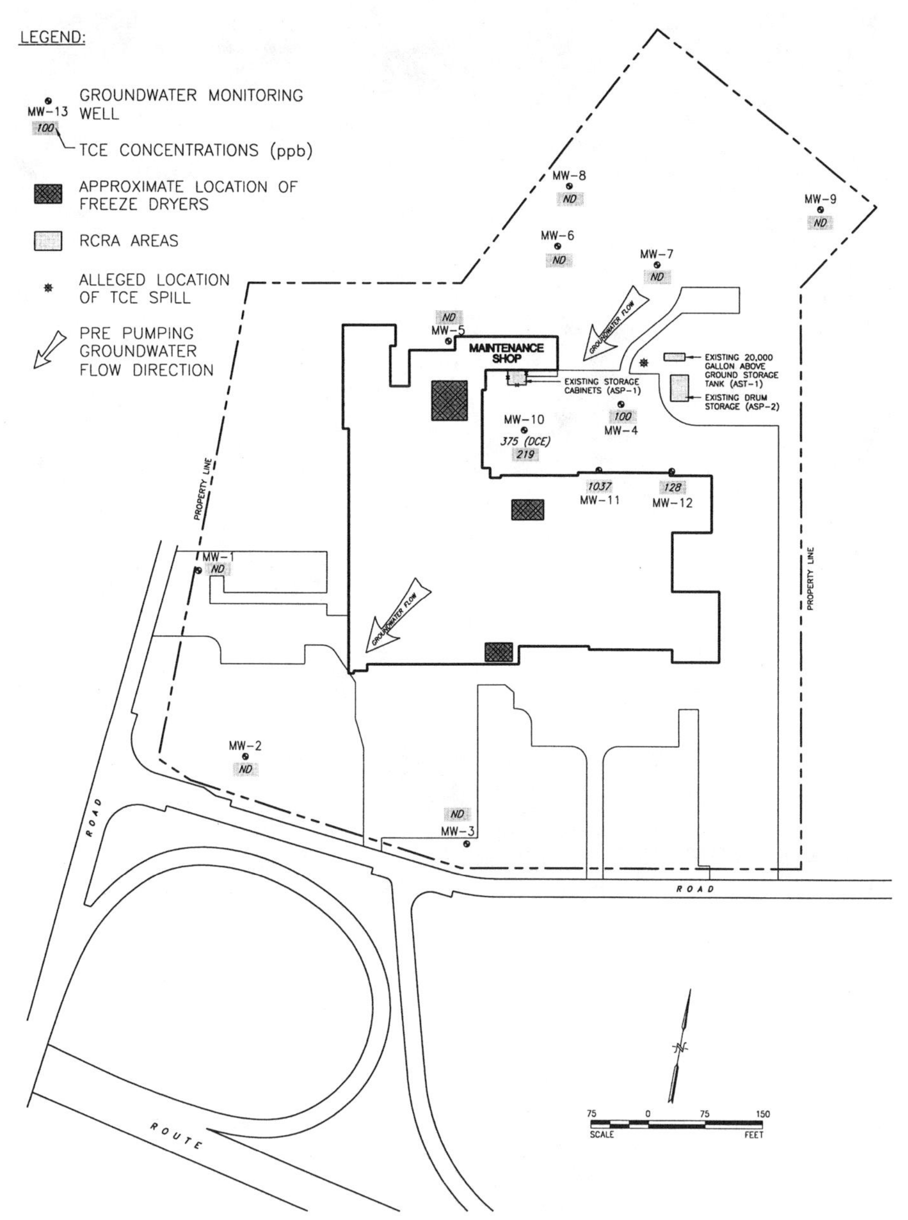

Figure 22.1 A facility map and initial groundwater data.

LEGEND:

BR-4 143 — 1987 APPROXIMATE SOIL BORING LOCATION AND TCE SOIL CONCENTRATION – ppb

BX-21 37 — 1991 APPROXIMATE SOIL BORING LOCATION AND TCE SOIL CONCENTRATION – ppb

A 100 — APPROXIMATE SOIL GAS SAMPLE LOCATION AND OVA VALUE – ppm

OVA STUDY ALLEGED LOCATION OF TCE SPILL

REPORTED 1984 SPILL AREA

N

20 0 20 40

SCALE FEET

Figure 22.2 Analytical data from the "alleged spill" location.

in Figure 22.2.[7] TCE was found in the soil at BX-2, BX-21, and BR-4, which were all in the vicinity of the chemical waste storage tank. BR-17, which was directly in the "alleged spill" location, had no detected concentrations of TCE. In addition, the highest total volatile organic compound (VOC) concentrations measured in soil gases were also near the chemical storage tank (i.e., see OVA study area).

Although no evidence of TCE contamination in the "alleged spill" area could be found, Universal proposed cleaning up the groundwater down gradient of the "alleged spill" and removing all of the contaminated soil from below the chemical units and RCRA storage areas. At this point, the cleanup cost was estimated to be less than $275,000.

GROUNDWATER REMEDIATION: PHASE I

Universal Engineers designed and installed the Phase I groundwater recovery/treatment system in late 1987 directly down gradient of the "alleged spill" area to contain the TCE contamination in the northern portion of the site (i.e., to ensure that the plume would not move off site). The location of the Phase I extraction system with the estimated zone of capture of the TCE plume is illustrated in Figure 22.3.

As the groundwater extraction progressed, the groundwater monitoring data showed that consistently high TCE concentrations were occurring in monitoring well MW-10. As can be seen from Figure 22.3, MW-10 is not down gradient of the "alleged spill".[8] In addition to the high TCE concentrations, MW-10 also evidenced very high concentrations of dichloroethylene (DCE).[9] Meanwhile, MW-4 (the well directly down gradient of the "alleged spill") contained little or no DCE. These data suggested that the contamination being monitored by MW-10 had an earlier origin than the contamination monitored by MW-4. It is also important to note that the location of the Phase I extraction system isolates MW-4 from MW-10 (i.e., they are monitoring uniquely different contamination sources).

These data were interpreted by both Universal Engineers and the state regulatory agency as evidence that there was another source of contamination other than contamination from the spill incident of 1983. Indeed,

[7]The aerial photograph was taken from a U.S. Geological Survey database.

[8]Universal Engineers mapped the upper surface of the shallow bedrock. The mapping showed that a dense nonaqueous phase liquid (DNAPL; TCE) could not have moved from the "alleged spill" location into the vicinity of MW-10.

[9]A daughter product of TCE.

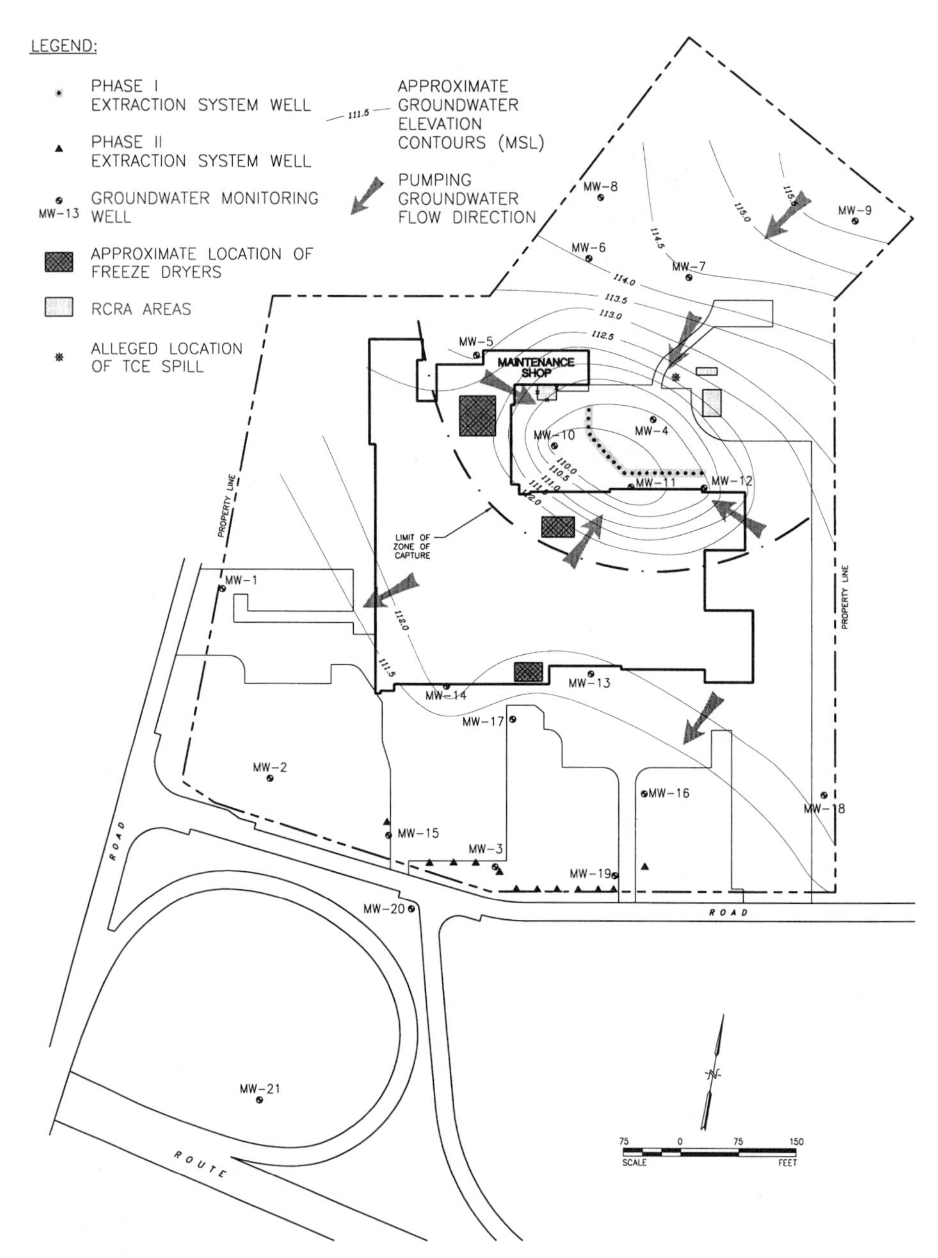

Figure 22.3 A map of the Phase I extraction system and zone of capture.

Universal reported that a separate and completely independent plume had been discovered northwest of MW-10.

To further confuse the situation, evidence now surfaced (from still another old employee) that there had been a spill of TCE[10] within one of the buildings sometime in 1978, while the facility was under the ownership of Argon's predecessor General Laboratories.[11] In spite of this knowledge, Argon and its consultant never attempted to investigate "under the building" and, as such, never verified if, indeed, there was a source of TCE contamination in this location (i.e., up gradient of MW-10).

LAWSUIT 1

Based on the 1978 TCE spill, Argon sued General Laboratories in early 1988 for the existing soil and groundwater contamination. The lawsuit between the parties was quickly and quietly resolved with a small payment to Argon in exchange for total indemnification of General Laboratories by Argon.

General Laboratories hired a well-known expert who examined the available data, hypothesized a phantom model,[12] and prepared a declaration to the effect that the 1978 spill could not have caused the contamination. The fact that the expert's phantom model was just that, and had little, if any, relevance in ascertaining the source, no less the extent, of the contamination, the declaration, essentially unevaluated and unchallenged by the company's consultant, won the day for the former owner (General Laboratories).[13]

In part, this result could be attributed to the cleanup estimate prepared by Universal (i.e., $275,000). Universal assumed that the TCE from the 1983 spill had "all dissolved in the groundwater" and, as a result, would be cleaned up within "a few months to a year." The settlement was in the approximate amount of Universal's Phase I cost estimate.

By 1989, the facility transfer had taken place (to Xenon Corporation), with the former owner (Argon) assuming the cleanup responsibility (and

[10]The spill was discharged directly into the building floor drains.

[11]The location of the TCE spill occurred near the freeze dryers adjacent to the maintenance shop.

[12]The expert only produced maps of the model's output. There was never a description of the model or any validation of the model's predictions.

[13]It is fair to point out that one should never doubt that "science fiction," when presented with conviction and eloquently articulated, can and often has been accepted as "science."

obviously the associated cost) and with Universal Engineers continuing to perform site remedial work.

GROUNDWATER REMEDIATION: PHASE II

In late 1989, groundwater monitoring began in the southern portion of the site. Universal Engineers produced a map of the TCE plume with a modified estimate of the zone of capture for the Phase I groundwater extraction system (see Figure 22.4). As a result of the continuing movement of the plume, Universal Engineers developed a Phase II remediation plan, which was approved by the state regulators. The Phase II groundwater extraction and treatment system was estimated to cost $200,000 in capital costs and would require some $8000 per month in operating costs.[14]

By the time the Phase II groundwater extraction system was ready to operate (early 1990), the Phase I system, which was expected to operate for less than a year, had been running for well over two years at an accumulated cost in excess of $1,000,000.[15]

THE INSURANCE LITIGATION

With costs running far beyond expectations, Argon now went on the offensive and sued its insurance carrier for recovery of costs resulting from the "1983 accidental spill of TCE." In 1983, Argon's insurance policy would only pay for contamination that was "sudden and accidental." At the time of the lawsuit, it was the opinion of Universal's project manager of the Argon site that the "TCE contamination that had been identified on site originated from multiple sources north of MW-10 and MW-4 but that the vast majority of the TCE contamination was the result of the 1983 'alleged spill.'"[16]

With battle lines drawn, the insurance company's counsel retained a forensic consultant to hopefully bring order out of chaos and shed the appropriate investigative light on the claim. Although clearly disturbed by the magnitude of the claim, the insurance company was anxious to deter-

[14]The location of the Phase II extraction wells are shown in Figure 22.3.

[15]By 1994, the combined Phase I (still running) and Phase II remedial costs exceeded $4,000,000.

[16]Although the forensic consultant was accused by Argon's counsel of coining the term "alleged," it was pointed out that Universal, in its report to the state, had actually introduced the term (is one therefore to assume that Universal did not totally buy into the spill theory—and if that was the case, why did it then locate the remedial system to intercept an "alleged spill"?).

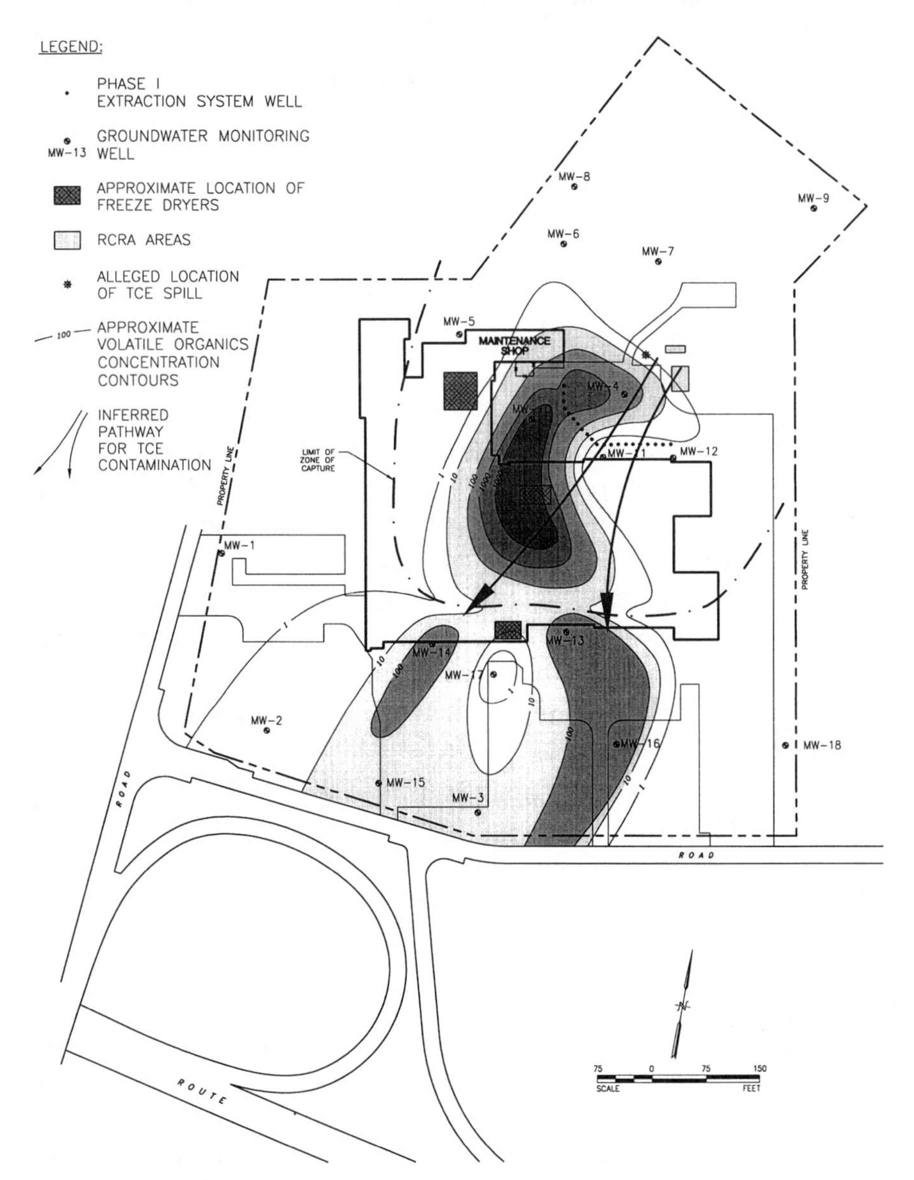

Figure 22.4 A facility-wide TCE plume map.

mine whether or not the claim was valid and, indeed, made an early but futile attempt to settle without going to war (see Sidebar 22.1).

The Forensic Investigation

The forensic investigation focused on three key areas: (1) sources of contamination, (2) distribution of the contamination, and (3) type of contamination.

Sources of Contamination

Based on the site operations and employee depositions, the TCE contamination could have originated from spills and leaks associated with each chemical unit, as well as at the waste storage areas (specifically AST-1 and ASP-1), the maintenance shop (which used TCE), the TCE spill to the floor drain, and the 1983 "alleged spill."

Based on the contamination monitored at MW-10, the source could have been the TCE spill in the building or spills/leaks at the drum storage area ASP-2. Except for shallow soil borings in the location of ASP-2 (showing no TCE contamination), there were no deep soil borings around or through the building floor near the (interior) TCE spill or at the maintenance shop. All of these locations are up gradient of MW-10 (based on the gradient established by the extraction wells). Thus, at this point in the investigation, there were no source areas identified by Universal Engineers (i.e., by default).

The groundwater data at MW-4 suggested that there was an up gradient source. As a result, the source area could have been either AST-1 and ASP-2 (several shallow soil borings identified TCE around AST-1) and/or the 1983 "alleged spill." However, no TCE was ever found in the shallow soils of the area identified by Argon employees who witnessed the spill. No deep soil borings were ever completed at this location. If 20 gallons of TCE had spilled onto a fairly sandy soil (with a shale bedrock at 15 feet), it would be expected that a DNAPL would be found at or near this location. Once again, Universal made no investigation in the area to determine whether or not a DNAPL existed.

Contaminant Distribution

According to Universal Engineers, the distribution of the TCE contamination (see Figure 22.4) across the site resulted from the 1983 "alleged

Sidebar 22.1 The Games Counsel Play

The chess game between opposing counsel is always interesting to observe. At one point, Argon's counsel proposed a settlement meeting in Denver, to which the insurance company's expert was invited. While at the airport, the expert received a call canceling the meeting. This issue came up in the first settlement conference, when the owner of Argon admonished the insurance company and its counsel for failing to meet with Argon to resolve the case issues. The judge questioned counsel for the insurance company, and she described the aborted Denver meeting, pointing out that Argon's counsel had canceled the meeting at the last minute. Imagine the surprise and shock when Argon's owner stated that he had no knowledge of such a meeting. You had to be there to see the "hang dog" expression on Argon's counsel's face—he had never informed his client of the meeting—apparently the whole idea was to rattle the insurance company's counsel.

This was not the only attempt to resolve the case. After the initial forensic investigation, which suggested an alternate-source theory to the spill, a conference call was initiated with opposing counsel, as well as Universal's project manager. When the alternate-source theory was presented during this call, opposing counsel went on a rampage, insisting that he, rather than the technical experts, understood all of the technical issues and that therefore the insurance company's expert was not only wrong but also dishonest it its approach.

spill." However, when the TCE contamination plume was compared to the direction of groundwater flow, there were several problems with Universal's theory. How did the dissolved TCE contamination move up gradient of MW-10? If the TCE was dissolved in water, why was TCE still present on the site?

One possible explanation was that the DNAPL from the 1983 "alleged spill" moved along the soil/bedrock interface to a location up gradient of MW-10. However, Universal's bedrock map showed a high bedrock ridge between MW-10 and MW-4, making it impossible for a DNAPL to move anywhere near MW-10. As a result of the groundwater flow direction and bedrock topography, the source of the TCE contamination that was being monitored at MW-10 could not have originated from the 1983 "alleged spill."

Type of Contamination

Finally, the historic distributions of volatile organic compounds (VOCs) in MW-4 and MW-10 were uniquely different. The data for each well is given in Table 22.1.

These data clearly showed that the area being monitored by MW-10 was, at best, only being partially remediated, while the area up gradient of MW-4 was, indeed, showing improvement. These data also showed that the concentration of TCE in MW-10, although fluctuating over several sampling intervals, was generally high and was not strongly influenced by the ground extraction system. Universal chose to explain these higher concentrations in MW-10 as having been caused by lower groundwater levels; however, MW-4 did not show a similar trend. The more likely result is that there still remains trapped TCE in the soil pores in the vicinity of MW-10 (i.e., in other words, a DNAPL).

Summary

All of these data suggest that there are at least two sources of TCE in the groundwater. Of these two sources, the source up gradient of MW-10 is the major source (estimated to be at least 90%, based on the VOC ratio shown in Table 22.1). The only information lacking is actual boring and sampling data up gradient of MW-10, as well as an analysis of the area directly below the 1983 "alleged spill."[17] The collection of these data would either prove or ruin Argon's position on the insurance coverage.

An Attempted Settlement

Reflecting on the fact that ongoing investigations and trial preparation could/would be costly, the insurance carrier called for a settlement conference. At this meeting, Argon made the first presentation to the mediator. Argon's new expert, who had just been retained a week earlier, offered the opinion that the TCE contamination on the Argon property was consistent with a spill having its origin at the point of the 1983 "alleged spill." (Surprise! Surprise!)

[17] As discussed in Chapter 1, the lack of key technical information is a classic example of why experts will have a difference of opinion.

Table 22.1 Groundwater Data Summary (ppb)

	MW-10			*MW-4*			
Date	*TCE*	*DCE*	*VC*[a]	*PCE*	*TCE*	*DCE*	*VC*
Apr 87					100		
May 87	219	375			24		
Jan–Mar 88	705	758			131		
Apr–Jun 88	1,480	2,260			1,320		
Jul–Sep 88	245	690	21		320		
Oct–Dec 88	440	1,300	33		600	8	
Jan–Mar 89	430	1,100	25		230	4	
Apr–Jun 89	630	11,000	46		720	8	
Jul–Sep 89	1,650	9,200	54		3,600	65	
Oct–Dec 89	420	1,100			230		
Jun 90	750	6,500	1,500	84	2,200	89	
Sep 90	1,500	9,200	1,700	18	410	23	
Feb 91	380	5,200	1,500	9	270	13	
Jul 92	1,500	7,480	1,900	21	570		
Apr 93	500	2	44	19	480		
Jul 93	310	12	71	5	75		
Jul 94	220	1,003	240	11	150	6	
Oct 94	64	173	56	2	23	1	
Jan 95	360	627	300	4	48	4	
Apr 95	1,600	11,095	2,900	1	13	1	
Jul 95	220	1,023	350	5	50	3	
Oct 95	656	8,311	820		5	12	3
Jan 96	1,050	5,676	1,280		72	4	
Jul 96	369	990	206	3	26	2	
Oct 96	835	626	80	12	167	8	
Jan 97	1,070	623	149	5	72	4	—
Totals	17,603	86,324	13,275	199	11,906	255	3
		(117,202)			(12,363)		

[a] Vinyl chloride is a decomposition product of TCE and DCE (and also tetrachloroethylene [PCE]).

Notes: (1) DCE is reported as the sum of all DCE isomers.

(2) Chemical ratios:

	MW-10	MW-4
TCE/DCE	1:4.9	59.8:1
DCE/VC	6.5:1	85.0:1

(3) Total VOC ratio for MW-10:

[MW-10 / (MW-10 + MW-4)] × 100 = 90%

Argon's expert based this opinion on his calculated rate of TCE transport in groundwater (i.e., TCE would move at a rate 279.8 feet per year). He also argued that the reason a DNAPL had not been found in the area of the 1983 "alleged spill" was because the incorrect sample preservation method was used. Because the insurance company's forensic consultant had to provide his report (consisting of site maps, plume plots, and other graphics) prior to the meeting, Argon's expert also took a considerable amount of time during his presentation to point out all of the technical errors on each graphic.

The forensic expert for the insurance company began his presentation by pointing out to the mediator that the graphics criticized by Argon's expert were developed entirely from data or portions of maps created by Argon's engineering consultant (Universal Engineers).[18] With that embarrassing moment out of the way, the forensic expert opined that the 1983 "alleged spill" could not have contributed more than 10% of the contamination and that the real problem was under the building and up gradient of MW-10.

The insurance company's forensic expert also compared for the mediator the rate of groundwater flow calculated by Argon's expert to an earlier calculation completed by Universal Engineers. As previously noted, Argon's expert calculated the rate of TCE movement in the groundwater to average 279.8 feet per year. However, Universal Engineers had previously calculated the average rate of TCE movement to be 50.6 feet per year (i.e., prior to any litigation). As a result, the question remained, who do you believe or were they both wrong? Both experts agreed that a DNAPL had not yet been identified in the location of the 1983 "alleged spill." However, the fact that a DNAPL had not been found was not relevant since *no* TCE had been found in the soil in any form (i.e., gas phase, adsorbed, or dissolved in soil water). The fact that a DNAPL was not identified was simply misleading.[19]

The real issue in this litigation has been the lack of information about contamination up gradient of MW-10 and deep soil borings beneath the 1983 "alleged spill" area. The insurance company's expert challenged Argon to collect this information. Argon's expert claimed that this information would be collected in the next two to three months (by this time, even the state regulatory agency insisted that this had to be done).

[18]All of the graphics clearly referenced the source material from Universal Engineers.

[19]This is not a chicken-and-egg argument. If a DNAPL was present, the other forms of TCE in the soil would be present.

Argon simply refused (or was not prepared) to even discuss multiple sources of site contamination and allocation percentages. Argon insisted that its "one-spill theory" was sufficient to explain the contamination. Faced with numerous unanswered questions, there appeared to be no resolution in sight. As the meeting ended, the issue of a "canceled settlement meeting" (see Sidebar 22.1) was aired.

The Depositions

Having arrived at a settlement impasse, the well-traveled path now led to depositions of the key technical players (i.e., Universal's project manager for the Argon site, Argon's expert, and the insurance company's expert). The deposition process was uneventful in the context that all of the opin-

Sidebar 22.2 The Document Shell Game

Several months prior to the deposition of Universal's project manager, the insurance company's forensic expert had traveled to Universal's headquarters to review and copy the relevant technical project files. These files consisted of approximately 35 cartons of documents, which represented both technical and financial records. However, as a result of questioning by counsel, additional cartons of documents began to appear. By the end of the deposition, almost as many documents were produced during the deposition as were made available during the initial discovery.

Discovery, in and of itself, is a complex process in that if the appropriate request is too narrow, the production will often be equally narrow. On the other hand, if the discovery request is overly broad, the plethora of documents can create a management and review nightmare. In this particular case, the careful preparation of deposition questions—a collaboration between counsel and the forensic expert—"uncovered" the additional documents, which either were truly "forgotten" or were held out of the initial discovery. The excuse made at the deposition was that these documents had been "misfiled" and, further, the project manager had not realized that other staff members who had worked on the project had their own files.

ions previously expressed by all parties remained unchanged. At the deposition of Universal's project manager, however, there was found to be (perhaps not surprisingly) a major discrepancy involving prior document productions. This is discussed in Sidebar 22.2.

New Data

The critical site information that should have been obtained almost eight years earlier was finally collected. The newly collected site information up gradient of MW-10 is illustrated in Figure 22.5. These data clearly demonstrated a major source of contamination under the building. This source

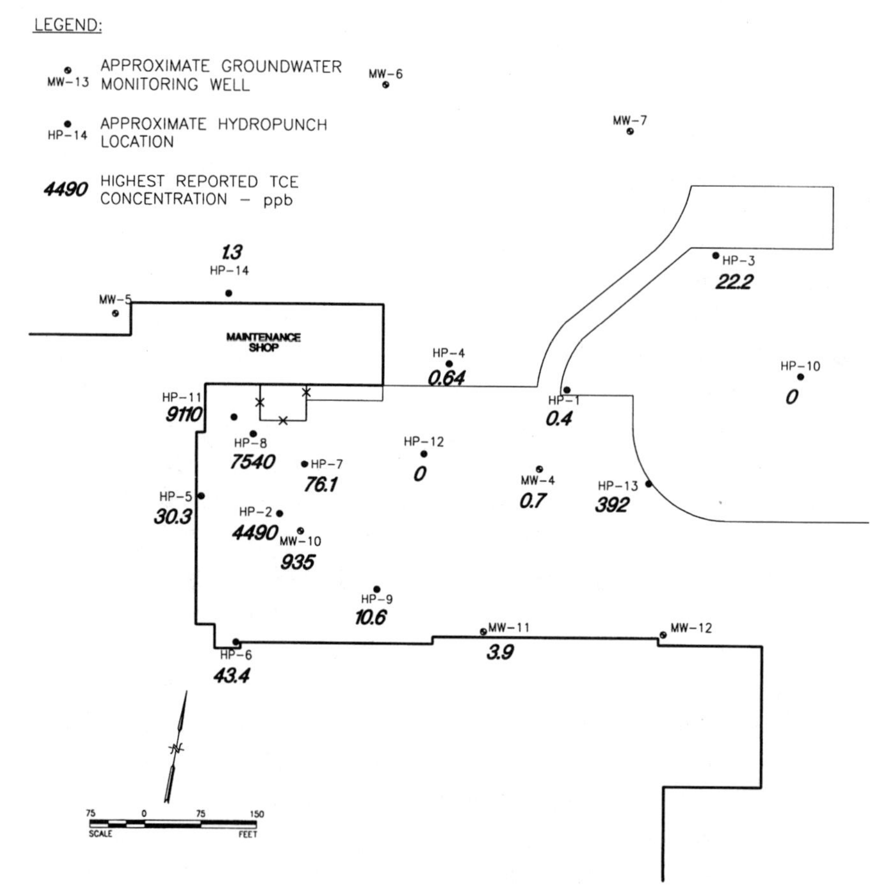

Figure 22.5 Hydropunch and monitoring well data near MW-10.

was totally independent of the 1983 "alleged spill." In addition, a "DNAPL search" in the area of the 1983 "alleged spill" never turned up the presence of a DNAPL, even when the "proper" sampling method was used. Ultimately, the parties reached a settlement (lest we forget, even winning costs money).

LESSONS LEARNED

1. It helps to know the most probable remedial solution, as well as the estimated cost, prior to making a claim for damages. Had Argon's consultant conducted a thorough investigation, Argon's first lawsuit could well have yielded a much larger settlement and the insurance issue may never have surfaced.
2. Do not install a remedial system unless you know the sources of contamination. By remediating the wrong source, years of groundwater extraction and treatment were almost completely ineffective and at an enormous cost to Argon.
3. Do not stick to a litigation theory (i.e., a sudden and accidental event caused the contamination) if you cannot support it technically (unless your client is willing to gamble on the outcome).

Chapter 23

An Act of God

The southern United States is one of the largest producers of phosphate fertilizers in the world. In the Red Ridge region, which contains an abundance of this natural resource, Red Ridge Fertilizer Company converted phosphate ore into calcium superphosphate by digestion with sulfuric acid. The resulting sludge and wastewater, which contained fluoride (F^-), was routinely discharged into lagoons, both on site and off site. The Red Ridge operations were purchased by National Fertilizer Company (NFC) in the early 1960s. When NFC ceased operations in 1983, the U.S. Environmental Protection Agency (EPA) investigated the site.

The site investigation revealed extensive fluoride contamination of both the soil and the groundwater. The EPA required NFC to remediate the site, at an estimated cost of approximately $14,000,000. As a consequence, NFC submitted to its commercial general liability (CGL) insurance company a claim for the cost of the remediation. Upon review of the claim file, the insurance company refused to pay the claim.

National Fertilizer Company sued its insurance carrier to recover its anticipated remedial costs and retained an expert to address soil and groundwater contamination issues, with emphasis on the source of the contamination. After an extensive review of the company's documents and site investigation reports, the expert concluded that the groundwater contamination was the result of "an act of God." In other words, the fluoride contamination was unexpected and unintended.

SITE HISTORY

Red Ridge Fertilizer Company began manufacturing in early 1940 and operations continued until 1983 when the facility ceased operations. During its 43 years of operation, the facility manufactured "normal superphosphate" fertilizer [$CaH_4(PO_4)_2 \cdot H_2O$]. The manufacturing process combined the source rock containing fluorapatite, [$Ca_3(PO_4)_2]_3 \cdot CaF_2$, with sulfuric acid ($H_2SO_4$). When these two compounds were combined, normal superphosphate and calcium sulfate ($CaSO_4$) were formed, with hydrofluoric acid (HF) as a byproduct.

Because HF is a hazardous gas, its evolution during the sulfuric acid digestion step was controlled by scrubbing the off gases. The resulting HF-contaminated scrubber wastewater was subsequently discharged to the settling ponds[1] (see Figure 23.1). Beginning in 1956, a sodium silica fluoride (Na_2SiF_6) plant was added to the facility.

In 1959, the city of Crescent Dunes noticed that the water from its municipal water supply well registered a declining pH, coupled with increasing concentrations of both sulfate and fluoride. The city's well was located approximately 6000 feet southeast of, and hydraulically down gradient from, the Red Ridge Fertilizer site. Concern over the declining water quality caused the city to hold a public meeting to discuss the problem.

The meeting was attended by the Red Ridge Fertilizer plant manager, the manager of City Water Works, and the city's sanitary engineer. At the beginning of the meeting, the Red Ridge Fertilizer plant manager stated that the position of the company was such that it definitely did not want to pollute the water in this region. However, having made that statement, he further pointed out that the company did not want to go out and spend a lot of money (addressing this issue) before the company felt that it was necessary. The plant manager further stated that, although approximately 12,000 gallons per day of liquid waste was discharged at the plant, during periods of dry weather the water disappeared very quickly, suggesting that the wastewater was "evaporating."

The city's sanitary engineer felt that if 12,000 gallons per day of wastewater was being pumped at the plant, some of it must be getting into the ground; it could not all evaporate. The manager of the Water Works stated that there was a strong indication that the wastewater from the Red Ridge Fertilizer plant must be going underground and therefore polluting the city water. Based on these comments, city officials considered allowing the wastewater from the fertilizer plant to be discharged to its sewage treatment plant, with the proviso that the waste would not be harmful to the treatment system.

[1]This aerial photograph is from a U.S. Geological Survey database.

Figure 23.1 Site wastewater ponds.

In late 1962, the plant connected to the city's sewer. A pretreatment step, prior to discharge, consisted of mixing the wastewater stream with lime, forming a lime slurry, which was then discharged into settling ponds (at a pH range between 7 and 10). The ponds, acting as settling basins, allowed the sludge to settle out and the "clear effluent" was then discharged to the city's sewer.

In 1963, the plant was sold to NFC, which continued plant operations with no changes in the process. In 1983, all plant operations ceased.

In 1984, the EPA conducted an investigation of the NFC facility that indicated that the on-site soil and subsurface waters were contaminated with elevated levels of fluoride. In January 1988, the state conducted a groundwater assessment at the site, which revealed that site contaminants, primarily fluoride and sulfate, had polluted the groundwater.

Based on this finding, the EPA required a complete remedial investigation and feasibility study. At the completion of these studies, the EPA issued a Record of Decision (ROD), in 1991, that required the following remedy:

- Excavation and solidification/stabilization of approximately 26,215 cubic yards of contaminated sludge and soils from site sludge ponds
- Consolidation of all stabilized sludge and soils into one new lined pond
- Construction of an RCRA cap over the pond
- Construction of a slurry wall around the RCRA cap
- Implementation of institutional controls to include security fencing, access, and deed restrictions

The estimated present worth cost for this remedial action was $14,367,000.

A CGL INSURANCE CLAIM

The site investigation data showed that the fluoride contamination of the groundwater was associated with the disposal ponds. In spite of this information, NFC submitted a claim to its insurance carrier for the cost of the remediation. When the insurance company refused to pay, NFC sued the insurance company to recover the anticipated remedial costs. In order for the fertilizer company to prevail in its suit, it had to demonstrate that the soil contamination and, thus, the groundwater contamination were either unexpected and unintended and/or sudden and accidental. As a consequence, NFC retained an environmental expert to review the relevant site information. Having reviewed all of the site-related information, including interviews with employees, the expert expressed the following opinions:

- Based on deposition testimony, there remains a question as to whether the ponds at the site were lined.
- None of the waste disposal or byproduct handling practices was intended to allow the uncontrolled release or discharge of chemicals (i.e., fluorides) to the soil or groundwater.

- The practices and procedures were consistent with the practices and procedures of similar manufacturing companies with respect to the storage, handling, and disposal of chemicals and byproducts.
- In the absence of evidence to the contrary, it is reasonable to assume that sudden and accidental events have caused the conditions necessitating remedial actions.
- Sudden and accidental releases, standing alone, would have necessitated the heretofore listed remedial actions.
- The sudden and accidental releases of fluorides resulted from hurricanes in June 1953, September 1953, September 1956, October 1959, June 1966, and September 1975. Under these conditions, the ponds' ability to contain the water would have been exceeded, causing overtopping.

In summary, the expert opinions were that (1) seepage from the ponds did not contaminate the groundwater, but overtopping of the ponds contaminated the groundwater; and (2) it was industry practice to use ponds that were subject to overtopping. In other words, hurricanes, or acts of God, created the groundwater contamination.

AN OPPOSING OPINION

Counsel for the insurance company retained a forensic expert to evaluate the site documents, as well as the opinions of the NFC expert, with the objective of producing an opposing opinion. After reviewing and evaluating all of the available information, including employee depositions, the forensic expert delineated the following facts.

Fluoride Toxicity and Chemistry

Toxicity

Fluoride has always[2] been known as a health hazard. For example, the hazards of hydrofluoric acid from the production of phosphate fertilizers were known since 1867 [1]. An article from that time period reported that when the phosphate ore is added to sulfuric acid a highly volatile and corrosive liquid, which is hydrated hydrofluoric acid, is dangerously pungent and irritating and that a minute drop of it upon the skin produces a

[2] "Always" can be considered in the context of "well over 100 years."

painful sore; thus, the vessels containing it need to be handled with great caution.

In addition to the plant hazards to workers [2], the release of fluoride into the environment such that it would contaminate water supplies has also been reported and therefore known to be a hazard. Obviously, this was the reason that the 1942 Public Health Service Drinking Water Standards [3] stated that fluoride in excess of 1.0 part per million (ppm) shall constitute grounds for rejection of the water supply. In other words, the presence of fluoride above 1 ppm in drinking water supplies would render the supply unfit for consumption (i.e., the water resource would be considered "damaged").

The hazards of fluoride were also known to the plant workers. One employee, who began working at the facility in 1951, stated that he was told that the gas from the wet mix operation would be harmful to his health. So, as a safety measure, masks were required to be worn. Another worker, during the same time period, was injured by exposure to hydrofluoric acid and was advised by his doctor that he would be better off if he quit.

These data suggested that during the tenure of Red Ridge Fertilizer Company (i.e., 1940–1962) and NFC (1963–1983), it was known that fluoride was a toxic chemical and that its release to the environment would contaminate water resources. In spite of this knowledge, both companies continued to routinely discharge the HF wastewater into earthen ditches and ponds on the property.

Wastewater Chemistry

From 1940 to 1956, untreated wastewater containing HF was discharged into a ditch that emptied into the facility ponds. From 1956, and until 1983, some of the HF was removed from the wastewater by reclamation as a step in the production of sodium silica fluoride (Na_2SiF_6). This process included the following steps:

1. When calcium fluoride is present in rock phosphate, the following reaction occurs with sulfuric acid:

$$CaF_2 + H_2SO_4 + 2H_2O = CaSO_4 \cdot 2H_2O + 2HF$$

2. Because most rock phosphates contain silica, the following reaction occurs:

$$SiO_2 + 4HF = SiF_4 + 2H_2O$$

3. Silicon tetrafluoride then reacts with water to form hydrofluosilicic acid:

$$3SiF_4 + 4H_2O = 2H_2SiF_6 + Si(OH)_4$$

4. Sodium silica fluoride is then formed by the addition of sodium chloride according to the following reaction:

$$H_2SiF_6 + 2NaCl = Na_2SiF_6 + 2HCl$$

However, not all of the HF acid is removed from the wastewater stream. Under the best of conditions, only a portion of the HF can be removed. This is illustrated by a 1965 study that evaluated the recovery of fluorosilicate from the gypsum settling pond water using the same process used in the sodium silica fluoride process. In this study, the average pond water analysis showed that there was approximately 5001 ppm fluoride. After recovery (with the resulting wastewater redirected to the ponds), test results showed that about 45% of the fluoride remained (i.e., not recovered), and of that amount about half of the fluoride is associated with the gypsum cake. The amount of fluoride that would later go into solution (from the gypsum cake) was unknown. The fluoride recovery method evaluated in the study was not implemented.

In 1962, the facility connected to the city sewer (the wastewater pretreatment step was described earlier). The "clear effluent" was then released to the sewer. Effluent chemistry is compared in Table 23.1 for effluents prior to and after the implementation of the pretreatment process.

These data show that, even after pretreatment, there remained a significant amount of fluoride in the wastewater to be discharged to the municipal sewer.

In 1967, the discharged wastewater was released to the city sewer at a pH of 4. At this pH, it was not possible to achieve the 10-ppm F^- discharge criteria. At the average wastewater discharge of pH 4, the fluoride concentration stood at a measured 2500 ppm. Given the thousands of ppm fluoride in the wastewater effluent, it was apparent that any seepage from the ponds would contain fluoride at a concentration well above drinking water standards (as well as above any natural fluoride in the aquifer).

As shown previously, similar reactions could also occur with soil in the ponds, with calcium, magnesium, sodium, and potassium cations reacting to form fluorosilicate. All of these compounds are well above the solubility of calcium fluoride (CaF_2). Thus, the limiting concentration of fluoride in

Table 23.1 Effluent Chemistry Before and After Treatment

	Water Analyses (mg/L)	
Constituent	*1957 E Pond*	*1961 Clear Effluent*
pH	0.42	10.1
Total acidity	40,000	NA
F^-	2,680	832
Cl^-	26,200	NA
SO_4	2,600	NA
PO_4	271	NA
Ca	120	NA
$Ca(OH)_2$	NA	1,000
Na	4,080	NA
K	10	NA

NA, not analyzed.

alkaline ponds would be controlled by CaF_2, which has a reported solubility of approximately 7.8 ppm, which is still above drinking water standards.

In summary, this basic chemical information was known by the chemical engineers who were employed by both fertilizer companies.[3] As a consequence, these companies understood that fluoride-contaminated wastewater was being discharged into facility ponds well above drinking water quality standards and therefore any leakage or seepage from these ponds (should it occur) would contain substantial amounts of fluoride, resulting in damage to the resource. The question then becomes, did Red Ridge Fertilizer and NFC know that the wastewater would contaminate the groundwater?

The Facility Ponds

A 1954 U.S. Department of Agriculture Soil Survey describes the soils at the site as consisting of loose to very friable gray sands and very sandy subsoils with sandy clay substrata below 30 inches with percolation rates

[3]People often forget that in order to operate a plant that depends on process chemistry, chemists and chemical engineers must be integral to the facility. Since "chemistry" is "chemistry," the only missing ingredient is the "application" of chemical principles throughout the operations.

usually greater than 10 inches of water per hour. This fairly rapid rate of percolation was verified by plant employees (who discussed the observed rapid disappearance of water when placed on the ground).

A soil chemist who worked at the facility commented that the wastewater discharged to the ponds leached into the ground just like rainfall and that the fluoride would move into the soil with the percolating wastewater. Another employee remarked that the north pond never needed to be emptied since the sand would allow the water to seep out of the pond.

Based on the depositions of numerous employees, the ponds had all been made by digging into the ground with a clam bucket to a depth ranging from 8 to 15 feet. There were never any liners in the ponds or ditches. To these observers, none of the ponds was engineered or designed, nor were any drawings or specifications found that addressed pond design. The occurrence of a fence line[4] through one of the ponds clearly demonstrates both the haphazard nature of how these ponds developed and the absence of pond liners (see Figure 23.2).[5] One employee reported a small fish pond that had an asphalt bottom.[6]

All of the employees reported that several times during the year various ponds would be cleaned out to remove the acid sludges, by dredging the pond bottoms. The aerial photograph also indicates that the berms around the ponds were made of soil and/or excavated sludge from the ponds.[7] Several employees also reported on the use of dynamite in the ponds (i.e., to increase pond percolation by breaking up the material caked on the bottom). With the dredging of the pond bottoms aided by the help of dynamite, the leaching of wastewater through the pond bottoms was enhanced. Although a sludge blanket forming on the bottom of the pond over time would reduce the percolation out of the pond, there was no data or likely engineering/scientific reason why percolation would have been totally eliminated.

Further evidence supporting percolation from the ponds came from a review and analysis of climatological data for the city of Crescent Dunes. Figure 23.3 shows the 30-year mean monthly precipitation and potential evapotranspiration for the city.[8] These data demonstrate that, on average,

[4]In the photo in Figure 23.2, the fence line has been computer enhanced.

[5]It is hard to imagine a "lined pond" with a fence line and fence posts running through it.

[6]To use this testimony to infer that "based on deposition testimony, there remains a question as to whether the ponds at the site were lined" is intentionally misleading.

[7]A simple method of berming a pond or lagoon is to take the dredged materials and bulldoze them to the sides, thus building up the berm.

[8]The data and graph were generated by the state's climatologist.

Figure 23.2 Wastewater pond fence line location. In the original aerial photograph, the fence posts are visible. The fence line has been added to this image to show its location.

the monthly precipitation is always greater than the potential evapotranspiration. In other words, the ponds are always filling up with rainwater. The conclusion was that most of the pond wastewater was being lost by percolation and not by evaporation. In addition, none of the employees ever saw the ponds overflow, even during rainy periods. All of this information clearly demonstrated that the ponds leaked and therefore contaminated the groundwater.

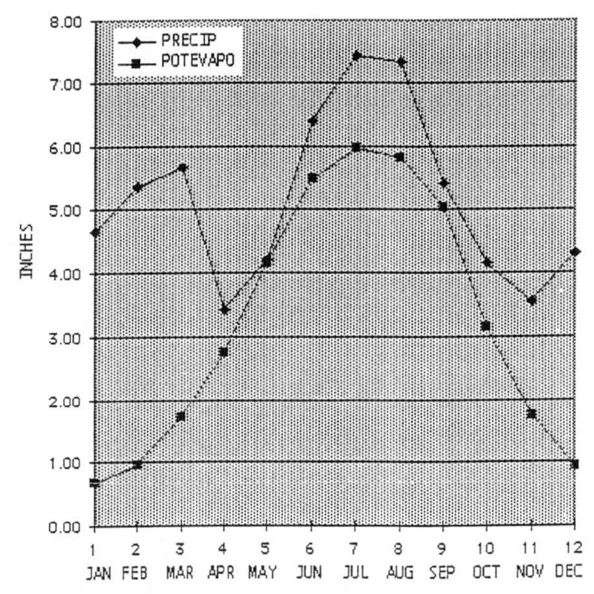

Figure 23.3 City of Crescent Dunes mean monthly precipitation and potential evapotranspiration.

This conclusion is further supported by the groundwater data collected by the city of Crescent Dunes. The well monitored by the city was approximately 6000 feet directly down gradient of the wastewater ponds. The monitoring well data for this well are given in Table 23.2.

Because no other city wells contained fluorides (and the local aquifer from which city water was obtained did not contain fluoride), this well was closed by the city of Crescent Dunes in 1958. The well water was considered contaminated, even though the fluoride content was below the 1942 U.S. Public Health Service standard for rejection.

The fluoride plume in the late 1980s is shown in Figure 23.4. This fur-

Table 23.2 City Monitoring Well Data (mg/L)

	Constituent			
Date	*pH*	SO_4	F^-	*Na*
8/1/57	4.61	25.7	0.33	5.4
9/5/57	4.76	21.5	0.26	5.3
3/15/58	4.52	13.2	0.40	4.8
6/20/58	4.39	20.2	0.39	4.6

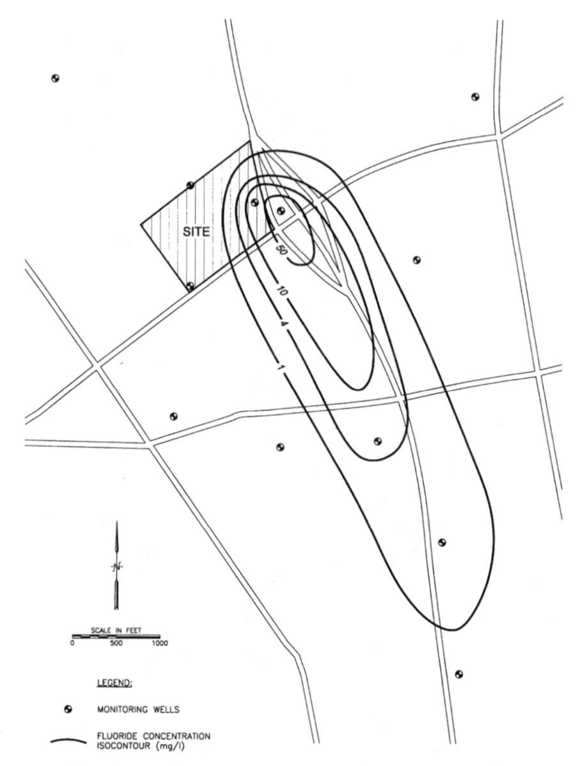

Figure 23.4 Groundwater fluoride plume.

ther supported the fact that the fertilizer wastewater ponds had and were contaminating the drinking water aquifer. By utilizing unlined ponds (state of practice), both Red Ridge and Fertilizer and NFC expected and intended that the ponds would seep and that the chemicals contained therein would reach the groundwater, thereby causing the contamination. Otherwise, recalling that the company had stated that it definitely did not want to pollute the water, it could have lined the ponds (state of the art), preventing the contamination.

State of the Art[9]

When an industry stores or discharges liquid industrial waste or industrial solid waste containing water-soluble chemicals to soil, the pre-1960s

[9]What is the difference between state of the art and state of practice? State of the art is what you know you can do if you desire to do so (i.e., spend the money); state of practice is what you have been doing and continue to do (because it is usually cheaper).

literature clearly demonstrates that seepage will pollute groundwater. During this same time period, the literature also demonstrates that seepage could have been significantly reduced or eliminated by using engineered liners for storage and disposal facilities, as shown by the following examples.

- In 1939, it was reported that "[v]arious materials for sealing the lagoon bottom were considered, including concrete, bitumels, bentonite and clay. . . . Clay was chosen because of its cost. . . . To have sealed it by other methods, such as application of bentonite, would have cost at least $15,000" [4].
- A 1940 article states that "[o]ccasionally, storage pits [for oil field brines] are constructed of wood or steel and sometimes earthen pits are lined inside with gunite, concrete, asphalt or basal sediments. The general practice is to construct the pits by the cheapest method. Little attention is given to selection of most suitable material, proper mixing or compaction of the embankments. . . . These pits are usually pervious to salt water [wastes]. Large quantities of salt water seep through the bottoms into subsurface formations and pollute the underground waters" [5].
- The story remains unchanged in the mid-1950s when it is reported that "[a]dsorption lagoons are specifically designed to permit seepage of liquid into the ground. This seepage ultimately reaches a stream or an underground water stratum; hence adsorption lagoons should not be employed for toxic wastes. . . . All lagoons act as absorption lagoons to some extent unless they are built of an impervious material like concrete, or lined with asphalt or other sealer" [6].
- By 1960, "[s]tudies and tests conducted by several universities and by the U.S. Agricultural Research Service have shown that buried plastic film linings, particularly vinyl and polyethylene, seem to meet the requirements [of an ideal liner] better than other materials" [7].

Based on the literature, both fertilizer companies had every opportunity to appreciate the consequences of unlined ponds and to provide appropriate liners in plant operation, storage, and disposal areas to contain their chemicals. The only issue was cost.

Because the data so clearly support the conclusion that the groundwater contamination from the ponds was expected and intended, the opinion given by NFC's expert appears to intentionally twist the site facts away from the ponds to so-called sudden and accidental events (i.e., the hurricanes). Recalling his fourth opinion in which he stated that, "In the absence of evidence to the contrary, it is reasonable to assume that sudden and accidental events have caused the conditions necessitating remedial actions"—obviously there was sufficient "evidence to the contrary," so that it was not "reasonable to assume" that sudden and accidental events caused the site contamination. The contamination was from the expected and intended use of unlined ponds.

This demonstration alone should be sufficient to rebut the NFC expert's opinions. However, since the expert was relying on the hurricanes as the cause of the site contamination, this opinion also need to be rebutted.

An Unsupported Theory

In order to establish an opinion that individual hurricanes caused a pond to overtop its berm requires, at a minimum, knowledge of the pond wastewater elevation relative to the pond's freeboard,[10] the rate of water loss from the pond, and the rate of precipitation into the pond. Obviously, if the input rate is greater than the output and the freeboard is less than the net increase in the pond water elevation, the wastewater will overtop the pond berm.

Predicting Pond Elevation

In order to predict the elevation of any pond (liquid), it is necessary to know the following:

- What was the pond water level on the day of the hurricane? Since the ponds were reported to have been from 8 to 15 feet deep, there could have been more than ample stormwater storage capacity. Several employees reported that the ponds were periodically dry. Since there are no pond elevation data, the freeboard would have been unknown.
- The pond elevation would be subject to both wastewater input and output functions. For example, what were the percolation and evaporation rates (thus, the drop in pond elevation). What was the wastewater input and discharge to the sewer?

[10]The distance between the pond water elevation and the top of the berm is the freeboard. This is a fundamental property of an engineered pond that is usually taken into account so that stormwater input (i.e., a hurricane) will not exceed the designed freeboard.

National Fertilizer Company's expert failed to present any of this information.[11] Therefore, he presented no scientific basis by which to establish a pond's freeboard. As a result, even if the expert knew the amount of precipitation from a given hurricane, an overtopping event could not be predicted (even taking into account wind factors[12]). He did, however, have precipitation data or at least he thought he had.

Interpreting Rainfall Data

As shown earlier, the expert alleged that the hurricanes in June 1953, September 1953, September 1956, October 1959, June 1966, and September 1975 caused the overtopping and associated damage. The precipitation data for the city of Crescent Dunes are compared to the dates of the hurricanes in Table 23.3.

These data demonstrate that three of the hurricanes (June 1953, October 1959, and June 1966) would have added little or no water (maximum up to 0.18 foot of water) to the pond. Therefore, half of his opinion about the impact of the hurricanes was completely wrong. The storms of September 1953 and September 1975 would have added 0.76 to 0.59 foot of water to the pond. For the September 1953 storm, the average precipitation would have been equal to approximately 0.25 foot per day. Over this 3-day period, the net output (seepage) was unknown. The September 1975 storm had a significantly less daily contribution over the 10 days of the hurricane. These data suggest that the hurricanes of September 1953 and September 1975 had little impact on the water level of the ponds, over the duration of the storms.

The only hurricane that had a potential impact would have been the storm of October 1959. During that storm, 0.97 foot of water would have been added to the ponds in one day. Therefore, there is only one potential "act of God" that may have caused the groundwater contamination (assuming that, even after seepage, insufficient freeboard existed). However, the city of Crescent Dunes had already determined two years earlier that the groundwater aquifer was contaminated. In addition, no plant employee ever reported overtopping of the ponds or flooding of the plant site.

Finally, the pattern of soil fluoride contamination is not consistent with "an act of God" theory. As discussed previously, the ROD remedial actions

[11]What is illustrated here is a basic failure to carry out a simple water "mass balance" on the pond.

[12]Which were not measured.

Table 23.3 Hurricane and Precipitation Data Comparison

June 1953 Precipitation Day	Amount	
		Hurricane duration June 2 to June 6
9	0.67	
12	0.98	
14	0.47	
15	1.51	
16	0.97	
22	0.54	
25	0.23	
26	1.23	
27	2.83	
28	0.67	
Total	10.10	Total precipitation from hurricane = 0

September 1953 Precipitation Day	Amount	
3	0.40	
4	1.46	
18	0.85	
19	3.12	
24	0.72	Hurricane duration September 23 to 28
25	4.69	
26	3.72	
Total	14.96	Total precipitation from hurricane = 9.13

September 1956 Precipitation Day	Amount	
1	0.30	
2	0.23	
7	0.96	
8	0.06	
16	0.01	
17	0.18	
18	0.06	
22	0.39	Hurricane duration September 21 to 30
23	2.57	
24	11.68	
25	0.09	
Total	16.53	Total precipitation from hurricane = 14.34

October 1959 Precipitation		
Day	*Amount*	
5	0.11	
6	0.89	Hurricane duration October 6 to 8
7	0.81	
8	0.54	
10	3.15	
11	0.56	
14	1.01	
15	0.24	
16	0.78	
17	0.09	
20	0.25	
21	0.09	
22	0.05	
28	2.16	
29	1.88	
30	0.02	
Total	12.63	Total precipitation from hurricane = 2.24

June 1966 Precipitation		
Day	*Amount*	
9	0.01	Hurricane duration June 4 to 14
15	0.39	
16	0.41	
17	0.05	
20	0.19	
21	0.15	
29	0.79	
Total	2.00	Total precipitation from hurricane = 0.01

September 1975 Precipitation		
Day	*Amount*	
6	3.50	
7	0.61	
11	0.22	
12	0.11	Hurricane duration September 13 to 24
17	1.47	
21	0.93	
22	3.52	
23	1.12	
Total	11.53	Total precipitation from hurricane = 7.04

required that the soils containing fluoride concentrations greater than 1500 ppm be excavated, solidified, and consolidated into a single pond, which would then be capped. Figure 23.5 shows the location of the former boundaries relative to soil samples that exceed 1500 ppm fluoride. This demonstrates that the vast majority of the fluoride contamination was directly under the ponds, as expected. The few areas of soil contamination outside pond boundaries are more likely the result of berm erosion and transport of fluoride sludges away from the pond boundaries. This erosion and transport would have occurred under normal precipitation and would not depend on high-precipitation events alone.

Based on Figure 23.5, it is clear that the groundwater contamination did not occur from the fluoride contamination outside the ponds. In fact, the vast majority of the soil fluoride contamination and the groundwater contamination resulted from the discharge of the fluoride wastewater to the unlined ponds.

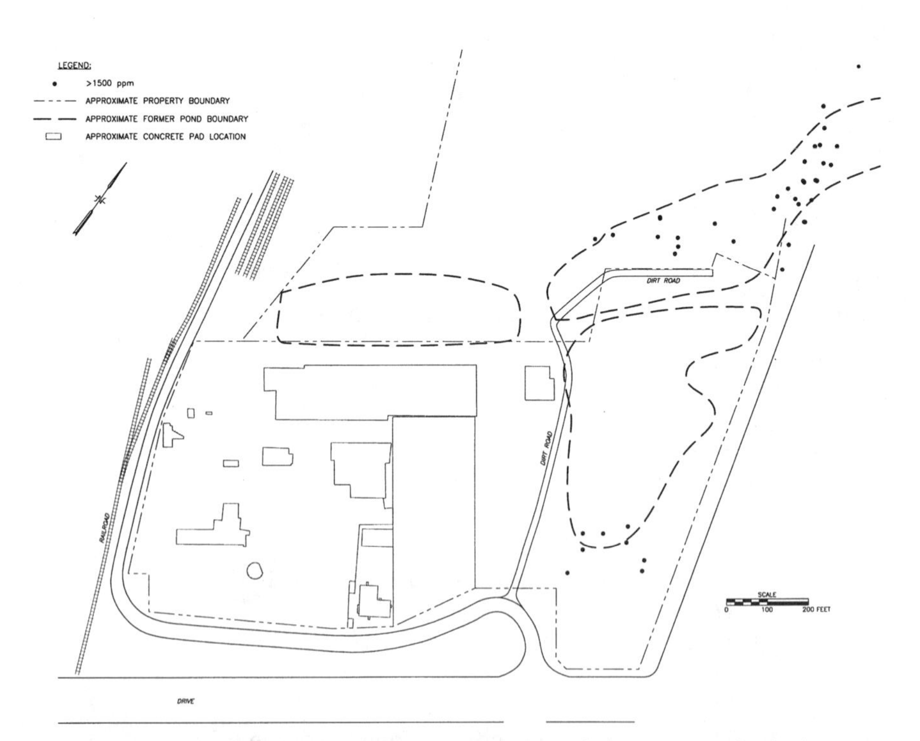

Figure 23.5 Soil fluoride concentrations that exceed cleanup standard relative to pond boundaries.

The opposition report containing the information discussed previously was finalized and forwarded to counsel for the insurance company. With the expert reports for both consultants completed, the deposition of each expert began.

DEPOSITIONS

The depositions began with the NFC's expert. A summary of some of the opinions he expressed during his testimony follows:

- After being shown numerous historical professional articles on unlined pits, ponds, lagoons, and landfills that were shown to have contaminated groundwater, he conceded that unlined ponds would pollute groundwater. However, it was his opinion that the pond bottoms and sides would have been "sealed" by the sludge, although he did not know the permeability of the ponds' sides or bottoms.
- As a Ph.D. engineer, he did not understand "freeboard." Therefore, he had no ability or data to support his opinion that the ponds overtopped their berms, other than the fact that there had been numerous hurricanes with high precipitation and winds that would have caused these sudden and accidental events.
- When shown all of the climatic data for the city of Crescent Dunes, he admitted that he had made a mistake. However, he did mention that it was not his fault since his staff did not collect appropriate climatic data for him to use in support of his opinions (or were the data omitted on purpose?).

The deposition of the insurance company's forensic expert was well supported, both by experience and by documentation, and thus was not impeached. A summary of some of the forensic expert's opinions given during his testimony follows:

- At this site, the unlined ponds, located in sandy soil and close to the groundwater, allowed the fluoride-contaminated wastewater to percolate into the soil and thus into the groundwater. Although a sludge blanket developed in the ponds over time, the bottoms were periodically disturbed (i.e., by excavation and blasting) so as to allow greater seepage than when the sludge blanket had formed. A sludge blanket, however, while having some effect on the perme-

ability, cannot be considered equal to an "engineered" liner. Further, employee testimony supported pond leakage.[13]
- There was no scientific basis or testimony to conclude that the ponds were ever overtopped due to a hurricane.
- Even on the odd chance that an overtopping event did occur, the amount of contamination resulting from such an overtopping (recognizing that the upper surface of water in the ponds contained the lowest concentrations of contaminants) would have been negligible when compared to the continuous seepage from the ponds.
- Properly engineered ponds include consideration of events such as storms when calculating freeboard.
- The fertilizer companies could have used state-of-the-art technologies that would have substantially reduced or eliminated the contamination.

Within days of the forensic expert's deposition, the case settled. The National Fertilizer Company's expert did not convince his client or counsel for the insurance company that the fertilizer company could prevail in court with "an act of God" opinion. As a result, the settlement was highly favorable to the insurance company.

LESSONS LEARNED

1. Experts must be careful to review all data collected by "staff" to ensure that there are no errors and/or data gaps.
2. Even under the most obvious site conditions, an expert can still concoct a seemingly "scientific" opinion that is intentionally misleading so as to support his client's truth.

REFERENCES

1. W. T. Brande and A. S. Taylor, *Chemistry*, 2nd American edition, Blanchard and Lea, Philadelphia: (1867).
2. P. Parrish and A. Ogilvie, *Calcium Superphosphate and Compound Fertilisers: Their Chemistry and Manufacture*, Hutchinson's Scientific and Technical Publications, Paternoster House, London: (1939).

[13]It is hard to argue something did not happen when people saw it happen and said so.

3. U.S. Public Health Service, "Public Health Service Drinking Water Standards, *Public Health Reports*, 58(3): 69 (January 15, 1943).
4. C. Lee, "Sealing the Clay Lining of the Lagoon at Treasure Island by Use of Sea Water," presented at meeting of Soil Mechnaics and Foundation Division of American Society of Civil Engineers, San Francisco, California (July 26, 1939).
5. B. Williams, "Salt of the Earth—a Pollution Problem," *Water Works Engineering*, 93(19): 1165 (September 11, 1940).
6. F. Grunham, *Principles of Industrial Waste Treatment*, New York: John Wiley & Sons (1955).
7. J. Anderson, "Vinyl Film Liner for Earth-Fill Reservoirs," *Civil Engineering*, 30(6): 42 (June 1960).

Appendix A

A Historic Review of Domestic Water Quality Criteria and Standards

This review is not intended to be exhaustive. Rather, it is provided to give a sense of the evolutionary changes in water quality criteria. Given the vast array of water quality criteria for domestic, aquatic, agricultural, industrial, and recreational uses, this review only addresses chemical domestic standards.

Water quality criteria specify concentrations of water constituents that, if not exceeded, are expected to result in a water environment that allows an identified use (i.e., use as a domestic water supply, use as a freshwater or marine ecosystem, use for agriculture, etc.). If water quality criteria are violated for any substantial length of time, an adverse effect on aquatic life, hazards to humans or other consumers of aquatic life, may result. Thus, when evaluating the potential impacts of chemicals on water quality, any adverse effects must be considered for all aquatic environments, not just hazards to humans.

Criteria, however, are not standards. Water quality standards connote a legal requirement for a particular water resource or for an effluent limitation. A water quality standard may use a water quality criterion as a basis for regulation or enforcement, but the standard can differ from the criterion based on site-specific conditions or hazards.

Water quality criteria are derived from experimental/*in situ* observations of the acute and chronic toxicity of aquatic organisms. Given the great variation among aquatic species, a specified concentration of a toxic compound will elicit a very different response for each organism. Historically, studies that began in the early 1900s only evaluated a relatively small number of compounds using a limited number of aquatic organisms. Given the vast array of organisms and chemicals, researchers resorted to using "indicator organisms" to assess general levels of toxicity. As toxicity studies have advanced over the years (which also included carcinogenic and teratogenic evolutions since the 1960s), it is no wonder that water quality criteria have changed and expanded with time.

CRITERIA AND STANDARDS

Report of Advisory Committee on Standards for Drinking Water Supplied to the Public by Common Carriers in Interstate Commerce, 1925

"The water should be . . . clear, colorless, odorless, and pleasant to the taste, should be free from toxic salts, and should not contain any excessive amount of soluble mineral substances nor of any chemicals employed in treatment. . . . Appropriate tests for the quantitative determination of physical and chemical characteristics are given in Appendix IV of this report together with the values which should ordinarily not be exceeded when these tests are applied."

Chemical Substance	*Limit Not to Be Exceeded (ppm)*
Lead	0.1
Copper	0.2
Zinc	5.0
Sulfate	250
Magnesium	100
Total solids	1000
Chloride	250
Iron	0.3

ppm = mg/L (milligrams per liter or parts per million).

U.S. Public Health Service, Drinking Water Standards, 1942

The presence of the following substance above the published limit shall constitute grounds for rejection of the water supply:

Chemical Substance	*Limit Not to Be Exceeded (ppm)*
Lead	0.1
Fluoride	1.0
Arsenic	0.05
Selenium	0.05

"The following chemical substances which may be present in natural or treated waters should preferably not occur in excess of the following concentrations where other more suitable supplies are available in the judgement of the certifying authority."

Chemical Substance	*Limit Not to Be Exceeded (ppm)*
Copper	3.0
Iron and manganese together	0.3
Magnesium	125
Zinc	15
Chloride	250
Sulfate	250
Phenolic compounds	0.001
Total solids	500

U.S. Public Health Service, Drinking Water Standards, 1962

These standards provide the following two types of drinking water limits: "Limits which, if exceeded, shall be grounds for rejection of the supply. Substances in this category may have adverse effects on health when present in concentration above the limit."

Substance	*Concentration (mg/L)*
Arsenic	0.01
Barium	1.0
Cadmium	0.01
Chromium (hexavalent)	0.05
Cyanide	0.2
Lead	0.05
Selenium	0.01
Silver	0.05

"Limits which should not be exceeded whenever more suitable supplies are, or can be made, available at reasonable cost. Substances in this category, when present in concentrations above the limit, are either objectionable to an appreciable number of people or exceed the levels required by good water quality control practices."

Substance	*Concentration (mg/L)*
Alkyl benzene sulfonate (ABS)[1]	0.5
Arsenic	0.01
Chloride	250
Copper	1
Carbon chloroform extract (CCE)	0.2
Cyanide	0.01
Iron	0.3
Manganese	0.05
Nitrate	45
Phenols[2]	0.001
Sulfate	250
Total dissolved solids (TDSs)	500
Zinc	5

One of the most important criteria in the 1962 standards is the CCE. According to these standards, the CCE is a "... practical measure of water quality and as a safeguard against the intrusion of excessive amounts of potentially toxic material into water. ... It is proposed as a technically practical procedure which will afford a large measure of protection against the presence of undetected toxic materials in finished drinking water." It is important to note that, by the very nature of the method, the CCE procedure underestimates the amount of potentially toxic organics in water. Thus, any detectable level of CCE organics is a conservative estimate of potential toxic materials in water.

[1]An anionic surfactant.

[2]The term "phenols" is understood to include cresols and xylenols.

Federal Water Pollution Control Administration, Water Quality Criteria, 1968

Surface Water Criteria for Public Water Supplies[3]

Constituent / Characteristic	*Permissible Criteria[4] (mg / L)*
Ammonia	0.5 (as N)
Arsenic	0.05
Barium	1.0
Boron	1.0
Cadmium	0.01
Chloride	250
Chromium (hexavalent)	0.05
Copper	1.0
Iron (filterable)	0.3
Lead	0.05
Manganese (filterable)	0.05
Nitrates plus nitrites	10 (as N)
Selenium	0.01
Silver	0.05
Sulfate	250
TDS	500
Uranyl ion	5
Zinc	5
CCE	0.15
Cyanide	0.20
Pesticides	
Aldrin	0.017
Chlordane	0.003
DDT	0.042
Dieldrin	0.017
Endrin	0.001
Heptachlor	0.018
Heptachlor epoxide	0.018
Lindane	0.056
Methoxychlor	0.035
Organic phosphates plus carbamates	0.1

[3]The Subcommittee recognizes that surface waters are used for public water supply without treatment other then disinfection. Such water at the point of withdrawal should meet Drinking Water Standards in all respects.

[4]Those characteristics and concentrations of substances in raw surface waters which will allow the production of a safe, clear, potable, aesthetically pleasing and acceptable public water supply which meets the limits of Drinking Water Standards after treatement.

Toxaphene	0.005
Herbicides	
2,4-D plus 2,4,5-T plus 2,4,5-TP	0.1
Phenols	0.001

	Radioactivity (pCi/L)
Gross β	1.0
Radium-226	3
Strontium-90	10

pCi/L = picocuries per liter.

Water Quality for Freshwater and Marine Organisms

Substances of Unknown Toxicity "All effluents containing foreign materials should be considered harmful and not permissible until bioassay tests have shown otherwise."

Industrial and Other Toxic Wastes "Wastes from tar, gas, coke, petrochemical, pulp and paper manufacturing, waterfront and boating activities, hospitals, marine laboratories and research installation wastes are all complex mixtures having great variability in character and toxicity. Due to this variability, safe levels must be determined at frequent intervals using flow-through bioassays of the individual effluents."

U.S. Environmental Protection Agency, Quality Criteria for Water, 1976

Substance	*Criteria (mg/L)*
Arsenic	0.05
Barium	1.0
Cadmium	0.01
Chromium	0.05
Copper	1.0
Iron	0.3
Lead	0.05
Manganese	0.05
Mercury	0.002
Nitrates	10.0
2,4-D	0.1
2,4,5-TP	0.01

Endrin	0.0002
Lindane	0.004
Methoxychlor	0.1
Toxaphene	0.005
Phenol	0.001
Selenium	0.01
Silver	0.05
Chloride	250
Sulfate	250
Zinc	5

U.S. Environmental Protection Agency, 1980

Toxic Pollutant	*Human Health Effects (μg/L)*	*Cancer Risk (1/10⁶)*
Benzene	0[a]	0.66 μg/L
Carbon tetrachloride	0[a]	0.40 μg/L
Chlorobenzene	488	N/A
1,2-Dichloroethane	0[a]	0.94 μg/L
1,1,1-Trichloroethane	18,400	N/A
1,1,2-Trichloroethane	0[a]	0.6 μg/L
1,1,2,2-Tetrachloroethane	0[a]	0.17 μg/L
Chloroform	0[a]	0.19 μg/L
1,1-Dichloroethylene	0[a]	0.033 μg/L
Toluene	14,300	N/A

μg/L = micrograms per liter (parts per billion).

[a] For these pollutants, EPA recommended that the human health criteria be set at zero, but noted that this level might not be feasible. Therefore, it also provided interim risk levels based on calculations that would permit one case of cancer per million people over a lifetime.

U.S. Environmental Protection Agency, Primary Drinking Water Standards, 1995

Substance	*Maximum Contaminant Level*	
Alachlor	2.0	μg/L
Aldicarb	3.0	μg/L
Aldicarb sulfoxide	4.0	μg/L
Aldicarb sulfone	2.0	μg/L
Atrazine	3.0	μg/L
Bentazon	18.0	μg/L
Benzene	1.0	μg/L
Benzo[*a*]pyrene	0.2	μg/L
Carbon tetrachloride	0.5	μg/L
Carbofuran	18.0	μg/L
Chlordane	0.1	μg/L
Dalapon	200.0	μg/L
1,2-Dibromo-3-chlorophane (DBCP)	0.2	μg/L
1,2-Dichlorobenzene (*o*-DCB)	600.0	μg/L
1,4-Dichlorobenzene (*p*-DCB)	5.0	μg/L
1,1-Dichloroethane (1,1-DCA)	5.0	μg/L
1,2-Dichloroethane (1,2-DCA)	0.5	μg/L
cis-1,2-Dichloroethylene (*c*-1,2-DCE)	6.0	μg/L
trans-1,2-Dichloroethylene (*t*-1,2-DCE)	10.0	μg/L
1,1-Dichloroethylene (1,1-DCE)	6.0	μg/L
Dichloromethane (methylene chloride)	5.0	μg/L
2,4-Dichlorophenoxyacetic acid (2,4-D)	70.0	μg/L
1,2-Dichloropropane	5.0	μg/L
1,3-Dichloropropene	0.5	μg/L
Di(2-ethylhexyl)adipate	400.0	μg/L
Di(2-ethylhexyl)phthalate (DEHP)	4.0	μg/L
Dinoseb	7.0	μg/L
Diquat	20.0	μg/L
Endothall	100.0	μg/L
Endrin	0.2	μg/L
Ethylbenzene	680.0	μg/L
Ethylene dibromide (EDB)	0.02	μg/L
Glyphosphate	700.0	μg/L
Heptachlor	0.01	μg/L
Heptachlor epoxide	0.01	μg/L
Hexachlorobenzene	1.0	μg/L
Hexachlorocyclopentadiene	50.0	μg/L

Lindane	2.0	μg/L
Methoxychlor	40.0	μg/L
Molinate	20.0	μg/L
Monochlorobenzene (chlorobenzene)	30.0	μg/L
Oxamyl (Vydate)	200.0	μg/L
Pentachlorophenol	1.0	μg/L
Picloram	500.0	μg/L
Polychlorinated biphenyls (PCBs)	0.5	μg/L
Simazine	4.0	μg/L
Styrene	100.0	μg/L
2,3,7,8-TCDD (dioxin)	0.00003	μg/L
1,1,2,2-Tetrachloroethane	1.0	μg/L
Tetrachloroethylene (PCE)	5.0	μg/L
Thiobencarb	70.0	μg/L
Toluene	1,000.0	μg/L
Toxaphene	3.0	μg/L
1,1,2-Trichlorobenzene	70.0	μg/L
1,1,1-Trichloroethane (1,1,1-TCA)	200.0	μg/L
1,1,2-Trichloroethene (1,1,2-TCA)	5.0	μg/L
Trichloroethylene (TCE)	5.0	μg/L
Trichlorofluoromethane (Freon 11)	150.0	μg/L
1,1,2-Trichloro-1,2,2-trifluoroethane (Freon 113)	1,200.0	μg/L
2,4,5-Trichlorophenoxy-proprionic acid (2,4,5-TP Silvex)	10.0	μg/L
Vinyl chloride	0.5	μg/L
Xylenes (*m,p,* and *o*)	1,750.0	μg/L
Total trihalomethanes	100.0	μg/L
Aluminum	1.0	mg/L
Antimony	0.006	mg/L
Arsenic	0.05	mg/L
Barium	1.0	mg/L
Beryllium	0.004	mg/L
Cadmium	0.0005	mg/L
Chromium	0.05	mg/L
Copper	1.3	mg/L
Cyanide	0.2	mg/L
Lead	0.015	mg/L
Mercury	0.002	mg/L
Nickel	0.1	mg/L
Nitrate (as N)	10.0	mg/L
Nitrite (as N)	1.0	mg/L
Selenium	0.01	mg/L
Silver	0.05	mg/L
Thallium	0.002	mg/L

Gross α	15	pCi/L
Gross β	50	pCi/L
Radium-226 and -228	5	pCi/L
Strontium-90	8	pCi/L
Tritium	20,000	pCi/L
Uranium	20	pCi/L

Appendix B

Selected Portions of State Pollution Control Regulations for California, Georgia, Illinois, Louisiana, Massachusetts, Michigan, New Jersey, Ohio, PennysIvania, and Texas

CALIFORNIA

1949 The Dickey Act: California Water Quality Control Act

Legislative History of Dickey Act

- The Dickey Act resulted from the legislative activities that began on January 10, 1947, when Assemblyman Dickey introduced House Resolution No. 27 to the California State Assembly. This resolution, which was adopted by the House on June 20, 1947, called for the creation of a Fact-Finding Committee on the Problems of Water Pollution resulting from the disposal of industrial wastes and sewage. The committee was "directed to ascertain, study and analyze all facts relating to procedures designed to prevent pollution of (1) the underground water flows, (2) the surface streams and (3) the lower reaches of the principal river systems, and coastal bays and shores; including but not limited to the operation, effect, administration, enforcement and needed revision of any and all laws in any way bearing upon or relating to the subject of this resolution."

- The committee was also directed to present its findings to the state legislature by the tenth day of the 1949 legislative session. The importance of this report was duly noted by Governor Earl Warren in his January 3, 1949 Annual Message to the Legislature. In that speech, Warren called attention to "the need for preventing the contamination of the waters of this State—the waters of our streams, our offshore waters, and our underground waters. These waters are being polluted to an alarming degree by industrial wastes and sewage disposal . . . [and] I trust that your committee studies of this subject will result in effective measures to meet this condition."

Dickey Act and Industrial Wastes

- The Water Pollution Control Act of 1949 was a direct result of the work done by Dickey's fact-finding committee. The act, which created the California State Water Pollution Control Board, defines "waters of the State" as "any waters, surface or underground" and "pollution" as any impairment of the quality of waters by "sewage or other wastes" that might "adversely or unreasonably affect such waters for domestic, industrial, agricultural, navigational, recreational or other beneficial use." "Other wastes" are defined as "any and all liquid or solid waste substance, not sewage, from any producing, manufacturing or processing operation of whatever nature" and clearly include all industrial wastes—both liquid and solid.
- The act required that "any person proposing to discharge sewage or other waste . . . shall file with the regional board of that region a report of such proposed discharge." The Regional Water Quality Control Board "after any necessary hearing, shall prescribe requirements as to the nature of such proposed or existing discharge with relation to the conditions existing from time to time in the disposal area or receiving waters upon or into which the discharge is made or proposed." This section effectively allowed the board to issue industrial ("other") waste disposal requirements either by describing the characteristics of the discharge or by describing the discharge in terms of the condition to be maintained in the receiving waters or disposal area—or by a combination of both of these methods. In a subsequent opinion, the term "disposal area" refers to indirect discharges onto the ground. Thus, the board could also impose requirements for indirect discharge of industrial wastes.

GEORGIA

1957 Georgia Water Quality Control Act

- It shall be unlawful for any person to discharge or permit to be discharged into any of the waters of this State any sewage, industrial wastes or other wastes which would adversely affect the health of the people of the State through the dissemination of toxic or radioactive substances.
- It shall be also unlawful to use any waters of the State for the disposal of sewage, industrial wastes, or other wastes so as to render such waters unsuitable for their then current uses.

1964 Georgia Water Quality Control Act

- An Act to create within the Department of Public Health a Division of Georgia Water Quality Control and a State Water Quality Control Board; to prohibit pollution, to define the method of obtaining permits for the disposal of sewage, industrial and other wastes in the waters of the State.
- The people of the State of Georgia are dependent upon the rivers, streams, lakes and subsurface waters of the State for public and private water supply and for agricultural, industrial and recreational uses.
- "Waters" or "waters of the state" includes any and all rivers, streams, creeks, branches, lakes, reservoirs, ponds, drainage systems, springs, wells, and all other bodies of surface or subsurface water, natural or artificial, lying within or forming a part of the boundaries of the State which are not entirely confined and retained completely upon the property of a single individual, partnership or corporation.
- "Pollution" means any alteration of the physical, chemical, or biological properties of the waters of this State, including change of the temperature, taste or odor of the waters, or the addition of any liquid, solid, radioactive, gaseous, or other substances to the waters or the removal of such substances from the waters, which will render or is likely to render the waters harmful to the public health, welfare, or harmful or substantially less useful for domestic, municipal, industrial, agricultural, recreational, or other lawful uses, or for animals, birds, or aquatic life.

- "Sewage" means the water carried waste products or discharges from human beings or from the rendering of animal products, or chemicals or other wastes from residences, public or private buildings, or industrial establishments, together with such ground, surface or storm water as may be present.
- "Industrial wastes" means any liquid, solid or gaseous substance or combination thereof resulting from a process of industry, manufacturing, or business or from the development of any natural resources.
- There is hereby created within the Department of Public Health a division to be known as the Division of Georgia Water Quality Control.
- It shall be unlawful to use any waters of the State for the disposal of sewage, industrial wastes, or other wastes.
- No person, without first securing from the board a permit, shall construct, install or modify any system for disposal of sewage, industrial wastes, or other wastes or any extension or addition thereto when the disposal of the sewage, industrial wastes or other wastes constitutes pollution.
- Any person desiring to erect or modify facilities or commerce or alter an operation of any type which will result in the discharge of sewage, industrial wastes or other wastes into the waters of the State shall apply to the Board for a permit to make such discharge as defined in this Chapter.

1964 Water Supply Quality Control

- The public and community water supplies. . . . their use and need for such water and the quality of such water is a major factor involving the health and welfare of all people in the State of Georgia.
- Waters or "Waters of the State" means and includes any and all rivers, streams, creeks, branches, lakes, reservoirs, ponds, drainage systems, springs, wells and all other bodies of surface or underground water, natural or artificial, of this State.

ILLINOIS

1911 Act, Rivers and Lakes Commission

- That the Governor of the State of Illinois shall forthwith after the taking effect of this Act, appoint a Rivers and Lakes Commission in and for the State of Illinois.

- It shall be the duty of said Rivers and Lakes Commission to see that all of the streams and lakes of the State of Illinois wherein the State of Illinois or any of its citizens has any right, or interest, are not polluted or defiled by the deposit or addition of any injurious substances, and that the same are not affected injuriously by the discharging therein of any foul or injurious substances, so that fish and other aquatic life is destroyed.

1921 Amendments to the Rivers and Lakes Commission Act

- It shall be unlawful for any person, firm or corporation, to throw, discharge, dump or deposit, or cause, suffer, or procure to be thrown, discharged, dumped or deposited any acids, chemicals, industrial wastes or refuse, poisonous effluent or dye-stuff, clay or other washings, or any other substance deleterious to fish life, or any refuse matter of any kind or description containing solids, substance, discoloring, or otherwise polluting navigable lake, river or stream in this State, or lake, river or stream connected with or the waters of which discharge into any navigable lake, river or stream of this State or upon the borders thereof, or any watercourse whatsoever.

1929 Act, Sanitary Water Board

- An Act to establish a Sanitary Water Board and to control, prevent and abate pollution of the streams, lakes, ponds and other surface and underground waters in the State.
- It shall be the duty of the Sanitary Water Board to study, investigate and, from time to time, determine, ways and means of eliminating from the streams and waters of the State so far as practicable, all substances and materials which pollute, or tend to pollute, the same and to determine methods, so far as practicable, of preventing pollution that is detrimental to the public health, or to the health of animals, fish or aquatic life.
- Pollution shall be regarded as existing in any said waters if the result of any discharge of any liquid or solid substance, the quality of said waters, is impaired for public water supply, bathing or recreational purposes; or if obnoxious odors result from such discharge near buildings, roads and lands occupied or used by human beings.

- No sewerage system which proposes to discharge into any of the aforesaid waters sewage or any liquid or solid substance of a decomposable or putrescible, acid or other character, which may cause pollution of any of the aforesaid waters of the State, shall be installed until a written permit for such sewerage system has been granted by the Sanitary Water Board.
- If the pollution of any of the aforesaid waters within the meaning of this Act is continued contrary to orders of the Sanitary Water Board, it shall constitute a nuisance which may be abated in actions commenced and maintained by the Attorney General.

1951 Sanitary Water Board, Revised Statute

- Whereas the pollution of the waters of this State constitutes a menace to public health and welfare, creates public nuisances, is harmful to wildlife, fish and aquatic life, and impairs domestic, agricultural, industrial, recreational and other legitimate beneficial uses of water, and whereas the problem of water pollution in this State is closely related to the problem of water pollution in adjoining states, it is hereby declared to be the public policy of this State to maintain reasonable standards of purity of the waters of the State consistent with their use for domestic and industrial water supplies, for the propagation of wildlife, fish and aquatic life, and for domestic, agricultural, industrial, recreational and other legitimate uses; to provide that no waste be discharged into any waters of the State without first being given the degree of treatment necessary to prevent the pollution of such waters; to provide for the pollution, abatement, and control of new or existing water pollution.
- "Pollution" means such alteration of the physical, chemical or biological properties of any waters of the State, or such discharge of any liquid, gaseous or solid substance into any waters of the State as will or is likely to create a nuisance or render such waters harmful or detrimental or injurious to public health, safety or welfare, or to domestic, commercial, industrial, agricultural, recreational, or other legitimate uses or to livestock, wild animals, birds, fish or other aquatic life.
- "Sewage" means the water-carried human or animal wastes from residences, buildings, industrial establishments, or other places, to-

gether with such ground water infiltration and surface water as may be present.

- "Industrial Waste" means any liquid, gaseous, solid or other waste substance or a combination thereof resulting from any process of industry, manufacturing trade or business or from the development, processing or recovery of any natural resources.
- "Waters of the State" means all accumulations of water, surface and underground, natural or artificial, public or private or parts thereof, which are wholly or partially within, flow through, or border upon this State or within its jurisdiction.
- "Stream standard" means such measure of purity or quality for any waters in relation to their reasonable and necessary use.
- The Board is hereby authorized to issue, continue in effect, or deny permits, under such conditions as it may determine to be reasonable for the prevention and abatement of pollution, for the discharge of sewage, industrial waste or other wastes, or for the installation or operation of sewage works or parts thereof.
- No person shall throw, run, drain or otherwise dispose into any waters of this State, or cause, permit, suffer to be thrown, run, drained, allow to seep or otherwise dispose into such waters, any organic or inorganic matter that shall cause pollution of such waters.

1964 Illinois Sanitary Water Board, Rules and Regulations SWB-8

- These Minimum Criteria shall apply to all waters at all places and at all times in addition to specific criteria applicable to specific sectors.
- Free from substances attributable to municipal, industrial or other discharges that will settle to form putrescent or otherwise objectionable sludge deposits; or which will form bottom deposits that may be detrimental to bottom biota (such as coal fines, limestone dust, fly ash, etc.); free from floating debris, oil, scum and other floating material attributable to municipal, industrial or other discharges; free from materials attributable to municipal, industrial or other discharges producing color, odor or other conditions in such degree as to create a nuisance; free from substances attributable to municipal, industrial or other discharges in concentrations or combinations which are toxic or harmful to human, animal or aquatic life.

LOUISIANA

1910 Acts No. 183, House Bill No. 19

- To protect rice planters and owners of the canals who use water for irrigation purposes against the pollution of the streams by salt water, oil and other substances, and also to protect the fish in said streams and making it a misdemeanor to contaminate said streams by draining or permitting the said water to be drained in said streams.
- It is hereby declared unlawful and a misdemeanor for any officer, manager, or employee of any corporation or any person acting for himself, or for any one else to knowingly and willfully empty or drain into, or permit to be drained from any pumps, reservoirs, wells, or oil fields into any of the natural streams or drains of the said State, from which water is taken for irrigation purposes any oil, salt water or other noxious or poisonous gases or substances which would render said water unfit for irrigation purposes or would destroy the fish in said streams.

1940 Acts No. 367, House Bill No. 994

- To create a Stream Control Commission to have control of the streams and waterways and coastal waters of the State and waste disposal therein and any pollution thereof with power to make rules and regulations governing the same; to prohibit the harmful pollution of any waters of the State and the coastal waters of the Gulf of Mexico within the territorial jurisdiction of the State of Louisiana; to regulate and control public and private waste disposal into any of the waters of the State.
- There is hereby created a Stream Control Commission herein referred to as the Commission, which shall consist of the Commissioner of Conservation, the President of the State Board of Health and the Attorney General of the State of Louisiana or their duly authorized representatives.
- That it shall be unlawful for any person to discharge or permit to be discharged into any of the rivers, streams, lakes or other waters of the State any waste or any pollution of any kind that will tend to destroy fish or other aquatic life or wild or domestic animals or fowls or be injurious to the public health or against the public wel-

fare in violation of any rule, order or regulation of the Commission, and any person so discharging any such waste or pollution into any of the waters of this State shall be deemed guilty of a violation of the provisions of this Act.

- In an emergency causing or likely to cause irreparable damage, or if the public interest requires, the Commission may issue a temporary order requiring that such waste disposal and such waste discharge or pollution be stopped and terminated.
- "Person" shall be construed to include any municipality, industry, public or private corporation, or co-partnership, firm or any other entity.
- "Waters of the State" shall be used to include rivers, streams, lakes and all other water courses and waters within the confines of the State of Louisiana and also the Gulf of Mexico and other bordering waters.

1948 Acts No. 386, R.S. 56: 362

- Pollution of Waters: In order to prevent the pollution of any of the waters of the state, the killing of fish, or the modification of natural conditions in any way detrimental to the interest of the State, no person shall discharge or permit to be discharged into any waters of the state, or into drains which discharge into such waters, any substance which kills fish, or renders the water unfit for the maintenance of the normal fish life characteristic of the waters, or in any way adversely affects the interest of the state.

1951 A Regulation Requiring the Submission of Reports for the Discharge of Industrial Waste and for the Construction or Alteration of Treatment Works. Adopted by the Stream Control Commission

- "Person" means any individual, public or private corporation, political subdivision, governmental agency, municipality, industry, copartnership, association, firm, trust, estate, or any other legal entity whatsoever.

- "Treatment Works" means any facility primarily designed and installed for the purpose of treating industrial wastes before final discharge or deposit into waters of the State.
- "Waters" shall be construed to mean public waters including lakes, bays, sounds, ponds, impounding reservoirs, springs, wells, rivers, streams, creeks, estuaries, marshes, inlets, canals, the ocean within the territorial limits of the State, and all other bodies of surface, natural or artificial, inland or coastal, fresh or salt, within the jurisdiction of the State of Louisiana.
- "Industrial waste" means any water-borne liquid, gaseous, solid or other waste substance or a combination thereof resulting from any process of industry, manufacturing trade or business, or from the development of any natural resource.
- "Other wastes" means garbage, refuse, decayed wood, sawdust, bark, lime, sand, cinders, ashes, offal, oil, tar, dyestuffs, acids, chemicals and all discarded substances other than industrial waste as defined in this Section.
- "Pollution" shall be construed to mean the discharging into any of the waters of the state any waste or any pollution of any kind that will tend to destroy fish or other aquatic life or wild or domestic animals or fowls or be injurious to the public health or against the public welfare.
- Requirements for the submission of reports. Any person intending to discharge industrial waste at any location in the State where such person is not now discharging industrial waste; any person intending to increase the quantity of industrial waste which he is discharging to State waters on the effective date of this Regulation; any person intending to construct a new outlet, or build, add to, or alter any treatment works for the handling of industrial waste, shall before starting such work, advise the Louisiana Stream Control Commission, Baton Rouge, Louisiana, in writing concerning his intentions, and shall supply to the Commission a general report describing the sewerage system which is proposed and the steps which will be taken to protect the waters of the State against new pollution or an increase in existing pollution.

1953 (As Amended) Rules Governing Disposal of Waste Oil, Oil Field Brine, and All Other Materials Resulting from the Drilling for, Production of, or Transportation of Oil, Gas or Sulfur. Adopted by the Stream Control Commission

- No oily fluids shall be discharged to, or allowed to flow on the ground, or be carried from the original lease in open ditches, or discharged or allowed to flow into any stream, lake or other body of water.
- No oil field brine shall be discharged into any stream, lake or other body of water, or into any ditch or surface drainage leading to any stream, lake or other body of water when it is determined by the Stream Control Commission that such discharge would adversely affect the palatability of a source of potable water to an appreciable degree, or would be deleterious to the public health . . . or whereby any lawful use of an such waters by the State of Louisiana, or by any political subdivision, or by any corporation, association, partnership, or person or any other legal entity may be lessened or impaired or materially interfered with.

1970 Acts No. 404

- "Water pollution" includes the introduction into state water bodies of any substance in concentration which results in the killing of fish or other aquatic life in numbers or in a manner materially detrimental to the interests of the state or renders the water unfit for maintenance of the normal fish or aquatic life characteristics of the waters, or in any way adversely affects the interests of the state in respect to its fish or other aquatic life.
- In order to prevent the pollution of any stream or other water body of the state, the killing of fish or other aquatic life, or the modification of natural conditions in any way detrimental to the interests of the state, no person shall knowingly discharge or knowingly permit to be discharged into any waters of the state, or into drains which discharge into such waters, any substance which causes "Water Pollution."

1970 Acts No. 499, House Bill 335

- An Act to prohibit discharge of untreated wastes into the Mississippi River; to provide a procedure for enforcement and for civil penalties; to provide for issuance of preliminary injunctions and to define certain terms used in the Act.
- "Persons" means any municipality, political subdivision, public or private corporation, individual, partnership, association or other entity.
- "Treatment works" means any plant or other works which accomplishes the secondary treating, stabilizing or holding of wastes.
- "Untreated wastes" means wastes which have not been treated in treatment works.
- "Wastes" means human or animal wastes and liquid industrial wastes.
- No person in this state shall willfully and intentionally discharge or cause to be discharged any untreated wastes into the Mississippi River.

MASSACHUSETTS

1897 Special Act, Chapter 510

- Pollution of Sources of Water Supply and Regulations of the State Board of Health.
- The State Board of Health shall have authority time to time to examine all streams and ponds used together with all springs, streams and water courses tributary thereto with reference to their purity, and shall have authority to make rules, regulations and orders for the purpose of preventing the pollution, and securing the sanitary protection of the same.
- Upon complaint to said State Board of Health by a board of water commissioners, or the president of a water or ice company, that manure, excrement, garbage, sewage, or any other matter is so deposited, kept or discharged as to pollute or tend to pollute the waters of any stream, pond, spring or water course used by a city, town, water or ice company as a source of water supply.
- After such hearing, if in its judgment the public health requires it, shall prohibit the deposit, keeping, or discharge of any such material, or other cause of pollution as aforesaid, and shall order any

person to desist therefrom and to remove any such material theretofore deposited. . . . But said board shall not prohibit the use of any structure which was in existence at the time of the passage of this act.

1899 Regulations of the State Board of Health

- No cesspool, privy or other place for the reception, deposit or storage of human excrement, and no urinal or water-closet not discharging into a sewer, shall be located, constructed or maintained within fifty feet of high water mark of Stony Brook Reservoir, so called, said reservoir . . . or within fifty feet of high-water mark of any reservoir, lake, pond, stream, ditch, water course, or other open waters, the water of which flows directly or ultimately into said Stony Brook Reservoir.
- No house slops, sink waste, water which has been used for washing or cooking or other polluted water, shall be discharged into the ground fifty feet, or upon the ground within two hundred and fifty of the high-water mark of said Stony Brook Reservoir, or into the ground within fifty feet or upon the ground within two hundred and fifty feet of high-water mark of any open waters flowing as aforesaid into said Stony Brook Reservoir.
- No garbage, manure, putrescible matter whatsoever shall, except in the cultivation and use of the soil in the ordinary methods of agriculture, be put upon the ground within two hundred and fifty feet of high-water.

Act of 1929, Chapter 181

- An Act Prohibiting the Discharge of Oils and Their Products Into or On Certain Waters and Flats.
- Whoever pumps, discharges or deposits, or causes to be pumped, discharged or deposited, into or on the waters of any lake or river or into or on tidal waters and flats, any crude petroleum or any of its products or any other oils or any bilge water or water from any receptacle containing any of the said substances, in such manner and to such extent as to be a pollution or contamination of said waters or flats or a nuisance or be injurious to the public health, shall be punished by a fine.

Act of 1935, Chapter 381

- An Act prohibiting the discharge of oils and their products, refuse and certain other matter into or on the waters and flats of Boston Harbor and its Tributaries.
- Whoever pumps, discharges or deposits, or causes to be pumped, discharged or deposited, any crude petroleum or any of its products, or any other oils, or any bilge water or water from any receptacle containing any of the said substances, or any other matter or refuse, into the waters or flats of the Boston Harbor, or any of its tributaries in such manner and to such an extent as to be pollution or contamination of said waters of flats or a nuisance or to be injurious to the public health, shall be punished by a fine.

Act of 1941, Chapter 388

- An Act Authorizing the State Department of Public Health to regulate pollution and contamination of inland and tidal waters.
- It shall from time to time after notice to all persons interested and a public hearing and subject to the approval of the governor and council, prescribe and establish rules and regulations to prevent pollution or contamination of any or all of the lakes, ponds, streams, tidal waters and flats within the commonwealth or of such tidal waters and flats; provided, that nothing in said rules and regulations shall adversely affect any industry or any municipal sewerage system existing on January first, nineteen hundred and forty-one, and that nothing contained herein shall affect other powers and duties of the department as defined by any general or special law.

1949 New England Interstate Water Pollution Control Compact

- The growth of population and the development of the territory of the New England states has resulted in serious pollution of certain interstate streams, ponds and lakes, and of tidal waters ebbing and flowing past the boundaries of two or more states; and such pollution constitutes a menace to the health, welfare and economic prosperity of the people living in such area; the abatement of existing pollution and the control of future pollution in the interstate waters

of the New England area are of prime importance to the people and can best be accomplished through the cooperation of the New England states in the establishment of an interstate agency to work with the states in the field of pollution abatement.
- The states of Connecticut and Rhode Island and the commonwealth of Massachusetts (the states of Maine, New Hampshire and Vermont when authorized and do join herein) are now bound and do agree.

Act of 1952, Chapter 501

- An Act of further regulating the discharge of injurious substances into waters used for fishing.
- If the commissioner determines that any marine fisheries of the commonwealth are of sufficient value to warrant the prohibition or regulation of the discharge or escape of sawdust, shavings, garbage, ashes, acids, oil, sewage, dyestuffs, or other waste material from any saw mill, manufacturing or mechanical plant, or dwelling house, stable or other building, which may, directly or indirectly, materially injure such fisheries, he shall thereupon give written notice of such determination to the commissioner of public health.

Act of 1956, Chapter 620

- An Act Establishing in the Department of Natural Resources a Water Resources Division and defining the powers and duties of said division.
- The deferred operation of this act would tend to defeat its purpose, which is to establish forthwith a Division of Water Resources in the Department of Natural Resources under the control of a Water Resources Commission which shall be the agent of the commonwealth in coordinating all activities of federal, state and other agencies in the conservation, development, utilization and disposal of water for the purpose of preventing loss of life and damage to property by erosion, floodwater and sediment in the watersheds of the commonwealth, and to obtain necessary financial assistance from the federal government, therefore it is hereby declared to be an emergency law, necessary for the immediate preservation of the public safety and convenience.

- There shall be established a Division of Fisheries and Game and a Division of Water Resources, within the Department, but not under the supervision and control thereof. The Division of Water Resources shall be under the supervision and control of the Water Resources Commission.

Act of 1957, Chapter 678

- The department shall take cognizance of the interests of life, health, comfort and convenience among the citizens of the commonwealth; shall conduct sanitary investigations and investigations as to the cause of disease. It shall have oversight of inland waters, including surface and subsurface waters, sources of water supply, and shall control the pollution or contamination of any or all of the lakes, ponds, streams, tidal waters and flats within the commonwealth and of the tributaries of such tidal waters and flats.

Act of 1966, Chapter 685

- An Act establishing a Water Pollution Control Division in the Department of Natural Resources.
- There shall be in the Department subject to the control of the Water Resources Commission a Division of Water Pollution Control.
- It shall be the duty and responsibility of the Division to enhance the quality and value of water resources and to establish a program for the prevention, control, and abatement of water pollution.
- The Division is hereby authorized to propose Water Pollution Abatement Districts consisting of more than one city or town, or designated parts of towns.
- No person shall make or permit a new outlet for the discharge of sewage or industrial waste or wastes, or the effluent therefrom, into any of the waters of the Commonwealth nor shall he construct or operate a new disposal system for the discharge of sewage or industrial or other wastes or the effluent therefrom into the waters of the Commonwealth without first obtaining a permit, which the director is hereby authorized to issue subject to such conditions as he may deem necessary to insure compliance with the standards established for the waters affected.

MICHIGAN

1929 Public Act No. 245

- An Act to create a stream control commission to have control over the pollution of any waters of the state and the great lakes, with power to make rules and regulations governing the same, and to prescribe the powers and duties of such commission; to prohibit the pollution of any waters of the state and the great lakes; and to provide penalties for the violation of this Act.
- For the purpose of carrying out the provisions of this act there is hereby created a stream control commission, hereinafter referred to as the commission, the commissioner of agriculture, and the attorney general.
- The commission shall be authorized to bring any appropriate action in the name of the people of the State of Michigan, either at law or in chancery, as may be necessary to carry out the provisions of this act, and to enforce any and all laws relating to the pollution of the waters of this state.
- The commission or any agent duly appointed by it shall have the right to enter at all reasonable times in or upon any private or public property for the purpose of inspecting and investigating conditions relating to the pollution of any waters in this state.
- The commission shall establish such pollution standards for lakes, rivers, streams and other waters of the state in relation to the public use to which they are or may be put, as it shall deem necessary. It shall have the authority to ascertain and determine for record and in making its order what volume of water actually flows in all streams, and high and low water marks of lakes and other waters of the state, affected by the waste disposal or pollution of municipalities, industries, public and private corporations, individuals, partnership associations, or any other entity. It shall have the authority to take all appropriate steps to prevent any pollution which is deemed by the commission to be unreasonable and against public interest in view of the existing conditions in any lake, river, stream or other waters of the state.
- It shall be unlawful for any person to discharge or permit to be discharged into any of the lakes, rivers, streams, or other waters of this state any waste or pollution of any kind that will tend to destroy fish life or be injurious to public health.

1944 Michigan Administrative Code

- A regulation on Minimum Standards for the Location and Construction of Wells Used for the Production of Untreated Public and Semi-Public Water Supplies, Other than Municipal Supplies.
- Source of Contamination: A ground water supply shall be obtained from a properly constructed and maintained well located so that the area within 75 feet of the casing or suction pipe shall be free from sources of possible contamination such as seepage pits, cesspools, outhouses, barnyards, septic tanks, disposal fields, county drains and other sources of contamination.

1949 Public Act No. 219

- An Act providing for the supervision and control by the state board of health over waterworks systems and sewage disposal systems, for the submission of plans and specifications for waterworks and/or sewerage systems and the issuance of construction permits therefore, for the supervision and control of such systems, for the classifying of water treatment plants, and sewerage treatment plants, for the examination, certification, and regulation of persons in charge of such water treatment plants and sewage treatment works, and providing penalties and defining liabilities for violations of this Act.

1949 Public Act No. 117

- An Act to create a water resources commission to protect and conserve the water resources of the state, to have control over the pollution of any waters of the state and the great lakes, with power to make rules and regulations governing the same, and to prescribe the powers and duties of such commission; to prohibit the pollution of any water of the state and the great lakes; to designate the commission as the state agency to cooperate and negotiate with other governments and agencies in matters concerning the water resources of the state.
- It shall be unlawful for any person to discharge or permit to be discharged into any of the lakes, rivers, streams, or other waters of this state any substance which is injurious to the public health or to the conducting of any industrial enterprise or other lawful occupation;

or whereby any fish or migratory bird life or any wild animal or aquatic life may be destroyed or to the growth or propagation thereof be prevented. . . . Any person who shall discharge or permit to be discharged any waste or pollution into any waters of this state, in contravention of the above provisions of this section, shall be deemed to violate the provisions of this act.

1965 Public Act No. 87

- An Act to license and regulate garbage and refuse disposal; and to provide a penalty for violation of this act.
- No person shall dispose of any refuse at any place except a disposal area licensed as provided in this act.
- Sanitary landfill operations shall be so designed and operated that conditions of unlawful pollution will not be created and injury to ground and surface waters avoided which might interfere with legitimate water uses.
- Water-filled areas not directly connected to natural lakes, rivers or streams may be filled with specific inert material not detrimental to legitimate water uses and which will not create a nuisance or hazard to health.

NEW JERSEY

1899 Chapter 41. An Act to Secure the Purity of the Public Supplies of Potable Waters in This State

- No sewage, drainage, domestic or factory refuse, excremental or other polluting matter of any kind whatsoever which, either by itself or in connection with other matter, will corrupt or impair, or tend to corrupt or impair, the quality of water of any river, brook, stream or any tributary or branch thereof, or any lake, pond, well, spring or other reservoir from which is taken, or may be taken, any public water supply of water for domestic use in any city, town, borough, township or other municipality of this state, or which will render, or tend to render, such water injurious to health, shall be placed in, or discharged into, the waters, or placed or deposited upon the ice, of any such river, brook, stream or any tributary thereof, or of any lake, pond, well, spring or other reservoir above the point from which any city, town, borough, township or other municipality shall or may obtain its supply of water for domestic use, nor shall such sewage, drainage, domestic or factory refuse, excremental or other polluting matter be placed or suffered to remain upon the banks of any such river, brook, stream or of any tributary or branch thereof, or any lake, pond, well, spring or other reservoir above the point from which any city, town, borough, township or other municipality shall or may obtain its supply of water for domestic use as aforesaid; and any person or persons, or private or public corporation, which shall offend against any of the provisions of this section shall be liable to a penalty of one hundred dollars for each offense; and each week's continuance, after a notice by the state or local board of health to abate or remove the same, shall constitute a separate offense.

1910 Chapter 215. An Act Prohibiting the Discharge of Sewage, Excremental Matter, Domestic Refuse and Other Polluting Matter into Fresh Water

- No person shall hereafter discharge or permit to be discharged into any fresh water any sewage, excremental matter, domestic refuse or other polluting matter. The term "fresh water" as used in this act shall be taken to mean and include all water commonly known as fresh and which may be used for human consumption, irrespective of whether such water shall be found in a stream where the tide ebbs and flows or not.

1940 Pollution of Waters

- No person shall discharge or permit to be discharged into any fresh water any sewage, excremental matter, domestic refuse or other polluting matter. The term "fresh water" as used in this article, shall mean and include all water commonly known as fresh water and which may be used for human consumption, whether or not such water shall be found in a stream where the tide ebbs and flows.
- No effluent from any seepage disposal system or any plant for the purification or treatment of sewage or industrial wastes shall be

discharged into any of the potable waters of this state, which, in the opinion of the department, is of such a character as will or may cause or threaten injury to the users of any of such waters.

OHIO

1925 House Bill No. 113

- No city, village, county, public institution, corporation or officer or employee thereof or other person shall provide or install a water supply or sewerage, or purification or treatment works for water supply or sewage disposal, or make a change in any water supply, water works intake, water purification works, sewerage or sewage treatment works until the plans therefor have been submitted to and approved by the state department of health.
- No city, village, county, public institution, corporation or officer or employee thereof or other person shall establish as proprietor, agent, employee, lessee, or tenant, any garbage disposal plant, shop, factory, mill, industrial establishment, process, trade or business, in the operation of which an industrial waste is produced, or make a change or enlargement of a garbage disposal plant, shop, factory, mill, industrial establishment, process, trade or business, whereby an industrial waste is produced or materially increased or changed in character, or install words [works] for the treatment or disposal of any such waste until plans for the disposal of such waste have been submitted to and approved by the state department of health. For the purposes of this act industrial waste shall be construed to mean a water-carried or a liquid waste resulting from any process of industry, manufacture, trade or business, or development of any natural resource.
- The state department of health shall study and investigate the streams, lakes and other bodies of water of the state and waters forming the boundaries thereof, for the purpose of determining the uses of such waters, the causes contributing to their pollution and the effects of the same, and the practicability of preventing and correcting their pollution and of maintaining such streams, lakes and other bodies of water in such condition as to prevent damage to public health and welfare.

1939 Ohio River Valley Sanitation Compact

- The compact was negotiated by representatives of the states of Illinois, Indiana, Kentucky, New York, Ohio, Pennsylvania, Tennessee, and West Virginia.
- The intent was to control future pollution and the abatement of existing pollution of the waters of the Ohio drainage basin—so as to protect and maintain the waters of said basin in a satisfactory sanitary condition, available for safe and satisfactory use as public and industrial water supplies after reasonable treatment, suitable for recreational usage, capable of maintaining fish and other aquatic life, free from unsightly or malodorous nuisances due to floating solids or sludge deposits, and adaptable to such other uses as may be legitimate.

June 19, 1951 Water Pollution Control Act of Ohio

- "Pollution" means the placing of any noxious or deleterious substances in any waters of the state which renders such water harmful or inimical to the public health, or to animal or aquatic life, or to the use of such water for domestic water supply, or industrial or agricultural purposes, or for recreation.
- "Industrial Waste" means any liquid, gaseous or solid waste substance resulting from any process of industry, manufacture, trade or business, or from the development, processing or recovery of any natural resource, together with such sewage as may be present, which pollutes the water of the State.
- "Waters of the State" means all streams, lakes, ponds, marshes, watercourses, waterways, wells, springs, irrigation systems, drainage systems, and all other bodies or accumulations of water, surface and underground, natural or artificial, which are situated wholly or partly within, or border upon this state, or are within it jurisdiction, except those private waters which do not combine or effect a junction with natural surface or underground waters.
- It shall be unlawful for any person to cause pollution as defined in section 2(a) of this act of any waters of the state or to place or cause to be placed any sewage, industrial waste or other wastes in a location where they cause pollution of any waters of the state, and any such action is hereby declared to be a public nuisance, except in such

cases where the water pollution control board has issued a valid and unexpired permit, or renewal thereof, as provided in this act.

- It shall be unlawful for any person to whom a permit has been issued to place or discharge, or cause to be placed or discharged, in any water of the state any sewage, industrial waste or other wastes in excess of the permissive discharges specified under such existing permit without first receiving a permit from the board to do so.

1963 House Bill No. 624

- No person shall cause pollution as defined in the Revised Code of any waters of the state, or place or cause to be placed any sewage, industrial waste, or other wastes in a location where they cause pollution of any waters of the state. Any such action is hereby declared to be a public nuisance, except in such cases where the water pollution control board has issued a valid and unexpired permit.
- No person who is discharging or causing the discharge of any sewage, industrial waste, or other wastes into waters of the state shall continue or cause continuance of such discharge after September 27, 1952, without first obtaining a permit therefor issued by the board, pursuant to rules and regulations to be prescribed by it.

1971 Solid Waste Disposal Act

- "Solid Wastes" means such unwanted residual solid or semisolid material as results from industrial, commercial, agricultural, and community operations, excluding earth or material from construction, mining, or demolition operations and slag and other substances which are not harmful or inimical to public health, and includes garbage, combustible and non-combustible material, street dirt, and debris.
- "Solid waste disposal" means final disposition of solid wastes by means acceptable under regulations adopted by the public health council under section 3734.02 of the Revised Code.
- The public health council shall adopt regulations having uniform application throughout the state governing solid waste disposal sites and facilities in order to assure that such sites and facilities will be located, maintained, and operated in a sanitary manner so as not to create a nuisance, cause or contribute to water pollution, or create a health hazard.

PENNSYLVANIA

The Clean Streams Law of 1937

- "Industrial waste" shall be construed to mean any liquid, gaseous, radioactive, solid or other substance, not sewage, resulting from any manufacturing or industry, or from any establishment, as herein defined, and mine drainage, refuse, silt, coal mine solids, rock, debris, dirt and clay from coal mines, collieries, breakers or other coal operations. "Industrial waste" shall include all such substances whether or not generally characterized as waste.
- "Pollution" shall be construed to mean contamination of any waters of the Commonwealth such as will create or is likely to create a nuisance or to render such waters harmful, detrimental or injurious to public health, safety or welfare, or to domestic, municipal, commercial, industrial, agricultural, recreational, or other legitimate beneficial uses.
- "Waters of the Commonwealth" shall be construed to include any and all rivers, streams, creeks, rivulets, impoundments, ditches, water courses, storm sewers, lakes, dammed water, ponds, springs and all other bodies or channels of conveyance of surface and underground water, or parts thereof, whether natural or artificial, within or on the boundaries of this Commonwealth.

1951 Act No. 315, P.L. 1304
"Local Health Administration Law"

- Bucks County Department of Health—Rules and Regulations Governing Dumps and Landfills.
- No person shall open, operate or maintain a sanitary landfill or dump in any area within the County of Bucks, Commonwealth of Pennsylvania, subject to the jurisdiction of the Department, who does not possess a valid written permit issued by the Department.
- Plans together with a descriptive report of the proposed sanitary landfill or dump shall be submitted in duplicate with each application. The plans shall show the location, area and topography of the site, the type of soil formations encountered, the depth of the proposed fill, the level of any ground water encountered, the location of access roads, the drainage of surface water, the distance from the nearest occupied buildings, the availability of acceptable cover ma-

terial and all other pertinent data required by the Department to clearly indicate the orderly development, operation and completion of the project.

- Sanitary landfills or dumps shall not be located on a site where the deposited refuse will intercept ground water or surface streams or where any drainage therefrom is likely to pollute any waters of the Commonwealth including underground waters.

TEXAS

Texas Water Agencies

1961 State Water Pollution Control Act

- An Act to establish a State Water Pollution control board, and to provide for the control, prevention and abatement of pollution of surface and underground waters of the State, . . .
- "Waters" shall be construed to be underground waters and lakes, bays, ponds, impounding reservoirs, springs, rivers, streams and creeks.
- "Waste" means sewage, industrial waste, and other wastes, or any of them, as herein below defined.
- "Industrial waste" means any water-borne liquid, gaseous, solid, or other waste substance or a combination thereof resulting from any process of industry, manufacturing, trade, or business.
- "Pollution" means any discharge or deposit of waste into or adjacent to the waters of the State, or any act or omission in connection therewith, that by itself, or in conjunction with any other act or omission or acts or omissions, causes or continues to cause or will cause such waters to be unclean, noxious, odorous, impure, contaminated, altered or otherwise affected to such an extent that they are rendered harmful, detrimental or injurious to public health, safety or welfare, or to terrestrial or aquatic life, or the growth and propagation thereof, or to the use of such waters for domestic, commercial, industrial, agricultural, recreational or other lawful reasonable use.
- Within twelve (12) months after the date upon which this law becomes effective, every person who upon such effective date is discharging or permitting to be discharged any waste into or adjacent to the waters of the State shall apply to the Board for a permit to continue such discharge if it is his desire to so continue.

1967 Texas Water Quality Act, Chapter 313

- An Act to establish the Texas Water Quality Board, prescribe its powers, duties, functions, and procedures and to provide for the establishment and control of the quality of the waters in the state and the control, prevention, and abatement of pollution; validating previous actions of the Texas Water Pollution Control Board; providing penalties.
- The Board follows the definitions of terms set forth in the State Water Pollution Control Act of 1961 (i.e. Waters, Waste, Pollution).

1967 Promulgated Rules of the Texas Water Quality Board

- The Board follows the definitions of terms set forth in the State Water Pollution Control Act of 1961.
- Texas Water Development Board: (a) underground waters: The Texas Water Development Board investigates all water quality matters concerning groundwater in the State and reports all findings as to water quality to the Board together with its recommendations in regard thereto. (b) injection wells FOR industrial and municipal wastes: The Texas Water Development Board administers Article 7621b with respect to issuing permits for injection wells for subsurface disposal of industrial and municipal wastes, other than salt water and other wastes arising out of or incidental to the drilling for or the producing of oil or gas.
- The permit shall; describe the location of each authorized place of discharge or deposit of waste; specify the maximum quantity of waste which may be discharged at any time and from time to time at each place of discharge or deposit; specify the quality, purity and character of waste which may be discharged at each place of discharge or deposit.

1969 Solid Waste Disposal Act

- An Act relating to the control of the collection, handling, storage, and disposal of putrescible and non-putrescible discarded or unwanted materials, including solid materials and certain materials in liquid or semi-liquid form, referred to in this Act as solid waste.
- "Solid waste" means all putrescible and non-putrescible discarded

or unwanted materials, including solid materials and industrial solid waste; as used in this Act, the term "solid waste" does not apply to waste materials which result from activities associated with the exploration, development, or production of oil or gas and are subject to control by the Texas Railroad Commission.

- "Industrial solid waste" means solid waste resulting from or incidental to any process of industry or manufacturing, or mining or agricultural operations including discarded or unwanted solid materials suspended or transported in liquids, and discarded or unwanted materials in liquid or semi-liquid form; the term "industrial solid waste" does not include waste materials, the discharge of which is subject to the Texas Water Quality Act.

1969 Water Quality Act—Revision

- Purpose of Policy is amended as follows: It is declared to be the policy of the State of Texas to maintain the quality of the waters of this state consistent with the public health and enjoyment, the propagation of and protection of fish and wildlife, including birds, mammals and other terrestrial and aquatic life, the operation of existing industries and the economic development of the state; to encourage and promote the development and use of regional and area-wide waste collection, treatment, and disposal systems to serve the waste disposal needs of the citizens of the state; and to require the use of all reasonable methods to implement this policy.
- "Pollution" means the alteration of the physical, thermal, chemical, or biological quality of, or the contamination of any water in the state that renders the water harmful, detrimental or injurious to humans, animal life, vegetation, or property or to the public health, safety, or welfare, or impairs the usefulness or the public enjoyment of the water for any lawful or reasonable purpose.
- "Permit" means an order issued by the board in accordance with the procedures prescribed in this Act establishing the treatment which shall be given to wastes being discharged into or adjacent to any water in the state to preserve and enhance the quality of water, and specify the conditions under which the discharge may be made.
- "To discharge" includes deposit, conduct, drain, emit, throw, run, allow to seep, or otherwise release or dispose of; or to allow, permit or suffer any such act or omission.

1969 Injection Well Act

- Definition of Terms are the same as those established in the Texas Water Quality Act of 1967.
- Industrial and Municipal Wastes: Applications to the Board;
- Before any person commences the drilling of an injection well, or before any person converts any existing well into an injection well, for the purpose of disposing of industrial and municipal waste, other than waste arising out of or incidental to the drilling for or the producing of oil or gas, a permit therefor shall be obtained from the board.
- Casing of Well: The board or commission shall require that the injection well be so cased as to protect all fresh waters from pollution by the intrusion of industrial and municipal waste. The casing shall be set at such depth, with such materials, and in such manner as the board or the commission may require.

Texas Railroad Commission

1919 Oil and Gas Conservation Laws

- The Legislature enacts a statute requiring the conservation of gas and oil, forbidding waste, and giving the Railroad Commission jurisdiction.
- Fresh water, whether above or below the surface, shall be protected from pollution, whether in drilling or plugging.

1934 Amended Rules and Regulations: Oil and Gas Conservation

- Fresh Water to Be Protected: Fresh water, whether above or below the surface, shall be protected from pollution, whether in drilling, plugging or disposing of salt water already produced.

1953 Amended Rules and Regulations: Oil and Gas Conservation

- Exploratory Wells: Any oil or gas well or well drilled for exploratory purposes shall be governed by the provisions statewide or field rules which are applicable and pertain to the drilling, safety, casing, protection, abandoning and plugging of wells, and all operations in connection therewith shall be carried on so that no pollution of any stream or water course of this State, or any subsurface waters, will occur as the result of the escape or release or injection of oil, gas, salt water or other mineralized waters from any well.

1967 Amended Rule 8, Conservation of Oil and Gas

- The Commission is of the opinion and finds that the amendment of Rule 8 of the General Conservation Rules of Statewide Application will afford a greater protection for the fresh water resources of the State, will reduce the threat of pollution from improperly handled brines and oil field brines, and will give notice to all operators of the responsibility placed upon them in the handling of salt water and oil field brines.
- All operators conducting oil and gas development and production operations are prohibited from using salt water disposal pits for storage and evaporation of oil field brines and mineralized waters.

1969 Amended Rule 8, Conservation of Oil and Gas

- The operator shall not pollute the waters of the Texas offshore and adjacent estuarian zones (salt water bearing bays, inlets, and estuaries) or damage the aquatic life therein.
- No oil or other hydrocarbons in any form or combination with other materials or constituent shall be disposed of into the Texas offshore and adjacent zones.

Index

Absolute pollution exclusion, 91–95, 97
Absorption, 78, 168, 175, 223, 237, 239, 527
Accident, 3, 9, 69–71, 77–79, 81, 233, 241, 497
Accidental, 91, 92, 96, 101, 214, 312, 317, 496, 497, 504, 513, 518, 519, 528, 533
Acid mine drainage, 22, 395
Acid rock drainage, 360, 362, 399
Activated carbon, 36, 65
Activated sludge, 36, 225
Administrative Dispute Resolution Act, 309
Administrative order, 120
Admissibility, 262–267, 272, 283–285, 288, 290, 293, 299
Admissibility of expert testimony, 262–265, 272, 284, 288
Adsorbed, 11, 237, 510
Adsorption, 36, 126, 128, 134, 143, 147, 527
Advocate position, 3
Aerial photographs, 106, 109, 112–114, 258, 296, 330, 332, 350, 487
Aerosol, 23
Affidavit, 7, 15, 341, 342
Agribusiness, 19, 44
Agricultural Plastics Company, 26
Agriculture, 19, 26, 27, 448, 522
Air:
  dispersion, 104
  emissions, 1, 21, 338
  quality, 52, 308
Algae, 224, 242
Allocation, 2, 86–89, 314, 315, 319–321, 323, 361, 407, 408, 421–423, 427, 429, 431, 433–436, 438, 439, 441, 449, 453, 511
Alternative dispute resolution (ADR), 301–303, 306–311, 313–315, 317–319, 322–324
American:
  Arbitration Association, 322
  Chemical Society, 35
  Society for Testing Materials (ASTM), 105, 110, 114, 115, 155, 220
  Society of Civil Engineers, 35
  Water Works Association, 26, 27, 35, 39, 40
Animations, 248
Antimony, 367, 374–376
Applicable or Relevant and Appropriate Requirements (ARARs), 59, 60
Aquatic:
  life, 218, 221, 225, 226, 229
  toxicity, 218, 221, 224, 229, 399
Aquifer, 124–127, 130, 133, 134, 140–143, 145–148, 259, 316, 411, 423, 458, 474, 479, 481, 521, 525, 526, 529
Arbitration, 100, 287, 301, 307, 308, 313, 315, 321–323
Archeologists, 151
Archives, 14, 109, 332, 333, 362, 374
Area Landfill, 486
Argon, 176, 497, 498, 503, 504, 506–508, 510, 511, 513

Arkansas, 104
Army, 50, 119, 120
Aromatics, 181, 225
Arsenic, 11, 21, 22, 36, 108, 349, 351, 360, 367, 369, 373–376, 389
Arsenopyrite, 370
Asbestos, 44, 63, 81, 82, 84, 97, 219
Ash, 29, 366, 460, 461
Asphalt:
  Institute, 26
  membrane lining, 26
  plastic liners, 26
Atomic absorption spectrometry, 168, 175

Bacteria, 28, 31, 34, 97, 225, 242, 243, 395
Bag-house, 23
Barium, 21, 224, 349, 350, 365, 366, 379, 387
Barrow pits, 485
Baryte, 350, 366
Battery manufacturing, 30
Beneficial use, 31, 448, 450
Bentonite, 26, 489, 491, 527
Benzene, 69, 78, 97, 169, 170, 272, 446, 457
Biochemical oxygen demand (BOD), 35–37, 229, 230
Biodegradation, 28, 37, 201, 206
Bio-oxidized, 29
Black ash, 366
Blast furnace, 367, 369, 370
Bodily injury, 71, 72, 83, 89, 92, 93, 95
Boiling point, 153–155
Boiling range, 153–155, 167, 179
Boring logs, 2, 356, 363, 378
Bottom line, 36, 43, 44
Breach of contract, 72
Burden of proof, 117, 266, 275, 282, 286
Business records, 296, 297

Cadmium, 21, 22, 349–351, 356, 360, 365, 378, 386, 388
Calibrate, 9, 124, 126, 128
California, 26, 33, 119, 224, 225, 230, 284
Cancer, 52, 217, 219, 221, 223, 224, 226, 228, 237, 266
Cancer potency slope factors, 237
Canning, 30
Carbon:
  carbon-14, 177, 178, 181
  disulfide, 167
  monoxide, 52, 69, 94, 96, 97
Carbonaceous, 351
Carbonic acid, 375
Carcinogen, 217, 219, 226, 229, 237
Cartographic, 109, 113
Cause test, 80, 81
CCE, 230
Cement, 26, 30
Chalcopyrite, 370, 375, 376
Chemical:
  degradation, 235
  fingerprinting, 151–153, 167, 168, 174, 176, 177, 180, 181, 201, 202, 204, 205, 207, 208
  oxygen demand (COD), 35–37
  storage area, 437
  use history, 407, 430, 431, 434, 438, 449
  waste, 20, 258, 467, 501
Chemicals of concern, 7, 234–239, 241–243, 319
Chlorinated:
  hydrocarbon, 116–118, 422, 480, 482
  solvents, 118, 422
Chlorination, 29
Chlorine, 41, 118, 178, 228
Cholera, 28, 33, 217
Chromium, 21, 220–225, 230, 235
Chronic, 2, 221, 222, 226, 228, 229
Claim, 66, 70, 72–74, 77, 85, 86, 94–97, 99–101, 116, 203, 244, 246, 262, 268, 318, 401, 407, 434, 450–453, 458, 469, 470, 472, 477, 480, 483, 485, 496, 497, 504, 506, 513, 515, 518
Claims-made, 72, 99–101
Clay, 22, 24, 26, 141, 351, 458, 491, 522, 527
Clean:
  Air Act, 52, 53, 58, 180, 309
  Water Act (CWA), 32, 50–52, 230, 262, 275, 309
  Water Act of 1972, 50

Climate, 15, 533
Closure, 54–57, 301, 310, 413, 470, 486, 487, 489, 491–496
Coal, 9, 19, 22, 29–31, 175, 181, 220, 369, 370, 373, 374, 455–458, 460, 461, 468
Coal tar, 175, 220, 457, 458, 461, 468
Co-elution, 169, 174, 181
Co-mediators, 306, 315
Commercial general liability insurance (CGL), 69–77, 79, 82, 84, 86, 89–91, 99, 101, 469, 515, 518
 coverage, 71, 75, 76, 99
 policy, 70, 72, 89, 99
Commingled, 2, 20, 115, 375, 422
Community, 19, 35, 215, 239, 240, 266, 270, 271, 274, 284, 487
Compliance, 32, 41, 42, 51, 56, 60, 65, 101, 152, 310, 494
Comprehensive Environmental Response, Compensation and Liability Act (CERLA), 25, 32, 53, 57–67, 76, 77, 110, 112, 114, 262, 269, 280, 281, 293, 309–311, 323, 355, 356, 471, 478
Concrete cores, 437
Conductivity, 125–127, 141, 142, 147, 170, 316
Containment, 24, 32, 55, 110, 477, 494, 495
Contaminant, 14, 93, 95, 96, 117, 123–127, 130, 133, 134, 136, 140, 144, 146, 229, 315, 319, 332, 399, 412, 421–423, 434, 435, 450, 506
Contaminant age, 435
Contamination, 3, 8, 9, 11, 31, 33, 51, 54, 59, 64–67, 70–73, 75– 78, 80–86, 90–92, 94, 98, 99, 101, 103, 104, 110, 111, 115–117, 120, 123, 124, 126–128, 147, 148, 175, 178, 179, 207, 208, 217, 224, 226, 229, 230, 258, 276, 280, 281, 291, 310–312, 318, 321, 331, 332, 338, 349–352, 355–357, 360–363, 376–379, 388, 390, 395–400, 405, 411, 414, 420– 423, 427, 429, 430, 434–439, 443–447, 449, 452, 455, 457, 458, 460, 470–473, 475, 476, 478, 481–483, 485, 489, 491, 493, 497, 498, 501, 503, 504, 506–508, 510–513, 515, 518, 519, 526–529, 532, 534
Continuous trigger, 82, 85, 86, 89, 312
Copper, 21, 22, 349–351, 356, 360, 365, 367, 369, 372–376, 379, 387, 388
Corporation, 26, 61, 269, 403, 473, 478, 497, 503
Corrective action, 54, 123, 412
Corrosion, 24, 372, 375
Cost of doing business, 485, 496
Covenants, 112
Criminal liability, 309
Crude oil, 178, 180, 181, 463
Cupola-type furnace, 369
Cyanide, 36, 460, 464

Dairies, 30
Damage, 21, 31, 44, 52, 59, 70–72, 76–78, 80–87, 89–96, 98–100, 214, 215, 226, 231, 312, 314, 317, 318, 323, 361, 403, 450, 495, 522, 529
Darcy's law, 28
Data gaps, 4, 218, 234, 235, 329, 362, 460, 534
Daubert, 262, 263, 265–273, 275–279, 282–290
DDT, 43, 108
Decompose, 177, 226, 228, 486
Deeds, 112
Deep injection well, 39
Deep pocket, 8, 350
Default values, 239
Defendant, 14, 64, 66, 88, 110, 111, 243, 264, 266, 271, 274, 276, 278, 280, 281, 319
Defense cost, 77
Defense Environmental Restoration Program (DERP), 117
Deferred maintenance, 44, 45
Degradation, 50, 203, 204, 206, 230, 235, 435
Degreaser, 119, 474
Delaney Amendment, 217
Delayed maintenance, 45

Dense nonaqueous phase liquid (DNAPL), 8, 480, 482, 506–508, 510, 513
Department of Justice (DOJ), 313
Deposition, 4, 9–12, 14–16, 104, 277, 283, 327, 329, 330, 339–345, 511, 512, 518, 533, 534
Dermal absorption, 237, 239
Detection limits, 234
Diagnosis, 273, 274, 276, 278, 296
Dichloroethene (DCE), 435, 501
Dichlorophenol, 39
Dickey Act, 33
Diesel:
  engines, 154
  oil, 403, 405, 411, 412
Diffusion, 129, 237
Dilution, 28
Dioxin, 321
Discharge, 23, 24, 28–32, 41, 42, 50–52, 65, 78, 92, 93, 95, 101, 126, 222, 318, 359, 360, 363, 390, 400, 405, 435, 448, 450, 517, 518, 520, 521, 528, 532
Disease, 28, 71, 214, 217, 218, 226, 228, 274, 486
Disposal, 1, 7, 8, 20–22, 25, 27–29, 31–33, 35–41, 53–56, 60–62, 64, 67, 81, 91, 93, 101, 103, 105, 106, 110, 113, 117, 119, 222, 226, 229, 234, 235, 309, 320, 355, 357, 362, 366, 369, 370, 375, 411, 412, 422, 442, 444, 450, 451, 463, 464, 470, 472–475, 477, 485–488, 491, 493, 518, 519, 527
Disposal practice, 1, 29, 39
Dispute review boards, 301
Dissolved oxygen, 242, 243
Distillation, 30, 36, 154, 155, 201
Ditches, 23, 26, 108, 113, 520, 523
Documentary evidence, 103, 291, 293–295
Domestic waste, 27
Dose-response, 216, 217
Dow, 26, 36, 43, 262, 265
Drain, 363, 437, 438, 506
Drinking water standard, 132, 225, 230, 521
Drosses, 374
Drugs, 151
Dry cleaning, 57, 65, 227
Dry wells, 30, 477, 479, 481
Due diligence, 67, 111, 241, 442
DuPont, 43
Dust, 22, 34, 69, 82, 97, 237, 365, 371–375, 453, 461, 486
Dyes, 37, 175, 221

E. Coli, 243
Earth:
  floors, 363
  pits, 24, 527
Effect test, 81
Effluent, 32, 52, 517, 521
Electromagnetic spectrum, 176
Electron capture detector (ECD), 170, 180
Electronic, 260, 421, 422, 427, 470, 473
Electroplating, 30
Elution time, 168
Emissions, 1, 21, 23, 271, 338
Endangered species, 314
Enforcement, 7, 32, 43, 50, 52, 53, 308, 313, 322, 495
Engineered, 24, 42, 486, 523, 527, 534
Enjay Chemical Company, 27
Environmental:
  audit, 66, 67
  forensics, 1, 2, 4, 6, 9, 18, 20, 70, 115, 123, 151, 233, 244, 245, 261, 336
  impairment liability insurance (EIL), 69, 71, 72, 99, 101, 469–472, 474–476, 480, 482
    coverage, 469
    law, 49, 50, 263
    policy, 72, 469, 482
  site assessments, 110, 112
Epidemic, 28, 33
Epidemiological studies, 213, 267, 274
Erosion, 21–23, 83, 487, 494, 495, 532
Escrow, 350
Ethics, 4
Ethylbenzene, 169, 170, 457

Ethylene:
dibromide, 170
dichloride, 170, 227, 229
Evaporation, 23, 201, 229, 524, 528
Evidentiary issues, 261
Excavation, 411, 444, 450, 451, 460, 465, 466, 479, 482, 518, 533
Exhibit, 246, 248, 259, 260, 344, 345, 347, 395
Expected and intended, 312, 526–528
Expert:
report, 15, 259, 341, 342
testimony, 214, 247, 262–265, 267–269, 271, 272, 275, 276, 278, 280, 282–286, 288–291, 295, 298, 327, 329, 346, 356
Explosives, 174, 175
Exposure:
pathways, 236, 239
theory, 82, 83, 312

Facilitation, 301, 303, 306, 315, 317
Fact-finding, 303, 323
False negative, 169
Fate and transport, 37, 235, 312, 315, 316, 319, 321, 332, 447
Feasibility study, 60, 491, 518
Federal:
Insecticide, Fungicide and Rodenticide Act, 309
Rules of Evidence, 264, 265, 267, 284, 288–290
Water Pollution Control Act, 32, 52
Feedlots, 30
Fertilizer, 30, 515, 519, 520
Fiber, 24, 30, 177
Financial assurance, 55, 56
Fire hazard, 25
Fish, 37, 44, 109, 221, 224–226, 229, 243, 471, 523
Floor drain, 23, 506
Flue dust, 373, 375
Fluoride, 22, 515, 516, 518–523, 525, 529, 532, 533
Forensic:
expert, 3, 10, 11, 70, 100, 124, 151, 152, 245, 330, 331, 333, 339, 343, 408, 411, 417, 420, 427, 433, 434, 438, 439, 444, 445, 449, 458, 480–483, 510, 511, 519, 533
investigation, 1, 13, 18, 71, 327, 356, 444, 458, 464, 470, 473, 493, 506, 507
process, 2, 5, 10, 17, 20, 262
Formerly Used Defense Sites, 117
Fortuitous event, 72
Free product, 411
Frye test, 264, 284
Fugitive, 22

Galena, 370
Garbage, 9, 28, 29, 34, 36, 39, 43, 125
Gas chromatograph, 154, 155, 167–171, 174, 176, 179–181
analysis, 167, 169, 179
trace, 155, 167, 168
Gas from coal, 9, 19
Gas migration, 485, 486, 489, 491–493
Gasoline, 94, 117, 153, 154, 167, 169, 170, 175, 179–181, 200, 204, 205, 442, 463
Gastrointestinal, 221, 227, 243
Gay Lussac Towers, 371, 372
General Electric Company v. Joiner, 262
Generator, 62, 456, 498
Geochemical, 20, 388
Glass manufacturing, 30
Glover Tower, 371–373
Gold, 356, 367, 374, 375, 407, 412, 420
Gradual, 25, 69, 82, 92, 101, 469
Graphics, 17, 248, 259, 329, 332, 345, 346, 510
Gravel pits, 485
Groundwater:
cleanup, 3, 38, 397, 411, 448, 498
contamination, 3, 9, 11, 65, 66, 98, 117, 120, 224, 226, 229, 321, 349–351, 355, 357, 360, 362, 363, 376, 388, 390, 395, 398, 399, 405, 411, 421, 423, 443, 444, 447, 455, 457, 470–472, 475, 482, 485, 491, 493, 497, 498, 503, 515, 518, 519, 527, 529, 532
elevation, 316

Groundwater (*continued*):
monitoring, 9, 54, 55, 125, 128, 315, 405, 411, 442, 451, 458, 474, 491, 501, 504

Half life, 177
Hall conductivity detector, 170
Harmful, 51, 82, 85, 242, 271, 516, 520
Hazardous:
chemicals, 20, 25, 67, 498
Ranking System, 58
Health:
hazard, 23, 276, 359, 519, 520
material, 60
science, 214, 215, 220, 230, 231
substance, 57, 71, 83, 111, 355
waste, 6–8, 27, 32, 53–57, 67, 76, 270, 280, 314, 411, 470, 471, 496, 498
Hearsay, 263, 291, 294–297, 299
Herbicides, 22, 30
Hexavalent, 221, 224, 225, 235, 474
Hidden, 40
High molecular weight compounds, 174
High pressure liquid chromatography (HPLC), 167, 174, 176
Historical documents, 7, 291, 296, 374
Historical knowledge, 6, 219, 222, 227
Historic maps, 103
Hoffman-La Roche, 43
Hot spots, 241, 408, 411, 436, 438
Human health, 53, 55, 59, 213, 214, 218, 220–222, 226, 227, 230, 233, 235, 238, 239, 241–244, 315, 447
Hydraulic conductivity, 127, 142, 316
Hydraulic Linings, Inc., 27

Illinois, 74, 97
Imhoff Tank, 28
incinerators, 308
Indicator, 37, 215, 247
Industrial:
facility, 236
park, 421–423, 473, 475
solid waste, 526
waste, 8, 27, 31, 33, 35, 37, 39–41, 43, 526
waste lagoons, 37, 39
Water and Wastes Journal, 26
workers, 236
Ingestion, 97, 221, 237, 243
Inhalation, 221, 223, 228, 237, 274
Injection well, 39
Injury, 70–72, 77, 78, 80–87, 89–95, 97, 99, 100, 214, 215, 231, 244, 279, 312
Injury-in-fact, 82, 84, 312
Innocent landowner, 64, 65, 110–112, 114
Insecticides, 174, 229, 318, 320
Insurance, 2, 3, 9, 14, 55, 69, 70, 72, 76–79, 86–89, 91–93, 95, 98, 99, 101, 109, 114, 152, 214, 231, 262, 310–313, 315, 327, 328, 332, 405, 408, 420, 469–477, 480, 482, 483, 485, 487, 492, 493, 496, 497, 504, 507, 508, 510, 511, 513, 515, 518, 519, 533, 534
Intentional, 44, 70, 78, 88, 214, 228, 317
Internal Revenue Service, 483
International Paper Company, 26
Ion exchange, 36
Irrigation, 26, 113, 448, 494
Isoparaffins, 181
Isotope, 176–178
Isotope ratio mass spectrometer, 176

JAMS, 307, 322
Joint defense group, 493, 496
Joint and several liability, 86, 87, 89
Judicial gatekeeper, 261

Known loss, 78, 79
Koppers, 43
Kriging, 388

Lagoon, 9, 26, 34, 42, 312, 320, 472, 474, 475, 527
Lakes, 21, 29, 34
Landfill, 6, 7, 29, 32, 91, 115, 141, 411, 412, 442, 485–489, 491, 493–496
Land use, 20, 114, 306, 308, 314, 323, 359

Lawsuit, 70, 73, 75, 88, 94, 241, 275, 320, 328, 400, 407, 444, 485, 503, 504, 513
Leach, 29, 34, 375, 395
Lead, 21, 22, 69, 82, 97, 104, 129, 140, 167, 170, 178–180, 200, 204, 205, 219, 271, 272, 274, 289, 306, 313, 320, 349– 351, 356, 360, 362, 365, 367, 369–379, 386, 387, 408, 496
  carbonate, 367, 370, 374, 375
  sulfate, 367, 377
  sulfide, 367, 369
Leaded gasoline, 167, 170, 175, 180, 204, 205
Leaks, 23, 357, 363, 422, 446, 463, 477, 506
Leases, 65, 112, 405
Lederle Laboratories, 43
Legislation, 31, 38, 43, 52, 53, 57, 114
Lender liability, 63
Liability, 32, 33, 56, 57, 60–67, 69, 71, 76, 77, 79, 80, 86–89, 91, 94, 97, 100, 105, 108, 110–112, 115, 116, 274, 309, 310, 315, 355, 469, 475, 515
Liner technology, 26, 27
Lipids, 175
Liquid chromatography, 167, 174
Lithophone, 350, 351
Litigation, 1, 2, 4, 6, 7, 9, 12–14, 16, 18, 20, 66, 73, 74, 147, 148, 151, 152, 233, 240–242, 245, 259, 261, 263, 270, 272, 285, 287, 302, 303, 306–308, 311–313, 322–324, 327– 331, 333–335, 338, 339, 342, 343, 346, 420, 438, 453, 472, 504, 510, 513
Livestock, 225
Love Canal, 57, 113
Lubricating oil, 154
Lumber mills, 19

Magnesium, 22, 367, 376, 521
Magnetite, 370, 374
Maintenance, 24, 44, 45, 55–57, 118, 243, 312, 338, 347, 456, 487, 491, 494, 495, 498, 506
Manganese tricarbonyl, 180
Manifestation, 82–86, 312
Manifests, 83, 109
Manufacturing Chemists' Association, 39
Manure, 208, 375
Maps, 14, 103, 106, 109, 114, 140, 330, 332, 445, 458, 460, 510
Mass spectrometer, 170, 176, 178
Material Safety Data Sheets, 25
Mattes, 374
Meat packing and rendering, 30
Mediation, 301, 303, 306, 307, 311, 313, 315, 317–321, 323, 324, 345, 346, 400, 403, 407, 408, 411, 413, 414, 420, 422, 431, 434–436, 438, 476
Metal finishing, 30
Metallurgical, 362
Metal plating, 30, 235
Metal refurbishing, 235
Methane gas, 486
Methyl cellosolve, 167
Methylene chloride, 430
Methyltriethyl lead, 167, 180
Michigan, 33, 224, 269
Microorganisms, 37, 222, 225, 230
Military, 105, 116–119, 121
Mill tailings, 21, 22
Mining, 21, 27, 362
Mini-trials, 301, 303, 307, 313, 323
Model, 5, 9, 104, 124–128, 140, 143, 144, 146, 147, 203, 204, 235, 286, 315, 319, 330, 363, 437, 480, 481, 503
Monitoring wells, 125, 128, 315, 405, 413, 423, 429, 442, 445, 446, 458, 489, 491
Monsanto, 43, 218
Municipal:
  treatment, 42
  waste, 7, 27
  water district, 104, 105

Naphthenes, 181
National:
  Academy of Sciences, 277
  Ambient Air Quality Standards, 52
  Contingency Plan (NCP), 59, 63, 310

National (*continued*):
  Pollution Discharge Elimination System (NPDES), 32, 51, 52, 109
  Priorities List (NPL), 58, 59, 67, 491
Natural gas, 19, 456
Natural resources, 19, 29, 44, 59, 63, 67, 215, 231
Navigable, 31, 50–52
Near-infrared wavelengths, 175
Negligence, 66, 88, 309
Negotiated Rulemaking Act, 309
Nematodes, 318
Neutrons, 176
New Jersey, 29, 94, 97, 99, 221
Nickel, 21, 22, 176, 243, 350, 365, 388
Nitrogen-14, 177
Nonaqueous phase liquid, 480
Non-binding, 302, 303, 323
Non-coercive techniques, 306
Nonferrous metals, 30
Nonflammable, 118
Non-selective detector, 170
Nuclear, 19
Nuisance, 33, 66, 90, 309, 355, 476, 495
Number of occurrences, 80, 86

Occupational exposure, 228, 274
Occurrence, 69, 70, 72, 77–81, 83, 84, 99–101, 107, 311, 356, 360, 461, 463, 472, 473, 523
Occurrence-based, 69, 70, 77, 79–81, 84, 99–101
Odor, 31, 34, 41, 101, 226, 230, 243, 486
Off-site, 71, 98, 99, 101, 110, 357, 359, 389, 390, 397, 423, 474, 475, 481, 501, 515
Ohio, 74, 224
Olefins, 181
On-site, 71, 101, 320, 374, 422, 423, 463, 474, 479, 481, 504, 515, 518
Open dumps, 28, 29, 32 485, 486
Operation and maintenance (O&M), 491, 492
Opinion, 3–15, 84, 246, 261, 262, 265–268, 275, 276, 278, 279, 283, 285–287, 289, 290, 303, 315, 316, 340–343, 360, 398, 412, 417, 445, 504, 508, 510, 519, 527–529, 533, 534
Ore, 29, 177, 178, 200, 350, 356, 362, 366, 367, 369, 370, 373– 375, 399, 400, 515, 519
Ore hearth, 367, 370, 374
Organic:
  carbon, 129, 144, 316, 461
  lead, 167, 170, 179, 180
Organism, 213, 220
Owned property exclusion, 71, 97–99
Owner/operator, 1, 67
Oxygenates, 180

Paint, 8, 30, 97, 148, 154, 176, 235, 241, 269, 349–352, 355–357, 360–363, 365, 366, 375, 377, 396, 398–400
Paracelsus, 219
Paraffins, 181
Part A application, 470
Particles, 23, 176, 373, 374, 386
Particulate, 1, 23, 52, 376
Peer review, 114, 265, 266, 270, 274, 282, 287
Pennsylvania, 33, 94, 113, 272
Perchloroethylene (PCE), 3, 65, 92, 143, 144, 225–230, 237, 316, 423, 429, 434–438, 477, 478, 489
Percolate, 35, 486, 533
Permit, 31, 32, 41, 42, 50, 52, 56, 109, 116, 312, 441, 470, 487, 489, 527
Persistence, 38, 316
Personal injuries, 275
Pesticides, 22, 30, 38, 44, 108, 213, 316, 320
Petrochemical plants, 19
Petroleum, 18, 19, 30, 39, 94, 114, 167, 175, 176, 179, 180, 208, 310, 349, 350, 442, 455, 458, 460, 461, 463, 464
Pharmaceuticals, 3, 30, 175, 262, 265
Phase I, 110, 112, 115, 349, 350, 356, 501, 503, 504
Phosphates, 242, 243, 520
Photo-ionization detector, 170
Photo-synthesis, 177
Physician, 214, 215, 274
Pigments, 175, 375

Pipe, 243, 433
Pit, 42, 116
Plaintiff, 14, 66, 243, 266, 269, 270, 273, 274, 278–282, 355
Plastic, 26, 27, 465, 527
Plume, 108, 116, 119, 123, 126–128, 130, 132, 133, 140–147, 389, 421, 422, 448, 470, 478, 480, 501, 503, 504, 507, 510, 525
Poison, 216, 217, 222
Pollutant, 32, 71, 91, 95–97, 152
Pollution Control Commission, 33
Pollution exclusion, 91–97, 312
Pond, 366, 456, 458, 460, 464–467, 472, 474, 475, 518, 521, 523, 524, 528, 529, 532–534
Poor quality, 31, 335, 337, 338
Population, 28–30, 59, 219, 224, 227, 379, 487
Porous media, 28, 128
Portland Cement Association, 26
Postclosure, 54, 55, 57, 495
Potable water, 226
Potential responsible party (PRP), 58, 60, 70, 73–75, 103, 110, 115, 116, 310, 311, 318, 320, 321, 422, 423, 430, 431
Practice, 1, 4, 6, 13, 14, 20, 24, 29, 36, 39, 42, 50, 64, 105, 111, 114, 115, 117, 146, 214, 217, 231, 235, 238, 239, 241, 268, 277, 279, 286, 289, 296, 312, 341, 344, 346, 347, 371, 519, 526, 527
Precipitation, 486, 494, 523, 524, 528, 529, 532, 533
Preferential migration pathways, 316
Presentation technologies, 248
Prior knowledge, 215, 470, 476
Pristane, 203, 204
Privy, 28
Process mineralogy, 4, 14, 373, 399, 460, 464
Profits, 36, 44
Property:
  damage, 71, 72, 76, 82–84, 87, 89, 92, 93, 95, 98, 100, 214, 312, 317, 318, 323
  transfer, 111
Pro rata, 87, 88
Protocols, 151, 205, 206, 313, 318, 463
Public:
  health, 7, 31, 40, 52, 59, 116, 217, 218, 223, 224, 229,520
  records, 110, 294, 296, 297
  welfare, 52
Pulp and paper, 30
Pump and treat, 482
Purdue Industrial Waste Conference, 41
PVC liners, 27
Pyrite, 356, 357, 360, 362, 370–376, 389, 390, 395
Pyrrhotite, 370, 371, 374

Quarries, 485

Radioactive, 19, 22, 117, 181
Rainfall, 21, 22, 242, 523, 529
Rate setting, 314
Raw materials, 20, 106
Reactivity, 25, 370, 520
Real estate, 67, 94
Rebuttal, 4, 120, 399, 414, 431
Receptors, 234–236, 239, 240
Record of Decision (ROD), 480, 492, 493, 518, 529
Recreation, 405
Recycling, 115, 320, 442
Refineries, 19
Refuse, 22, 29, 30, 50, 487
Regulation, 12, 25, 30–33, 50, 51, 494
Regulators, 34, 35, 51, 76, 240, 311, 346, 360, 420, 504
Regulatory agency, 41, 75, 112, 119, 234, 235, 237, 238, 240, 241, 346, 397, 414, 423, 430, 451, 501, 510
Release, 21, 23, 25, 56, 59, 63–66, 70, 71, 90, 92–95, 101, 111, 114, 116, 123, 124, 137, 203, 233, 239–241, 280, 281, 310, 315–317, 319, 395, 399, 435, 450, 460, 469, 472, 484, 518, 520
Remedial investigation (RI), 8, 59, 60, 349, 350, 351, 356, 360, 479, 482, 491, 518
Repositories, 109, 334
Reservoirs, 24, 26, 27
Resource allocation, 314

Resource Conservation and Recovery Act (RCRA), 8, 32, 53, 55, 56, 58, 67, 108, 120, 309, 311, 355, 469, 470, 496, 498, 501, 518
Risk, 58, 59, 79, 81, 85, 87, 213, 217, 233–244, 288, 293, 307, 308, 321, 342, 399, 412, 423, 448, 450, 471, 476
Risk assessment, 59, 213, 217, 233–243, 412
Rivers, 21, 29, 30, 34, 43, 50
Rivers and Harbors Act of 1899, 30, 50
Roasting, 367, 370, 372–374, 376
Routine maintenance, 312
Rubber processing, 30
Runoff, 24, 27, 31, 118, 486, 494

Salt, 527
Sanborn map, 351, 356, 376, 456
Sand filtration, 29
Sanitary, 23, 28, 29, 32, 33, 38, 40, 222, 331, 485–487, 493, 516
Scientific community, 215, 239, 266, 270, 271, 284
Scrubber, 516
Sediments, 233, 237, 242, 243, 360, 527
Seepage, 8, 9, 24, 26, 29, 90, 93–95, 101, 130, 519, 521, 522, 527, 529, 533, 534
Selenium, 22, 373, 376
Self-insured, 89
Sensitivity, 167
Septic tank, 29
Settlement, 4, 17, 58, 285, 306–309, 313–319, 321, 323, 329, 345, 352, 398, 407, 408, 418, 420, 434, 438, 439, 441, 450– 454, 476, 477, 481–484, 493, 495, 496, 503, 507, 508, 511, 513, 534
Sewage, 26–28, 31, 33, 39, 41, 96, 222, 225, 230, 243, 516
Sewage and industrial waste, 33, 39
Sewer, 28, 51, 109, 243, 405, 517, 521, 528
Silent Spring, 43
Silica, 69, 365–367, 376, 516, 520, 521
Silver, 21, 351, 356, 367, 370, 374, 375
Sintering, 367, 370
Site history, 103, 105–108, 110, 112–115, 121, 345, 360, 367, 374, 399, 473, 476, 493, 515
Site specific, 5, 9, 14, 408, 483, 493
Slag, 104, 351, 356, 360–363, 367, 369, 370, 373–375, 378, 379, 386, 387, 395, 396, 460, 461
Smelter, 104, 320, 356, 362, 370, 371, 373–375, 378, 390
Smoke, 34, 92, 93, 95, 96, 101, 104
Smoking gun, 116, 119
Soil contamination, 338, 351, 357, 377, 399, 405, 423, 434, 437, 438, 444, 471, 518, 532
Solar, 19
Solid waste, 1, 29, 32, 53, 109, 373, 485–488, 493, 526
  Disposal Act, 32, 53
  waste management, 109, 486, 487
Solvent, 36, 116–118, 167, 179, 422, 434, 437, 474, 475, 477, 478, 480, 481
Speisses, 369, 374
Sphalerite, 370
Spills, 23, 24, 32, 357, 360, 377, 387, 389, 396, 458, 463, 477, 498, 506
Spray evaporation, 23
Staff Industries, 26
  implementation plans, 52
  Pollution Discharge Elimination System, 52
  Water Quality Control Board, 225, 230
State-of-the-art, 14, 27, 36, 42, 43, 526, 534
Statistical analysis, 386
Storm, 9, 363, 529
Stormdrain, 359, 363, 396
Storm drains, 23
Streams, 21, 30, 34, 41, 125, 224, 234
Strict liability, 62, 66, 274, 309
Strontium, 21
Sulfides, 21, 22, 356, 367, 369, 370, 372–374, 376, 390, 395
Sulfur dioxide, 374–376
Sulfuric acid, 8, 350, 356, 357, 360, 362, 363, 365, 367, 371, 373, 375, 376, 395, 396, 399, 515, 516, 519, 520

Sump, 23, 103, 436–438, 445, 474, 475
Superfund, 33, 55, 58, 64, 217, 306, 309–311, 314, 315, 319–321, 491
Superfund Amendments and Reauthorization Act, 64, 310
Surplus capital, 44
Synthetic chemicals, 30
Synthetic liner, 489

Tactic, 11
Tailings, 21, 22, 366, 375
Talc, 274, 275
Tank, 24, 28, 29, 105, 108, 350, 445, 446, 501
Tanneries, 19
Tanning, 30, 221
Taste, 31, 41, 226
Technical neutrals, 314, 315
Testimony, 4, 9, 10, 14–17, 117, 124, 214, 245–248, 259, 260, 262– 265, 267–272, 274–278, 280, 282–292, 295, 296, 298, 327, 329, 330, 340–342, 344–347, 356, 518, 533, 534
Tetraethyl lead, 167, 180
Tetramethyl lead, 167, 179
Textiles, 30
Thematic exhibits, 247
Thin layer chromatography, 167, 174
Third-party coverage, 71
Thorium, 22
Tidal flats, 351, 366, 374, 375
Timber products, 30
Timeline, 106, 114, 248
Toluene, 169, 170, 278, 279, 430, 457, 498
Topographic maps, 109, 114
Tort damages, 72
Total pollution exclusion, 91, 95, 96
Toxicity, 1, 21, 25, 31, 36, 37, 41, 60, 176, 214, 218, 219, 221–224, 227–230, 237, 321, 395, 399, 519
Toxicological threshold, 217
Toxicologist, 214, 233, 238, 276
Toxicology, 213, 214, 216, 218, 219, 221, 227, 231
Toxic Substance Control Act (TSCA), 58, 309
Toxic tort, 2, 66, 215, 231, 262, 315
Trace metals, 21, 29, 349, 350, 356, 357, 360, 362, 363, 365, 366, 371, 377–379, 387, 395, 396
Tracer dye, 39
Transactional costs, 312
Transporter, 60, 61
Treatment, storage, and disposal facility (TSDF), 54, 56
Trench landfill, 486
Trespass, 66, 90, 309, 355
Trichloroethane (TCA), 227, 228, 230, 429, 477, 478
Trichloroethylene (TCE), 37,116–119, 121, 179, 225–230, 423, 429, 435, 438, 470, 472–479, 498, 501, 503, 504, 506–508, 510
Trigger, 74, 81, 82, 84–86, 89, 312
Trigger of coverage, 81
Trigger date, 312
Trimethylethyl lead, 167, 180
Trivalent, 221, 225, 235
Truth, 1–6, 8–10, 12, 13, 17, 18, 50, 294, 295, 534
Tutorial exhibits, 247

United States:
  Army Corps of Engineers, 119
  Bureau of Reclamation, 26
  Department of Agriculture, 26, 522
  Environmental Protection Agency (EPA), 7, 25, 32, 51, 52, 54, 58, 59, 63, 109, 119, 120, 151, 168, 180, 204, 219, 235, 239, 310, 313, 320–322, 379, 468, 470, 474, 479, 493, 518
  Geological Survey (USGS), 28, 109, 114, 356
  Public Health Service, 40
Uncalibrated, 9
Underground storage tank (UST), 105, 108, 123, 349, 408, 411, 413, 443, 444, 446, 449, 450, 452
Underground waters, 33, 527
Unexpected, 17, 78, 92, 312, 463, 492, 493, 515, 518
Unintended, 92, 312, 515, 518
Unlawful, 90
Unleaded gasoline, 180

Unlined:
lagoon, 42, 475
pond, 472
Unregulated, 31, 41, 113
Unvalidated, 9
Upjohn, 43

Vanadium, 176
Vapor:
degreasing, 118
dryer, 436, 437
phase, 153, 237, 448
Vessels, 24, 520
Videotape, 259
Vinyl chloride (VC), 43, 226–228, 230, 316, 435, 489
Vinyl plastic liners, 26
Violation, 75, 441, 442
Visuals, 245–248, 259, 260, 347
Volatile organic compound (VOC), 237, 241, 242, 423, 430, 431, 437, 474, 491, 501, 508

W. R. Meadows, 26
War Production Board, 119
Water:
district, 104, 105, 116
Pollution Abatement Committee, 40
Pollution Control Act, 32, 52
Quality Control Board, 225, 230
Quality Improvement Act, 32
resources, 21, 22, 28, 31, 32, 308, 520
table, 39, 40, 141
wells, 34, 104, 113, 470, 474, 489
Waters of the state, 33
Wavelengths, 175
Wells, 30, 34, 35, 41, 51, 104, 105, 108, 113, 125, 128, 145, 315, 318, 319, 359, 388, 389, 405, 413, 423, 429, 442, 445, 446, 458, 470, 473–475, 477–479, 481, 489, 491, 506, 525
Wetlands, 378
Wildlife, 221, 222, 225
Wisconsin, 74
World Health Organization, 38, 219, 277
World War II, 44, 105, 116–120, 216

Zinc, 21, 22, 104, 349–352, 355, 356, 360, 362, 363, 365–367, 370, 372–374, 376, 378, 379, 386–390, 396, 397